History of the Bumpers College:

Evolution of Education in the Agricultural, Food and Life Sciences in Arkansas

by Gary Zellar and Nancy Wyatt

Arkansas Agricultural Experiment Station
Fayetteville, Arkansas 72701

Published by
Arkansas Agricultural Experiment Station
University of Arkansas

Milo J. Shult, Vice President for Agriculture and Director
Charles J. Scifres, Associate Vice President for Agriculture

ISBN: 0-9622858-2-X

Arkansas Agricultural Experiment Station Special Report 194
CP3M899PM

Table of Contents

Acknowledgments

I want to thank Dean Charles Scifres, who inspired this project, for giving me the opportunity to bring to light the history of the Dale Bumpers College of Agricultural, Food and Life Sciences and trusting me with the responsibility of getting the facts straight. I hope he finds that the trust has not been misplaced.

I would like to take this opportunity to honor the memory of Dr. Stephen F. Strausberg, who began this project, but whose untimely death in 1997 left the work incomplete. I benefitted tremendously from the notes and research that Dr. Strausberg left behind. As a professor, he deepened my understanding of history through his lectures, seminars and casual conversations. While his research and the knowledge that he imparted made this a better book, any errors of fact or vagaries of interpretation in this work are entirely my own. I would like to think that he would be proud of the finished product.

My appreciation goes out to Dr. Jeannie Whayne, who introduced me to the possibilities of agricultural history as a rich field of study and had faith enough in my research and writing abilities to recommend me for this position.

I would also like to thank Howell Medders, the Director of Communication Services, who oversaw the project, for his patience and guidance in getting the book completed. Dr. Glenn Hardy aided tremendously in giving his time and sharing his recollections and insights as well as allowing me access to records from his personal archives. Special thanks go out to Lyell Thompson, Roy Rom, Glenn Hardy, Charles Caviness, Howell Medders and Charles Scifres, who read the manuscript and whose insightful comments and suggestions made it into a better book. Terri Waller, from the Arkansas Alumni Association, Glenn Hardy and Howell Medders deserve special credit for collecting the names and information for the appendices.

Without the aid and insight of Michael Dabrishus and Andrea Cantrell in Special Collections at Mullins Library, the "real story" might never have come to light. Their combined knowledge of the vast amount of material in Special Collections made my job much easier. I would like to thank Michael Dabrishus especially for guiding me through, and giving me access to, the University Archives. Andrea Cantrell and Ethel Simpson both gave their time unselfishly in many discussions regarding arcane intricacies of University politics in years gone by. I would also like to thank Vicky Jones at the Southwest

Collection Special Collections Library at Texas Tech in Lubbock Texas for her help in locating and copying material from the Seaman Knapp Papers and Bradford Knapp Papers.

I want to extend special appreciation to co-author Nancy Wyatt, for her patience, good humor and insights as the "project" was completed and transformed into a book.

Finally, I want to thank my wife Brooke, our daughter Violet and our son Garrison for their patience, understanding and inspiration that helped spur this book toward completion before the next millenium.

Gary Zellar
Fayetteville, Arkansas
July 1999

Preface

When I came to the University of Arkansas in the Spring of 1994, I asked for a copy of the history of the College. I was offered a history of the Experiment Station and told that *From Hills and Hollers*, a history of the poultry industry, was being written, but that there was no history of the College. Understanding the role of agriculture in the beginnings of the land-grant system and knowing that the University of Arkansas at Fayetteville was the state's land-grant institution, it seemed incredible that there was no written record of the people and contributions of the College of Agriculture and Home Economics, as the College was called at that time. It almost seemed as though resident instruction had lost its rightful place in the overall mission in agriculture and at the University of Arkansas. Evaluating enrollment trends over the past several decades and sensing faculty attitudes did indeed reinforce that feeling—not that teaching was considered to be unimportant but that it just was not at parity, in terms of philosophy as well as financial resources, with the other facets of the original tripartite mission—research and extension.

A lot has changed since my inquiries in March 1994 about the history of the College. As in colleges of agriculture nationwide, there has been a rededication to teaching by faculty and administration at the University of Arkansas, accompanied by a pervasive zeal for the student population and its welfare. This philosophical movement of resident instruction to parity with research and extension has re-established that balance critical to fulfilling our mission. Another critical change has been the broadening of this mission since the College's inception. Never wavering from its dedication to production agriculture, there is a clear understanding among the administration and faculty that colleges of agriculture in the new millennium are also in the food business and the people business and in the business of providing leadership in the applications of biological technologies in the broadest sense to the advancement of society. Raising the status of the Department of Home Economics to the School of Human Environmental Sciences was a step critical to the appropriate evolution of the College, truly offering the opportunity for students to pursue opportunities in professions such as hospitality management, interior design and apparel studies that, at least on the surface, are totally unrelated to agriculture. Perhaps the benchmark change was the rededication of the College to its core values and naming it in honor of Senator Dale Bumpers in 1995. This action truly renewed the College as the Dale Bumpers College of Agricultural, Food and Life Sciences (DBCAFLS).

The DBCAFLS Alumni Society is in its third year, and a renewed spirit permeates the College—a new spirit of community. The Alumni Society is providing that essential link among all the College's students, former, present and future. It further establishes a fundamental truth—the College doesn't "belong" just to the faculty or the administration or even the students; it belongs to all the people.

So a great deal has changed during a very short time. And the next 10 years or so will bring even greater change in the agricultural, food and life science industries and the roles that "colleges of agriculture" will be expected to play as society's demands for natural resources and their products steadily increase. The progressive industrialization of agriculture, the expected continuation of biological innovation at an even greater rate of change than we have experienced heretofore, an ever-increasing sensitivity to environmental integrity and the world politics surrounding the global food supply will continue to increase the expectations of graduates of the agricultural, food and life sciences.

Finally, the College, I believe, will continue to play a unique role in agricultural programs in Arkansas. Administered by the University of Arkansas-Fayetteville, the College is the lynchpin that connects the Division of Agriculture, which operates at the University of Arkansas System level, to the Fayetteville campus and its teaching mission. Both components of the Division, the Arkansas Agricultural Experiment Station and the Cooperative Extension Service, contribute heavily to the College. Most of the faculty in the College are faculty of the Experiment Station with joint teaching-research appointments, and some are also appointed in the Cooperative Extension Service. The glue that assures the success of this relationship is the students, both undergraduate and graduate.

Gary Zellar and Nancy Wyatt have effectively captured the evolution and growth of the College. A lay person in reference to agriculture, Dr. Zellar has presented, in unbiased and unabashed terms, the personalities and times that decided the course of the College. Dr. Wyatt's experience as faculty of the College brought tone and sometimes temper and chronology to the book as she carefully edited and contributed to it. Drs. Zellar and Wyatt, perhaps unknowingly, proved that history does indeed repeat itself. Their characterization of the rich tradition of the Dale Bumpers College of Agricultural, Food and Life Sciences strengthens my belief that it also has an incredibly bright future—that the College will indeed serve as the guidepost for the future of Arkansas. Thank you both, Drs. Zellar and Wyatt, for doing what needed to be done and doing it so well. There was no better time to do it than as we enter the new millenium.

Charles J. Scifres
Dean and Associate Vice President for
Agriculture-Research
July 1999

Chapter One

Roots of the College of Agriculture, 1871-1899

The history of the Dale Bumpers College of Agricultural, Food and Life Sciences is intertwined with the very founding of the University of Arkansas itself and the beginnings of the land-grant system of colleges and universities. When the Arkansas General Assembly established Arkansas Industrial University as a land-grant institution in 1871 under the terms of the 1862 Morrill Act, the legislators enjoined the Board of Trustees to organize the new University "according to the true spirit and intent of the acts of Congress."[1] The Board did indeed launch their educational experiment in the spirit of the Morrill Act by resolving that:

> the leading object shall be–without excluding other scientific and classical studies and including military tactics, to teach such branches of learning as are related to Agriculture and the Mechanical Arts in order to promote the liberal and practical education of the industrial classes in the several pursuits and professions of life...(and)....To prosecute experiments for the promotion of agriculture and horticulture.[2]

Yet, in spite of the board's injunction and the efforts of administration, faculty and supporters, agricultural studies failed to take hold during the University's first 30 years, and there were few students.

The reasons for this lack of interest were complex and mirrored many of the problems encountered in other early land-grant institutions. With the scientific study of agriculture still in its infancy, there was no coherent body of research to draw on; few professors were qualified to teach the subject matter, there was no established curriculum, and few books focused on agriculture. Public school systems had yet to develop to any great extent in the United States, and in Arkansas no such system existed until 1875. Thus, many of the students at the early land-grant universities, the "sons and daughters of workmen and farmers" who were supposed to be the special focus for the educational program envisioned in the Morrill Act, had received only a rudimentary education and were woefully unprepared to begin scientific studies.

Additionally, many of the agriculture programs at the early land-grant universities suffered from inadequate legislative appropriations. The situation in Arkansas was particularly acute. The state was still recovering from the devastating economic and political effects of the Civil War. Arkansas had been the site of numerous major battles and skirmishes that had destroyed much of its agricultural productive capacity, which was the state's principal source of income. Arkansas had also experienced vicious and protracted guerrilla warfare that divided the loyalties of communities and families and left personal and political scars that were slow to heal.

However, the major obstacle to the development of agricultural studies at the University of Arkansas and other land-grant institutions in this period was simply that few students elected to take the courses. While the National Grange and other state and local farmer's groups endorsed agricultural education in principle, most farmers distrusted the idea of "book farming" in practice. Students were also prejudiced against agricultural studies because at this early stage the courses were still associated with manual-labor educational theories that required them to work on the University farm in "practical application" of the book farming they learned in the classroom. In fact, the few students from rural areas who came to colleges and universities came to escape the drudgery associated with farm work and gain professional training that would liberate them from labor in the fields. In spite of these handicaps, efforts were made during the first 30 years of the University of Arkansas to establish and develop a program of agricultural studies that would provide the foundation for the creation of a College of Agriculture at the University.[3]

In the first of these efforts, the Board of Trustees sent an investigating committee to inspect the operations of two northern land-grant institutions in order to guide them in their efforts.[4] The committee first visited Illinois Industrial University at Urbana, established in 1867. They were suitably impressed with the progress made in Urbana during the four short years of its existence. The University had both Agricultural and Mechanical Colleges and an experimental farm of 160 acres as well as colleges devoted to the classics and other sciences. The building program then underway at Illinois and the expansive scope of academic instruction interested the Arkansas trustees. They also noted the agricultural education system taking shape at Illinois that joined scientific theory with practical application. An Illinois Industrial University circular proclaimed:

> The aim of this school is to educate scientific agriculturalists....The practical farmer...laughs at the notion of sending his son to learn (the arts of farming) from a set of scientific professors....The real object of scientific farming...is not to teach *how* to plough, but

> the reason for ploughing at all....So that he may understand the reason of the processes he uses and may intelligently work for the improvement of those processes.[5]

The work on the University farm caught the visitors' attention as well. Though they did not necessarily dwell on the experimental aspects, the trustees did notice that, "a practical and experienced farmer" managed the operation and supervised the student labor. Instead of requiring students to work on the farm as part of their practical education, as was the general practice in so-called "industrial education" at the time, labor on the farm was both voluntary and "remunerative." When the committee asked, "will experimental farming pay?" the Illinois faculty answered that indeed it would, "with proper management."[6]

The next stop for the committee was Michigan State Agricultural College in Lansing. Michigan State held the honor of being the first state-supported agricultural college in the nation, created in 1857. Indeed, the program at Michigan State, with its emphasis on experimental work and a curriculum that successfully joined the study of agriculture with science, had an enormous influence in the development of agricultural education in the early years of the land-grant system. Professor William James Beal, who began his tenure at Michigan State in 1871, was largely responsible for the direction of the program there. Beal had studied under Louis Agassiz at Harvard and adapted Agassiz's innovative approaches to scientific studies, which were then current in Europe, to the scientific study of agriculture in the United States. What

Small building south of the Agricultural Experiment Station Building that was used for many purposes over the years, including as an Entomology Building, Offices of the Director of the Station, Ag Chemistry, Ag Education, Ag Library and the Extension Service. See map, page 20, building number 6.

the trustees saw at Michigan State, then, was an institution that was on the very "cutting edge" of agricultural education, which was still in the embryonic stages of development in this country.[7]

While the committee was impressed with Michigan State's efforts in experimental agriculture, particularly their stock breeding program, and with their ability to join theory and practice in agricultural studies, the committee observed that the college was "almost exclusively agricultural in its character" and still (at that time) lacked a mechanical department. Unlike Illinois Industrial University, with its more cosmopolitan and voluntary curriculum, the students at Michigan State were required to enroll in "a full and complete agricultural course." In addition, work on the college farm was compulsory.[8]

On the basis of their observations at these seemingly dissimilar institutions, both of which were operating under the aegis of the Morrill Act, how then should the Board of Trustees proceed in organizing Arkansas Industrial University? In their report to the full Board of Trustees, the visiting committee recommended that the Board combine the best features of both institutions. To advance the understanding of scientific agriculture, the committee believed that the experimental work and the wedding of theory and practice in agricultural education seen at Michigan was the better system. However, in regard to student labor on the University farm, the board adopted the Illinois voluntary approach. The Board also followed the Illinois model of a more cosmopolitan and general curriculum for the University as a whole and made the selection of an agricultural course of study elective rather than compulsory. Thus, the Board fulfilled the General Assembly's injunction to organize the University in the spirit of the Morrill Act. While emphasizing "the agricultural and mechanic arts," the Board did not exclude either classical or scientific studies.[9]

The Board revealed its intention to emphasize agricultural studies at the new University in how they defined the "Courses of Instruction." The first of the four colleges mentioned in the Board's January 1871 deliberations is "the College of Agriculture." The College was then divided into two "schools": the School of Agriculture and the School of Horticulture. The College of Agriculture at the University of Arkansas existed only on paper until 1872 when the Board outlined a full four-year course in agriculture and hired the first professor of theoretical and practical agriculture, Richard Thurston, M.D.[10]

Thurston had no formal training in agriculture but presumably had practical experience from managing his farm in Van Buren. He received $40 a lecture to present 12 lectures on agriculture during the 1873 spring term. The topics for the lectures included such subjects as "the garden and the orchard," "how plants grow" and other more general discourses on farming and agriculture. In one of his lectures he extolled the virtue and necessity of applying scientific principles to farming:

> ...to gain...the greatest return from the soil for the least outlay of labor is the primary object of farming....this object can be best obtained only by a judicious use of natural and organic forces, the knowledge of which constitutes science in its broadest sense.

Thurston went on to say that a deeper understanding of physics, chemistry and biology as well as mechanics can be applied to agriculture and called on his students to join in establishing "a new order of scientific farming" to combat the "old fogyism...[that] drags the ponderous old system impelled by rules and appliances of an hundred years ago."[11]

It's not clear how many students attended Thurston's lectures and were convinced to come forward in the service of scientific agriculture, but as was true for many of the other land-grant institutions in this early period, the number of students enrolled in agriculture courses were few. In any case, Thurston did not return the next academic year. The next professor of agriculture would not be appointed until 1888, and the College itself disappeared until it was officially organized in 1905.

In October 1873 the Board took the initial steps toward organizing its experimental farm by electing T.L. Thompson as professor of theoretical and applied chemistry and superintendent of the University farm. Thompson came to the University with a B.S. Degree from Iowa State Agricultural College in Ames. In addition to teaching regular classes in organic, physical and analytical chemistry, Thompson also taught agricultural chemistry classes. Once again, it is not clear how many students, if any, were enrolled in Thompson's agricultural chemistry classes, but later reports indicate that interest was low.

While the 160 acres that made up the University grounds had all originally been a farm, only the western portion of the tract was set aside to "prosecute experiments for the promotion of agricultural and horticultural pursuits." Thompson took up the work on the farm with enthusiasm. He first cleared 40 acres of wooded land in order to increase the area under cultivation and had the four-acre orchard trimmed and pruned.

Anxious to begin experimental studies, Thompson conducted limited experiments with spring wheat. He wanted to see if farmers in northwestern Arkansas could successfully plant their wheat crop in the spring, when they would have more time to devote to it, instead of the fall, when they were busy harvesting other crops. He deemed the wheat experiments a failure because "heavy spring rains came on and blasted the crop in the beginning of its growth" and vowed to begin the experiment again next season. Thompson rated the vegetable garden and corn crops more successful.

After a survey of the 160 acres that served as both the campus and the University farm, Thompson recommended that the trustees commit themselves to a beautification program to coincide with construction of the new main building. He suggested ornamental and shade tree plantings on some of the farm land close to the building "as a corn field or potato patch would be incongruous with a fine building."

The farm work also included instructing students in "practical agriculture" and supervising their work. Thompson said that using student help was a problem because, "we can not always expect first-class labor from the students in school." Another problem with the work on the farm, according to Thompson, was the shortage of implements and the lack of tillable ground, which made it difficult to keep those students who wanted to work busy.[12]

The problem with keeping the student workers on the farm busy no doubt increased after the Board reversed its earlier stance on elective course work. In 1874, concerned with the lack of students in the agriculture course, they passed a resolution that required all beneficiaries, or students who attended the University free of charge, to take a course in agriculture and mechanics. So that the instruction would be grounded in the practical realm as well, labor on the farm was made compulsory for the beneficiaries.

Professor Thompson's untimely death in October 1875 left the University without a farm superintendent or a person qualified to teach the agriculture course. The board hired a student, John H. Bell, to run the farm and appointed another Iowa State graduate, H.L. Harvey, to teach chemistry. However, there was no attempt to fill the position in the Agriculture College, which was now called the Agriculture Department.[13]

An early postcard shows the University of Arkansas campus as it would have looked around the turn of the century. The building in the foreground is University Hall, known as "Old Main." To the left is Gray Hall. The building with four towers to the right is the Agricultural Experiment Station Building. The two buildings directly behind Old Main are the original buildings of the University of Arkansas. (Photo courtesy Photograph Collection, Special Collections, University of Arkansas Libraries.)

In January 1876, the Executive Committee of the Board issued a report on the Agriculture Department's status. The report conceded that efforts to give agricultural studies a prominent place in the University had so far come to naught. Even the required course for the beneficiaries was a failure because the students were mostly

> ...elementary scholars who required years of instruction to enable them to understand the proportions and mysteries of chemistry which is absolutely essential to scientific agriculture.

In June 1876, the Board appointed their secretary, J.F. Simonds, to the position of regent and asked him "to take control of the Agricultural and Mechanical Departments" and make a full report on their condition as well as that of the farm.[14]

Simonds' detailed account dealt mainly with the physical needs of the Agriculture Department in conducting experimental scientific agriculture and is replete with recommendations for carrying the work forward. His suggestions for the preparation and conditioning of soil in the experimental fields with the use of green manure crops, barnyard manure, lime, potash and gypsum showed a thorough knowledge of the most progressive thinking on the subject at that time.

Simonds' most enthusiastic recommendation was for the development of "a blooded stock Department" similar to the one that had won the visiting committee's praise at Michigan State. In Simond's opinion this was the only way the University could make the farm self-sustaining as well as earn it the praise and attention from the Arkansas farmers that it so desperately needed. Regarding the stock breeding program, Simonds remarked:

> We can and I think we ought to take the lead in this branch of husbandry, and say to all "come on," instead of remaining the butt of sly jokes concerning our blooded stock and Book farming.

He also recommended shipping crates of the farm's best apples to legislators in Little Rock with the boxes marked "University Apples" in order to draw favorable attention to the agricultural work at the University.[15]

In Simond's report, his most prescient observation, which countered the thinking of the Board and most of the public on the matter, dealt with what the true purpose of the University farm should be. He said: "I do not understand that (the University farm) is purely for profit to your treasury, which is to pay off all your University debts and redeem state script at par." Simonds went on to say that neither was the farm intended as an employment agency for "young men who are too dull to study and too lazy to work." Rather, he envisioned the farm as an educational tool to be used to illustrate the principles of scientific agriculture in the same way that the "Air Pump, the Electrical machine and the Telescope" are used to teach the fundamentals of physics and astronomy.[16]

Unfortunately, Simond's enthusiastic and pointed report resulted in neither more students in agriculture nor more appropriations from the Legislature. Following Simond's recommendations, the Board dropped the requirement that beneficiaries enroll in the agriculture course. Without the requirement, there were no students listed in the agriculture courses from 1877-79, and the board dismantled the Departments of Agriculture and Horticulture.[17]

The lack of agriculture students and, as some critics maintained, an absence of agricultural instruction at all finally had political and financial repercussions that briefly jeopardized the future of the University itself. In 1879 Washington County's Levying Court refused to raise the taxes needed to pay the interest on the bonds that the county purchased in order to bring the University to Fayetteville. Local political rivalries played a major role in bringing on this confrontation between the University's opponents and its supporters. However, one of the issues that the opposition chose to spearhead their attack bore directly on the progress of agricultural instruction at the University. The Levying Court said they had no obligation to raise the taxes because the University "was being conducted almost solely as a literary institution" in violation of the intent of the Morrill Act, which emphasized instruction in the agricultural and mechanic arts.[18]

In early 1880 the University filed a lawsuit in order to force the county to levy the taxes. The University argued that although there was no demand for agricultural courses, all science classes as taught at the University either directly or indirectly emphasized agriculture. Chemistry Professor H.L. Harvey testified to this effect. Judge Lafayette Gregg, one of the founders of the University and its leading supporter in Fayetteville, echoed Harvey's position but added that the practical work on the farm was of little value and a needless expense for the University. Other witnesses testified that agricultural instruction lagged at the University because the Legislature had not appropriated the necessary funds to develop the program. The University won the day in court, but nagging issues had been brought to the forefront and needed resolution before agricultural education could take its rightful place at Arkansas' land-grant university.[19]

Throughout the early 1880's officials at the University issued numerous reports on the bleak state of affairs in the Agriculture Department and the need to give agricultural studies a prominent place in the curriculum. Finally, after the 1885 Legislature rejected President George M. Edger's plea for $25,000 to underwrite agricultural and mechanical studies and to develop the University farm, the board appropriated $4,500 from the University's general funds for the contingency. It appears that Edger acted in step with growing public sentiment throughout the state. One letter printed in the *Arkansas Gazette* reflected the growing agrarian discontent:

> Let us have an agricultural University or none. We have only to banish the study of dead languages and other so-called higher branches...and see what Arkansas can do to rear scientific farmers.[20]

Acting in response to these agrarian political pressures, Edger and the board completely reorganized the University in 1886, dismissing all existing faculty and electing replacements. The board combined the Departments of Chemistry, Biology and Mineralogy under a new professor, George D. Purinton. The professor had a degree in chemistry from West Virginia and experience in college teaching in both Iowa and South Carolina and had been employed in private industry analyzing fertilizers. The board also made Purinton superintendent of the University farm with a small contingency fund to put the farm in order and begin experiments. In doing so, they hoped that students would be attracted to agricultural studies and Purinton could begin to build departments of Agriculture and Horticulture. Students did not appear for the courses, but Purinton did begin experiments and "set the farm in order."[21]

George D. Purinton, Superintendent of the university farm 1886-1887. (Photo courtesy Photograph Collection, Special Collections, University of Arkansas Libraries.)

Purinton contacted the United States Department of Agriculture for advice on how to proceed. He also enlisted the support of a local nurseryman, J.B. Gill of Springdale, who donated 400 fruit trees of different kinds in order to rejuvenate the orchard. A.B. Mayer Co., of St. Louis, provided material for chemical analysis as well as fertilizer for the orchard. The Engineering Department at the University helped lay out the plats for the experimental plots and the orchard. This cooperation among the College, federal authorities, local business and private industry anticipated the alliances that underscored many future successes in agricultural studies.

Purinton compiled the results of the experiments and published *Arkansas Industrial University Farm*, the first bulletin issued by the Department of Agriculture and Horticulture in May 1886.[22] President Edger hoped that this milestone would induce Arkansas' congressional delegation to support a bill pending in Congress that would provide federal funding for agricultural experiment stations.[23]

The bill, popularly known as The Hatch Act, was a landmark event in the growth and development of the land-grant system of colleges and universities. The Hatch Act became law March 2, 1887, and provided $15,000 annually to individual agricultural experiment stations, most of them associated with land-grant institutions, to fund research in scientific agriculture. The legislation not only provided badly needed funds, especially for states such as Arkansas, but also met the need for a body of scientific research that agriculture teachers could draw on. Another feature of the bill enjoined the stations to issue bulletins outlining the progress of their research and provided free use of the U.S. mail to disseminate the information to farmers and the public.

The importance of the Hatch Act for the future of agricultural education in the United States and in Arkansas cannot be overstated. It gave agricultural studies legitimacy at a time when they were floundering in Arkansas and provided groundwork on which they could grow in the future.[24]

Growing agrarian discontent in the nation that provided the support for the passage of the Hatch Act had a parallel movement in Arkansas. The focus for these agrarian political forces in Arkansas was the re-organization of Arkansas Industrial University along lines that would emphasize agricultural and mechanical studies. The result was the Barker Bill, passed by the Arkansas Assembly in March 1887. The legislation gave heavy emphasis to "practical studies in agriculture and mechanics" and outlined a complete course in agricultural studies. The bill also mandated that all male students, whether in classical, scientific or agricultural studies, should work at least three hours a day on the University farm and gave agriculture students 10 cents an hour, whereas classical students would earn only three cents an hour. Beneficiaries were required to enroll in either agricultural or mechanical studies while students taking only liberal arts courses of study paid tuition. In addition, the position of "Superintendent of Agriculture" was established, and the superintendent was required to submit detailed reports on the progress of the University farm and the Agriculture Department to the Legislature.[25]

Purinton published the second bulletin from the Arkansas Industrial University College of Agriculture in January 1887.[26] In it he outlined experiments conducted with oats and corn and the effects of using either commercial fertilizers or manure compost. Purinton submitted his report for the Agriculture Department in May 1887 in which he voiced the perennial lament on the lack of students. The primary reason for "the inherent unpopularity of this course of study grow(s) out of its association with the idea of manual labor." Purinton tendered his resignation in August 1887, no doubt influenced by the manual labor provisions contained in the Barker Bill.[27]

The board elected Albert E. Menke as Purinton's replacement. Menke's credentials included degrees from King's College in London, Harvard University, the Halle University in Germany and a Ph.D. in Chemistry from Cornell University. Menke's full title during his first years at the University was "Superintendent of Agriculture, Chemistry, Theoretical and Practical Agriculture, Mineralogy and Veterinary Art."

Although Menke's scientific background was prestigious, he had never spent any time on the farm. Undeterred, Menke proceeded to put the farm in order, and with the help of a newly appointed adjunct professor, S.S. Twombly, began instruction in agricultural chemistry, soil preparation, land reclamation, crop rotation, forage and fodder crops and a host of other subjects. For the first time since beneficiary students were required to take a course in agriculture in 1876-7, there were students listed the Agriculture Department.[28]

Albert E. Menke, Superintendent of the university farm 1887, Director of the Station 1888-1891. (Photo courtesy Photograph Collection, Special Collections, University of Arkansas Libraries.)

Agricultural studies at the University were given another boost in February 1888 when the Arkansas Assembly accepted the terms of the Hatch Act. The University used $4,000 of the Hatch funds to build a brick experiment station building and $4,000 to equip its laboratories. The Board appointed Menke Director of the Station in September 1888 and employed 12 staff members, three of whom were also professors at the University, S.S. Twombly, F.W. Simonds and Menke. Three other branch stations were also opened at Jonesboro, Pine Bluff and Texarkana in order to conduct experiments in the different climatic regions in the state. The branch stations also showed other parts of the state the benefits of scientific agriculture and thus encouraged support for the University.

In its first year of operation, the Agricultural Experiment Station issued seven bulletins covering topics ranging from the "cultivation of cotton and corn," to the "diseases of animals" and "tests of varieties of grapes and strawberries." As Stephen F. Strausberg pointed out in his work on the history of the Experiment Station, the early work at the Station was geared toward solving farmers' practical problems and increasing local production rather than investigating "underlying scientific principles." This orientation of the work at the Station combined with Menke's and Purinton's efforts to enlist the aid of farmer's organizations and industry laid the foundation for the development of extension work that would link agricultural teaching with research to serve the needs of Arkansas' farmers.[29]

In 1889 there were 70 students reported in the Agriculture Department. Of these, three were listed as "collegiates"; the rest were preparatory students. Menke asked that the two-year "Short Course" in Agriculture, first offered in the 1888-89 school year, be put on par with the Normal Course for training teachers. Once students completed the Short Course, he urged that they be awarded a diploma, but not a degree. Menke went on to say:

> Not that I think a degree or a diploma amounts to anything as affecting the value of the course but since we occupy an anomalous position, being dependent on the popular vote and since the newspapers continually ask, "how many agriculture graduates have you?" It seems to me that this will be an answer that will please the people (and) the press and will work no injury to the student.[30]

At this point in the Agriculture Department's development, Menke's ear was tuned to the political winds as well as practical reality.[31]

During the 1890's the Department of Agriculture took on an identifiable shape, though it was not until after the turn of the century that it came to be recognized as the College of Agriculture. While student enrollment could be described as a "roller coaster ride," with recognizable ups and downs, distinct areas of study under different professors and instructors appeared that would coalesce into separate departments in the College. Efforts to draw student and public interest in agricultural studies also continued but with more success than had been seen in earlier times.

The following table gives some idea of student enrollment in the Agriculture Department or School, as it was sometimes known, during the 1890's.

Enrollment in the 1890's.

Year	1890	1891	1892	1893	1894	1895	1896	1897	1898	1899
Preparatory	66	33	49	29	33	23	16	1	-	-
Collegiate	3	3	3	-	3	-	1	7	7	19

The agricultural preparatory students in 1898 and 1899 were not accounted for in a separate category; instead they were lumped together with other "science" preparatory students. With this in mind, it could be said that the total number of Agriculture students during this period is larger than a glance at the table would indicate.[32]

In 1890 the Department of Agriculture offered classes in agriculture, horticulture, stock breeding, stock feeding, veterinary anatomy, veterinary science, dairying, chemistry and mineralogy, leading to either a Bachelor of Science in Agriculture or a certificate for the completion of the Short Course. Menke and G.L. Teller, an adjunct professor whose primary assignment was as a physics and mineralogy professor, taught the agriculture and mineralogy segments of the course. Jerome F. McNeil, professor of geology and biology, provided instruction in horticulture. R.R. Dinwiddie, the Experiment Station veterinarian, taught veterinary science.

W.F. Bates, the farm foreman, and J.M. Moore, assistant foreman, took charge of "practical agriculture." Bates taught dairying as well as the preparatory course in agriculture. Courses in horticulture and entomology leading to a Bachelor of Science degree were also offered in the Science Department under McNeil. These courses were directed toward training professional sci-

The Agricultural Experiment Station Building, built in 1888 at a cost of $4,000, was later used for vocational teacher training, as a general extension building, music building and editorial services building. It was razed in 1972. It stood on Maple Street where the entrance to the Alternative Pest Control Center is now located. See map, page 20, building 5. (Photo courtesy Photograph Collection, Special Collections, University of Arkansas Libraries.)

entists for posts at the experiment stations opening across the country. The agriculture students were also required to take general classes such as biology, botany and zoology in the Science Department as well as the usual English, mathematics and political science classes required of all students at the University.[33]

In 1891 the Board dropped the compulsory labor requirement and allowed beneficiaries to enroll in liberal arts studies free of tuition. While the Board's decision to return to an elective curriculum and a "voluntary and remunerative" system of student labor removed the onus of agriculture students working as "quarry slaves," these actions probably accounted for the drop in enrollment in the Agriculture Department from 66 students to 33. Nonetheless, Menke observed that prizes offered by the Board to agriculture students raising the best crops had "a good effect in stimulating the interest of the students and resulted in keen competition." Menke heartily recommended that such prizes be continued. Menke also praised the work of W.F. Bates in providing dairying instruction and advised that this feature of agricultural studies be given special emphasis. The "blooded stock" in the University dairy herd had won prizes in stock judging competitions the previous year, and Menke was anxious to capitalize on the favorable publicity that would draw attention to the Department of Agriculture from across the state.[34]

In 1892 the Board authorized further cash prizes. In line with expanding the dairy operations, $25, $15 and $10 were offered for the best five pounds of butter made by students in the dairying classes. Menke asked the Board for $2,500 for barns and other physical improvements, including equipment for the dairy operation. Menke also recommended that the Board appropriate money to fund a series of "Farmers' Institutes" to be held at various locations across the state. The Agricultural College at the University of Wisconsin had initiated such a program in 1886, and it had done much to draw the interest of Wisconsin's farmers in the work of the College and the Experiment Station.

The Board approved Menke's plan in November 1893 and drew $250 from the general University funds and $250 from the Experiment Station monies to underwrite the project. They instructed the Experiment Station personnel and the "professors at AIU" to devote to the events whatever time was necessary in order to make them a success and solicited the support of Arkansas' Commissioner of Agriculture, W.G. Vincenheller, as well. In the spring of 1894 the first Farmers' Institutes were held in Washington, Arkadelphia, Searcy and Monticello. Papers were read by University President John L. Buchanan and by Menke and McNeil. Robert L. Bennett, the Station Director at that time, and C.L. Newman and W.F. Bates, agriculturalists at the Station, also presented papers and gave demonstrations.[35]

Menke was pleased with these first adventures into agricultural extension education. Through personal contact with the "scientific agriculturalists" at the University, the farmers themselves could judge the value of the program and would be more likely to support the University and perhaps send their sons and daughters there. With a teaching program providing training for scientific agriculturalists (albeit, a "Short Course"), an active Experiment Station producing a body of research and with the beginnings of extension activities taking the knowledge from the classrooms and research directly to the people, Arkansas Industrial University now had the three elements in operation that are credited with making the land-grant system a success. However, a few bumps in the road had to be eliminated before the program at Fayetteville could be judged a success.

The Depression of the 1890's had its effects on student enrollment at the University. After 1894, and through the remainder of the decade, student numbers declined by as much as 20%. In an effort to shore up enrollment in the Agriculture Department, further prizes were offered agricultural students. In 1894 the Board offered awards of $50, $30 and $20 to agriculture students for composing the best essays on horticulture. The essays would then be read at the commencement exercises. These were considerable sums in a period of economic depression.

While the number of students continued to decline, Menke and President Buchanan both noted with optimism that the number of students studying agriculture at Arkansas exceeded that of any other Southern university and was equal to the number of collegiates at the University of Wisconsin.[36]

R.L. Bennett, Dean and Director of the Station 1891-1903. (Photo courtesy Photograph Collection, Special Collections, University of Arkansas Libraries.)

In 1896 the Board separated the Agriculture Department from the Chemistry Department. Robert L. Bennett, the Director of the Experiment Station since 1891, was made Superintendent of Agriculture and the Farm and the head of the new department. The new department had four sub-departments roughly analogous to the departments in the Experiment Station. The Agriculture Department, supervised by Bennett, offered classes in soils, farm crops, farm machinery and buildings, animal breeding and rural economics. The Horticulture Department was headed by the Station's horticulturist, J.T. Stinson, and featured classes in plant propagation and breeding, fruit culture, vegetable gardening, landscape gardening and forestry. G.L. Teller, chemist at the Station, instructed in the Agricultural Chemistry and Meteorology Department. R.R. Dinwiddie, the Station's veterinarian, headed the Animal Pathology and Bacteriology Department and gave classes in animal physiology, anatomy, nutrition and reproduction as well as farm hygiene and bacteriology. The Board enforced this marriage of research and resident teaching with the injunction that:

> The Experiment Station officers shall in no way diminish the amount of Experiment Station work that has heretofore been required of them because of their duties as teachers.

This reversed an earlier Board stance taken in 1891 that directed that the teaching faculty and Experiment Station staff be composed of different persons. In this early period then, the stage was set for what would continue to be a dilemma in colleges of agriculture throughout the country: if Experiment Station and teaching faculty were joined, how much time should the individuals devote to research and how much to teaching?[37]

At the end of 1897, the Board elected a new Professor and Superintendent of Agriculture, C.L. Newman. Newman held a degree in agriculture from Alabama A&M, taught agriculture at the University of Tennessee and had worked as an agriculturalist at the branch experiment stations in Pine Bluff and Camden since 1893. Newman set to work immediately improving the grounds, re-seeding pastures and repairing buildings as well as teaching the agriculture course. Students taking the agriculture course could earn either a

C.L. Newman, Superintendent of Agriculture 1897. (Photo courtesy Photograph Collection, Special Collections, University of Arkansas Libraries.)

four-year Bachelor of Science degree or a Bachelor of Scientific Agriculture, which was basically the same curriculum with more agriculture-related classes. The Short Course in agriculture also remained an option.[38]

The Horticulture Department was given a breath of new life at this time. The Board appropriated $3,500 for building a new greenhouse and a forcing house and to equip new laboratories. In addition to the regular four-year course in horticulture that led to a B.S. degree, J.T. Stinson suggested to the Board that a ten-week "winter course" be established for "farmer's sons who have but little time at their disposal and who want to better acquaint themselves with agricultural principles." Lectures would be given on soils, agricultural chemistry, economic entomology, plant spraying, market gardening and the care of fruits adapted to Arkansas.

Stinson made another suggestion reminiscent of J.F. Simonds' 1876 proposal to send out crates of "University apples." He recommended that the department send out "choice varieties of fruits" raised as nursery stock by students on the University farm to prominent men in the state. In this way new varieties could be introduced and tested, and attention would be drawn to the work at the University. This plan was later carried out beginning in 1899 under the supervision of W.G. Vincenheller, the Vice-Director and chief pomologist at the Experiment Station, who held Farmers' Institutes across the state focusing on large fruit culture.[39]

As the nineteenth century ended, the 30-year struggle to establish agricultural studies at the University was nearly complete. There was an established faculty trained in scientific agriculture from the leading agriculture colleges and universities from across the country. Regularly scheduled college and preparatory courses attracted an ever-increasing number of students. The Experiment Station, with its four branch stations and the University farm, had done much to gain recognition for the accomplishments of Arkansas' land-grant university in the realm of scientific agriculture. Extension activities undertaken by the Station personnel and the agriculture faculty with farmers in the state increased the visibility of agricultural studies as well. Linking agricultural studies with manual labor theories of education had fallen by the wayside, and the University catalog descriptions of the agriculture courses stressed that labor was required from agriculture students only in order to demonstrate techniques or to carry out experiments. This factor more than any other is credited with the growing popularity of the classes with students.

During the first 30 years at the University of Arkansas, agricultural education had grown through fits and starts. Economic, political and educational conditions in Arkansas were not yet ripe for the maturing of agricultural studies. After building a foundation during this period of experimentation and testing of waters, such studies were poised to undergo a period of consolidation and sustained growth during the first two decades of the twentieth century.

Endnotes

1 John Hugh Reynolds and David Yancey Thomas, *History of the University of Arkansas* (Fayetteville: University of Arkansas, 1910), 52-53.

2 Arkansas Industrial University, *First Report and Catalogue of the Arkansas Industrial University*: August 1873 (Fayetteville: Democrat Book Inc, 1873), 15-17. Hereafter cited as *Catalog*, followed by year and page number.

3 Reynolds, 18-19; Charles E. Kellogg and David C. Knapp, *The College of Agriculture: Science in the Public Interest* (New York: McGraw Hill, 1966), 11-12; Alfred Charles True, *A History of Agricultural Education in the United States: 1785-1925*, United States Department of Agriculture, Miscellaneous Publication No. 36 (Washington D.C.: GPO, 1929), 114 -116; Robert A. Leflar, *The First 100 Years: Centennial History of the University of Arkansas* (Fayetteville: University of Arkansas Foundation, 1972), 59-60.

4 Kellogg and Knapp, 11; Reynolds and Thomas, 69-70.

5 University of Arkansas Board of Trustees, *Minutes of Meetings: September 18, 1871-February 27, 1891* (Fayetteville: University of Arkansas Library Special Collections Microfilm Publication, 1986), 14 (hereafter cited as *Minutes*, followed by page number).

6 *Minutes*, 11-15; True, 184-187.

7 True, 133-135.

8 *Minutes*, 16-17.

9 *Minutes*, 18 19; Reynolds, 53 54.

10 *Minutes*, 119, 90; Arkansas Industrial University, *First Report* (Little Rock: Little Rock Printing Co., 1873), 23-24.

11 Arkansas Industrial University, *Second Report: 1873-4* (Little Rock: Little Rock Printing Co., 1874), 223-227.

12 *Minutes*, 151, 184-186; Reynolds, 94-95.

13 Harrison Hale, *University of Arkansas-1948* (Fayetteville: University of Arkansas Alumni Association, 1948), 151-152; *Catalog: 1874-75*, 18, 103.

14 University of Arkansas Archives, *Board of Trustees Records,* Box 1 File: "Reports of Executive Committee, January 20, 1876," and Box 2: File: "Regents Report, 1876" (Special Collections, Mullins Library, University of Arkansas Fayetteville, uncataloged collection), hereafter cited as *Trustees Records* followed by box and file designation.

15 *Trustees Records*, Box 2: File, "Regent's Report, 1876."

16 Ibid.

[17] *Catalog: 1875-76* (Fayetteville: Democrat Book and Pub., 1876), 119.
[18] Reynolds and Yancey, 84.
[19] Ibid, 86.
[20] *Arkansas Gazette*, July 17, 1885, 6:1.
[21] *Trustee's Records*, Box 2, "President's Report," 15-16.
[22] Arkansas Agricultural Experiment Station, *Arkansas Industrial University Farm.* Bull. 1, University of Arkansas, 1886.
[23] Arkansas Industrial University, Department of Agriculture and Horticulture, *Bulletin No. 1: May 1886* (Little Rock: Mitchell and Bettis, 1886), 1-7; *Trustees Records*, Box 2, "President's Report, 1886"), 18-20.
[24] Kellog and Knapp, 12-13; Reynolds and Thomas, 283-284; True, 208-209, 221-222.
[25] *Arkansas Gazette*, July 17, 1885, 6:1, March 30, 1886, 6:1, March 11, 1887, 3:3-4, June 23, 1887, 6:1; Leflar, 48-49; Hale, 57-58, Reynolds and Thomas, 141-44.
[26] Arkansas Agricultural Experiment Station, *Oats and Corn*, Bull. 2, University of Arkansas, 1887.
[27] Arkansas Industrial University Department of Agriculture and Horticulture, *Bulletin No. 2: January 1887* (Little Rock: Mitchell and Bettis, 1887), 1-4; University Archives, Office of the President, *Dean's Reports*, Box 1, "Agriculture, Chemistry and Physics: 1882-97." Hereafter cited as *Dean's Reports* followed by box and file designation.
[28] Leflar, 57-58; Reynolds and Thomas, 270, 278, 273; Hale, 62-63.
[29] Reynolds and Thomas, 289; Stephen F. Strausberg, *A Century of Research: Centennial History of the Arkansas Agricultural Experiment Station* (Fayetteville: Arkansas Agricultural Experiment Station, 1989), 6-10.
[30] *Deans Reports*, Box 2, Agriculture Department, 1890-97, "Report of the Agriculture Department" (no date, but internal evidence indicates May 1889), 2-3.
[31] Hale, 153; Reynolds, 142; *Trustee's Records*, Box 2, "President's Report, 1889," 5-6; *Minutes*, August 8, 1889, 211.
[32] *Catalogs, 1890-99*, passim.
[33] *Catalog, 1890-91* (Little Rock: Woodruff Printing, 1891), 53-61.
[34] *Dean's Reports*, Box 2, "Agriculture Department 1890-97," Menke's Report, November 25, 1891, 1-2; *Arkansas Gazette*, September 15, 1889, 9:1.
[35] *Catalog, 1893-94*, 89; *Dean's Reports*, Box 2, "Agriculture Department, 1890-97," Menke's Reports, 1892, 1-2 and November 12, 1894, 1-2; *Trustee's Records*, Box 2, "Committee of Agriculture Report," 1-6.
[36] Reynolds, 157, 547; *Trustee's Minutes*, 312, 324; *Dean's Reports*, Box 2, "Agriculture," Menke's 1894 Report, 1-2; *Trustee's Records*, Box 2, "President's Reports," Buchanan's 1894 Report, 1-3.
[37] *Catalog, 1896-97* (Fayetttevile: n.d., n.p.) 54-58; *Board Minutes*, 348-349, 229.
[38] Reynolds, 501-2, 275; *Catalog: 1897-98* (Little Rock: Democrat Press, 1897), 104-105.
[39] *Dean's Reports*, Box 4, "Horticulture Department Report, 1897"; *Board Minutes*, 368-69: *Catalog: 1897-98*, 105-106.

Chapter Two

The College of Agriculture Begins: 1900-1920

By 1900 the economic, educational, social and political conditions that inhibited the growth of agricultural education at the University of Arkansas and other land-grant institutions during the nineteenth century had changed. These changes ushered in the University of Arkansas College of Agriculture in 1905, and they were responsible for the College's continued growth and consolidation for the following two decades.

Agriculture overall was in its most prosperous condition since the Civil War. Production expanded and diversified through the first two decades of the twentieth century, and farmers enjoyed an unprecedented era of high and stable prices. From 1900 to 1910, prices on the three key agricultural commodities—corn, wheat and cotton—rose an average of 48 percent. Since Arkansas was primarily dependent on agricultural products for its state revenue, the farmer's prosperity was in part reflected in increased legislative appropriations for the University of Arkansas. By 1907 state expenditures for the University had more than doubled the amount spent in 1895. Federal aid during this period also increased. The second Morrill Act of 1890 increased the federal funds available to land-grant institutions by $25,000. The Nelson Amendment raised the amount again by $5,000 in 1907. The Adams Act of 1906 similarly increased the support for experiment stations by $15,000.[1]

Other federal legislation in this period widened the scope of the land-grant college's educational mission and provided new career avenues for graduates from colleges of agriculture. The Smith-Lever Act created the United States Agricultural Demonstration and Cooperative Service in 1914, funding extension programs for farmers. Land-grant institutions administered the extension programs and provided the expertise. The Smith-Hughes Act of 1917 provided for agricultural and vocational education in secondary schools and for training of teachers for those positions.[2]

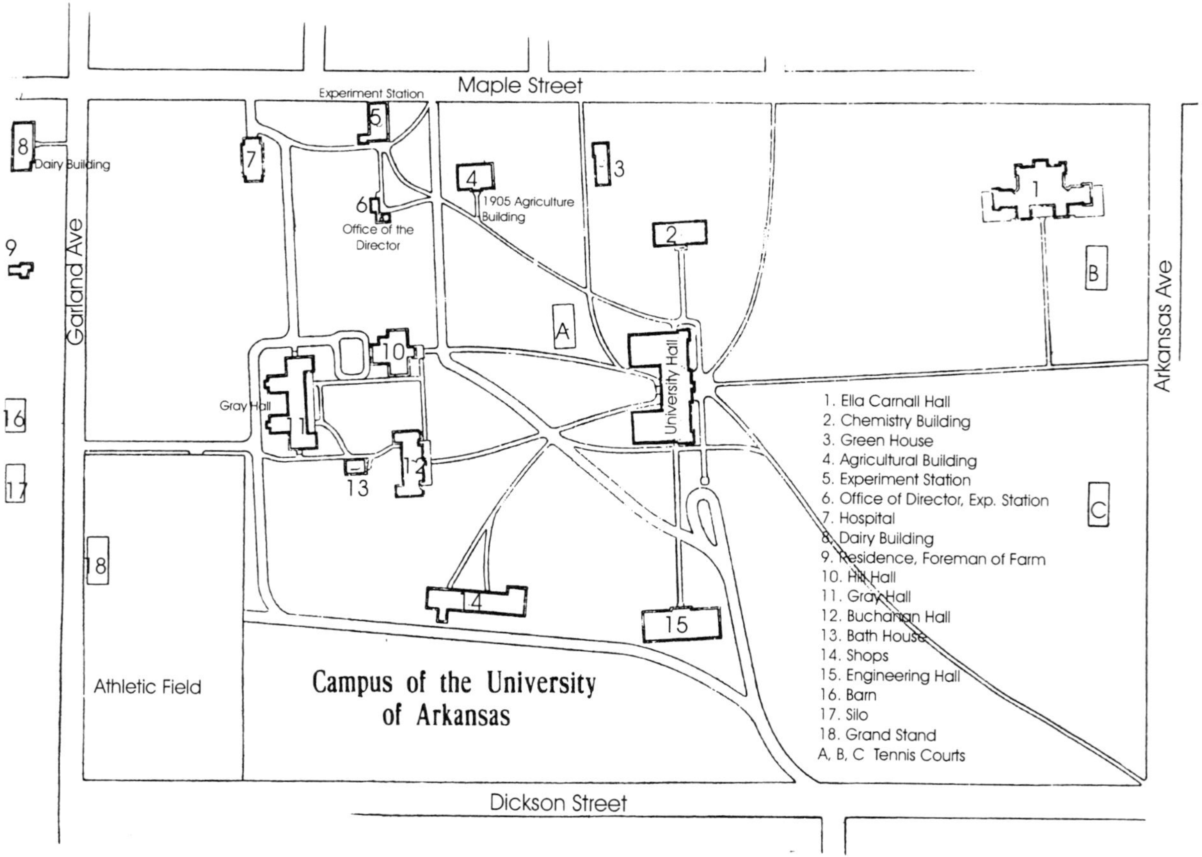

The University of Arkansas campus in 1908-1909. The 1905 Agriculture Building (4) would still have been considered "new" at this time. This is the building that is currently the Agricultural Annex, just north of the current Agriculture Building. The Dairy Building (8) had just been completed on the corner of Garland and Maple (where Silas Hunt Hall is now located).

Along with improved economic conditions in the agricultural sector and the growth in appropriations came improvement and growth in public education across the nation and in Arkansas in particular. When Arkansas Industrial University was established in 1871, the town of Fayetteville and Washington County were among the few locales in the state that could claim any sort of public school system. The state still had a long way to go before a statewide system of public schools would reach into some rural areas of the state, but by the turn of the century the number of accredited high schools whose graduates were eligible to begin collegiate studies at the University had increased dramatically. This trend was given a boost in 1908 through the initiation of a statewide campaign to increase the number and quality of public schools in Arkansas. From this larger pool of high school graduates, with an increasing number from rural areas, the number of students enrolling at the University doubled in the years between 1900 and 1910. After 1910 the growth in numbers was steady until another boom period began in the 1920's.[3]

Agricultural education in the land-grant system had matured in the 40 years after Congress passed the first Morrill Act in 1862. The agricultural colleges turned out increasing numbers of scientifically trained researchers, professors and agricultural experts. The Association of American Agricultural Colleges and Experiment Stations (AAACES), established in 1887 following the passage of the Hatch Act, took the lead in establishing curriculum standards and suggesting models for organizing and administrating agricultural colleges. The Association's annual meetings became a forum for the discussion of various issues pertaining to agricultural education and drew attention to the needs and the promising possibilities of an expanded agricultural education system. The AAACES was not only an effective lobby but did more than any other organization to enhance the professional image of agriculture teachers and researchers and to lay the pedagogical foundations for agricultural studies.[4]

One indication of the maturation of agricultural studies can be seen in the proliferation of textbooks, journals, technical manuals, Experiment Station bulletins and other agricultural literature in the years leading up to, and just beyond, the turn of the century. In the early days of the land-grant colleges and universities, there were but a handful of books on the scientific principles of agriculture; by the mid 1890's the number had grown but was still fewer than 100. In 1906 the Association's agricultural textbook bibliography contained 389 titles, and the quantity and the quality of the literature continued to grow throughout the period.[5]

Student interest in agricultural studies was reflected in a steady growth in enrollment in agricultural courses after 1905 at the University. The sharpening of student interest can be seen most clearly in the organization of the Agricultural Society in 1908 and the formation of a chapter of the Alpha Zeta national agricultural honor fraternity in 1917. The agriculture students brought attention to their activities and included the entire campus and community in them when they initiated "Agri Day" in 1915, which soon became a major campus and community event.[6]

In the nineteenth century the focus of political agitation surrounding agricultural studies at the University had been whether such studies were necessary and how to foster them. New problems arose as agricultural studies at the University of Arkansas moved into the twentieth century and became firmly established. The concern became who would determine the priorities for the College and how to derive the most benefit from the programs initiated through the institution. With increased appropriations, a multiplicity of clienteles in teaching, research and extension work and a more complex administrative structure, the College of Agriculture could scarcely avoid political entanglement. The College's success was in a large part determined by the effectiveness of its administrative leadership in navigating the political landscape inside the College and University and outside the campus with the state Legislature, the public at large and the federal government.

In 1900 the Agriculture Department reduced its sub-departments from the four listed in 1897 to two, Agriculture and Horticulture. C.L. Newman, the Superintendent of Agriculture, taught the agriculture course, and Ernest Walker replaced J.T. Stinson as horticulture professor. Walker represented the new breed of academically trained scientific agriculturalists being generated by the land-grant system during this period. He held a B.S.A. degree from Cornell, one of the most prestigious agricultural colleges in the nation, and had taken course work in botany at Harvard. After an active career in teaching and research at Cornell and Clemson College in South Carolina, Walker accepted the position at the University of Arkansas and was a major figure in the development of horticultural studies and research as well as con-

A photograph of part of the campus, circa 1910. The view is from the south looking north. The small building to the left is the office of the Director of the Experiment Station. The 1905 Agriculture Building is to the right. Barely visible through the trees behind the small building are the spires of the Agricultural Experiment Station Building. (Photo courtesy Photograph Collection, Special Collections, University of Arkansas Libraries.)

tributing to entomological studies. Walker was also active in the state Horticultural Society and other agricultural organizations in the community.

Stinson's resignation in 1899 to become Director of the Missouri Experimental Fruit Station ended his official connection with the University, but his new career kept him in contact with his colleagues in Arkansas. He developed a program that equipped a railway car with fruit sprayers, microscopes and informational literature and visited towns and farm communities along the rail line. Through this method, which was later adopted in Arkansas and other states, Stinson made direct contact with a wider audience with information about the Experiment Station's work. He later took a position as an agriculturalist for the Missouri Pacific Railway, showing that new careers in private industry were opening to agriculture graduates.[7]

In the 1901-02 school year, the Agriculture Department cataloged 12 collegiate and 34 preparatory students in the agriculture and horticulture classes. While the numbers were not large in comparison to those in other University departments, they represented a growing interest in agricultural studies, particularly among students at the collegiate level. At this time Newman was making efforts to recruit students from other departments who, while not intending to take a degree in agriculture, might be interested in taking agricultural classes "as a part of a liberal education." It's not known how many students elected to enroll in agriculture classes on this basis, but this trend would become more noticeable in the years after the establishment of the College and was indicative of patterns established at other land-grant institutions where agricultural colleges developed within the context of a "liberal arts" university.[8]

A major milestone in the development of agricultural studies was reached in 1904 when the University awarded the first Bachelor of Science in Agriculture to Hugh Ellis Morrow. Morrow, a Fayetteville resident, began his studies in 1899. His course work was originally geared toward earning a B.S. degree in chemistry, but he switched to agriculture in his junior year and joined 30 other graduates in other fields at the June 1904 ceremony. Morrow parlayed his agriculture degree into a position as a professor in the Chemistry Department, a post he held for the next 16 years.[9]

Hugh Ellis Morrow. (Photo courtesy Photograph Collection, Special Collections, University of Arkansas Libraries.)

The next year there were 10 collegiate students in agriculture and a total of 71 students taking either agriculture or horticulture classes. Presumably, the majority of these students were enrolled in the classes as part of a liberal arts education.

In the 1904-05 academic year, the University administration and the staff in the Agriculture Department began steps toward reorganizing the department into a College. The whole process was part of a larger movement in the land-grant system promoted by the AAACES aimed at standardizing the agricultural education curriculum and organizing colleges into specialized departments in order to make more efficient use of the accumulated scientific knowledge then at hand. The founding of the College of Agriculture at the University was also caught up in a political wrangle at both the state and university levels. It would not be the first time that politics would be a determining factor in the development of agricultural studies at the University, nor would it be the last.

Newman, as the Superintendent of Agriculture, seemed to be the logical choice to head the new College, given his well-regarded service in the department and his scientific credentials. However, he was passed over for the Dean's post in the new College of Agriculture. The Trustees fired Newman in 1904 as Superintendent and Professor of Agriculture. According to Newman, his dismissal was purely political. He had been heard criticizing the abilities of Experiment Station Director W.G. Vincenheller, whom he believed was a political appointee of Governor Jeff Davis. Furthermore, Newman let it be known loudly and publicly that he distrusted the "political trickery" of Governor Davis.

The Board then elected George A. Cole as the professor of agriculture and left the Superintendent's job vacant. Cole had taught for 12 years in the Preparatory Department, was the principal there for one year and previous to that was principal of several public schools in Washington County. Cole had no formal training in scientific agriculture as such but had spent his first 15 years on his family farm in Ohio. Cole was joined by the Director of the Experiment Station, W.G. Vincenheller, in bringing the plans for organizing the College before the trustees and the Legislature.

Some at the University considered both Cole and Vincenheller political choices rather than "scientific men," and to a certain extent this was true. Vincenheller had no formal training in agriculture either, except for a short stint in the 1890's as a "special student" at the University. He did serve as Arkansas' Commissioner of Mines, Manufacturing and Agriculture for six years in the 1890's, during which time he was indeed involved in politics. Following his tenure as Commissioner, he was employed as a pomologist at the Arkansas Experiment Station and later appointed as Vice-Director. In this role Vincenheller took the lead in promoting fruit culture throughout northwestern Arkansas and became popular with both farmers and businessmen in the area. In 1903 the Trustees appointed Vincenheller Director of the Experiment Station to replace Robert L. Bennett, who had resigned in protest over political interference in the scientific operation of the Station.[10]

After his appointment as Director, Vincenheller became more concerned with the lack of funding for the Station than with political interference. At this time, Jeff Davis, Arkansas' flamboyant "populist" governor, began gearing up a campaign for the United States Senate. The governor attempted to garner support in northwestern Arkansas by pushing for larger appropriations for the University. In early 1905, with the field ripe for increased funding requests, Vincenheller went, hat in hand, to Arkansas Sen. W.P. Fletcher, who headed the legislative committee in charge of University appropriations. The result of this meeting was a $35,500 state appropriation for the Experiment Station, which nearly matched the Station's $42,000 in federal funds. Previous to this the state had contributed only marginally to finance Station operations. In the bargain Fletcher inserted in the appropriation legislation that the money should be used for new buildings and maintenance and that "instruction be provided in practical agriculture, horticulture, entomology, veterinary and kindred subjects."[11]

W.G. Vincenheller, Dean and Director of the Station 1903-1908. (Photo courtesy Photograph Collection, Special Collections, University of Arkansas Libraries.)

The 1904-5 University Catalog contained a special notice to the effect that, "new and thorough courses in agriculture...will be established at the Experiment Station beginning with the September, 1905 term."

With Vincenheller appointed Dean and Experiment Station Director, the College was divided into seven departments following recommendations of the AAACES Committee on Instruction. Vincenheller's dual appointment as Dean and Director followed the trend at other land-grant institutions. The arrangement had its advantages in coordinating the resident teaching and research work with the limited resources available. But, as both agricultural research and teaching grew more complex, priorities diverged, and this required a Dean with nimble administrative skills to sort matters out.

The resident teaching personnel included two professors and six instructors. All held concurrent positions with the Experiment Station, except one, G.A. Cole. Only two had association with agricultural studies at the University before 1904-05. Dinwiddie had been the Experiment Station veterinarian since the Station was created in 1887, and Walker, as mentioned above, had only recently come to Arkansas. So, in a sense, the College of Agriculture was something new under the sun.[12]

Agriculture Faculty 1904-1905.

Department	Staff	Degrees	Tenure at UA
Agricultural Chemistry	John H. Norton	B.S.A.,B.S.	1904
Agronomy	G.A. Cole, Professor	B.S., M.A.	1904**
Animal Husbandry	Robert W. Wade	B.S.A.	1905
Dairy Husbandry	Victor A. Hooper	Dairy Graduate*	1905
Entomology	Charles Frederick Adams	B.S.A., A.M., M.D.	1905
Horticulture	Ernest Walker, Prof.	B.S.A	1900
	J.L. Hewitt, Instructor	B.S.A.	1905
Veterinary Science	Robert R. Dinwiddie	M.D., V.S.	1887

* Two-Year Degree
** Began in Agriculture Dept.

Along with a Dean, a College and six new teachers, agricultural studies also received something new that was a symbol of the arrival of the College of Agriculture, something that everyone could see. Out of the funds earmarked for agricultural studies the Trustees authorized the construction of two new buildings, a $9,000 Agriculture Building and $5,000 to erect and equip a new Dairy Building. The Agriculture Building, built during 1905-06, was a 54-ft x 68-ft two story brick structure with a basement. With limestone trimming and a slate roof and equipped with steam heat and electric lights, the building was "modern" in every detail and made an attractive home for the new college. The first story housed classrooms, a library, reading rooms and a lecture room. The second story accommodated the Dean's office, a bulletin room and another library, and the entire west side was set up as a 27-ft x 50-ft exhibit hall. The basement contained propagation rooms as well as storage facilities.

The new, two-story, 41-ft x 73-ft Dairy Building, completed in 1908, replaced the 30-ft x 41-ft, single-story "Dairy Barn" built in 1890. The first story held a modern creamery and dairy and cheese rooms. On the second story were offices, a library, laboratories, reading rooms, a lecture room and a milk testing laboratory. The basement contained a refrigerated cheese curing room and other work rooms.[13]

With these new facilities the College began a "semi-commercial" dairy operation. The dairy bought milk from local farmers because the University farm herd was still not large enough to provide the milk needed for instruction in butter and cheese making. The College then offered processed milk, butter and cheese for sale to the public at low prices through the University Creamery. Over the course of several years, this operation was built up until by 1909 the Creamery generated $20,000 in income for the College.[14]

The College's physical plant now included these two new buildings, the Experiment Station building, a forcing house, greenhouses and stock barns and various shops on the University farm. While Vincenheller and others still considered the farm acreage inadequate and they tried to persuade the Legislature for appropriations for expansion, for the moment they had to settle for

The first Agriculture Building was built in 1905 at a cost of $9,000. Over its long life it has been used as a Home Ec Building, the University Infirmiry, and (its current role) the Agricultural Annex. It stands with its back to Maple Street directly north of the current Agriculture Building. See map, page 20, building 4 (Photo courtesy Photograph Collection, Special Collections, University of Arkansas Libraries.)

The Dairy Building, also known as the Animal Industry Building, was completed 1908 at a cost of $5,000. It was located at the corner of Garland and Maple where Silas Hunt Hall now stands. The building housed equipment for pasteurizing and bottling milk, buttermaking and ice cream manufacturing as well as the offices of the Animal Industry Department, classrooms and labs. See map, page 20, building 8. (Photo courtesy Photograph Collection, Special Collections, University of Arkansas Libraries.)

the 160 acres originally pledged for use as a farm that was now principally occupied by the University campus. Still, progress had been made toward a physical setting for the College.

As in previous years, the new College's course of study offered students a four-year program leading to a B.S.A. degree and a two-year course, now called a "Special Course," in which the student earned a certificate. In the collegiate course all students in their first two years, and part of the third, took classes in horticulture, animal husbandry, dairying, agronomy, entomology and veterinary science in addition to the English, mathematics and science required for all students. When students were in their junior year, the College required each student to choose a specialty for intensive study in one of the seven departments. Agriculture students then spent 17 of their 30 total hours of course work on the declared specialization during their junior year and 24 hours out of 30 in their senior year. The course work closely followed the standards recommended by the AAACES and matched programs offered in many other land-grant colleges.[15]

In keeping with the philosophy of developing the College within the framework of a liberal arts university, the College also offered agriculture classes of a more general nature. A general horticulture class was geared for those students working toward a teacher's certificate at the University and also targeted a growing interest among the middle class in beautifying their homes with decorative plants and shrubbery.[16]

A new program, named the "Short Winter Course," was patterned after similar successful programs at other land-grant colleges. The course's objective was to bring farmers to the College during their winter slack time for practical instruction in the latest research results. Such courses had proven to be quite a public relations coup for the Agriculture College at the University of Wisconsin, and they had the added advantage of extension work that did not have to leave the campus. The College held the first "Farmer's Course" from January 6-18, 1907.[17] Twenty-five people attended the two-week course. Nearly all the College and Experiment Station faculty and staff gave lectures or demonstrations on such subjects as orchard care, Texas tick fever eradication, dairying, animal feeding and nutrition, soil fertility and cotton production, to name but a few. Although the attendance at the course was not as large as the organizers had hoped, the program grew in the coming years and became an eagerly anticipated annual event that brought much favorable attention to the College.[18]

Whereas in former times agricultural classes and faculty offices had been scattered throughout the campus or held in the Experiment Station's cramped quarters, now students, faculty and research personnel had a home of their own and had the use of their own lecture halls, libraries and reading rooms. This allowed a greater cohesion and the development of stronger academic and social ties. In early 1906 the staff and faculty at the College made strides toward cementing their relationship by forming the Research Club. Ernest Walker, J.L. Hewitt and Victor Hooper took leading roles in the new club. They held meetings every two weeks and discussed topics of interest, read papers and encouraged original research in the field. While membership of the Research Club was confined to faculty and staff, it was but a short step further to the creation of an agricultural student organization.[19]

In October 1908 agriculture students took the first steps toward organizing what became known as the Agricultural Society. Students in the Agricultural Club (as it was originally known) elected M.B. Oates as the club's first president, along with H.H. Holtzclaw, vice-president and J.H. Southworth, secretary-treasurer. Ernest Walker and Albert K. Short, animal husbandry professor, and Rufus J. Nelson, agricultural education professor, two new appointees in the College of Agriculture, assisted the students in organizing the club. The Agricultural Society became an important feature in the College of Agriculture, helping coordinate social functions and providing a forum in which a wide range of issues could be addressed.[20]

Following the 1908 Farmer's Course, Dean Vincenheller embarked on a tour of the state in order to spread the gospel of "modern methods of farming." In doing so, he picked up the agricultural extension work that the University began in 1894 in which he participated as Agriculture Commissioner and again in 1899 when he was Vice-Director at the Experiment Station. Both of these programs had been canceled because of lack of support. Vincenheller was now on more solid ground. The AAACES had formed a special committee that promoted the development of Farmers' Institutes and other extension activities across the nation. The USDA also joined in the movement through providing funds, personnel and information through their Farmer's Cooperative Demonstration Extension program, which was begun by Seaman A. Knapp. Armed with information on improved cotton and corn varieties recently developed at the Experiment Station and provided with $4,000 from the USDA, Vincenheller spoke at over 26 Farmers' Institutes in a two-month period. He demonstrated new methods and promised to distribute samples of the new seeds to farmers the next fall. There is no way to determine how successful Vincenheller was at getting the farmers to adopt new ideas. But the Farmers' Institutes were well attended and became a permanent feature of the College's extension program.[21]

Following Vincenheller's return from his tour of the state, troubling rumors regarding mishandling of funds at the Experiment Station and the College began to surface. In June 1908, the Board appointed a sub-committee to investigate alleged improprieties in leasing additional land for Experiment Station use and in handling the funds generated by the College's dairy operation. The case hung fire through the summer until the Board's August meeting. At that meeting the sub-committee announced that they still were not prepared to issue their report. However, newspaper accounts said that the investigation had accumulated 60 pages of testimony and compiled a six-page summary of its findings already. Vincenheller wrote out his resignation at this time, citing ill health and exhaustion brought on by his administrative duties as his reasons for doing so. However, the Board did not accept his resignation until November, and the results of the sub-committee's investigation were never made public.[22]

C.F. Adams, Dean and Director 1908-1913. (Photo courtesy Photograph Collection, Special Collections, University of Arkansas Libraries.)

Following Vincenheller's resignation, the Board appointed Charles F. Adams, head of the Entomology Department since 1905, as Acting Director of the Experiment Station. In June 1909 they made his appointment permanent and elected him College Dean and Experiment Station Director. Adams was only 32 when he assumed the position of Dean. Although he was young, his research and teaching experience was expansive. He held degrees in medicine and agriculture and had done extensive research and teaching in such varied fields as histology, marine biology and entomology.

During his tenure from 1908-1913, Dean Adams tried to push a research-oriented agenda at the College. He attempted to have the teaching loads reduced and research funds increased. In these efforts he was unsuccessful and appeared to be bucking the tide of increased student numbers in agricultural classes and a blossoming interest in the College's agricultural extension program.[23]

In January 1909, the campus newspaper's agricultural news editor reported that there were "over 300" students in agriculture classes. While this was undoubtedly wishful thinking (the numbers were closer to 150), it goes to show that by this time, someone involved in agricultural education at the University could dare to hope that such numbers were possible when only a few years before the agriculture students could be counted on the fingers of one hand. The faculty, including professors, adjunct professors and assistants, had grown from two in 1904 to 22 in 1909.

In the 1908-10 school years, the College expansion continued; two new departments were added to accommodate the interest in agricultural education (Department of Agricultural Education) and extension work (Department of Extension Education).

The Department of Agricultural Education was headed by Rufus J. Nelson, who graduated in 1907 with an M.S. in Agriculture from the University. Nelson, along with Adjunct Professor J.M. Wilson, focused instruction on subjects needed by students intending to teach agriculture at the high school level. Such vocational education classes, targeted especially for rural areas of the state, were part of the state-wide effort at educational reform that sought to bring educational facilities to the most remote parts of Arkansas. This new demand for secondary school agriculture teachers opened another career path for those studying agriculture at the University. The demand would increase in the years after WWI, when vocational agriculture classes became a regular feature of secondary school curricula across Arkansas.[24]

The College's role in agricultural extension work became increasingly important in the years between 1909 and 1914 when the terms of the Smith-Lever Act transformed agricultural extension work. George A. Cole, who resigned his position as head of the Agriculture Department (which replaced the Agronomy Department in 1907) in 1908, became the Superintendent of Farmers' Institutes and the head of the Department of Extension in 1909. Cole's popularity among farmers made him the natural choice for the job. That year he was elected president of the Arkansas Farmer's Union and also the president of the Arkansas State Fair. Cole organized the College's agricultural extension work into three branches: the Farmers' Institutes, Boy's Corn Clubs and correspondence classes. In his efforts, Cole was free to solicit the aid of any of the College faculty.

From single meetings with interested groups of farmers to weeks of interaction and instruction as seen in the Farmer's Winter Course, the program of the Farmers' Institutes took many forms. The Extension Department adopted J.T. Stinson's "rolling experiment station" extension method in 1911. Railroad companies, vitally interested in farmers' patronage, equipped a train with sprayers and equipment and staffed it with College and Experiment Station personnel who offered demonstrations and lectures to interested fruit farmers at various locations along the line. The so-called "Educational and Demonstration" train also featured the use of improved pruning implements and orchard heaters provided by local hardware dealers. Ernest Walker was quick to point out that the expensive heaters were not generally recommended for orchards in northwestern Arkansas.

While businesses, railroads and the College all had a mutual interest in bringing new information to Arkansas farmers, the kind of information imparted depended on its source. Generally speaking, the College's agricultural experts tried to keep their talks and demonstrations understandable but still based on scientifically proven principles. As a rule, they wanted to avoid the "promotional" approach sometimes used by agribusiness concerns and railroads. Walker's comments regarding the efficacy of equipment promoted by the hardware dealers brings into focus the potential for conflicts in the cooperative extension partnership between the College and business.[25]

H.S. Mobley, president of the local Farmer's Union and a USDA special field agent for boy's and girl's club work, aided Cole in putting together the second phase of the College's extension program. The idea of organizing Boy's Corn Clubs and Girl's Canning Clubs in order to draw attention to the Farmers' Institutes began in Illinois in 1902. The first Boy's Corn Club in Arkansas was organized in 1908. The clubs proved so popular that Seaman Knapp organized a special department within the USDA's Cooperative Demonstration program in 1908 to promote their activities. Nationwide, by 1909 there were over 150,000 boys and girls in these clubs, the forerunners of the 4-H program.

The third part of the extension program at the College included correspondence courses. The College offered 16 courses in various agricultural subjects such as elementary agriculture, vegetable gardening, principles of canning and preserving, soils, dairying, farm management and home and school sanitation. The courses were available to anyone "who desires an agricultural education and who can–not...attend college." Farmers interested in improving the farm operation, farm wives seeking to beautify the farm home or enrich farm life, boys and girls preparing for college, home gardeners or teachers preparing to teach agriculture were all potential correspondence students. The College also encouraged Farmer's Union locals to organize classes and promised to send teachers to visit the classes occasionally in order to give personal instruction.[26]

In the first three months of 1913 alone, the College's extension activities tallied 45,438 lectures to farmers, 60,000 letters sent, 105,000 bulletins printed, 120 newspaper articles run and 3,124 boys and girls enrolled in clubs. The work expanded to even greater proportions after the Legislature appropriated $7,500 for agricultural extension work in 1913-14.

Extension activities promoted by the College that focused on farm women, called the "Home Demonstration" program, took on increased importance after a Home Economics Department was created within the College in 1913. Since agriculture and home economics are closely related subjects, focusing on farm and home production as integrated units, it was a logical progression to bring home economics under the purview of agricultural studies. The union of the subjects mirrored the conditions on most farms where the work of the home and the farm are "indissolubly connected." Food and nutrition, bacteriology and entomology and other related subjects hold a common scientific interest for both fields of study, so it seemed fitting that home economics should develop within the realm of agricultural education.[27]

As early as 1908, Mrs. Rufus J. Nelson, the wife of agricultural education professor Nelson, began a series of lectures on domestic sciences on Saturday mornings at the Agriculture Building. Nelson, assisted by Walker, Hewitt and Hooper, was trying to "arouse interest in the subject" in order to convince the Board of Trustees to fund such a department at the University. Reportedly there was a domestic science class with 31 students in 1909, but it is not clear whether the class was held in the College of Agriculture or the School of

Students at work in an entomology classroom, circa 1919.

Education. In any case, the Home Economics Department began in the College in the 1913-14 school year. The Board elected Sarah Petit, who held a B.S. in domestic science from Columbia University, as the head of the new department and Mrs. R.C. Thompson, with a B.S. from Kansas State, as instructor. The department offered full courses in either domestic science or domestic arts.

Originally the department and classes were housed in Peabody Hall. Tucked off in the corner of the basement, the department had neither the space nor the equipment necessary to serve the needs of the students. Consequently, many students requesting work in the department were turned away in the first years of its existence. Requests for increased appropriations for the department were routinely turned aside by either legislators or the governor. After Governor George W. Hays vetoed $8,000 appropriated by the 1915 Assembly for improvements and equipment in the Home Economics Department, the students condemned him, "to eat soggy biscuits for the rest of his days."

Unfazed by the barriers to increased enrollment, the home economics students and faculty quickly established themselves by forming the Home Economics Club in 1914 and hosting and serving at numerous University and College of Agriculture events. By the 1917-18 year, the Home Economics

Department provided seven of the 18 graduates from the College of Agriculture. So, in spite of early difficulties, the department grew rapidly and would go on to form a vital and integral part of the College.[28]

With the addition of three new departments, especially Home Economics, the student enrollment in the College expanded. University President John C. Futrall, a tireless advocate for agricultural studies, said in his 1914-15 report that student enrollment in the College of Agriculture was "increasing more rapidly than any other division in the University." According to Futrall, there were 93 students listed as working toward degrees in the College at this time; if the number of students just taking agriculture classes were added, then the total number enrolled in the College would triple. Futrall praised the new Dean of the College of Agriculture, Martin Nelson, for his efforts in trying to juggle the College's research, resident teaching and extension responsibilities. Nelson took over the Deanship after C.F. Adams resigned in July 1913. Nelson had earned both his B.S.A. and M.S. degrees from the University of Wisconsin College of Agriculture. Nelson came to the University of Arkansas in 1908 as Professor of Agronomy and head of the Department where he was deeply involved in both teaching and research.[29]

Martin Nelson, Dean of the College 1913-1920. (Photo courtesy Photograph Collection, Special Collections, University of Arkansas Libraries.)

As Martin Nelson stepped into his new job as Dean, the faculty of the College was in the midst of a statewide extension campaign. College faculty and staff oversaw 40 Farmers' Institutes that averaged one week in length. The institutes were held in 26 counties, and over 9,000 farmers attended the meetings. Following the end of this campaign in September 1914, the so-called "Profitable Farming Campaign" inaugurated by the Profitable Farming Bureau, a group of Little Rock bankers and businessmen, barnstormed the cotton regions of the state. Twenty faculty and staff from the College devoted a total of 450 hours to the Profitable Farming Campaign before the faculty had to wind up their involvement in January 1915.

Although the extension work was a tremendous help to the farmers in the beleaguered cotton country, it came at the expense of the resident teaching responsibilities at the College. Dean Nelson said that during the extension campaigns, classroom instruction suffered and that most of the College faculty and staff had been unable to take vacation time.[30]

In May 1914 the Smith-Lever Act created a national Agricultural Extension Service. It arrived in Arkansas just in time to keep the agricultural extension work at the College from overextending itself. Governor George W. Hays accepted the terms of the act in June. The Arkansas Assembly passed an enabling act in January 1915 and appropriated $45,260 to match the federal seed money pledged for the program that went beyond the initial $10,000 appropriated by Smith-Lever. The terms of the act called for a cooperative arrangement between the Extension Services of the USDA and the College of Agriculture. The College would administer all local, state and federal funds allocated for the program and agreed to cooperate in all federal extension work conducted by the USDA. The act completed the third side of the "Agricultural Triangle" of resident teaching, research and extension created to carry out the mission of the land-grant system.

The over $100,000 in funding pledged to the Extension Service program from federal, state and local monies for the 1914-15 biennium dwarfed any amount previously used in Arkansas for agricultural extension. The program grew rapidly as many involved in agriculture saw the great possibilities in being able to improve farming operations as well as the lives of farmers through extension activities. The farmers themselves were slower to accept this infusion of "book farming." But it would not be many years before the county agent was a familiar figure with most farmers, who would regularly seek the agent's advice.[31]

The creation of the Extension Service took much of the responsibility for extension work from the College of Agriculture's resident teachers and researchers, but it didn't take all of it. The teachers and researchers still contributed with lectures, demonstrations and consultations, as the Smith-Lever Act intended. In the 1915-16 year alone, the College's resident faculty gave 460 lectures, 37 demonstrations and 30 consultations.

In most states the Dean of the agriculture college was also Extension Director, and the Extension Service headquarters were located on the campus of the land-grant university. In Arkansas a different arrangement developed in the first years of the program. The Extension Director was made a separate officer from the Dean of the College. Though officially answerable to the Dean as a member of the College faculty, the Director also had considerable administrative leeway to chart an independent course. L.C. Lassetter, the first Extension Director, had an office in Fayetteville, where he was officially headquartered. However, Lassetter spent most of his time in Little Rock, where the USDA's Farmer's Cooperative Demonstration Extension Service program was headquartered before the Smith-Lever Act was passed.[32]

Bankers and businessmen were also keenly interested in agricultural extension work. Consequently, a group of Arkansas bankers and businessmen formed the Profitable Farming Bureau (PFB), an organzation that would assist in getting the latest information out to farmers. The Profitable Farming

Campaign, inaugurated by the PFB in 1914, brought businessmen, bankers and the College's agricultural experts together to provide advice for the benefit of the state's farmers.

Unfortunately, the alliance between the businessmen and scientific agriculturalists fell apart over who should decide what kind of information should be given to the farmers. An extension worker involved in the campaign observed that bankers had been trying to get farmers to do their bidding for years and that for just as long, farmers had been ignoring them. The businessmen, led by the PFB's E.J. Bodman, a Little Rock banker, felt that the College and the Extension Service were not moving quickly or energetically enough to help improve farming conditions in Arkansas. On the other hand, some in the College and Extension Service felt that Bodman and his associates promoted farming practices that did not have proven scientific value and did more good for the businessmen selling seed or fertilizer than for the farmer.

E.J. Bodman, a Little Rock banker, head of the Profitable Farming Bureau. (Photo courtesy Photograph Collection, Special Collections, University of Arkansas Libraries.)

In June 1915 Bodman pressured President Futrall and the Board of Trustees to replace Dean Nelson and Extension Director Lassetter with people "more energetic...more renowned" and more in step with the PFB program. Though Nelson remained Dean of the College, Lassetter was downgraded to Assistant Director in charge of extension specialists, and J.H. Miller, an extension officer from Kansas A&M College, was appointed Dean and Director of Agricultural Extension Service.

Bodman handpicked Miller as Lassetter's replacement and dubbed him, "the best man in extension work in the United States." However, Bodman got more than he bargained for in Miller. Within months of his appointment in June 1915, Miller and Bodman were at odds. In an August 1915 letter to Bodman, Miller not only disputed some of the Bodman's recommendations on scientific grounds, but he also questioned the right of Bodman and his committee to direct agricultural extension work in Arkansas at all:

> In what way is the responsibility of successful agriculture in Arkansas placed upon your Committee? Why should you, a busy banker, continue to hold this terrible responsibility of directing agricultural interests after you were able to have the University follow your advice and select a Director of Extension work?

Miller also objected to the charge that he was "moving too slow" on the agricultural front, the same complaint that Bodman had leveled against Lassetter.

> (In) my ten of years of experience in this work...I have never made an announcement, or promoted a plan or an idea, or recommended a method or a seed or agricultural proposition whatever that I ever *had to take back* [emphasis in the original]. I cannot stay in this work if I am to be pushed, crowded and directed into things before I am ready.[33]

Such frank talk stung Bodman's concern for the importance of the PFB's agenda, and he began to lobby for Miller's removal. Although Futrall tried to stick by Miller for a season, he eventually gave in to Bodman, and the trustees fired Miller in October 1916. During the interim siege Bodman went far afield to find Miller's replacement; indeed he went to the very top. Bodman commented on Miller in a February 1916 letter to Bradford Knapp, the United States Director of the Extension Service for the South, and then offered Knapp the job:

> ...a man cannot continue to act as he has and hold out and unless he changes, there is likely to be a change. Besides, as I wrote you before, we would like to see you at the head of Extension in the State. You would find a bunch of loyal men back of you, men who have seen the light and would lend every ounce of support that they have.[34]

Later events would reveal the irony in Bodman's remarks after Knapp came to Arkansas as Dean of the College.

The controversy continued on into the 1920's and had an unsettling effect on the College as Bodman and the PFB brought political pressure to bear in an effort to influence appointments in the College and Extension Service as well as the direction of agricultural studies.[35]

Fortunately, involvement of agriculture students in the College progressed on a smoother course. With increased student enrollment in the College of Agriculture, both the Agricultural Society and the Home Economics Club grew as well. On the Wednesday before Thanksgiving in 1915, the Agricultural Society and the Home Economics Club sponsored the first "Agri Day" celebration. The day was declared a holiday from classes, and the agriculture students sponsored a parade, a picnic and a "barn dance" in the evening. Agri Day, which was later moved to the springtime to take advantage of warmer weather, became an established tradition on the campus and was an eagerly anticipated community event in Fayetteville.[36]

Student involvement took on a patriotic flavor after the United States entered World War I in April 1917. The agriculture and home economics students both contributed to the war effort. The men enlisted or were drafted into the armed forces and participated in the "Food Will Win the War" agricultural production campaigns. The women organized and cooked from "wheatless" "meatless" and "sugarless" menus on and off campus and helped

The first Agri Day was held the day before Thanksgiving in 1915. Pictured here are "home ec girls," attired in white with matching caps, who served coffee and "agri specials," special cakes for which the recipes were given as souvenirs. They also served 300 people at a barbeque at noon. Later reports commented on the "riotous expense" of the floats for the Agri Day parade that year: $90. (Razorback, 1916)

disseminate information on the subject across the state. Home economics students and faculty also participated in the various campaigns to raise agricultural production and organized classes in food and nutrition as it related to the war. The University gave class credit for agriculture students who left the classrooms for the trenches or for those who went back to the farm to increase production for the war effort.[37]

In the midst of the war effort, Congress passed the Smith-Hughes Act. In September 1917 the State Board of Education designated the University of Arkansas as the state institution qualified to receive the funds made available for training secondary school vocational agriculture teachers under the terms of the Smith-Hughes Act. The legislation was another milestone in the history of agricultural education that ranks with the Morrill Act, the Hatch Act and the Smith-Lever Act in importance.

The act's provisions opened career opportunities for students in the College of Agriculture as public school teachers. Also, it brought agricultural education into public schools and exposed students at an early age to the ideas of scientific agriculture. More importantly, the Smith-Hughes Act provided money to hire teachers and start programs for vocational education. This was a welcome development, especially in rural states such as Arkansas where any appropriations for school support were needed.

President Futrall often pointed to the deficiencies of Arkansas' public schools as one reason why agricultural studies at the University did not attract more students. In later years one national study found that 22 percent of the students from the Smith-Hughes schools went on to college, and of those, 36 percent chose agricultural colleges. The Smith-Hughes funds also went to support night schools, short courses and other extension activities as long as qualified Smith-Hughes teachers were part of the program. The Smith-Hughes programs molded a new generation's interest in agricultural education and can take much of the credit for increased enrollment in agricultural colleges in the 1920's. With this upsurge of interest, the Agricultural Education Department in the College took on a new significance and would continue to play a key role in the College in the years to come.[38]

Enrollment 1905-1920.

Year	1905	1906	1907	1908	1909	1910	1911	1912
Students	10	13	19	13	13	23	26	58
Graduates	1	3	1	1	4	1*	-	-

Year	1913	1914	1915	1916	1917	1918	1919	1920
Students	124	93	139	113	81	78	98	147
Graduates	5	10	3	6	18	10	10	8

*M.S. awarded.

As was the case with the enrollment in agriculture courses in earlier years, the number of students in the College of Agriculture from 1905 until 1920 was also a bit of a roller coaster. The number of students showed a slow and but steady increase until 1912 when it jumped from 26 to 58 students. In 1913 it jumped 114 percent to 124 students and see-sawed up and down until WWI, when army enlistments kept enrollment below 100. Enrollment stayed below 100 before it reached 147 in 1920. In the period between 1912 and 1920, the enrollment in the College of Agriculture averaged 14% of the total at the University.

> *If there is any arena where capitalistic enterprise and environmental integrity come face to face, it is in agriculture....If there ever was a center of thought and activity that combines the most forward technology with a love of the land, in every aspect of that term, it is agriculture!*
>
> *JoAnne Sears Rife, BSHE 1956, MS 1971*

Following the dramatic increase in students after 1911-12 and particularly after the creation of the Home Economics Department, the College made perennial pleas for more space and funds for maintenance. The building erected in 1905 that was received with such acclaim was now deemed inadequate. The College had more than 10 times the number of students, and the faculty and staff had tripled. The addition of extension responsibilities and training Smith-Hughes teachers all contributed to overcrowding.

The need to expand the University farm acreage was also among the College's foremost needs. The original 160 acres pledged for the University farm shrank to 40 acres as the campus expanded over the years. Dean Nelson described most of the farm acreage as "thin, rough and hilly" and unsuitable for experimental farming. In his 1918 annual report, President Futrall put the purchase of additional land for a farm as the top priority for the University. He recommended that the Legislature appropriate enough money to buy a farm of at least 500 acres and went on to say, "Until it is secured and well equipped, it will be an impossibility for the College of Agriculture to do for the state the service which it is expected to do and should do."

By 1919 the enrollment in the College climbed back toward the century mark with 98 students. In addition to offering a special four-week short course for returning veterans, the College prepared vocational training for disabled soldiers that aided greatly in helping make the transition from soldier to civilian a smooth one. One other new feature was the purchase and renovation of a six-room "Practice Cottage" on Dickson Street for the Home Economics Department.

Finally, in April 1919, the Trustees announced the purchase of 423 additional acres for the University Farm one and a quarter miles north of the campus. The Trustees also said they were considering buying an additional 175 acres in the near future. This was indeed welcome news for the supporters of the College of Agriculture. However, the Board ignored the requests of Dean Nelson and President Futrall for a new Agriculture Building, a building for the Home Economics Department, increased salaries for teachers and other requests for maintenance. The season certainly appeared to be ripe for a showdown with the Legislature over these additional requests. Unfortunately, Futrall, Nelson and the Extension Service personnel were deeply embroiled at the time in a struggle with E.J. Bodman and the PFB over the direction of the College and the future of the University itself.[39]

The Home Economics Practice Cottage, 1919. (Razorback, 1919)

The PFB organized a major campaign during the summer of 1918 when Bodman led a group of bankers, businessmen and politicians on a tour of agricultural farms and facilities in Missouri, Iowa and Wisconsin. The tour included a comparative inspection of the extension services and agricultural colleges at Iowa State and the University of Wisconsin with Arkansas. Arkansas' programs did not compare well with the well-developed and well-financed operations in Wisconsin. Bodman used the information to underscore his contention that the leadership at Arkansas' College of Agriculture and Extension Service was "less than inspiring" and needed rejuvenation. He overlooked the fact that the taxpayers and state legislators in both Iowa and Wisconsin had liberally supported their land-grant institutions for many years.[40]

The PFB campaign intensified in the fall of 1918 after Bodman hired H.M. Cottrell, a self-proclaimed "agricultural expert," to guide the publicity for the upcoming spring 1919 campaign. In his speeches and newspaper articles, Cottrell mixed information gleaned from Extension Service circulars and Experiment Station bulletins with his own overstated claims for success and promotional bombast. While the general public was captivated by Cottrell's promise that Arkansas farmers could increase their annual income by $300 if they followed the PFB program, many scientific-minded agricultural experts felt that Cottrell's statements were deliberately misleading. Cottrell, Bodman and the Extension Service personnel originally planned the 1919 campaign as a cooperative effort. Once the project got underway, some Extension Service agents and specialists complained that Cottrell and Bodman were railroading the Extension Service into endorsing the PFB's extravagant claims. In the midst of this friction, Bodman once again pressed Futrall to dump Lassetter for a PFB-endorsed Extension Service Director. This time he wanted Cottrell. Futrall would not accommodate him, though, especially after receiving negative reports about Cottrell's earlier activities in Colorado.[41]

H.M. Cottrell. (Photo courtesy Photograph Collection, Special Collections, University of Arkansas Libraries.)

In June 1919, Bodman and Cottrell launched a concerted effort to replace Futrall, Nelson and Lassetter with more compliant University officials. The showdown came at a specially arranged Board of Trustees meeting at the office of Governor George H. Brough in Little Rock on June 21. Bodman and the PFB presented their case for replacing the officials and revamping Arkansas' Extension Service. However, Futrall and various University and Extension people (but not Dean Nelson) presented evidence of the PFB's political interference in the Extension Service and outlined several examples of the misinformation promulgated by Cottrell and others. Futrall's strong presentation carried the day, and the Trustees supported him. Part of the evidence he used was a lengthy letter from Bradford Knapp criticizing the PFB's tactics. Some of Knapp's remarks echoed J.H. Miller's comments four years earlier:[42]

> Members of the Profitable Farming Committee ought to know by this time that I have never approved of their employment of an agricultural man to conduct agricultural campaigns....[W]hat I have had to say upon all occasions has been that business men have no business to be giving agricultural instruction to farmers any more than farmers should be giving banking instructions to bankers.[43]

The PFB's conduct was not the only target for Knapp's criticism. He also scrutinized the behavior of officials of the College of Agriculture and Extension.

> I was shocked...to find that the Extension forces in which the University of Arkansas College of Agriculture and this Department are so vitally interested, permitted themselves to be whipped into line in the campaigns which were ill-advised and stood mute when misinformation was being peddled out to farmers.

Knapp issued rules of conduct for Extension Service personnel in regards to such campaigns and threatened censure for any employees who failed to follow the guidelines in the future.[44]

That fall, after the political climate had cooled somewhat, Futrall offered Bradford Knapp the Dean's post in the College of Agriculture. It is unlikely that Futrall was motivated by Bodman's charges of incompetence in the College in seeking Knapp for the post. But having a man of Knapp's reputation at the head of the College opened many possibilities. In any case, Nelson accepted a post as Vice-Dean with equanimity, and Lassetter was retained until a successor was found. After extended discussions with Futrall and the Trustees over what his role would be, Knapp accepted the offer in November 1919 and began his new duties in January 1920.

This somewhat surprising turn of events even received Bodman's measured blessing. Given his earlier pledge of cooperation, there was little else he could do. Knapp accepted Bodman's offer but told him bluntly that any cooperation would be on his terms only. That being said, Knapp went on with his mission to "transform agriculture in Arkansas" using the University of Arkansas College of Agriculture as his vehicle. Knapp was as good as his word; as the College moved into the 1920's, an era of great change and expansion was in store.[45]

Endnotes

[1] True, 199-200, 221, 233, 273; Reynolds and Thomas, 154, 161, 175; Leflar, 71; Strausberg, 22.

[2] True, 288-89, 291-93.

[3] True, 291-93; C. Vann Woodward, *Origins of the New South: 1877-1913* (Baton Rouge: Louisiana State University Press, 1966), 400-403.

[4] True, 206, 210-19.

[5] Ibid, 230.

[6] *University Weekly*, 21 October, 1908, 1:3; 18 November 1915, 1:1; *University of Arkansas Cardinal, 1915,* 176.

[7] Reynolds and Thomas, 275, 533-34, 527-28.

[8] *Catalog, 1901-02*, 110-13; True, 216-17; John W. Jenkins, *A Centennial History: A History of the College of Agricultural and Life Sciences at the University of Wisconsin-Madison* (Madison: College of Agricultural and Life Sciences, 1991), 55, 58-59.

[9] Reynolds, 497.

[10] Clifford L. Newman Biography, Box 2: File 26, George A. Cole Biography, Box 1: File 30, *Reynolds and Thomas Research Papers*, Mullins Library Special Collections, Fayetteville. Hereafter cited as *Research Papers.* Futrall to "Mamma," September 9, 1908, Box 2: Folder 6, Futrall to "Father," May 4, 1913, Box 2: Folder 8, *John Clinton Futrall Papers*, Mullins Library Special Collections, Fayetteville; Reynolds and Thomas, 276; Leflar,74-75.

[11] Reynolds and Thomas, 276.

[12] *Catalog, 1905-06*, 139-148.

[13] Ibid, 31-33.

[14] Arkansas General Assembly, *Report of the Joint Committee to Visit the University of Arkansas* (Little Rock: n.p., 1911), 103-104, 114-15. Hereafter cited as *Joint Committee Report.*

[15] *Catalog, 1905-06*, 139-140, 118.

[16] Ibid, 146.

[17] Jenkins, 47-48; True, 278-79.

[18] *University Weekly*, January 16, 1907, 1:2; *Catalog, 1906-07*, 125-26.

[19] *Catalog, 1905-06*, 60.

[20] *University Weekly*, October 21, 1908, 1:3, October 28, 1908, 4:3, November 11, 1908, 2:2-3.

[21] *University Weekly*, February 13, 1908, 2:3; *Trustees Minutes*, 412; Reynolds and Thomas, 150-51.

[22] *Trustee Minutes*, June 9, 281; August 28, 285-86; *Arkansas Democrat*, August 29, 1908, 7:5, 8:5; *Arkansas Gazette*, August 30, 1908, 16:3.

[23] Reynolds, 437-38; Strausberg, 27-28.

[24] *Catalog, 1909-10*, 153-36; Woodward, 404-406; True, 345.

[25] *Catalog, 1909-10*, 154; *Arkansas Sentinel* (Fayetteville), December 16, 1909, 3:1, January 2, 1910, 8: 2-3; *University Weekly*, December 1, 1911, 2:3, Dec. 15, 2:2.

[26] *Catalog, 1909-10*, 154; Hale, 197-98; True, 393-94.

[27] True, 269.

[28] *University Weekly*, October 21, 1908, 1:3, November 12, 1914, 2:3; *Dean's Reports*, Box 2, File: 1908-09, 37; *University Bulletin* 8(15):37 (December, 1914).

[29] *University Bulletin*, 2-3; Hale, 107; Reynolds, 499.

[30] *University Bulletin*, 4-5; "Report of Extension Work in Arkansas, July 1, 1914-June 30, 1915, 1-2," *University of Arkansas Cooperative Extension Service Records, 1914-88*, Box 7, File: History, University of Arkansas Mullins Library Special Collections, Fayetteville. Hereafter cited as *E.S. Records*.

[31] *E.S. Records*, 1-3; Hale, 194-95.

[32] *University Bulletin* 10(21):56-57 (January 1917); *E.S. Records*, 3.

[33] Leflar, 105-106; J.H. Miller to E.J. Bodman, August 17, 1915, Box 13, File: Profitable Farming Bureau, *Dean's Reports*.

[34] Bodman to Bradford Knapp, February, 18, 1916, Box 1, Folder 1, *Bradford Knapp Papers*, Southwest Collection, Special Collections Library, Texas Tech University, Lubbock, Texas. Hereafter cited as *Knapp Papers*.

[35] Leflar, 106-120.

[36] *University Weekly*, November 18, 1915, 1:1, January 21, 1915, 1:3; *University of Arkansas Razorback*, 1917, n.p.

[37] *University Weekly,* October 25, 1917, 3:5, November 8, 1915, 3:1, December 13, 1917, 1:1, January 10, 1918, 1:3, February 21, 1918, 1:2.

[38] *University Weekly*, September 28, 1917, 3: 4-5; *University Bulletin* 10(21):9 (January 1917), 12(19):6-7 (December 1918); True, 369-74, 382.

[39] *University Weekly*, January 9, 1919, 1:1-3, February 6, 2:4, February 13, 4:4, April 10, 1:5; *University Bulletin* 19(21):33 (January 1917).

[40] Leflar, 108-109.

[41] Leflar, 110-12; W.L. Carlyle to Futrall, August 23, 1919, Box 2, Folder 3, *Knapp Papers*; Bodman to University Trustees, n.d., Box 13, File: Profitable Farming Bureau, *Dean's Reports*.

[42] Leflar, 112-120.

[43] *University Bulletin* 13(8):85 (July 1919).

[44] Ibid, 86-88.

[45] Futrall to Knapp, October 11 and 30, November 24, December 10 and 18, 1919, Box 1, Folder 1; Knapp to Futrall, December 29, 1919; Knapp to Bodman, December 29, 1919; Futrall to Knapp, January 1, 1919, Box 1 Folder 3; Box 2, Folder 1 Knapp to Nelson, January 2, 1920, Box 2, Folder 2, *Knapp Papers*.

Chapter Three

Growth and Consolidation: 1920-1930

The early 1920's was not the most propitious time for "transforming Arkansas' agriculture." Within months of taking office in January 1920, Dean Bradford Knapp was confronting disaster spreading throughout the Arkansas cotton country. Wartime demand, which had brought years of unprecedented high prices between 1915 and 1919, was followed by a worldwide contraction in the cotton market. Cotton that brought 40 cents per pound in June 1920 was selling for 6 cents in early 1921.

The sharp dive in the price of Arkansas' principal agricultural commodity brought economic distress in other sectors as well. Corn and wheat prices also plummeted to new lows not seen since the 1893 Depression. The fruit orchard industry, promoted in northwestern Arkansas since Purinton began the first experiments on the University farm in 1886-87, was virtually wiped out during the 1920's due to a number of factors. These included the imposition of grading standards, the poor quality of Arkansas apples, growing varieties unsuited to Arkansas conditions, exorbitant freight rates and plunging prices. Although cotton and other commodity prices recovered in the latter half of the 1920's, shaky economic conditions continued throughout the decade and reflected a general malaise in American agriculture.[1]

The problems confronting Arkansas agriculture were not confined to the economic sphere. The difficulties did not lie within the traditional scope of agricultural studies as they developed in the nineteenth century. Scientific findings and improved practices developed at the land-grant colleges coupled with increased mechanization on farms and improved transportation had gone a long way toward solving production problems on the nation's farms. However, to effect a positive change, Knapp and the staff at the College would also have to address structural, social, political and even racial issues. Rising sharecropping and tenancy rates, falling value of farm acreage, shrinking farm sizes, an increasing rate of farm mortgages and out-migration from rural to

Bradford Knapp - Dean 1920-1923. (Photo courtesy Photograph Collection, Special Collections, University of Arkansas Libraries.)

urban areas were among the problems facing Arkansas' farmers in the 1920's. African-American farmers were particularly troubled as racism compounded the effects of many of these dilemmas. If the appearance of the boll weevil in large numbers in Arkansas in 1921 and the devastating Mississippi Flood of 1927 are factored in, it is a wonder that any farmers in Arkansas survived the 1920's at all.[2]

In spite of conditions on the farms across the state, the enrollment in the College of Agriculture doubled during the decade. When the students arrived at the University, they found that agricultural studies had a different emphasis than in earlier times. Instead of concentrating on educating farmers who would then return to the farm and employ improved practices, the College now trained teachers, county agents, home demonstration agents, farm managers and agricultural experts for industry and business concerns. As University President Futrall pointed out, one scientifically trained practical farmer who returns to the farm has a beneficial, yet limited, influence. However, the agriculture graduate who goes on to a leadership position as a teacher, county agent or in agribusiness has a wider influence. To train these new leaders, Futrall sought and successfully corralled some of the foremost figures in American agriculture to head the College. Likewise, Bradford Knapp, and his successor, Dan T. Gray, did their utmost to attract outstanding teachers and researchers. They realized their goal in large measure, in spite of the low salaries for college teachers budgeted by the Legislature.[3]

The cry for more and varied research was also answered during this period as additional federal appropriations were made under the Purnell Act of 1925, and Branch Experiment Stations were established in various parts of Arkansas. The Legislature also made funds available to the University through various funding measures such as taxes on cigarettes, a severance tax on Arkansas' natural resources (a tax on digging bauxite or pumping oil from the ground) and an increased mil tax for the University.

The College enlarged the scope of its extension activities with the establishment of the annual Farmer's Week at the University in 1919. The program was encouraged and fostered throughout the 1920's and became one of the major events of the year for the College, the University and the community at large. The social activities of the agriculture students continued during this period as well and left their mark in the traditions of the University and community.

All and all, the 1920's were a decade of growth and consolidation for the College of Agriculture at the University of Arkansas. Yet, all the while, the fractious nature of Arkansas politics was never far from the surface in the development of the College. During Bradford Knapp's term as Dean, it was not only near the surface, but it affected Knapp's every move.[4]

Before taking over his duties, Knapp made a very careful assessment of the conditions in Arkansas that he had to work with, agricultural, educational and political. He consulted a former colleague from the USDA, W.L. English, who had observed the situation in Arkansas from his post as the agriculturalist with the St. Louis-San Francisco Railroad. English agreed with Knapp that agriculture in Arkansas was "on the verge of an era of wonderful development" and that there were "possibilities for more real constructive work in that state...than any place." However, he also cautioned Knapp that the situation was "loaded with dynamite" and told his friend that he had only a "fifty-one percent chance of success."

Early on in the negotiations leading to Knapp's accepting the Deanship, Futrall told Knapp that Bodman's agitation for the PFB had subsided. In fact it appeared to Futrall that some of the stockholders in Bodman's bank were dissatisfied with the amount of time that Bodman devoted to the PFB, and they clamored for him to resign his agricultural activities and spend more time with bank business. In any case, Futrall felt that Cottrell and the misinformation that he spread were the real problem. Experts in the College and Extension Service had called Cottrell's credibility into question on numerous occasions following the confrontation in the Governor's office earlier in the summer. By that fall, C.W. Watson, the head of the state Extension agents, asked that Cottrell submit his articles to Extension Service specialists in order to review them for accuracy before publication. This was a far cry from the situation that prevailed during the PFB's spring campaign when Extension Service people meekly stood by on the same stage as Cottrell while he made his fantastic claims unchallenged. While Knapp made certain that people knew where he stood on certain issues, he also assured everyone through his plain speaking that cooperation, rather than confrontation, was his style. He made this point in letters to Bodman, colleagues at the College and various Arkansas officials and leaders in agriculture.[5]

In his discussions with Futrall and others at the College, Knapp made clear his commitment to improving the teaching, research and extension capabilities of the College. From what he could see, Knapp believed that the Extension Service was the strongest leg of the "agricultural triangle" at the College. So, in the beginning of his administration, he concentrated on bringing better researchers and teachers to Arkansas and provided for more advanced training for them. With low salaries a perennial problem at the University, Knapp traded on his reputation as one of the leading figures in agri-

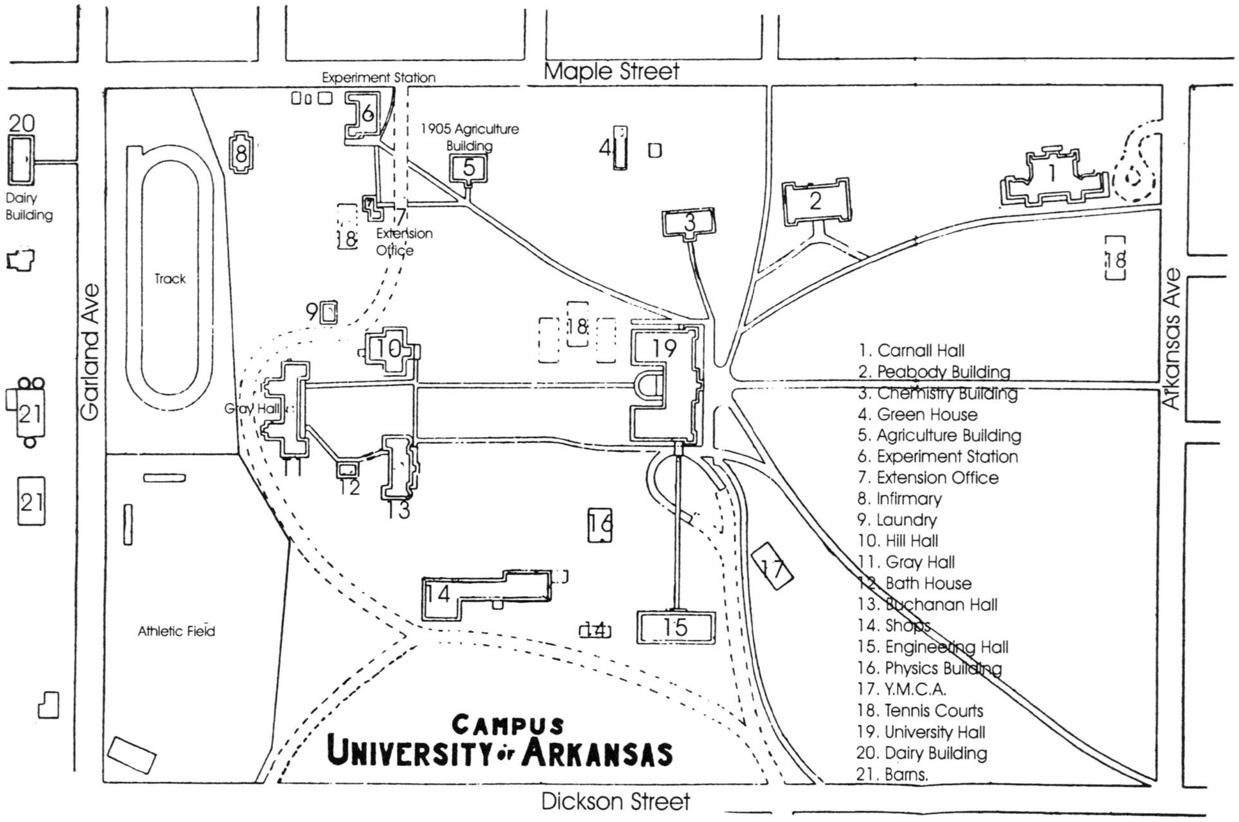

The University of Arkansas Campus in 1922-1923. The current Agriculture Building has still not been built. The athletic field and track are still on the upper campus, where the Fine Arts Center and Waterman Hall will later be built. Gray Hall is located where Mullins Library will be built. The barns are at the site of the present-day Student Union.

cultural education in order to draw people to the College. He also began a campaign for higher salaries for professors and researchers.

It was one of Knapp's lasting achievements that he was responsible for bringing Barnett Sure to Arkansas. Sure became one of the leading figures in agricultural chemistry, and among other things, the co-discoverer of Vitamin E and an internationally recognized expert in nutrition.

Sure came to Arkansas equipped with a Ph.D. For other faculty, Knapp arranged leaves of absence so that they could complete their doctoral studies. W.J. Baerg, who became one the world's leading experts on poisonous arthropods, was granted leave in 1921 as was plant pathologist H.R. Rosen.

Another notable faculty member that Knapp recruited was Dwight Isely, an entomologist, who was brought to Arkansas to fill in for Baerg while he was on leave. Isely remained at the University until his retirement and earned his own international reputation for his research on cotton pest control. Deane Carter in Agricultural Engineering was yet another Knapp appointee whose work and reputation had a positive impact on the College.[6]

One other item discussed by Knapp and Futrall was the creation of a Department of Agricultural Economics with Knapp as professor and head. Knapp made this one of the contingencies of his coming to Arkansas, and Futrall readily agreed. Knapp tapped A.D. McNair, a veteran Extension Service farm management specialist, to aid him with teaching the classes in the new department. Knapp believed strongly that the study of agricultural economics and marketing needed more attention. He outlined his views in a statement released in the student newspaper, the *University Weekly*, upon assuming his official duties on January 15, 1920:

> I am deeply interested in practical problems. Not only production but finances for farmers and their great problems of farm management, distribution and marketing must be studied and worked out along practical lines.

His views were in line with the latest recommendations of the curriculum committee of the Association of State Universities and Land Grant Colleges (ASULGC). But Knapp's concern went beyond the study of rural economics. He also envisioned expanding the department to studying and teaching about Ozark native crafts and industry and Arkansas rural life in general.

The new Department of Agricultural Economics did not catch on immediately, and, as it turned out, McNair had little spare time to devote to teaching. McNair, known by his nickname "Lespedeza" McNair because of his advocacy of planting this legume as a nitrogen-fixing, soil-building cover or forage crop, could devote only one-quarter of his time to teaching.[7]

Knapp also organized the Department of Agricultural Engineering during the 1920-21 term. The new department was under the joint supervision of the College of Agriculture and the College of Engineering. The creation of this department reflected the growing mechanization of all phases of production on the farm, especially in the years after WWI. While it is true that Arkansas and some other sections of the South were slower to adopt farm mechanization than farming operations in other parts of the country, there was still a keen interest in the subject. This was especially true after Knapp brought Deane Carter from Iowa State University in 1921 to head the Department. Carter's classes became very popular, and he took an active interest in developing and supporting the Agricultural Society and the Alpha Zeta chapter on campus. Carter also built his professional reputation and enhanced the image of agricultural engineering with publications and projects that ranged in scope from tractor mechanics to building sturdy, low-cost, yet attractive housing for rural Arkansans.[8]

Knapp also took steps toward streamlining cooperation between the College and the four State District Agricultural Schools at Jonesboro, Monticello, Magnolia and Russellville. The cooperation included coordinating scholastic work and certifying instruction at the district schools so that students from the district schools could enter the College with advanced standing. This arrangement not only helped to increase enrollment at the College but also provided students who were better prepared to undertake agricultural studies. In addition, the arrangements with the College raised academic standards at the district schools and improved their pedagogical foundations. The institutions also coordinated livestock breeding work in order to advance the bloodlines of their livestock with the least expense. Knapp also convinced the district schools to participate in the experimental work done by the College, which aided greatly in conducting a variety of projects adapted to different climate and soil regions found in Arkansas.

While the cooperation succeeded with three of the district schools and helped account for part of the rise in enrollment in the College during the 1920's, competition typified the relationship with the Russellville district school. This competition would result in an intense political effort to move the College of Agriculture to Russellville in the early months of 1923 and would have enormous consequences for Knapp and his tenure at Arkansas.[9]

One of the first efforts undertaken by Knapp as he assumed his duties as Dean was to publish an Extension Service Circular outlining his "Safe Farming" program for Arkansas farmers. After a careful consideration of world economic forces that influenced the price of cotton, the Dean cautioned farmers not to gamble their security by speculating in cotton futures. In Knapp's approach, which he had developed while directing extension work for the South, he advised farmers to make sure they had ample acreage devoted to food, feed and forage crops before turning their attention to cash crops such as cotton. The Dean also suggested that farmers pay greater attention to developing "safer" cash crops than cotton, i.e., crops that were not as susceptible to the vagaries of international market fluctuations.[10]

Knapp also took steps to put the 423 acres of the new University farm in shape. All departments pitched in to help. With neither the money nor the machinery necessary to "handle the farm properly," everyone did the best they could with what was available. Making adjustments here and taking short-cuts there, gradually the wet portions of the farm were drained, and improvements were made. The acreage was divided among various departments with 73 acres going to Horticulture, 75 to Agronomy, 10 to Entomology and Plant Pathology and the balance to Animal Husbandry and Dairying. Fifty-five acres were set aside for planting additional trees, adding to the orchard that was already in cultivation.

In all, the College spent $13,800 in making improvements on the farm in three years. Most of the money came from the general operating budget for the College rather than a special appropriation from the Legislature. With the improvements made and the hard work of the College personnel, the new farm made a profit of $37,000 from cash sales of farm products in the years between 1920 and 1923. Not included in the balance sheets was the incalculable educational and research benefit derived from the experiments conducted at the farm and the benefit of this information to the farmers across the state.[11]

My first year at Arkansas had many firsts because I did not come from a farm background. I suppose I will always remember my first Agri Day; it was less than outstanding. My lot was to show an animal. I was introduced to an old Hereford cow. She knew more about the operation than I did. She did her best to show me how but finally gave up because of my ignorance. Her revenge was finally stepping on my toe. She must have enjoyed this.

Charles L. Duff, BSA 1945

Knapp was also solidly behind an effort to establish branch experiment stations in other parts of the state. To insure that the College's experimental work would have applications that could benefit the state's diverse agricultural and climatic regions, Knapp recommended that the College establish branch stations at various locations. The Legislature had already made tentative moves in that direction, providing an appropriation for the purchase of a cotton experiment station somewhere in the central part of the state. However, political infighting over the exact location of the station delayed its establishment for many years. Knapp also recommended building a station in Arkansas' rice belt and one in southwestern Arkansas devoted to truck and fruit farming. As an adjunct to increasing cooperation with the district schools and with a view to increasing the College's visibility throughout the state, Knapp also suggested setting up "field stations" at each of the schools. The field stations could then share experiments with the researchers at the

College as well as conduct their own research with enlarged facilities. Knapp's plans for increasing research capabilities and the College's visibility throughout the state did not come to fruition during his term. However, he did lay the foundations for their success in the future.[12]

Knapp felt secure with his ability to manage the Extension Service in Arkansas. After all, along with his father, he had played a key role in developing the institution in the South. Futrall and Knapp discussed in detail the future plans for the Extension Service, and an understanding regarding the role of Extension Director Lassetter appears to have been reached. Futrall was reluctant to demote Lassetter to Vice-Director of Extension to coincide with Martin Nelson's reassignment as Vice-Dean of the College. He felt that Knapp would have his hands full with reorganizing the College resident teaching and research program and could not devote the time necessary to properly oversee the Extension Service's burgeoning operations. Lassetter continued with his title of Director and remained headquartered in Little Rock, although Knapp believed that the Director should stay in Fayetteville under closer scrutiny.

Given Knapp's experience with administering extension work and his reproving Lassetter's conduct during the 1919 PFB campaign, it is surprising that he agreed with Futrall's suggestion. Perhaps Knapp took to heart confidential advice from his friend W.L. English that he should wait at least six months before taking any "radical moves" in reorganizing the College's operations. In any case, Lassetter resigned in July 1920 to take a position in Atlanta as editor of *Progressive Farmer* magazine, and the Board named Milton T. Payne as his successor. Little is known about Payne and Knapp's feelings regarding his selection as director. It must be assumed that Payne had Knapp's approval, at least in the beginning. However, as events unfolded over the next two years, Knapp would consider Payne's appointment his worst mistake.[13]

Part of Knapp's campaign to transform Arkansas agriculture and the College's role in that transformation involved bringing together the perennially fractious political elements in Arkansas for a common goal. One well-wisher, congratulating Knapp on taking the job in Arkansas, mentioned that it would be a "relief from matters political" to be in a state that was "unanimously Democratic"; therefore, "there will be no worries as to how to trim one's political sails every two years." Knapp had left Washington in the midst of a change of administrations with the potential for a particularly contentious political battle within the USDA. However, the idea that he would find relief from political matters in Arkansas was one that had little basis in reality, as he would soon find out.[14]

During his first 18 months on the job, Knapp worked unceasingly to build bridges between the College and the people of Arkansas. He succeeded in bringing most of the bankers and businessmen into his orbit through a series of conferences and meetings with various groups. His efforts culminated with an open invitation for every bank in the state to send a representative to Farmers' Week in Fayetteville for special meetings. He also encouraged the growth and development of cooperative marketing organizations among farmers and businessmen. While he looked with favor on most of the activities undertaken by the newly formed Farm Bureau offices around the state, he always stressed that the organizations should have the farmers' interest foremost in mind, rather than that of the bankers and businessmen. In fact, he pointedly told some businessmen that the way that he envisioned cooperative marketing working was bound to hurt the middlemen involved in agricultural commodities, at least in the beginning, or until the farmers were getting fair and stable prices for their goods.

Knapp's credibility and visibility were given a boost after U.S. President Warren Harding summoned him to the Washington Agricultural Conference of 1922 to meet with other agricultural experts and discuss the agricultural crisis looming in America. After his return from the nation's capital, he and an old colleague from the Extension Service Office in Washington, J.A. Evans, made an extensive tour of the state in early 1922 to promote Knapp's "Safe Farming" campaign. An inveterate and almost compulsive writer, he also reached out to the farmers through print. Herbert Bowden, the editor of *Arkansas Farmer and Homestead* magazine, gave Knapp his own column in the journal, and a state newspaper, the *Arkansas Democrat,* also gave the Dean frequent space and covered his activities as news items. In addition to issuing his "Safe Farming" circulars and University and Experiment Station Bulletins, Knapp also contributed frequently to the Extension Service newsletter, *The Extension Cooperator.*

Knapp was even quite a hit with the student body in the College. When Dean Knapp was seen sporting a five-gallon, light tan Stetson hat on occasion, the male agriculture students quickly adopted the custom. Soon senior and junior agriculture students were wearing the lofty lids in imitation of their leader, and the "Agri specials" became a tradition that readily identified the agriculture students in any crowd and gave them a group identity long after Knapp had gone on to other fields of work.[15]

While all the publicity and goodwill generated for the College was important to him, Knapp despaired that most of it was due to the personal regard that people held for him rather than for the work of the College. The chairman of the Arkansas House Budget Committee, S.A. Hoover, touted Knapp as "one of the greatest men in America today." Yet, Governor T.C. McRae, also a Knapp supporter in principle, vetoed the appropriation bill that would have given the funding needed for a new agriculture building and an increase in teachers' salaries. Good will did not necessarily translate into political capital.[16]

Beginning early in 1921, various factions in the state began agitation for removal of the University, or at least the College of Agriculture, from Fayetteville. Similar calls to re-locate the University to a more central location in the state had failed in 1909, 1915 and 1919. In early 1922 Governor McRae suggested that the state should buy Camp Pike, an inactive Army base outside of Little Rock, and transfer the Colleges of Agriculture and Engineering there. General John J. Pershing, the Army Chief of Staff, torpedoed the proposal, saying that the Army might need the base in the future. However, by the end of 1922 another concerted push for the removal of the Agriculture and Engineering portions of the University was gaining steam. This plan called for combining the two colleges with the District Agricultural School in Russellville. A combination of factors gave this effort a greater chance for success.[17]

One factor was a heating up of state politics with the rise of the Ku Klux Klan as a force in Arkansas politics during the 1922 election cycle. While the College and Knapp did not become central issues in the campaign, the reactionary and anti-progressive tone voiced by some candidates carried over to agricultural issues.

> *When I enrolled as a freshman at the UofA, I fully intended to be an agri teacher in some high school in Arkansas. I worked for the Horticulture Department, Plant Pathology and Agronomy Department at "the farm," and I worked three summers at the Cotton Branch Experiment Station, Marianna. These jobs sold me on agri research. After doing my practice teaching, I knew that research was for me!! Not teaching!!*
>
> *R. Wallace Williams Jr., BSA 1942*

Peripheral issues involving the Farm Bureau and its relations with the Extension Service and Knapp's involvement with the founding of the Service itself did influence local elections to a degree. Dean Knapp caught the brunt of much of the criticism. One charge that greatly offended Knapp, and which he sought immediately to correct, was that the agricultural extension work initiated by his father and then carried out by himself was controlled by the Rockefellers as part of an international conspiracy by bankers to enslave farmers. According to George P. Sands, president of the Arkansas State Farmer's Union, the Rockefellers operated the Extension Service through their charitable foundation, the General Education Board, and supposedly called the shots for the Extension Service program, with the Knapps acting as educated dupes.[18]

From the other direction, some of the bankers and businessmen in the Farm Bureau painted Knapp's cooperative marketing proposals as so much agrarian radicalism, un-American in complexion and only slightly less dangerous than the Bolshevism that had triumphed in Russia. Knapp despaired of being able to correct the many distortions of his record and of his motives. He wrote to I.O. Schaub, his old friend at the Extension Service in Washington D.C.:

> There is so much politics in Arkansas that I would consider a change if it were an improvement. Things are going alright with me but there is always politics in this state and I get tired of the constant misunderstanding of my motives and unfair criticism of my best intended acts.[19]

At the time Dean Knapp had to deal with attempts to move the College, he was also troubled with dissension in the ranks. Declaring that teaching and research work was in "splendid shape," Knapp described the Extension Service, and in particular Director Payne, as the weak link in the agricultural triangle. He complained that Payne was, among other things, a poor administrator. Knapp also accused the Director of insubordination in attempting to run the Extension Service as "his own show" from Little Rock, separate from the College, all the while ignoring the plans and policies formulated by Knapp. Amid these growing tensions, and after the 1922 elections, the College removal faction began gathering their forces for the upcoming 1923 legislative session.[20]

Removal advocates, led by Rep. Reese Caudle and Steele Hays of Pope County, claimed they intended no injury to either Knapp or the staff at the College. However, Hays, a Farmer's Union man, borrowed much of the rhetoric for his speeches from the Farmer's Union anti-Rockefeller propaganda that indicted Dean Knapp. They said their fight was with "the University crowd" and corrupt legislators who kept the benefits of agricultural and mechanical education out of reach of the mass of Arkansas' citizens. Citing an educational survey of the University done by the U.S. Department of Education in the early months of 1921, the removalists claimed that the conditions that prevailed in 1921 still existed. To an extent this was true concerning antiquated buildings and poor equipment. But Knapp and the staff at the College of Agriculture had done much to correct most of the shortcomings that were within their power to correct, given that the Legislature and Governor considered University appropriations fair game for cost-cutting exercises.[21]

Knapp joined with President Futrall and the "University crowd" in opposing the College's removal to Russellville. Dean Knapp made speeches against removal and even drew up an Experiment Station Bulletin in defense of the College of Agriculture staying in Fayetteville. But his heart was not in the fight. He felt that the battle was between the various political factions in the state over issues that had little to do with improving the lot of Arkansas' farmers or the fate of agricultural education. Early on in the removal battle he said in confidence to his old friend O.B. Martin, who still was working with the Extension Service in Washington, D.C., that he was, "not in the least opposed" to moving the College, and would prefer re-locating to Little Rock if the removal faction won out. However, he also said that he would "steer a safe course" and try to stay above the fray.[22]

Staying neutral proved impossible for Knapp as the battle between Russellville's backers and Fayetteville's patrons raged in the Legislature from January through March 1923. Knapp publicized the growth of the College and its many-faceted services to the state and the farmers of Arkansas and emphasized what he considered phenomenal work given the meager resources allotted by the Legislature.

Before the removal battle finally ran out of steam, one of the most strident of removal forces spokesmen, Rep. W. Leroy Lee of Yell County, collapsed and died on the House floor during a speech. The removal bill itself was declared dead and then revived several times before finally being "indefinitely postponed" with the possibility of appearing as an initiative on the 1924 general election ballot. President Futrall and the "University crowd" won the removal battle, and Knapp had done his part. The College was secure and has not again been faced with a serious challenge to Fayetteville being its legitimate location.[23]

But the "dreadful siege" took its toll on Knapp, and he began serious consideration of "greener pastures" for his talents. Two events helped propel him on his way. The removal controversy dust had barely settled when President Futrall sent Knapp a directive. Futrall advised the Dean that the Department of Agricultural Economics, one of Knapp's creations, should be closed and along with it the title of professor of agricultural economics. Futrall complained that the department was a fiction, with no classes and no instruction. With Knapp's busy schedule that put him in different parts of the state and the country more regularly than he stayed at home in Fayetteville, Futrall saw little hope that the University could have a professor of agricultural economics or a department any time in the near future. He did not question Knapp's credentials for teaching such a class, only the University's need for a faculty title with no department. Futrall closed his letter with a warning to Knapp: "I hope you will not press the point of retaining in the University either the name of a Department of Agricultural Economics or of reserving the title for anyone on the faculty."[24]

In response, Knapp quoted Futrall's letter of December 1919 in which Futrall said that "it is perfectly satisfactory that the new Dean should be put down as Head of the Department of Rural Economics and Marketing" as a condition for Knapp's coming to Arkansas. However, Knapp did not press the issue of retaining the department or the title, only the overwhelming necessity of teaching and researching the subject. Knapp left no doubt about his passion for the issue, saying that he did not care about titles, "but I do care about the field of our work and (its) usefulness in the state." Knapp then gave his own warning to Futrall:

> If your letter means merely that the title is embarrassing but that the (Dean) may go right ahead dealing with these many economic questions, then we can get along. But if it means the subject is...to be

> removed from consideration (and the Dean) cannot deal originally with Agricultural Economics questions at all times freely...then we have very little left...and the damage is irreparable.[25]

Why was this seemingly unimportant detail of a "quasi" department so important to Futrall that he would risk Knapp's resignation over winning this point, particularly after he had promised Knapp the position in writing as a condition of his coming to Arkansas? Since Futrall prized loyalty to the University as an institution, perhaps he was disenchanted with Knapp's lukewarm efforts during the removal battle. However, there were other issues at work that came between these two champions of Arkansas agriculture. Futrall's vision of an improved agricultural environment in Arkansas was basically a conservative one, and the president had a well-honed reputation as a penny-pincher. Knapp's innovative approach to finding solutions did not sit well with Futrall's conservatism or the Legislature's miserly ways. The Dean himself felt stifled in his efforts. He was disenchanted with the possibilities for transforming Arkansas agriculture; his optimistic assessment at the beginning of his tenure did not survive his three-year journey through Arkansas politics.[26]

> *I was better at doing than in academics. The work on the UofA farm that I did and saw others do gave me confidence in agri work. One day in a meats class we were to butcher a steer. The instructor handed me a sledge hammer to kill the steer, which I did in one lick. Most of the other class members were pretty surprised. One fainted.*
>
> *William L. Bush, BSA 1949*

The final straw for Knapp came in June 1923 in his showdown with Director M.T. Payne over control and direction of the Extension Service. Knapp had tried for nearly a year to "let things work themselves out," but to no avail. He said he had tried to be "soft-hearted" in dealing with Payne's failure to follow directives or to submit plans of work to him for approval. This tack had served only to give the Director more opportunity for insubordination. Knapp called publicly for Payne's resignation and for the Extension Service headquarters to be located in Fayetteville. While the information did not reach the newspapers, Knapp let it be known that if the Payne did not resign by July 1, he would resign himself rather than continue to work under such conditions. In a letter to Carl Hollis, a banker and a Payne supporter, Knapp spelled out his position:

> I do not think you would relish my asking you to retain in your bank an employee who failed to keep you advised regarding important matters of the bank's business or who worked against you as an of-

ficer of the bank. If I cannot get a better man than Payne I shall be glad to relinquish my place.[27]

After a lengthy two-day Board of Trustees meeting June 11 and 12 in which the fate of the Extension Service was the main topic discussed, the Board decided that the Extension Service would remain in Little Rock. Payne did not submit his resignation, nor did the Board request it. The *Arkansas Gazette* newspaper reported that Payne, under a contract of an indefinite period, would remain with the Extension Service "for the present." Although Payne eventually did resign on July 31, it was not soon enough for Knapp. He told a friend, "I am convinced that they intend to throttle me here, but they are not going to do it." During the last two weeks in August, Knapp took his first vacation in three years, which included a week-long stay in Stillwater, Oklahoma. After returning from vacation, he resigned his post September 7 to take a position as President of Oklahoma A&M College.[28]

Across the state there was an immediate outpouring of regret and a chorus of voices asking Knapp to reconsider. E.J. Bodman and Gov. Thomas C. McRae were among those who asked Knapp to remain in Arkansas. Bodman told Knapp that a group of Arkansas businessmen and bankers had agreed to pool their resources and match Oklahoma's salary offer. Gov. McRae endorsed the plan as well. Knapp graciously declined and said the issues were not just the salary. What Arkansas really needed, he said, "is more funds for the Agriculture program, not the Dean's salary."

One thing that Knapp did not mention was that, if he accepted the offer, he would be making more than President Futrall, which would be sure to further estrange the two men. Knapp's salary had always been a contentious issue with the other Deans and faculty members at the University. Knapp told his brother, Arthur, that he was keeping quiet about the real reasons for his leaving Arkansas, but there is little doubt that his deteriorating relationship with Futrall played a major part in his decision to leave. In his new position he would have politics aplenty to contend with, as the situation in Oklahoma was at least as fractious, if not more so, than Arkansas. At least in his new position he would not have to answer to John Clinton Futrall.[29]

In spite of the discouraging circumstances that surrounded Knapp's leave-taking and the political battles he had endured while in office, his brief tenure in Arkansas had more positive effects than the length of his stay would imply. Enrollment increased dramatically from 98 in 1920 to 231 in 1923. The College had gained national recognition while Knapp was there. But more than that, Knapp had improved the research and academic reputation of the institution by bringing in scholars and teachers of the first rank and raising academic standards. While he had not accomplished all that he wished in the way of building up the physical needs of the College, much of what he wanted was out of his reach anyway, as the state was in a disastrous financial condition during the early 1920's. Knapp had built the foundation for many of the future accomplishments of the College, and his resignation set the stage for some immediate changes that had long-term benefits as well. In all, Knapp's tenure was successful in many ways, not the least of which is that he set into motion the modernization and consolidation of the College in a way that only people with a vision for the future can do.[30]

Without missing a step, Futrall summoned the ever-faithful Martin Nelson from his vacation in the Rocky Mountains to take over as interim Dean while a search was made for Knapp's replacement. The College faculty all signed a letter to Futrall asking that a thorough search be done to find someone as fully qualified as Knapp with the ability to "command and hold the respect of the businessmen and farmers of the state of Arkansas." The search began immediately and was quickly narrowed down to a few candidates, the leading one being Dan T. Gray, who had been the Dean of the College of Agriculture at Alabama Polytechnical Institute (which later became Auburn University) since 1920.[31]

Futrall invited Gray to visit Fayetteville's campus at the end of October. At a meeting with the Board of Trustees on November 3, Gray, Futrall and the Trustees discussed at length the role of the Extension Service and its relationship with the College.

After the Board elected Gray the new Dean of Agriculture, they went on to make the changes in the Extension Service that Knapp had been clamoring for for years. First, the Dean was made the Director of the Extension Service. The Associate Director, who had charge of the state work, was made clearly and definitively an administrative subordinate. In an effort to improve the professional quality of the county agents, all future Extension Service appointments had to be college graduates. The appointment of an Assistant Director was put off until Dean Gray assumed his duties in January in order that he could have full input in the selection process. In the meantime, T. Roy Reid, who had filled the post since Payne's resignation in July, continued as interim Assistant Director. The Trustees finally decided that the Extension Service headquarters would remain in Little Rock.[32]

Dan Gray - Dean 1924-1939. During his 15 years as Dean, enrollment of the College almost tripled and the faculty increased from 26 members to 61.

Futrall announced in early November that Gray had accepted the Dean's position and would begin his duties in January. Gray was eminently qualified for the job and had already garnered a reputation as a major figure in Southern agriculture. Many farmers with livestock recognized Dan Gray's name, as he had served as the animal husbandry editor for the widely read farm journal, *The Southern Ruralist,* for many years. A native of Missouri, Gray earned both his B.A. and B.S. degrees from the University of Missouri and held an M.S. from the University of Illinois where he did his graduate work in animal husbandry and nutrition. He went on to become the head of the Animal Husbandry Department at Alabama in 1908 before taking the same position at North Carolina A&M in 1913. During the war Gray had charge of all animal husbandry work in the South for the U.S. Extension Service. He also served as the secretary for the Association of Southern Agricultural Workers from 1913-1920 and was president in 1920-21. Gray returned to Alabama Polytechnic Institute as Dean of the Agriculture College in 1920 where Futrall had contacted him.[33]

After Futrall's announcement, E.J. Bodman wrote to Gray to announce that H.M. Cottrell would pay him a visit. Curious about why a private citizen would want to "interview" him regarding his suitability for a job he had already accepted, Gray wrote to T. Roy Reid and asked him to provide some information on Bodman and Cottrell. Reid replied at length describing their activities and the controversy surrounding their promotional campaigns in Arkansas. Reid advised Gray to keep Cottrell and Bodman at arm's length. When Cottrell arrived in Alabama, Gray politely greeted the "agricultural expert" and then promptly turned him over to associates. When Dean Gray arrived in Arkansas he handled Bodman in a similar way. However, Dean Gray's relationship with Bodman did undergo a change after Bodman was appointed to the Board of Trustees in 1925.[34]

Dan T. Gray was the perfect person to fill the Dean's post in the aftermath of Knapp's resignation and the removal battle. He worked assiduously to heal the scars that were left from the conflicts of the past year. As well as being an affable diplomat, he was a forward-looking proponent of modern agriculture that served the needs of a diverse clientele ranging from businessmen and bankers to sharecroppers and truck farmers. Dean Gray got on well with people from all backgrounds, but he was no pushover, as Bodman and his associates in the PFB soon found out.

The rest of the 1920's were more settled than the first three years of the decade. The agricultural crisis in Arkansas continued, but at least politics at the state and University level simmered down to a manageable level. While the enrollment in the College fluctuated from a high point of 265 in 1924 to a low of 164 in 1928, the research and academic facilities and the work undertaken by the College expanded and consolidated during this period.[35]

Enrollment 1920-1929.

Year	1920	1921	1922	1923	1924	1925	1926	1927	1928	1929
Enrollment	98	147	176	231	265	240	209	182	164	202
Graduates	6	8	11	18	18	25	34	29	41	22

During Gray's initial months on the job, he was busy evaluating the College's program and drafting a plan for the agricultural development and prosperity of the state. He also began a study of the College's curriculum in order to bring it into line with guidelines from the ASULGC. During this time the Dean had very little time for teaching, but as he told President Futrall, he was interested in putting together a course for the fall term that would include an overview of all facets of agriculture and the problems facing farmers. Gray made three recommendations to Futrall:

1. Strengthen the faculty and encourage more interaction among resident teachers and Extension Service personnel;
2. Establish a Branch Experiment Station system; and
3. Establish a building program for the College.

Gray emphasized that the faculty were scattered everywhere and not a single department was housed in rooms originally designed for its use.[36]

Pending the Legislature's approval for any building program, in the fall of 1924 the overcrowding at the College was relieved somewhat after the south end of the old men's dormitory, Gray Hall, was turned over to the College. The Departments of Agricultural Engineering, Farm Management (formerly Agricultural Economics), Agricultural Education, Entomology, Plant Pathology and Horticulture had used rooms in the northern part of the building since 1921, and these rooms were remodeled. New offices, lab spaces and classrooms were equipped, and the Departments of Agricultural Chemistry and Veterinary Science and Bacteriology moved in.[37]

That fall Dean Gray also began teaching his classes in the newly organized General Agriculture Department. The dean sought to implement recent recommendations made by the ASULGC calling for overview courses in agriculture. The recommendations were made to correct a trend toward over-

Gray Hall - The south end of this men's dormitory was turned over to the College. Among the departments housed completely or partially in the building were Ag Chemistry, Farm Management, Ag Education, Entomology, Plant Pathology, Horticulture, Bacteriology, Veterinary Science and Ag Engineering as well as the office of the *Arkansas Agriculturist.*

specialization in subject areas that left some agriculture graduates well versed in their areas of study but essentially ignorant of the wider foundations of agricultural studies. Gray was assisted in the department by Extension Service specialists, Smith-Hughes teachers and regular members of the College faculty.[38]

The College's support for excellence in research took several forms as the 1924-25 school year began. Barnett Sure's ground-breaking research in the role of vitamins in human nutrition was beginning to attract notice across the nation and would soon come to international attention as he refined his investigations in his new lab at Gray Hall. J.W. Read, the head of the Agricultural Chemistry Department, and Stella Palmer Hotz, the head of the Home Economics Department, published research results of their own in a 1924 Experiment Station bulletin, *Vitamins, Health and Daily Diet.*[39] The bulletin went into a second edition because of the heavy demand for the volume. The College also aided W.J. Baerg in completing his Ph.D. research on venomous insects by arranging for him to study tarantulas and scorpions in the Panama Canal Zone during part of the 1924-25 term. The National Research Council and the Governor of the Panama Canal Zone arranged for Baerg to take up research at several locations and labs in Panama. James Ayres Clark, a 1918 B.S.A. graduate from the University of Arkansas, who had worked in Panama for six years at a lab in Ancon, aided Baerg in his research.[40]

In another research development, the University established a program for the study of rural economics and sociology in the Economics and Sociology Department in the College of Arts and Sciences. A.W. Jamison, from the University of Illinois, was brought in to begin the research. While the program would include some teaching, it was mainly intended as a research project conducted through the Experiment Station. Plans were made to expand the program in cooperation with other Southern states as part of a long-term regional project. Futrall had completed his plans of dismantling the program in the College of Agriculture shortly following Knapp's departure in the fall of 1923. In a series of letters to Acting Dean Nelson, Futrall had made his intentions in that regard clear. However, the trend was moving against Futrall on the suitability of agricultural economics studies remaining in the curriculum of the College of Agriculture.[41]

The year 1925 was a benchmark year for the College of Agriculture. Many currents came together in 1925 that not only signaled the arrival of a new era in the history of the College, but also represented a maturing of agricultural studies at the University. The principal events included an expansion of the University farm, passage of the Purnell Act by Congress and the establishment of three branch experiment stations. Finally, and not least, the Legislature appropriated funds for the construction of a new agriculture building.

In January the University finalized arrangements that added another 100 acres to the University Farm. The deal was in the works since Futrall had

The entomology insectary was located behind the horticulture/plant pathology greenhouse and head house. The buildings were located where the Student Union now stands.

asked Nelson to evaluate the additional land that adjoined the farm earlier in the summer and give him a report. Nelson said that the so-called "Davis farm" (about 60 acres) part of the acreage was by far the most desirable, but the other portions of the parcel were less so and needed a considerable amount of work to make them suitable for experimental work. In spite of the shortcomings, the acreage was bought, boosting the size of the farm to just over 523 acres. Other improvements were also made on the farm, including moving the large barn on the site to a more suitable location, building cottages for the farm workers, laying tile in fields to improve drainage and painting and repairing existing buildings and fences.[42]

On February 24, 1925, Congress passed the Purnell Act, which supplemented the Hatch and Adams Acts for funding Experiment Station research work. The bill provided an additional $30,000 in research funds for each state the first year with additional increases annually until the total amount reached $60,000 by 1930. Besides providing additional money at a time when funding was difficult to come by, the act was significant because, for the first time, agricultural research funds were specifically targeted for areas other than production agriculture. The Purnell Act's special focus was agricultural economics and sociology and home economics. With this focus, new funding and the recommendation of the ASULGC, President Futrall realized the logic for such studies to reside in the College of Agriculture. Futrall might easily dismiss a Dean, but it was more difficult to refuse the potential $60,000 contained in the Purnell Act. Accordingly, the College established a Department of Rural Economics and Sociology begining in the 1925-26 school year.[43]

The "new" department had four professors: Dean Gray, the veteran A.D. McNair and two new professors, who would contribute significantly to understanding Arkansas' agriculture and rural environment as well as to the College's reputation as a major research and teaching institution. C.O. Brannen came to Arkansas from his job as an agricultural economist for the USDA, where he was a specialist in land tenure and agricultural finance from 1919-25. Brannen began his education at North Texas Normal College and went on to earn both his B.A. and M.A. From Peabody College in Nashville, Tennessee. Brannen also did graduate work in political science at Columbia University before coming to Arkansas. His research monograph, *Relation of Land Tenure to Plantation Organization*, was considered an important and groundbreaking study and was published by the USDA as a separate bulletin in 1924.[44]

J.A. Dickey came to the University from his position as an instructor in rural sociology at Cornell University in New York. Dickey was born in North Carolina and earned his B.A. from Elon College. He went on to study at the University of North Carolina under the tutelage of E.C. Branson and Howard Odum, two pioneers in the study of the South and Southern rural culture. One of his early studies in rural sociology, done with Branson, *How Farm Tenants Live*, dealt with one of the most pressing problems in the rural South, the increasing numbers of tenants and sharecroppers on Southern farms. He also worked on such problems as rural health and extension education for rural areas. Dickey would go on to earn his Ph.D. with research undertaken in Arkansas. Brannen and Dickey started the ball rolling in the department shortly after arriving in Arkansas. Dickey completed a study, *The Relation of Industry to Agriculture in Arkansas*, that was published as a University of Arkansas bulletin in 1926.[45]

Other research was also supported with the Purnell funds, including research in home economics. Madge Johnson, a professor in charge of research in the Home Economics Department, started a research project studying and assessing labor-saving devices and conveniences in farm homes. Johnson's research grew out of the "Country Life" movement from the turn of the century, which sought to make rural life more attractive and more modern without sacrificing "traditional" rural values such as thrift and self-help. The idea of making rural life more attractive took on a new meaning as concerned people (mainly academics and urban dwellers) sought ways to stem the floodtide of migration from rural areas to the cities during the 1920's. The earlier research in the Home Economics Department with diet and nutrition continued at an accelerated pace with the infusion of Purnell funds.[46]

The 1925 Arkansas Legislature, goaded by the federal largesse contained in the Purnell Act, finally appropriated $30,000 for the establishment of three branch experiment stations. The people of Lee County and its county seat, Marianna, had previously donated 160 acres as the site for a cotton experi-

The 185-acre fruit and truck station near Hope, Arkansas, was one of three branch experiment stations established in the 1920's. The Principal Experiment Station Administrator commended Arkansas' branch stations as "one of the most progressive ideas that I have observed in station organization in recent years" and singled out the Hope Station as "a thing of beauty as well as an adequate research facility."

ment station. Likewise, 160 acres was donated east of Stuttgart for a rice experiment station, and 185 acres was donated outside Hope for a fruit and truck experimental farm. All together the land provided by the people of these counties was valued at $40,000. When combined with the funds from the Legislature, the people's donation made sure that the branch station project was off to a good start.[47]

After pleading with the Legislature for years, the University and the College of Agriculture finally got the necessary funds to begin a building program on the Fayetteville campus, the first in 20 years. In February a legislative inspection committee visited the campus and returned to Little Rock with a recommendation that the Legislature approve funding for the building program. The Legislature passed the appropriation, and Governor Thomas C. McRae signed the measure in March 1925. The tremendous boost given the College in these first months of 1925 cannot be overstated. The University farm, that had been "but rocky hillside" six years earlier, now was an extensive experimental farm of 525 acres. The three branch experiment stations added another 500 acres for agricultural research in a variety of soil, crop and climate conditions.

In addition to the physical expansion of the College, the Legislature also raised faculty salaries to a level "making them commensurate with average university salaries" around the nation. Now the College would be able to hold the teachers and researchers that Knapp and Gray had so carefully selected and expand the research and teaching program to a level equal to that of any educational institution in the South.

Private industry and other organizations also chipped in to support the program at the College. Besides the National Science Council's support for Baerg's work, the Eli Lily Company and the American Medical Association provided additional underwriting for Sure's work on vitamin E. Farm equipment companies also donated machinery and farm building and housing materials to Deane Carter's Agricultural Engineering Department.[48] A short poem written in 1925 by one of the staff describes, in a less serious way, how the faculty contributed to the growth and development of the College:

What They Do
by C. Woolsey
(*With Apologies to Everyone*)

Baerg wants to know what Katy Did
Sachs wants to preach on soil
Read writes about Cyanamid
And A.D. talks about toil.

Stout sings about the little chick
And Wiggins of how to spray
Dvorachek's job has got its kick
While Nelson makes the hay.

Young heals sick plants with all his might
While Rosen talks Mosaic
And Cooper prays both day and night
For an apple less prosaic.

Rapp shouts about squash and peas
B. Sure tells why they're good
I'm "Sure" We'll all be glad to "B"
Herbivora del Cud

Why cows leave home we want to know
So Mason works this out
And Martin wants the steer to low
In pastures here about

Jake works with the cotton day and night
And Mac adores the soy

While Bleeker's bugs give us a fright
Which gives him plenty of joy,

Schilling doctors old "bossy" cow
And doesn't lose a minute
While Carter makes a model barn
Although there's nothing in it.

Barr raves about the kitchen sink
And Isely chases the bugs
While Jacobson leads us to think
Soil robbers are all thugs.

Miss Galloway hands out the books
And makes us bring them back
While Mrs. Taylor and her help
Know all about Home Ec.

Willbanks handles milk and cream
And makes it into butter
While Alex haunts the dairy barn
Of sheep and goats he'll mutter.

Holloway makes men of boys
And teaches them to teach
While Clayton handles ag ec.
And lives among statistics.

Casey makes seeds germinate
And Gus puts out the bunk
While Bell keeps everything in shape
And handles all the junk.

But after all is said and done
And each has had his say
It takes someone to back each one
And that man is Dean Gray.[49]

If the attendance at Farmers' Week held in Fayetteville by the College each August is any indication, Arkansas' farmers were also taking a new interest in supporting the College. Five hundred farmers came to town for the first Farmer's Week in 1919; by 1926 an estimated 5,000 farmers and their families attended. Many agriculture colleges held such events in the 1920's,

Automobiles owned and driven by Arkansas farmers who drove from all parts of the state for Farmers' Week, 1926. Farmers attended from 74 of Arkansas' 75 counties and spent an entire day inspecting experimental plots on the 550-acre farm. (Arkansas Agriculturist January 1927.) Attendance at the annual Farmers' Week reached a high of 6,077 in 1932 before the event was discontinued due to budget constraints.

but the one at Arkansas was the largest, both in attendance and variety of subjects addressed. But the days were not just devoted to so much "agricultural preachment." There was also horseshoe pitching and an old time fiddlers contest, an "Agricultural Pageant" as well as other amusements. Four days of lectures and demonstrations were followed by an enormous picnic and barbecue held at the newly expanded and improved University farm north of town. In this way the College transmitted genuine information as well as good will to the farmers of the state.[50]

The agriculture students were also busy expanding their organizations and helping develop the College. In June 1924 the agriculture students launched their own newspaper, *The Arkansas Agriculturist.* Published each month throughout the school year, *The Agriculturist* kept the students, faculty and public at large informed about the academics, research and social life at the College. The Agricultural Society and the Home Economics clubs were very active sponsoring and hosting campus events during this time as well. The Agri Day celebration, which had been switched to a date in April in order to take advantage of warmer weather, had grown in size and variety over the years. The 1926 Agri Day began with the usual oversize footprints of "Old Man Agri" appearing all over town guiding the way to the campus and the celebration. The agri parade, which kicked off the day's festivities, had grown to

Footprints of "Old Man Agri" (*Arkansas Agriculturist*)

include 50 major floats by 1926. The parade was followed by a luncheon served by the Home Economics Club on shingle plates. In addition to the educational and informative agricultural displays put up in the Agriculture building, the day also featured an "Agri show" filled with spoofs and skits. The evening was devoted to the annual Agri Ball where, keeping in step with the land-grant college mission, gingham dresses and overalls replaced gowns and black ties.

In addition to growth in Alpha Zeta, the agricultural honors fraternity, agriculture students were busy in "fields" other than academics on the campus. During the 1920's agriculture men were also active in athletics. Brad "Bo" Scott, an agriculture student and president of the Agri Day Association, was also captain of the 1926 Razorback football team. Lloyd Dhonau, the business manager for the *Agriculturist*, quarterbacked the team. Seven other agriculture students also manned the trenches on the football field, and others were active in track and other sports.[51]

Funding for a new Agriculture building was the icing on the cake for the students and faculty who had struggled in cramped quarters for years. They could scarcely believe their ears when they learned that the new building was planned to have over seven times the room of the old building. The *Arkansas Agriculturist* declared: "From now on when we are asked what college we are in, instead of saying, 'I am an Agri, now laugh, dammit!' Let's say, 'I am in the College of Agriculture and proud of it.'"[52]

Construction on the new $350,000 Agriculture building began in October 1926, as President Futrall shoveled the first concrete for the building's foundation. The building was completed and dedicated in June 1927 in a ceremony that featured an address by honored guest U.S. Secretary of Agriculture W.M. Jardine. Before his appointment as Secretary of Agriculture, Jardine served on the committee appointed to survey the University in early 1921. He must have been pleased that the recommendations he made in 1921 regarding the College of Agriculture were finally being followed. R.R. Dinwiddie, a former veterinarian and animal science professor at the University, who had been with the Experiment Station from its beginning in 1887 until he retired in 1914, also was invited to speak. E.J. Bodman, the former head of the Profitable Farming Bureau and now on the University Board of Trustees, was on the podium as well. This was truly a sign of how far the College and Dean Gray had come in healing the antagonism of the earlier years of the decade and moving ahead.[53]

The new, three-story, 257 x 105-ft building was originally conceived as the south side of an agricultural quadrangle. The first floor housed the Agronomy Department and a specially designed horticulture laboratory used for cotton research. Other labs in the building were also specifically designed for various agricultural experiments, unlike the makeshift arrangements that had persisted in earlier times. One professor had described the earlier lab con-

Construction on the current Agriculture Building began in October 1926. The building, designed in English Collegiate style, cost $350,000 to build and was dedicated June 1927. It was originally conceived as the south side of an agricultural quadrangle. Photograph is of the south side of the building during construction.

ditions as like "trying to fish with a bent pin." With the new labs and equipment, the research could now forge ahead unimpeded by archaic equipment or crowded conditions.

The second floor of the new building was home to the Dean's offices, the agricultural library, offices of the agricultural editor, and the Departments of Plant Pathology and Rural Economics and Sociology. The library occupied the entire east wing of the second floor, and with new books and journals ordered, Dean Gray said that it would be the finest of its kind in the South. The third floor contained the Departments of Horticulture and Entomology and three general classrooms. The classrooms had seating for 425 students, and two of them could be converted to use for general meetings.[54]

The old building, renovated and refitted, became the exclusive domain of the Home Economics Department. The change was overdue. For the greater part of the 1920's, home economics students made up the majority of the students in the College. Yet, they had been crowded into the cramped and poorly lit basement of Peabody Hall. After their relocation to the former Agriculture Building, their space expanded dramatically. The first floor contained the food laboratory, equipped with two new kitchens and the Home Economics Club room. The second floor had the department offices and the clothing laboratory. A large lecture room, a room fitted for teaching home nursing and additional storage space, a luxury unknown in the old location, were located on the third floor. The basement housed the research laboratories.

Along with a new location, the Department also had a new head. Henrietta K. Burton came to Arkansas from New York where she was doing graduate work at Columbia University. Burton's areas of expertise were child welfare and community extension and nutrition work. The Home Management House maintained by the department got a new manager that year also. Marian L. Normington, with her M.A. from the University of Chicago, had spent two years in nutrition research in the laboratories of the Home Economics Bureau in Washington D.C.

Whereas only a few years previously all the work in the Department of Home Economics was relegated to teaching and extension work, by 1927, thanks to the Purnell Act, research in the department was developing rapidly. Research Director Madge Johnson undertook a study of rural homes in Arkansas. Diet and nutrition studies running concurrent with Barnett Sure's work were again underway. A study of trends in savings account use was also being conducted. In another sign of growth and development in the department and coinciding with the establishment of a graduate school at the university, the first five graduate students in home economics were enrolled in the spring 1928 term.[55]

In line with the emphasis on research in the area of agricultural economics, Dean Gray initiated a course in agricultural business beginning in the fall 1928 term. In doing so Gray made the University of Arkansas College of Agriculture the first university in the South with such a program. The course was open to agriculture and home economics majors alike. Gray emphasized in his description of the course that women as well as men should be able to prepare themselves for the new careers opening in agribusiness. The College offered the course to give students a better understanding of the social and

A "practice meal" in the Home Economics Department, located at this time in the basement of the Peabody Building. The time period is estimated as 1920-1930. (Photo courtesy Photograph Collection, Special Collections, University of Arkansas Libraries.)

economic forces underlying the growing and marketing of farm products. Recognizing that there was a trend toward corporate or large-scale farming, the subjects covered were intended to train agricultural marketing specialists, farm managers, agricultural banking experts and people interested in running their farms in a more business-like fashion. In addition to classes in farm taxation, agricultural cooperation, farm finance and marketing, the course also introduced classes in the economics of food and clothing and the sociology of rural life. Among some of the career avenues open to graduates of the course were positions as farm managers, in the growing field of agricultural cooperative work and in the government with either the U.S. Bureau of Agricultural Economics or in a state experiment station. County agents could also benefit from such training, as could Smith-Hughes teachers.[56]

In 1929 another new course of study was added to the Horticulture Department, which became the Department of Horticulture and Forestry after Dean Gray hired Robert P. Holdsworth to teach and research the use and maintenance of the farm woodlot and the growing of timber as a cash crop. Holdsworth held degrees from Michigan State Agricultural College and the Yale University School of Forestry. Later Holdsworth received a fellowship at the Royal College of Forestry in Sweden. He combined these sterling academic credentials with practical experience as a forester in Montana and Washington.

Concerning the need for forestry studies in Arkansas, Gray said:

> Out of thirty-three million acres of land in Arkansas it is estimated that seventy percent is potential forest land. Yet, Arkansas has done little for its forests in an organized way. It is one of the six states in the Union not having a state forest service.

With the creation of the new department, Gray hoped to address this deficiency by training professional foresters, providing a body of forestry research particular to Arkansas and focusing public attention on the need for a state forest service.[57]

The importance of agricultural economics and the attention given to the research done in that field during the 1920's accelerated to new levels in the 1930's. However, this did not mean that research, teaching and extension work in production agriculture ceased. During the 1920's the outlines of the research in developing more productive and disease-resistant food and fiber crops took shape, leading to the so-called "green revolution" of the 1950's and 1960's. J.O. Ware was busy breeding new strains of cotton more suitable for Arkansas soils. Breeding experiments with corn and soybean were also underway. J.R. Cooper experimented with new apple varieties, and C.B. Wiggans worked with grapes. Plant pathologists V.H. Young and H.R. Rosen worked to find the cause of cotton wilt and fire blight disease in apple trees and identify disease-resistant strains of potatoes. Martin Nelson, now the assistant director of the Experiment Station and head of the Agronomy Department, made progress in experiments with fertilizers and soil types. The work of H.E. Dvorachek and R.M. Smith in livestock and poultry breeding, feeding and care was also very important. Smith's research helped promote the growth of the poultry industry in Arkansas, which would become a vital part of the state's agricultural economy. W.J. Baerg and Dwight Isely worked with control of insect pests such as the boll weevil and red spiders in cotton and the corn rootworm and corn stalk beetle. Isely was especially interested in biological and cultural control of insect pests, avoiding the use of poisons such as strychnine and lead arsenate.[58]

The Extension Service experienced phenomenal growth during the 1920's. By the end of the decade nearly every county had at least one county agent or home demonstration agent. The African-American Extension Service force grew during the decade as well with a force of over 20 agents by 1929 plus two Movable School agents who traveled the state in a one-and-a-half-ton panel truck.

Relations between the Extension Service and the College proper became harmonious after T. Roy Reid was named assistant director and Gray took over as Dean in 1924. Communication remained open and fruitful through the remainder of the decade and into the 1930's as the two men remained at their posts and viewed their work together as a partnership rather than a contest. They would need all the harmony they could muster to brace them for the Great Depression and World War II.[59]

People in Arkansas hardly noticed the stock market crash in October 1929. It appeared to them to be the economic disaster that they had been experiencing throughout the decade, only on a national scale. The effects of the crash and the economic depression that followed were felt deeply, however. The College of Agriculture felt the effects as well, but the College was prepared to weather the crisis and lend assistance wherever necessary, thanks to growth and consolidation that had taken place during the 1920's. Through the efforts of President Futrall, Bradford Knapp, Dan T. Gray, numerous faculty members, students, legislators and the people of Arkansas, the College was in a condition to serve the needs of the people in their time of crisis. A period of growth and consolidation was over for the time being; the challenge of the Great Depression and World War II lay ahead.[60]

Endnotes

1 Carl Moneyhon, *Arkansas and the New South* (Fayetteville: University of Arkansas Press, 1997), 134-135.

2 Moneyhon, 136; Strausberg, 42-43, 45-47.

3 *University Bulletin*, 10(21):33; *University Weekly*, March 22, 1920, 1:4; True, 300.

[4] Moneyhon, 145-146; True, 317; University of Arkansas, "First Annual Short Course for Farmers Program" (Fayetteville: College of Agriculture, 1919), Special Collections, Mullins Library, University of Arkansas-Fayetteville.

[5] W.L. English to Knapp, October 30, 1919, and Futrall to Knapp, October 6, 1919, Box 1, Folder 1; also Futrall to Knapp, January 1, 1920, Box 1, Folder 3; and D.C. Welty, Arkansas Commissioner of Agriculture to Knapp, December 16, 1919, Box 1, Folder 2; also Knapp to Lynn W. Osborn, January 2, 1920, and Knapp to Martin Nelson, January 2, 1920, Box 2, Folder 2. All in *Bradford Knapp Papers.*

[6] *University Bulletin*, 15(12):18, "Educational Survey of the University of Arkansas: Summary," hereafter cited as "Educational Survey"; Knapp to Nelson, January 2, 1920, Knapp to Osborn, January 2, 1920, Knapp to Frank Godchaux, June 27, 1920, Knapp to J. A. Evans, January 30, 1923, in Box 2, Folder 2, *Bradford Knapp Papers*; *University Weekly*, Oct. 7, 1920; *Arkansas Democrat*, July 23, 1920, 9:3; *Arkansas Traveler*, Oct. 6, 1921, 1:4; Strausberg, 45-46.

[7] True, 308; *The Agricultural Triangle*, 1(5):3; *University Weekly*, May 6, 1920, 1:5; Futrall to Knapp, Dec. 10, 1919, Box 1 Folder 1; Knapp to Bodman, April 18, 1919, and Knapp to Futrall, May 5, 1919, Box 2, Folder 2, *Bradford Knapp Papers.*

[8] *Catalog, 1920-21*, 132-133; Deane G. Carter, *The First Twenty Years of Agricultural Engineering: 1921-1941* (Fayetteville: Deane G. Carter, 1966), 9-12.

[9] *Dean's Reports*, 1923-24, Box 1, File 3 (although labeled 1923-24, the report is actually 1922-23); Leflar, 168-171.

[10] Bradford Knapp, *Safe Farming in Arkansas.* Arkansas Extension Service Circular No. 85 (February 1920), Fayetteville: College of Agriculture, University of Arkansas, 3, 7-8

[11] *University Weekly*, April 15, 1920, 1:5; *Facts about the College of Agriculture University of Arkansas*, University of Arkansas Experiment Station Bulletin No. 182 (Fayetteville: University of Arkansas, 1922), 11-12, hereafter cited as *Facts.*

[12] *Facts*, 13-14; *University Bulletin*, 14(18):7-8 (December 1920).

[13] Futrall to Knapp, October 11, and November 24, 1919, W.L. English to Knapp, October 30, 1919, Box 1 Folder 1; Futrall to Knapp, January 1, 1920, Box 1, Folder 2; Knapp to J.A. Evans, January 29, 1923, Box 2, Folder 2, *Bradford Knapp Papers.*

[14] T.M. Jeffords to Knapp, December 31, 1919, Box 1, Folder 3, *Bradford Knapp Papers.*

[15] *University Weekly*, April 22, 1920, 1:4; May 5, 1920, 1:5; *Arkansas Democrat* (Little Rock), July 23, 1920, 12:1; *Arkansas Agriculturist* (December 1929), 5; Knapp to Godchaux, June 27, 1920 and Knapp to J.S. Wannamaker, November 1, 1920, Box 2, Folder 2, *Bradford Knapp Papers.*

[16] *University Weekly*, January 21, 1921, 1:1, April 7, 1921, 1:6.

[17] Leflar, 167-168.

[18] Moneyhon, 142-144; Knapp to A.C. Davis, May 5, 1922, and Knapp to J.D. Eldridge, January 20, 1922, Box 2, Folder 2, *Bradford Knapp Papers.*

[19] Knapp to A.B. Hill, August 8, 1923, and Knapp to Fayetteville Democrat, August 8, 1922, Box 2, Folder 2; Knapp to Silas Bent, September 12, 1922, and Knapp to I.O. Schaub, April 26, 1922, Box 1, Folder 2; Knapp to A.C. Davis, May 5, 1922, Box 2, Folder 2, *Bradford Knapp Papers.*

[20] Knapp to O.B. Martin, January 10, 1923, and Knapp to I.O. Schaub, January 10, 1923, Box 2, Folder 2, *Bradford Knapp Papers.*

[21] Leflar, 168-169; "Why the Agricultural College Should be Separated from the University of Arkansas and Moved from Fayetteville," in *Miscellaneous Material on the Suggested Removal of the Agricultural College from Fayetteville to Russellville,* Special Collections, Mullins Library, University of Arkansas-Fayetteville.

[22] Ibid.

[23] Leflar, 170-171; *Arkansas Democrat,* January 25, 1923, 1:6; January 21, 2:6; February 1, 1923, 1:5-6, 4:4-5;

[24] Knapp to Harry Bellows, May 15, 1923, Futrall to Knapp, May 16, 1923, Box 1, Folder 3; Knapp to Futrall, May 21, Knapp to Carl Hollis, June 8, 1923, Box 2, Folder 2, *Bradford Knapp Papers.*

[25] Ibid.

[26] Knapp to J.A. Evans, January 30, 1923, Box 2, Folder 2, *Bradford Knapp Papers.*

[27] Knapp to J.A. Evans, January 30, 1923, and Knapp to Carl Hollis, June 8, 1923, Box 2, Folder 2, *Bradford Knapp Papers.*

[28] *Fayetteville Daily Democrat,* June 12, 1923, 1:6; *Arkansas Gazette,* June 13, 1923, 1:7; *Arkansas Democrat,* June 13, 1923, 2:5.

[29] Bodman to Board of Trustees, September 9, 1923, Box 2, Folder 3; McRae to Knapp, Telegram, September 9, 1923, Knapp to McRae, September 9, 1923, Box 1, Folder 3; Knapp to Arthur (Knapp), September 17, 1923, Knapp to Clarence Ousley, September 10, 1923, Box 2, Folder 3, *Bradford Knapp Papers.*

[30] Hale, 159.

[31] Nelson to Futrall, September 10, 1923, Futrall to Nelson, September 14, 1923, Box 11, File: "Nelson," *Dean's Reports*; *Arkansas Alumnus* (June 1924), 5; *Agricultural Triangle* (October 1924).

[32] Futrall to Nelson, October 30, 1923, Box 11, File: Nelson, *Dean's Reports; Trustees Minutes,* November 3, 1923.

[33] Strausberg, 48; *Arkansas Agriculturist* (October 1928), 7; *Arkansas Alumnus* (January 1924), 6; *Agricultural Triangle* (October 1924), 10.

[34] T. Roy Reid to Robert Leflar, June 4 and July 18, 1966, Box 4, File 4, *Leflar Papers.*

[35] *Arkansas Agriculturist* (October 1928), 7; Hale, 159-160; Moneyhon, 138, 144-145.

[36] *Agricultural Triangle* (October 1924), 11; Gray to Futrall, June 2, 1924, Box 12: File "Gray," *President's Records*; 1923-24 Report, Box 13, File "Agriculture, 1923-24," *Dean's Reports.*

[37] Carter, 8; *Agricultural Triangle* (November 1924), 27.

[38] *Catalog: 1924-25,* 155; True, 306, 308-310.

[39] J.W. Read and S. Palmer, *Vitamins, Health and the Daily Diet,* Bulletin 184, University of Arkansas, 1923.

[40] *University Bulletin* 18(17):8, 10-11; *Arkansas Alumnus* (October 1924), 8.

[41] *Arkansas Alumnus* (November 1924), 5; Futrall to Nelson, September 26, 1923, October 25, 1923, Box 11, File: Nelson, *Dean's Reports*; True, 317, 305-306.

[42] Nelson to Futrall, October 10, 1924, and July 19, 1924, Box 11, File: "Nelson," *President's Records*; *1924-25 Report,* 1-2, Box 13, File "Agriculture 1924-5," *Dean's Reports.*

[43] True, 317; *Catalog: 1925-26,* 185-186.

[44] *Arkansas Alumnus* (January 1926), 3; C.O. Brannen, *Relation of Land Tenure to Plantation Organization: With Developments Since 1920* (Fayetteville: University of Arkansas Agricultural Experiment Station, 1928), ii-iii.

[45] *Arkansas Alumnus* (January 1926), 4; J.A. Dickey, *The Relation of Industry to Agriculture in Arkansas*, University of Arkansas Bulletin, 19 (June 1926).

[46] *Trustee's Records 1925-26 Report*, 1-2, Box 2, File "President's Reports": David B. Danbom, *Born in the Country: A History of Rural America* (Baltimore: Johns Hopkins U.P., 1995), 167-168, 196-197.

[47] Ibid.

[48] *Trustees Records, 1925-26 Report* and *1927-28 Report* 1-2, Box 2, File: "Presidents Reports"; *Agricultural Triangle* (May 1925), 6; *Arkansas Agriculturist* (January 1926), 51.

[49] *Agricultural Triangle* (February 1925), 5.

[50] *Arkansas Agriculturist* (October 1926), 8-9, (September 1927), 3.

[51] *Arkansas Agriculturist* (April 1927), 157, (October 1927), 20; *Arkansas Alumnus* (May 1928).

[52] *Arkansas Agriculturist* (February 1928), 88.

[53] *Arkansas Agriculturist* (October 1927), 9; *Arkansas Alumnus* (September 1927), 7-8; *Trustees Records 1927-8 Report*, "Home Economics," 32-33, Box 2, File: "President's Reports."

[54] *Arkansas Agriculturist* (October 1927), 25, 37; *Arkansas Alumnus* (June 1927), 4-7; *Educational Survey*, 18.

[55] *President's Reports,* "1927-28 Report, Home Economics," Box 2.

[56] *Arkansas Agriculturist* (October 1928), 7, November 1929, 10; *Arkansas Alumnus* (September 1928), 2.

[57] *Arkansas Agriculturist* (October 1929), 7, 16; *Arkansas Alumnus* (October 1929), 2.

[58] Strausberg, 61-64; *Arkansas Agriculturist* (February 1925), 70, (November 1926), 35, (November 1927), 38 (May 1928), 149, (March 1930), 18.

[59] Gary Zellar, "H.C. Ray and Racial Politics in the African American Extension Service Program in Arkansas: 1915-1929," *Agricultural History* (Spring 1998), 72(2):436, 444-445; Mena Hogan, "A History of the Agricultural Extension Service in Arkansas" (M.A. Thesis, University of Wisconsin, 1940), 280-282.

[60] Moneyhon, 147.

Chapter Four

The Great Depression and World War II: 1930-1945

The economic, social and political consequences of the Depression and World War II transformed agriculture in Arkansas. It was not quite the transformation that Bradford Knapp, Dan T. Gray, J.C. Futrall and others had talked about in the 1920's. Nevertheless, it was a change so complete that a farmer who had slept through the 16 years between 1929 and 1946 would scarcely have recognized on awakening the farming operations that dominated the landscape. College professors, researchers and students played a key role in the transformation. Sometimes they worked actively as agents of change; at other times forces beyond their control buffeted them. Always, they were involved, seeking to improve the lives of all Arkansans through agriculture.

The economic and social dislocations of the Depression and the New Deal programs drove many farmers from the land, especially tenants, sharecroppers and farmers with small acreages. Whether government policy makers intended it or not, the New Deal programs and agencies favored the large planters and commercial operators. One result of the New Deal in agriculture was a concentration of land holdings in the hands of large-scale operators who farmed commercial quantities of the principal agricultural commodities. Agricultural mobilization for WWII, with its frenzied application of science, machinery and labor needed to fuel the arsenal of democracy, brought prosperity not seen since the WWI. The wealth fell into the hands of those who had reaped the lion's share of the benefits of the New Deal programs.

The College of Agriculture provided expertise and advisors for the myriad of New Deal agencies that came to Arkansas to battle the Depression crisis. The New Deal changed the relationships among the farmers, the College and the federal government. The biggest change for the College, besides expanding teaching, research and extension services to serve different clienteles, was

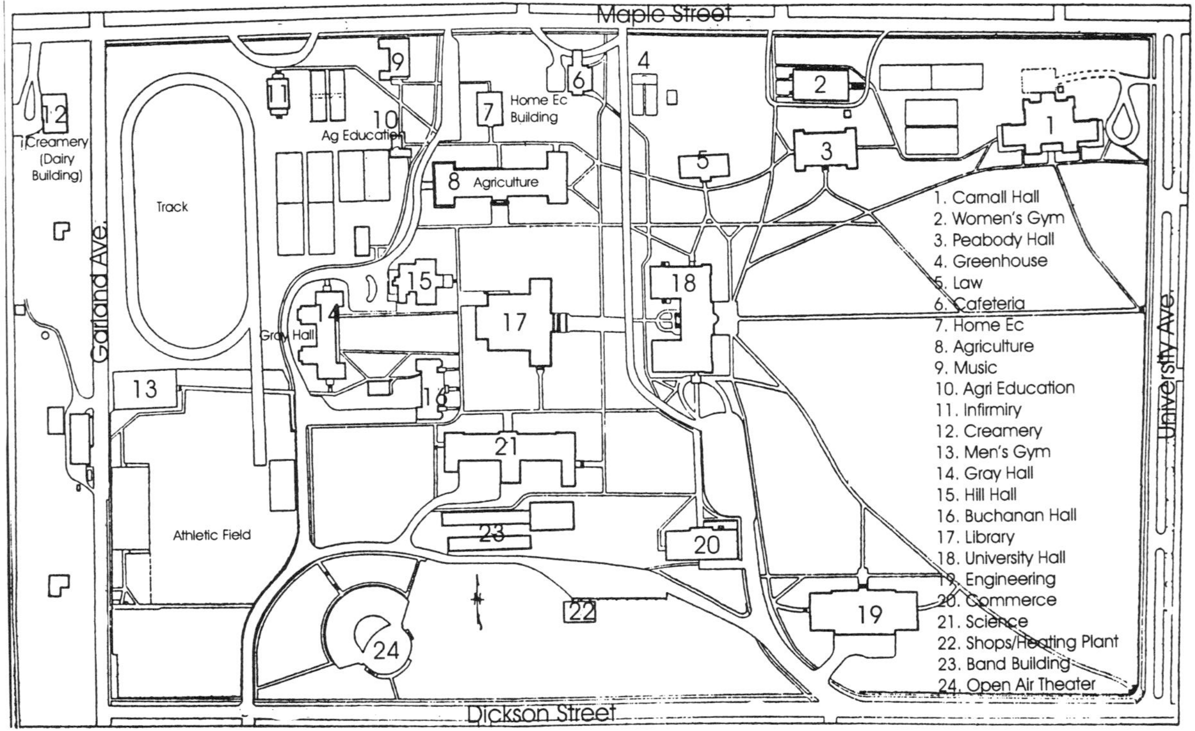

The University of Arkansas campus in 1936-1937. The new Agriculture Building (8) has been built, and the 1905 Ag Building (7) has been turned over to the Home Economics Department. The Agricultural Experiment Station Building has been relegated for use as a Music Building (9). Note the addition of the "Open Air Theater" to the landscape and Campus Drive, which cuts across campus from Dickson to Maple Streets behind Old Main. The Engineering Building (19) was built at the same time as the Agriculture Building. On the site of the current Plant Sciences Building is the university cafeteria (6).

that the College faculty was now called upon to help make and carry out public policy through their participation in the new federal programs and agencies.

In the late 1930's, Dean Gray was concerned that the New Deal agencies were taking over the role previously played by the College in developing an agricultural program for Arkansas. In at least one sense, Gray's fears were founded in fact. The number of federal agents and officials operating in Arkansas far outnumbered the College's personnel. However, the University of Arkansas College of Agriculture had educated most of these federal agents, and they owed at least their alumni allegiance to the College, if not their paychecks.

By the end of WWII, with the shift to commercial farming firmly established, the successful commercial farmer, if not a college graduate himself, depended on expert advice to a greater degree than ever before to stay competitive. In addition to the work done by the agricultural economists and rural sociologists, whose accomplishments took on greater importance during the Depression, the scientists and engineers at the College were also busy. During these years, particularly during the war years, advances in plant and animal breeding, progress in farm mechanization and strides in weed and insect control all contributed to the revolution in agricultural production that came at the end of the war. By the end of the war, the College was serving a different clientele and serving them in a different way.

During the years between 1930 and 1946, enrollment in the College of Agriculture see-sawed as conditions changed in society at large. In the early years of the Depression, with resources scarce and family incomes low, the number of students reached a low of 181 (1934). After New Deal student aid programs took effect, the numbers climbed to a record enrollment of 630 in 1940 before falling again during WWII as armed service enlistments and the draft took precedence over education. Whatever the enrollment numbers were, the agriculture and home economics students remained involved in college and community life. They established new service, honor and social fraternities and sororities throughout the period, adding to the cohesiveness in the College.

Enrollment 1930-1940.

Year	1931	1932	1933	1934	1935	1936	1937	1938	1939	1940
Men	127	118	116	98	115	179	268	318	378	399
Women	97	94	92	83	93	101	137	161	200	231
Total	224	212	208	181	208	280	405	479	578	630

The number of faculty members decreased in the early years of the Depression as budget issues became a concern, salaries were cut and no new faculty appointments were made. After the establishment of work/relief New Deal programs and the passage of the Bankhead-Jones Act, the situation improved as the additional funds were used to hire new faculty to relieve the overcrowded classrooms caused by the enrollment boom that accompanied the New Deal student aid programs established in 1935-1936. Salary cuts and crowded classrooms aside, the faculty of the University of Arkansas College of Agriculture remained one of the finest in the South, if not the nation, as they continued to provide excellence in instruction, world-class research and unstinting public service.

The Depression struck Arkansas hard. With its population 80 percent rural in 1930 and the majority of the farm incomes dependent on cotton, the fall in cotton prices from 17 cents a pound in 1929 to a new low of five cents in 1933 had a devastating impact. Other agricultural commodities followed suit. Farm incomes dropped 60 percent between 1929 and 1933. Tens of thousands of farms were lost because of delinquent taxes. Nearly one-third of the privately held land in the state was either delinquent in paying taxes or forfeited for failure to pay in 1933. Even for those farmers who still had collateral, credit from banks to pay taxes was not available. Over 50 percent of the banks in Arkansas failed between 1929 and 1931, and many of those in rural areas remained closed.[1] Arkansas farmers were dealt a double blow as the worst drought in memory struck the state from June 1930 through February 1931. One-third of the cotton crop was destroyed by the withering heat along with 70 percent of the corn yield. By February 1931 the Red Cross estimated that it was feeding one-third of Arkansas' population.[2]

> *I was determined to pursue a college education even in the face of depression–early thirties. You see I was 21 yrs. old and on my own with little to no finance. So, in the summer of 1932 I went to Ark. Tech. to work out fees for the fall semester. I did all kinds of work for 15 cts per hr. My brother and I rented an upstairs flat doing light housekeeping (batching), and by gall I made it. But let no one try to tell me how difficult it is going to college in the depth of our greatest depression.*
>
> *Dr. Edward M. Halsey, BSA 1942*

Throughout these crises in the early 1930's, the personnel in the College of Agriculture took the lead in providing aid for beleaguered farmers. The ASULGC appointed Dean Gray to head a commission to petition President Herbert Hoover for drought relief in December 1930. However, Gray's efforts were stymied by President Hoover's philosophical objections to direct aid to the stricken farmers. After assessing the political consequences of withholding aid, Hoover reluctantly agreed to extend federal loans, but in the face of the deepening economic disaster spreading through the countryside, most farmers were unable to secure loans at any level. T. Roy Reid, the Assistant Director of the College Extension Service, and Connie Bonslagel, the head of the Home Demonstration program, along with the local agents, played a key role in dispensing the Red Cross aid. However, the resources were stretched thin, and many farm families received barely enough food to sustain them above starvation levels. African-American farm families got even less attention as racism and planter's fears of "outsiders tampering with their labor supply" aggravated their suffering.

Hoover was no more willing to extend farm relief beyond what could be done through his Federal Farm Board than he was to provide federal assistance during the drought. What money was available was once again issued to those farmers able to secure loans. County agents were the conduit for these meager funds as well as the small amounts made available through the Reconstruction Finance Corporation beginning in early 1932.[3]

The University and the College of Agriculture felt the economic tremors as well. The University budget was cut 22 percent between 1931 and 1933. Faculty salaries were reduced eight percent in 1932-33 and another 14 percent in 1933-34. The Board of Trustees hired few new faculty and made no replacements for those who died or retired.[4]

The Farmer's Union renewed agitation to abolish the Extension Service, calling the program a wasted expense. While most farmers saw the county and home demonstration agents as the only government officials interested in their welfare, state and local revenues could no longer support their activities, and the Extension Service was reduced dramatically. The African-American Extension Service was the first to feel the cuts, and their numbers were reduced by 50 percent by 1932. Those who remained had their salaries reduced to a bare subsistence level. The Experiment Station federal funding was locked in to the land-grant funds and previously legislated levels, but state funding was curtailed significantly. Dean Gray defended the expenditures requested from the state by quoting a Department of the Interior survey done in 1930:

> The University of Arkansas Agricultural Experiment Station occupies a respected place among the agricultural experiment stations of the United States, and has accomplished more of direct money value for the industry than it could have accomplished by training five times the number of dirt farmers as it has....The State's investment in agricultural research at the University is repaid a thousand fold annually.[5]

Apparently the farm folk of Arkansas agreed with this assessment if their attendance at the annual Farmer's Week held at the University and the Visiting Days at the Branch Stations can gauge their interest in the work. For example, 4,821 farmers and their families attended the 1930 Farmer's Week in August, in spite of the drought and the economic distress in the countryside. The week's activities were highlighted by a pageant entitled "The Ten Commandments of Agriculture by Seaman A. Knapp" written by a local playwright, Annie Foster Stark.

The attendance at Farmer's Week continued to grow in these early years of the Depression, with over 5,000 visiting in 1931 and a record 6,077 appearing at the 1932 festivities before President Futrall canceled the event in 1933 due to budget constraints. The Branch Station Visiting Days were equally popular; 2,000 attended the Fruit and Truck Station Day at Hope in June 1930; 2,500 were at the Cotton Station at Marianna in September; and 450 came to the Rice Station outside of Stuttgart in October. After the cancellation of Farmer's Day, visiting days were continued at the Branch Stations and drew tens of thousands of visitors throughout the 1930's.[6]

The dark early days of the Depression did not dampen the enthusiasm of the College's student body. In spite of budget cuts, financial difficulties and decreasing enrollment, the agriculture and home economics students found ways to keep the Agri Day activities going, if only at a reduced level. Few students begrudged the five-dollar fee assessed all agriculture students used to finance the Agri Day activities. Though not many could afford the fee, everyone pitched in and found creative ways to raise the needed money.[7]

In addition to maintaining existing organizations such as the Agricultural Society, the Agri Day Association, the Home Economics Club and the Alpha Zeta honors fraternity, the students were also busy organizing new clubs, fraternities and sororities that added to the cohesion in the College. A chapter of the Alpha Gamma Rho agricultural social fraternity had its beginnings in an AGR Club organized in 1930. Fifteen agriculture students under the leadership of John Stevens formed the earliest chapter. The first presidents, James Niven and C.B. Guilland, recruited other agriculture students into the chapter, which grew to include 39 members by 1934. Dean Gray also helped with the organization, and Deane Carter, head of the Agricultural Engineering Department, served as its sponsor. The AGR national organization granted the University of Arkansas organization a charter in the spring of 1934.

The Home Economics students took initial steps toward establishing a Phi Upsilon Omicron chapter on campus in the early 1930's, but the organization did not take off until 1940. The students also organized college 4-H clubs for both the men and women on campus in the early 1930's. In addition to mentoring 4-H members in secondary schools throughout Arkansas, the clubs also organized cooperative 4-H houses close to the campus. These were a great help in defraying student living expenses and strengthened social cohesion among the agriculture students. The girls' 4-H Cooperative House, organized in 1932, was the first of its kind in the country. Their cooperative method of reducing college student living expenses was imitated on many campuses across the nation.

A college chapter of Future Farmers of America (FFA) was organized at this time, and later the members also started a cooperative house. The plans

The girls' 4-H Cooperative House, established at the University of Arkansas in 1932, was the first of its kind in the country; their cooperative method of reducing living expenses for college students was imitated on many campuses across the nation. Cost per semester to live in the house: $50. A new girl's 4-H Cooperative House (pictured here) was built in the 1950's. The residents relocated to the new building from their previous home at 753 West Dickson. (*Arkansas Agriculturist*, January 1946)

that governed these cooperative ventures allowed students to pay room and board, all or in part, with either canned goods or produce brought from the farm. The members shared other expenses as well as chores. One enterprising student brought a cow to school for two semesters, drank three pints of milk a day himself and sold the rest of the milk to buy other food items. Other students sold sandwiches or baked goods and took up what odd jobs were left in town to help meet expenses. By the 1936-37 school year, there were over 100 agriculture and home economics students living in the cooperative houses, or about one-quarter of the College's student enrollment.[8]

> *One spring I purchased the sweet corn crop produced at the farm from Professor Cooper. I sold it to the various dorms and sorority houses and the VA Hospital. I made a good profit.*
> *J.E. McNeil, Jr., BSA 1946*

The students also benefitted from scholarships provided by local 4-H, FFA and other agricultural clubs. Women's Home Demonstration clubs throughout the state were very active in providing what funds they could for needy students with scholastic ability. The campus Alpha Zeta chapter created a fund in 1932, sponsored by agronomy professor C.K. McClelland, to aid members in carrying on their schooling. Scholarships were available from private donors as well. International Harvester gave out two $500 awards each year to deserving students. The Chilean Nitrate of Soda Educational Bureau offered a $2,000 scholarship toward graduate study for investigations into the proper uses of fertilizers on cotton, corn and oats.[9]

As in earlier years the agriculture students contributed more than their share to athletics at the University. The College supplied many of the leading players for the 1932 Razorback football team. The sub-captain, Jim Edmundson, from Maysville, Oklahoma, played on the varsity squad for three years. Edmundson belonged to the Alpha Zeta fraternity where he was censor and was involved in many other campus activities. Three other agriculture students held positions on the varsity team, and there were two others on the reserve team. Another agriculture student, Louis "Red" Johnson, played right end and was the 1933 team captain.[10]

The students in the College certainly did their part to maintain the integrity of the institution during the hard times in the early phases of the Depression. Meanwhile, Dean Gray and the faculty were doing their part to uphold the reputation of the College and shepherd it through the crisis. Reactionary forces in the Legislature led a series of attacks on the faculty of the University, portraying them as a pack of "rich professors" being coddled at the public trough while they polluted the minds of Arkansas' youth with "foreign and northern" ideologies. The opponents of the College also dragged up the same rhetoric employed in the 1880's, that the College was not doing its job in training farmers. Dean Gray responded with a patient, yet forceful, reply:

> During these hard and difficult times the people of Arkansas are examining more closely than ever before the work of the various educational institutions of the state, the qualifications of the faculties to do satisfactorily the educational work delegated to them, and comparing this work and these faculties to similar work and faculties of neighboring states. This is entirely proper and should be done. Fortunately, the faculty of the College of Agriculture of the University is in the position to remain perfectly happy when the closest and most exacting analysis is made....In both scholastic attainments, as measured by advanced degrees, and the practical and scientific results (of research), the faculty of this College stands high when compared with the best faculties in America. Our faculty is a compliment to Arkansas, but not to Arkansas alone. The faculty is a credit to the educational institutions of the nation.[11]

Gray went on to say that 38 percent of the faculty held doctorates from "the leading universities of America and Europe" and 60 percent held master's degrees, with many of them having done advanced work toward their Ph.D.'s. Only Cornell University, considered by many the premier agricultural university in the United States, could boast of a higher percentage of faculty with advanced degrees. In the South, the University of Arkansas College of Agriculture had no equal. Left unsaid in Gray's rebuttal was the outstanding fact that two "rank foreigners on the faculty," (as the legislators described them) Barnett Sure, originally from Lithuania, and M.C. Kik, from Holland, had made great strides in improving and guarding the health of Arkansans as well as people around the world through their research on vitamins and nutrition.

A survey done by the Higher Education Commission of the United States Office of Education in 1931 buttressed Gray's remarks. The Commission noted that the College's job was no longer to instruct farmers in farming practices; that function was now carried out by the Smith-Hughes schools and the Extension Service. The job of the College was to train scientists, teachers, Extension Service workers and people going into agricultural industries. And in this field, the Commission felt that the University of Arkansas College of Agriculture was doing its job admirably well, especially in agricultural research.[12]

Dean Gray also had a good reason to be proud of the curriculum offered at the College. The course work was geared toward providing the undergraduate students with a solid foundation in the general principles of scientific agriculture rather than focusing on specialized fields of study. Specialization was generally reserved for graduate study. In this way the College prepared its undergraduates for careers in extension work, vocational agriculture, farm management, general farming, homemaking and other pursuits that required a solid understanding of a wide range of subject matter. However, as Dean Gray was careful to point out, that did not mean that the College was so inflexible that a student wishing to undertake specialized work would not be allowed to do so. In these cases the head of the department and the Dean would work with the student to formulate an individual course of study tailored to the student's area of interest.[13]

This flexible program allowed students to get the most value from their education. One illustration of the program at work involved Helen Morgan, the first woman to enroll in the forestry program. Morgan, from Ouachita County, had grown up on a farm that had 450 acres of well-managed timber. While her family had strived to insure that their woodlands were managed as a "renewable resource," such was not the case in much of the surrounding privately held acreage in the area. Morgan, who was studying for a career in Home Demonstration, wanted to return to her home locale equipped to show the benefits of improved forestry practices in enriching rural home life. According to Morgan, not only could forestry provide farm income in a steady and renewable way, but improved forestry practices could beautify the homestead, attract wildlife and prevent soil erosion, thus adding to the land value. Morgan went on to a career in Home Demonstration in Ouachita County after her graduation in 1931.[14]

The flexibility offered for home economics and agriculture majors could also be seen in the number of women enrolled in meat-cutting classes in the

Department of Animal Industry. While "butchering" would seem to be primarily a job for men in some circles, many times on the farm such gender distinctions fell by the wayside when there was a job to be done. In any case, 13 home economics majors enrolled in the class during the 1930-31 term in order to widen the scope of their studies in food science, nutrition and home demonstration. The Extension Service had emphasized a "Live at Home" program for many years in their work across the state in order to help alleviate the financial distress in rural areas. During the Depression the idea took on an added meaning. This program stressed home canning and preservation of foods. A key part of this focus was the preservation and curing of meat for home use. In the class the women also learned things hands-on that were applicable to food science and nutrition. What cuts of meat were the most nutritious and which ones were the most economical, learned from the viewpoint of the butcher, were valuable lessons for consumers in such times when every penny counted.[15]

> *[My] advisor enrolled me in an "Animal Husbandry Class." Seated on back row, surrounded by 70 male students. I was the lone female. Having heard rumors of some of the jokes this Prof. told, I rushed to his desk to inform him I was there. He was surprised, but said, "It's o.k., but there has never been a female in this class." The next day there still wasn't a female in that class.*
>
> *Wanda Richards Underwood Spivey, BSHE 1941*

During the early years of the Depression, before the New Deal programs began to make funds available for equipment and buildings, the various departments struggled along and made do with what was available. In 1932 W.L. Bleeker in the Department of Bacteriology and Veterinary Science had to turn students away because there wasn't enough lab equipment to serve the numbers wanting to take the class. However, in the next term he broke the classes into smaller sections so that everyone would be sure to have the necessary equipment. Even though this took time away from his research work, Bleeker made adjustments in his work schedule and asked Margret Smith, a researcher from the Home Economics Department, to teach some of the labs. V.H. Young in the Plant Pathology Department reported a general scarcity of equipment, including microscopes, slides and charts, needed to carry on routine work. J.R. Cooper, in the Horticulture and Forestry Department, said that though the student attendance and performance was "satisfactory," his teaching program was handicapped by a scarcity of microscopes. In the Entomology Department Baerg lamented that some research work would have to be reduced due to the budget reductions. Yet, in spite of the inadequate facilities, the teaching and research remained at a high level.[16]

The excellence in the teaching was reflected in part by the number of undergraduates going on to graduate work at the College and at other universities across the country. In the 1931-32 term there were seven graduate students in the College of Agriculture. Six were studying in the Rural Economics and Sociology Department, and one, Clyde R. Benbrook of Fayetteville, the recipient of the Chilean Nitrate of Soda scholarship, was working in the Agronomy Department. By 1932, 24 former undergraduates of the College had gone on to do master's or doctoral work at some of the most prestigious agricultural institutions in the country, including Cornell, Harvard, Iowa State, Wisconsin and Michigan State. Furthermore, many of these former Arkansas agricultural students returned to Arkansas as either teachers or researchers in the College in the 1930's.[17]

During this period of severe budget constraints, one important addition to the physical plant at the College was made in early 1932. Through thoughtful planning and the use of materials already available, Deane Carter, the head of Agricultural Engineering, provided the plans and engineering know-how to develop the first Poultry Experimental Farm. Located on 23 acres adjacent to the Main Experiment Station north of Fayetteville, the farm experimental work was supervised by R.M. "Chicken" Smith. Smith came to the College in 1927 and conducted experiments in cross-breeding, feeding and housing poultry. In conjunction with the Arkansas Poultry Improvement Association, Smith was instrumental in training poultry inspectors and in developing a statewide inspection program to protect poultry flocks from the spread of disease. From the research done by Smith and others, the College hoped to provide financially strapped Arkansas farmers with an alternative money-making enterprise. In doing so they helped lay the foundations for what would later become the most important agricultural industry in northwestern Arkansas.[18]

This was during the heart of the Great Depression, like bank holidays and poor economic conditions generally. The opportunity to work on and off campus just to stay in school for four years was a great experience. Sometimes the rate of pay was as little as 10 cents per hour. Many students had to drop out.... Generally the atmosphere was great. Guidance from the faculty was superb. Even under the poor economic conditions that existed, the College was able to place each graduate in some sort of employment.

Buford Poe, BSA 1935

The Department of Rural Economics and Sociology, under its head, C.O. Brannen, made major research contributions during these early years of the Depression. Studies begun in the late 1920's, by Brannen and others, on farm credit, land tenure and taxation were expanded and updated to meet the new crisis. The Department of Rural Economics and Rural Sociology was working in a field that was at the center of the crisis gripping the nation. Accordingly, the department expanded both the number of personnel and the scope of its work. As already noted, the bulk of the graduate work done in the College took place in rural economics and sociology. The teaching and research staff included nine teachers and researchers plus a clerical staff of half a dozen to help make available the research studies on questions of such vital interest to all. The department also set up cooperative research projects in conjunction with other departments at the College and the University, the most important liaisons being with the Agricultural Engineering and Home Economics Departments in the College and with the School of Business Administration and the Economics Department. However, for all the activity in the College toward solving the problems of the Depression, the work seemed for naught because of reluctance of state and federal governments to take action.[19]

The Arkansas state government was mired in hopeless gridlock during this period as Governor Harvey Parnell called one special legislative session after another, all the while proclaiming that Arkansas needed no new legislative palliatives. Parnell told the Assembly, "This is not the time to try out fantastic schemes nor visionary panaceas for all our ills." The same situation prevailed at the national level as President Hoover lost the confidence of the people and Congress appeared reluctant to act. Apparently the people felt differently about legislative palliatives and visionary panaceas, as they elected Franklin Delano Roosevelt by a landslide in 1932, though in his campaign, Roosevelt had promised little beyond a pledge to take action. After Roosevelt's election in 1932 and the beginning of the New Deal in the spring of 1933, the nation's agricultural system and the role that the federal government played in it would be transformed. Likewise, the College of Agriculture's mission took on a new direction that would transform its relationship with the people of Arkansas.[20]

The inauguration of Roosevelt in the spring of 1933 and the initiation of the New Deal were greeted with a collective sigh of relief. For the people of the nation, and especially the farmers, it was as if a spell had been broken. Their plea for action on their behalf was answered in the flurry of legislation and executive orders that accompanied the historic first "100 Days" of the Roosevelt Administration. The programs formulated and the agencies created to carry them out during this first phase of the New Deal agricultural planning were a mixture of both the visionary and the practical impulses found in Roosevelt's "Brain Trust." The agricultural programs were a blend of progressivism, old-style agrarianism, McNary-Haugenism[21] and hard-nosed politics, and no one was quite sure what the end results would be. If the New Deal policy planners agreed on one thing, however, it was that something must be done—and the sooner the better.[22]

The Agricultural Adjustment Administration (AAA), the Farm Credit Administration (FCA), the Credit Commodity Corporation (CCC), the Federal Emergency Relief Administration (FERA), the Public Works Administration (PWA), the Civilian Conservation Corps (another CCC) and the Civil Works Administration (CWA) were all among the so-called "alphabet agencies" created in the early New Deal to address the crisis. The implementation of these programs and the impact that they had on Arkansas' agriculture was intimately related to the work and development of the College of Agriculture as well. The effectiveness of the programs depended on local cooperation and the input of local expertise. The faculty and staff at the College, particularly in the Department of Rural Economics and Sociology, were the logical choice to help formulate and carry out the program objectives. In this way the College took on a role beyond its original mission of resident teaching, research and extension: that of helping to formulate public policy.

The work of the AAA, the linchpin of the New Deal agricultural program, was carried out under the direction of the College Agricultural Extension Service. T. Roy Reid, the Assistant Director of the Extension Service, was designated to direct the program in Arkansas. County Extension agents helped supervise the programs in their locales. The AAA's first objective, to increase farm incomes by raising prices on basic agricultural commodities through limiting the amount of acreage in production, began in the spring 1933 "plow-up campaign." Because most of the cotton and rice crops in Arkansas had already been planted by the time the AAA legislation became operational, the only conceivable way to limit production was for farmers to plow under part of their crops already in the ground. In addition to raising prices for commodities, the AAA program also added to farm incomes through direct cash payments from the government to farmers. In return for signing contracts agreeing to plow under a certain amount of acreage (usually one-third), the farmer received payments that reflected the parity price of the crop, minus the market price, multiplied by the average production per acre. The county agents, along with committees formed from local farmers and planters, oversaw the operations. Under the program 97,000 Arkansas farmers plowed under over one million acres and received payments worth $12 million.

Farm incomes doubled and cotton prices rose to 10 cents a pound by the fall of 1933. However, the benefits from AAA acreage reduction were lopsided. The program favored large planters and commercial operators at the expense of small farmers, tenants and sharecroppers. While AAA regulations stipulated that planters must share the payments in equal measure with their tenants and sharecroppers, in reality few of the latter received the money due them. The 1933 plow-up campaign did not displace many tenants or sharecroppers because contracts for the year had already been made. However, in the 1934-35 season and in ensuing years of the acreage reduction program, tens of thousands of these landless farmers, who formed the backbone of the Southern agricultural system, were forced off the land. Many of them were replaced by tractors, which their landlords bought with AAA crop reduction payments. In attempting to solve one problem, the New Deal planners had created another. College of Agriculture personnel were called upon to address the problem of the displaced tenants and sharecroppers as well.[23]

While most of the New Deal relief programs formulated in Washington were tailored to the concerns of the urban unemployed, by the fall of 1933, the planners realized that rural relief was also needed.[24] Harry Hopkins, Roosevelt's director of the CWA, called upon Dean Gray and C.O. Brannen to survey the needs of the rural unemployed in Arkansas and make recommendations. Brannen also was made director of the FERA survey on rural relief. In 1934 he was appointed state director of a project collecting data on tax delinquency and land values. Orville J. Hall, an assistant professor in the Department of Rural Economics and Sociology and a graduate of the College, was drafted as assistant director. FERA appointed another member of the department, T.C. McCormick, as director for a project making profiles of rural relief families in Washington, Lee and Cleveland Counties. Dean Gray gave B.M. Gile a four-month leave of absence in the fall of 1933 to work with the Commodity Credit Corporation. Gile amplified his previous ground-breaking research done in the early 1930's dealing with agricultural credit in Arkansas. Many of the findings from Gile's previous work had found their way into the New Deal farm credit legislation, making Gile the perfect person for the job. W.E. Mountcastle, a graduate of the College, directed a project under the Division of Subsistence Homesteads, which later became part of the Resettlement Administration. The goal of the project was to resettle displaced farmers and unemployed people in subsistence colonies in rural areas on small homesteads where they could raise their own food and be provided with at least part-time jobs in the area, a goal that appealed to the utopian planners in the New Deal. One of the offshoots of this project became known as Dyess Colony, named after the Arkansas FERA Director, W.R. Dyess, a fervent advocate of resettlement projects, who died in a plane crash in 1935. The resettlement projects were one of the most controversial of all the New Deal initiatives, extolled by supporters as a visionary solution to the problem of rural poverty and blasted by critics as the beginnings of Soviet style collectivism in America.[25]

In early 1934 Dean Gray was also made chairman of the Arkansas Land Utilization Commission. The commission's job was to study how farm lands were used in Arkansas, catagorize them and, if possible, formulate a statewide plan for more efficient use of lands that had been "retired" under the AAA program. In June 1934 Dean Gray took a year's leave of absence to become regional director of the AAA Land Use Survey. In this capacity Gray supervised the study of sub-marginal lands in Arkansas, Louisiana, Oklahoma and Mississippi in order to evaluate their possibilities for grazing, timber, wildlife management or recreational use. Operating on the premise that unemployment and poverty in Southern rural areas were due to overpopulation and underutilization of land resources, the land surveys sought to match rural populations with alternative methods of generating income and livelihood. President Futrall tapped C.O. Brannen to act as temporary dean during Gray's absence.

The New Deal also drafted many other College personnel from the teaching, research and extension branches of the College during this period to provide expertise in matters ranging from landlord-tenant relations to soil erosion control. It was a heady time for the College's staff, whose advice for many years had been accepted only grudgingly by farmers and ignored by most government policy makers.[26]

After his appointment as Acting Dean, Brannen tackled his overloaded agenda with enthusiasm. In addition to his work with the AAA, FERA and other New Deal agencies, Brannen had to coordinate the teaching, research and extension work at the College and tailor the College's program to the needs of the new agenda in agriculture. Budget issues were also a primary concern. For while FERA and PWA grants and loans offered some relief and even the possibilities for expanding research, teaching and improvement of facilities, the core of the College's budget remained tied to state appropriations, which were practically non-existent. University President Futrall was constantly on the lookout for ways to reduce expenses, and though he was always an advocate of agricultural education, he also had to regard the needs of the University as a whole.

One of Brannen's first actions as Acting Dean was the creation of a course on the new federal programs. The course, called "federal agricultural programs," was intended to acquaint students in the College with agricultural problems beyond those they had experienced first hand and provide them with an understanding of how the new federal programs operated to address the problems. Ultimately, the course gave the students the training necessary to participate in the programs, which many of them would do in the ensuing decade. As already noted, the bulk of the staff in the Department of Rural Economics and Sociology were on leave during the 1934-35 term working with various federal agencies. There was a greater call for the classes in this department, with over 200 students enrolling during the 1934-35 term, nearly doubling the 1932-33 numbers. Brannen arranged to have J.W. Jamison from the Economics Department take over some of the teaching duties. Graduate students also took over some of the classes. Increases were in store for other departments as well, but not until 1935-36 and thereafter.[27]

In 1934 the University received loans from the PWA for construction of a new library and a chemistry building. The Board's building committee wanted to use some of the remaining PWA funds to build additional dormitory facili-

Two greenhouses, built in 1935, with Public Works Administration funds.

ties. However, Brannen urged that the money be used to build additional greenhouses, which were badly needed by the Departments of Plant Pathology and Horticulture. At first Futrall was reluctant to use the funds at all, suggesting that the money be allowed to revert back to the federal treasury, thus avoiding the necessity of paying back additional loans, which would save the students money in the long run. Brannen eventually prevailed, and the two greenhouses were completed in the late summer of 1935.

Futrall was constantly on guard against what he considered useless expenses. In one incident Futrall questioned the need for the University to sign a National Recovery Administration (NRA) agreement for the operation of the University creamery. Fearing that adhering to the NRA codes and price schedules would increase the cost of the creamery's operations, Futrall asked Brannen for an explanation. Brannen replied that since the creamery was already in compliance with NRA standards, the College lost nothing by signing the agreement. Because the creamery did a considerable business with local consumers and merchants, Brannen wanted to avoid the embarrassment of not flying the NRA "blue eagle" (the symbol of the NRA, which stores and businesses either posted in their windows or flew on flagpoles to show that they had signed the "voluntary agreements" with the NRA on price and wage controls).

Part of Futrall's economy campaign included cutting travel expenses as well. The president decried using precious appropriations for travel to "useless and redundant" federal conferences. While he backtracked somewhat from characterizing the meetings as "useless" and reluctantly admitted that perhaps the College should be represented, he asked that Brannen restrict the number of College people travelling.

Faculty trips to professional meetings drew Futrall's special ire. Futrall turned down C.K. McClelland's request to attend the National Soybean Association meeting in the fall of 1934, saying that the College of Agriculture had more travel expenses than all other divisions in the University combined. After Brannen pointed out that McClelland was the Association's president and that attending the meeting would help with finding substitute crops for cultivated land under acreage restriction, Futrall relented, but he allowed McClelland only travel expenses and only for a reduced number of days.[28]

Dean Gray returned from his leave in the summer of 1935, and he too found a full agenda awaiting him. National attention was drawn to the plight of the tenants and sharecroppers displaced by the AAA acreage reduction program in Arkansas. The activities of the Southern Tenant Farmer's Union (STFU) in trying to organize the landless farmers focused attention on the negative aspects of the AAA program and how the federal efforts failed to address the needs of those at the bottom of the agricultural ladder. In response President Roosevelt created the Resettlement Administration by executive order in early 1935 to deal with the problem. T. Roy Reid was appointed the director of the program in Arkansas, and C.C. Randall, one of the district agents for the Extension Service, was named his assistant. Later that summer, Connie Bonslagel, the director of the Extension Service Home Demonstration program, was named as an assistant director.

Both the federal government and the state of Arkansas established commissions to study the tenant problem in Arkansas. College of Agriculture personnel played key roles on both commissions. Dean Gray, C.O. Brannen, T. Roy Reid, C.C. Randall, Connie Bonslagel and H.W. Blalock participated in the proceedings. W.T. Wilson, a graduate from the College and now a professor in the Department of Rural Economics and Sociology, made a survey of potential rehabilitation clients for the resettlement projects that became an important guide for determining who should be eligible for the program. J.G. McNeely, who came to the College in 1936, also did an important study of landlord and tenant relations that covered such topics as the fairness of contracts, supervision of tenants, distribution of crops and livestock, tenant migration and the change in tenant status. H.W. Blalock and K.H. Hunter's study

Connie Bonslagel, left, state home demonstration agent (with Mattie Melton and Esther Kramer) made the first official Extension air trip on May 12, 1931, from Little Rock to Rogers. (*Arkansas Agriculturist,* December 1931)

on tenant conditions in Arkansas received national attention and gave impetus for revamping the Resettlement Administration and recasting it as the Farm Security Administration (FSA) in 1937. T. Roy Reid resigned from his position as head of the Extension Service in 1937 to take over full-time administration of the FSA in Arkansas, and the Board appointed C.C. Randall as his replacement.[29]

College personnel undertook other studies concerning the needs of rural people at this time as well. Isabella Wilson, the head of the Home Economics Department, joined with W.H. Metzler of the Rural Economics and Sociology Department to conduct a study of the health of rural people and their access to health services. The massive project began in 1936 and continued through 1940. Wilson and Metzler found that rural health conditions were very poor in the state; rural people not only were afflicted with high rates of disease, but they had no access to medical care. These findings were instrumental in the development of legislation tailored to the needs of Arkansas' rural population.

In another joint project, undertaken by personnel from the Departments of Agricultural Engineering, Home Economics and Rural Economics and Sociology, researchers evaluated the status of rural housing in Arkansas. This study also drew national attention. Drawing on findings from the housing survey, Deane Carter helped design plans for low-cost housing for rehabilitation clients accepted by the FSA. Using low-cost native materials and unskilled labor lowered the costs for building new homes or refurbishing homesteads for rehabilitation clients.

> *Agri Dean W.R. Horlacher, who taught a course in Extension methods, told us that he had rather hire a "B" or "C" student who had participated in many activities than an "A" student who did nothing but study, but if the person could participate and make A's it was great.*
>
> *Dr. Janive Segraves Blanchard, BSHE 1944*

The Agricultural Engineering Department also did significant work on soil erosion in conjunction with the U.S. Soil Conservation Service (SCS). Perhaps the most significant study done in conjunction with the Rural Economics and Sociology Department, whose results were overlooked by many of the rural planners and policy makers at the time, had to do with the impact of mechanization in the Delta area of Arkansas. The findings in "Physical and Economic Inputs Required for Field Power on Delta Lands" showed the benefits of using mechanization over hand labor on expanded acreage.[30]

All the rehabilitation and resettlement efforts undertaken by the federal agencies could not change the unalterable fact that machines were replacing people in the commodity-producing agricultural sector. The trend, which began following the plow-up campaign in 1933, after planters and commercial operators bought tractors with their crop reduction payments, would continue at an accelerated pace during WWII and signaled the end of the tenant-sharecropping system in Southern agriculture.[31]

While the tenant-sharecropping system faded away under the influence of New Deal programs, a lasting legacy was the rural electrification program undertaken by the Rural Electrification Administration (REA). Roosevelt created the REA by executive order in 1935, and Congress legislated the agency the following year. Agricultural Engineering, in conjunction with the Electrical Engineering Department, drew up plans for this effort to improve the lives of rural Arkansans. Deane Carter and E.E. Karns issued an Extension circular on farm electrification in 1937 that provided farmers with basic information on wiring and then connecting their farms to the rural cooperative power lines just beginning to reach into the Arkansas countryside. Isabella Wilson, Deane Carter and Orville Hall collaborated in a 1937 study, "Cost, Use and Trends

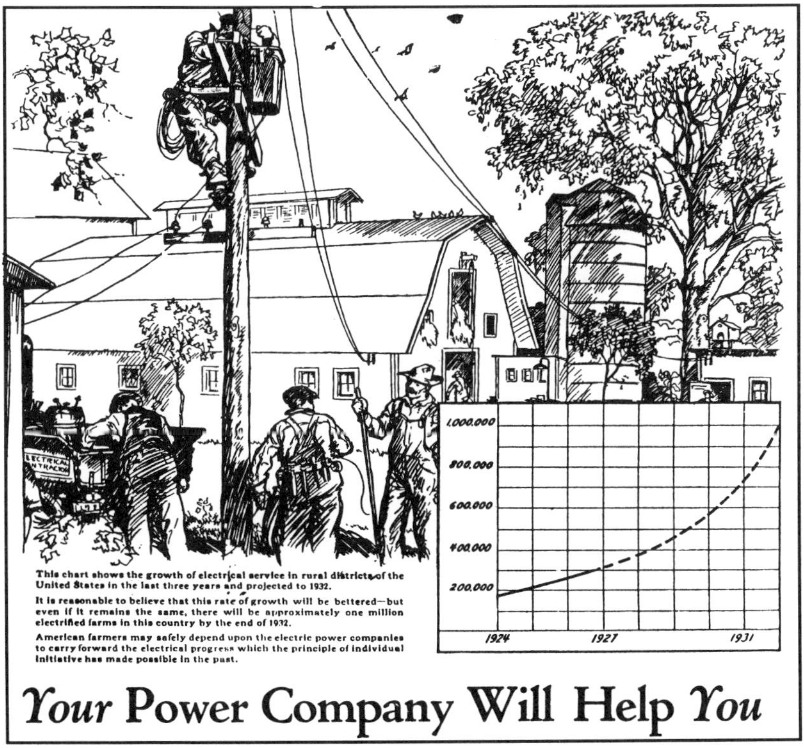

In the late 1920's, the push for rural electrification had begun, as can be seen from this portion of a 1927 ad in the *Arkansas Agriculturist.* In 1935 President Roosevelt created the Rural Electrification Agency, and members of the College faculty contributed to the effort through publications and research.

in Rural Electrification," that helped set standards for the blossoming number of rural electric cooperatives in the state. In the year 1937 alone there were more miles of electric power lines and more homes connected to them than all previous years combined. Eight thousand farm homes were connected in 1937, and similar gains were made each year until the war intervened and construction of more lines was interrupted.[32]

The research and extension work at the College received a boost from the federal government after Congress passed the Bankhead-Jones Act in 1935. An additional $1 million was made available for 1936 with additional increases yearly until the amount reached $6 million in 1940. Research into improved varieties and maintaining soil fertility was targeted in the act as well as the social and economic problems affecting agriculture. In spite of the acreage restrictions on cotton production and other commodities in the United States, research continued on improved cotton varieties that drew international attention. Agriculturalists from China, India and Peru visited the Experiment Stations to learn from the research conducted on new cotton varieties by L.M. Humphrey and studies done on cotton diseases undertaken by V.H. Young. C.K. McClelland's experiments on new soybean, oats and forage crops were of particular interest to farmers looking for different crops to plant on restricted acreage. McClelland penned a poem lamenting cotton's sagging reputation among government policy planners.[33]

In Defense of Cotton

Some say, and seem loathe to stop,
"How wasteful is the cotton crop,
How fertile farms it soon despoils,
And robs them of their richest soils."
This hue and cry must have an end,
When reason doth the crop defend.

If one a second thought would take,
He'd find himself in a mistake—
Year in, Year out, man plants the crop,
Although his friends advise a stop.
Cotton has no thought or aim,
And at man's feet must lie the blame.

If seed upon the cotton land be shaken,
Little from the soil is taken.
Lint alone robs not the farm,
Nor does the future generation harm;
It's elements from air and water come,

Built by energy from the sun.

The loss that causes such vexation,
Comes from faulty crop rotation,
Repeated stirring of the soil,
Burns out the humus just like oil.
Of more plowed down there is great need,
Since "germs" upon this substance feed.

One should small grains in the cotton grow,
Or vetch and clovers widely sow.
Were such crops valued at their worth,
And turned completely 'neath the earth,
The land would not now be cheated,
Nor humus ever be depleted.

Another loss that should not be,
Which takes the soil from hill to lea,
Which gullies every farm and field
And cuts down every acre yield,
Which goes on e'en to the ocean
Is what is known as soil erosion.

With terrace, cover crop, and ditch,
The run-down soils would soon be rich,
The streams would carry no more silt,
And better soils could then be built.
Soil depletion is a shame,
But cotton alone is not to blame.[34]

With the Bankhead-Jones appropriations offsetting some of expense of research, the College was able to open a new Livestock and Forestry Branch Station in Batesville in the spring of 1937. This effort was in line with the results of the land utilization studies undertaken by the College during the 1930's, which emphasized turning some of the acreage devoted to row crops over to grazing and timber management in order to raise farm incomes.

The College's branch station system and the facilities built by the Agricultural Engineering Department to house the operations of the stations were singled out for special praise by a USDA inspection team in 1938. Noting that the College coordinated the operations of the Branch Stations with the Main Station and with Extension activities in a remarkably efficient way, B. Youngblood, the Principal Experiment Station Administrator nationally, called the program, "one of the most progressive ideas that I have observed in station organization in recent years." Youngblood touted the Hope Branch Station as, "a thing of beauty as well as an adequate research facility."[35]

Not only did the New Deal programs address the research and building needs at the College, but they also were a great benefit to the students. Beginning in 1933 the federal relief programs provided jobs for students needing money to meet college expenses. Out of the 181 students enrolled in the College during 1934-35 term, FERA employed 32 students, 11 women and 21 men. The average monthly wages amounted to $15; this was not a princely sum, but with the help of the agriculture student's cooperative enterprises, the jobs were a great benefit.

In addition to library, clerical, laboratory and janitorial work, the students also gained useful experience in helping conduct the various surveys. FERA funds also helped in the creation of a nursery school associated with the Department of Home Economics. The students earned wages in helping at the school as well as class credit in child development classes. The program also benefitted dual-income families, as both parents sought work in increasing numbers to make ends meet during the economic hard times. The National Youth Administration (NYA), created by the Roosevelt administration via executive order in the summer of 1935, also was a great aid to the students. Under the guidelines set down to govern eligibility for the NYA programs, caseworkers paid special attention to students from rural locales. Agriculture students made up the largest single group employed in NYA jobs in Arkansas during the 1937-38 term. For example, six NYA student assistants were assigned to Barnett Sure's agricultural chemistry lab to aid in research.

While it is not known exactly how many students in the University of Arkansas College of Agriculture held jobs funded by the New Deal agencies, in the University as a whole, the number was estimated to be over one-third. The federal aid to students, especially after the NYA programs were initiated in Arkansas in 1936, accounted for the tremendous enrollment increases in the College in the years between 1935 and 1940. Enrollment jumped from 280 to 405 between the 1936 and 1937 school years alone. The climb continued for the rest of the decade before reaching a peak of 630 in 1940.[36]

> *The National Youth Administration (NYA) provided on-campus jobs for me. The man who supervised me was a helpful friend.*
> *Katherine Kennedy Hill, BSHE 1938*

The New Deal agencies and other expanded federal agricultural programs offered an additional benefit for graduates from agriculture colleges. Jobs awaited college-trained agricultural experts. Whereas most graduates before the advent of the New Deal were interested in jobs in the Extension Service or as vocational agriculture teachers, after 1933 their opportunities widened. The SCS offered places for 120 graduates of the College of Agriculture conducting soil erosion surveys across the country. As the number of federal agencies multiplied, so did the need for trained people to administer and implement policies. Out of the 88 June 1938 graduates, 15 found jobs with the AAA, 12 with the SCS and 11 with the FSA. One was working as a Works Project Administration (WPA) regional supervisor, and the rest found work with Extension Service or as Vocational Agriculture teachers.

In the 1939 spring term, the College offered an entire course devoted to studying the FSA. Through class lectures students learned the development and aims of the program. FSA officials, many of them former faculty members, came in to brief the class on the results of the program as well as how to conduct surveys, interview clients and evaluate their needs. The course was the only one of its kind in the country. However, there was a need for such a class in Arkansas, as the state was the center for much of the FSA's resettlement and rehabilitation work.[37]

With the increasing federal participation in, or as some would say, dominance of, agricultural programs came the feeling that the federal government was encroaching unnecessarily in areas that the land-grant colleges of agriculture had previously regarded as their special field. Dean Gray discussed the situation with President Futrall on numerous occasions beginning in the spring of 1938. Gray felt that the work of the College was being overshadowed by the federal initiatives and was coming up second best in the affections of the people of Arkansas. What was particularly galling for Gray were the federal attempts to dictate policy rather than cooperate, as had been the historic relationship between the land-grant institutions and the USDA. The gist of Gray's complaints, which Futrall seconded, was that the College and

Home Management House, 545 Oakland Ave. Home Economics students, usually four at a time, lived in the house for eight weeks for three semester hours of credit. Cost for the eight weeks was about $55.50. (*Arkansas Agriculturist*)

agencies such as the FCA, AAA, FSA and SCS were parallel and competing agencies. The federal agencies had as many as five times the number of personnel in Arkansas as the College had with the Extension Service and the College faculty combined.

In the early stages of their discussions, President Futrall had passed off any action on the College's part to bring the situation into balance. However, Futrall changed his attitude toward the issue when he read press reports put out by federal officials that the College was to be a partner in a rehabilitation project at Wedington Gap, 16 miles west of Fayetteville. The president immediately shot off letters to both Gray and Brannen demanding an explanation. Miffed that he had not been consulted in the matter, and even more that the federals were trying to crowd the College into a project it could not sustain, he told Gray that the University could not afford the $15,000 a year to hold up their end in the bargain. The scope of the project included resettlement housing, forest and wildlife management and livestock and pasture management. But the real selling point of the project for the federal officials and many of the local residents was a 100-acre lake with recreational facilities and vacation cabins located above the lake on a scenic ridge. Futrall chided Gray for falling into the expensive public relations trap laid by the federal officials, saying that the University, as an educational institution, had no business "running a hot dog stand." Gray pleaded ignorance of the matter, although he had served as chairman of the planning board that evaluated the project. Brannen did not immediately reply, which implied that perhaps he was the mysterious "College official" who the federals said had approved the College's participation.[38]

The land-grant institutions and the federal government reached an understanding of sorts on the issue of the proper relationship between the two at the historic Mt. Weather Conference in July 1938. The agreement stressed the importance of localizing the input in the development and execution of farm programs. State planning land-use committees, composed of local and state officials, were set up with federal funds to gather information and coordinate activities. For a season, anyway, the federal-state-local cooperative relationship was restored. Above all, the Mt. Weather Agreement emphasized that, to be effective advocates, the land-grant institutions must be able to stand above politics, a great idea in theory, but much more difficult to put into practice, especially during an election year in Arkansas.[39]

Governor Carl Bailey won his 1938 re-election contest easily, but he was still smarting from his defeat in a 1937 special election to replace U.S. Sen. Joseph T. Robinson, who had died in office. After his re-election Bailey announced menacingly, "My friends may expect the best of me, and my enemies may expect the worst."

The governor began immediately to strengthen his position for the next election and proceeded to clear state offices of officials whom he deemed not sufficiently loyal to him during the campaign. One of his special targets was personnel at the University of Arkansas. Dean Gray read the handwriting on the wall and announced his resignation in a letter to Futrall at the end of October 1938. Although there were other issues involved, such as the Dean's administrative relationship with Futrall, which had deteriorated since the flare-up over the Wedington project, Gray said he wanted to retire while he still had his health and "while the program of the college of agriculture is strongly organized on an independent, unhampered and non-political basis." Futrall accepted Gray's resignation but asked him to stay on until June 1939 and to delay any announcement of his action until a later date.

Gov. Bailey next targeted the Assistant Director of the Extension Service, C.C. Randall, who was summarily fired by the Board of Trustees in May 1939. The Board of Trustees had been under Bailey's control since the Legislature approved his plan to increase the number of Trustees on the Board in January. The reason given for Randall's dismissal was his failure to cooperate with federal agencies. Randall effectively refuted this point when he showed the Board copies of letters from the various federal agencies praising his very cooperation. Randall told reporters his firing was politically inspired because of his refusal to solicit donations in his office for Bailey's 1938 campaign. Bailey did not reply to the accusation but instead labeled Randall a tool of the Farm Bureau, an organization the Governor had accused on many occasions of being against the interests of the small farmer. The Farmer's Union was the Governor's base of support in the agricultural community, and since the 1920's the Union had been fighting a losing battle with the Farm Bureau over who would sit at the head of the table of farm organizations. While it was true that the Farm Bureau and the Extension Service were closely allied in Arkansas, Randall's activities on behalf of the tenant farmers had more than once set him on a collision course with the Farm Bureau.

There was an immediate public outcry over Randall's dismissal. The entire Extension Service force, minus one member, signed a petition on Randall's behalf. Deane Carter, whom the Board appointed as Gray's successor, protested the action and said he could not take up the post under "the present circumstances." Apparently the Board agreed and announced that W.R. Horlacher, the head of the Animal Science Department, would take up the duties as Dean beginning in June. It was rumored that Futrall would be Bailey's next target. However, the governor's "house-cleaning" campaign ended on a tragic note as President Futrall was killed in an automobile accident on the way home from Little Rock after a meeting of the Board of Trustees in September 1939.[40]

Although some criticized Futrall for his "autocratic" administrative style, throughout his 25-year presidency he was a steadfast advocate for the College of Agriculture. All the major accomplishments in the College that had to pass through the Board of Trustees or the Legislature had Futrall as a proponent, and the College's reputation as one of the top agricultural institutions in the country owed much to the interest in agricultural education shown by the flinty, yet perceptive, former classics professor and football coach .

Dan T. Gray's contributions to the growth, development and reputation of the College were equal to those of Futrall, but Gray's accomplishments occupied a different sphere. Gray achieved much for the College and the farmers of the state with his advocacy before boards, committees, commissions and professional groups. But in his role as "the Dean," he had taken a direct and active interest in the welfare of the students and the farmers of the state. During his 15 years with the College, enrollment almost tripled, new and innovative courses were introduced, and a strong foundation curriculum guided College of Agriculture graduates into careers in teaching, public service, research and business. The faculty, with one the finest records of scholastic achievement in the nation, increased from 26 members to 61, new departments were added, and the staff brought in and sustained by Gray conducted path-breaking research. The new Agriculture Building was constructed, and the plans for a new Home Economics Building were approved. In addition, the branch station system, also touted as one of best in the nation, was established and had grown to five stations with the Poultry Station included. It was indeed a record in which Gray could take just pride.

The twenty-fourth annual Agri Day honored the departing Dean with the largest celebration in recent years. Thirty floats, four bands and thousands of students and townspeople paid tribute to Gray with the theme, "A Progressive Agriculture for a Progressive State." From the staff at the College, Gray and his wife received an all-expenses-paid vacation in Mexico as well as a glowing letter of appreciation for Gray's many years of service to the College. At the June graduation ceremonies, the University awarded the Dean an honorary Doctorate in Science as a tribute to his unstinting service. However, Gray's greatest achievement could not be measured in enrollment numbers, buildings constructed or honors received; his enduring legacy was that he led the College through some of the most trying times and yet left it prepared to face the further challenges that were in store.[41]

In 1940 the University was under new leadership with J. William Fulbright as President. The College had W.R. Horlacher as Dean and H.E. Thompson as Assistant Director of the Extension Service. Fulbright and Thompson, identified as Gov. Bailey's appointments, did not survive after Bailey's defeat in the 1940 gubernatorial election. However, Horlacher continued at the helm in the College of Agriculture throughout the war years.

Horlacher, born and raised in Indiana, had earned his B.S. and M.S. degrees from Kansas State in Animal Science. He received his doctorate from the University of Wisconsin in 1923. Horlacher then went on to a teaching and research career at Texas A&M University until coming to the University of Arkansas College of Agriculture in 1936 as professor and head of the Animal Industry Department.[42]

W.R. Horlacher - Dean 1939-1946. (Photo courtesy Photograph Collection, Special Collections, University of Arkansas Libraries.)

As the Depression faded under the production imperatives of the war, the College had to look at the challenges of a world at war with a different focus. Through the three branches of the College, teaching, research and extension, Horlacher took steps to increase agricultural production in Arkansas to meet war-time needs.

In preparing for full war production, government planners wanted to avoid the mistakes made during WWI. Strict social and production controls were not established by the government as had been the case during WWI. Instead, the rationing and liberal price controls imposed benefitted most farmers. While price controls on manufactured goods that farmers had to buy were instituted early on, such controls on agricultural goods the farmers produced were not mandated until 1943. Another benefit for farmers came as production controls on agricultural commodities were loosened or suspended while price supports were maintained. In some cases producers of major commodities were earning 110% parity on the goods they marketed. Under these conditions rice and wheat production jumped 50 percent, and net farm income rose an astonishing 156 percent in the first years of the war.

The rapid switch from agricultural production limits to the need for full production accelerated the changes begun during the Depression. Farm sizes increased, farms were concentrated in fewer hands, and mechanization escalated. In spite of the rural rehabilitation programs of the FSA, tenants, sharecroppers and other "marginal" farmers left the countryside in droves, seeking higher-paying livelihoods in urban-centered war industries. This development led to a major labor shortage in rural areas. Though agricultural labor wages more than tripled between 1933 and 1943, they came nowhere near equaling industrial wages found in the cities. The demand for more livestock, fruits and vegetables and fiber and seed oil crops to feed and clothe the millions of men in arms and lubricate their weapons and machinery did more for the idea of diversified farming than the combined efforts of almost 30 years of extension work. Government-funded experimental work expanded crop and livestock breeding, leading to the development of new varieties that would fuel the "green revolution" after the war. Work with chemical pesticides and growth hormone-based herbicides also began during the war years, but their general use in agricultural production would wait until after the war when chemical companies were released from defense priorities. All the elements for the great post-war production revolution were in place during the war years, but the first priority was to win the war, and the University of Arkansas College of Agriculture contributed unselfishly to the effort.[43]

War preparation plans began even before America's formal entry into the war in December 1941. The United States delivered over 2 billion pounds of food and other agricultural commodities to Great Britain in 1940, and Arkansas farmers contributed their share. Arkansas, as an agricultural state located hundreds of miles inland, was described by defense planners as the "Heart of America's Defense." Horlacher, C.O. Brannen and others in the College worked with the State Planning Board and the federal Office of Production Planning to help meet production goals of unprecedented magnitude.[44]

Working through the Extension Service and the Experiment Stations, the College encouraged production of soybean, peanuts and other oil-bearing crops. "Arksoy," a variety developed by C.K. McClelland at the University Experimental farm, was considered particularly useful as an oil-bearing crop. It helped that Arkansas farmers had been encouraged to plant the crop as an alternative to cotton during the AAA acreage restriction programs. Experiments were begun on the viability of producing cork in Arkansas to relieve shortages of the essential war material. Farmers were forced to make dramatic cutbacks in their use of nitrogen fertilizers, as nitrates were needed for munitions production. Experiment Station and Extension Service personnel advised farmers, as they had for years, to plant more nitrogen-fixing crops such as hairy vetch, lespedeza and clover. Research expanded on finding better varieties of these "green manure" crops as well as soil tillage methods that would make the most of the plants' nitrogen-fixing qualities. Extension staff also distributed information formulated at the Experiment Station on how to handle farmyard manure so as to retain its plant nutritive value.

The increased need for animal fats and proteins to fuel the men in arms also encouraged the growth of the livestock and dairying industries in Arkansas. Butter production was especially encouraged as a true "Victory food." The researchers studied safer ways of handling dairy products as well as more efficient methods of drying and preserving eggs and dairy products. The Experiment Station staff responded to the call for livestock production through experimenting with improved breeding programs to produce animals that would reach market weight more rapidly and provide more meat per net weight. R.M. Smith's poultry breeding experiments were successful in this regard, as were other experiments conducted at the Experiment Stations with hogs and cattle. Food preservation and protection of nutritive contents were a special concern to armies on the move. Barnett Sure and M.C. Kik turned their research talents to these special war-time needs.[45]

In addition to the government funding for experimental agriculture research during the war years, the grants in aid from foundations and private companies increased significantly and set a pattern for further developments in this area for the post-war years. In 1942 the General Education Board provided $32,500 and $150,000 grants for research into forestry economics and to study land tenure problems in the Southwest. The board also provided $900 for teaching in rural economics and sociology, a $3,200 grant to study transportation problems related to marketing agricultural goods and an additional $2,800 to study land tenure in Arkansas.

In order to better coordinate the research efforts, in 1943 C.O. Brannen was named head of the new Bureau of University Research. Brannen brought together all phases of research at the University, both publicly and privately funded. The Bureau was especially concerned with planning the post-war transition to a peace-time economy. Extensive surveys of economic and social conditions in selected Arkansas cities and counties were made to help determine how the production slowdown from a full wartime economy might affect Arkansas. Rather than dictate programs through recommending legislation, the Bureau attempted to maintain a neutral stance by simply providing policy makers with important facts necessary to make a smooth transition.[46]

> *My first 2 ½ years were during World War II. At one time we had about 100+ girls and about 10 boys in the college! The presence on campus of 1500 troops made a difference. I am not sure they were taught any classes by agri professors; we were never in class with them. Their singing as they marched was very nice to hear; we got off the sidewalks when we met them....In 1945 in anticipation of the return of the first married students, we studied how to feed two people for $1.00 a day.*
>
> *JoAnne Sears Rife, BSHE 1956, MS 1971*

The Extension Service's "Live at Home" strategy of producing sufficient food and forage for home use took on a new dimension during war-time. Over 50,000 additional Extension agents were trained nationwide between December 8, 1941, and April 1, 1942. At over 4,000 meetings, agents were given special training in all lines of important work, including food preservation, home gardening and other ways to increase production. The "Victory Garden" campaign familiarized many city dwellers with agricultural extension work for the first time.

The influence of the Extension Service in bringing information to farm families and city dwellers during the war-time emergency cannot be overestimated. The Extension Service reached a total of 211,547 Arkansas farm families in the 1941-42 year alone. A total of 17,257 farm men, women and youths also participated directly in planning the war-time agricultural production campaign through membership on county planning committees organized by the Extension Service in 2,000 communities across the state. The Extension Service also enlisted the aid of over 17,000 farm women and men as neighborhood leaders charged with distributing essential information. And 62,500 4-H Club members signed pledges for all-out production of food and were responsible for producing 2.5 million pounds of pork, 1.5 million pounds of beef, 1.3 million pounds of poultry, 1.4 million gallons of milk and millions of bushels of irish potatoes, sweet potatoes, peanuts and soybeans during the 1944-45 year alone.[47]

The resident teaching faculty also did their part for the war effort. By 1943 eleven faculty members were on military leaves of absence in the various armed services. Many others left their faculty positions to join the government agencies responsible for planning and directing war-time agricultural production. While student enrollment declined dramatically after enlistments and the draft, the faculty was also reduced to a skeleton crew as no new faculty were hired to replace those who left.

In addition to their regular teaching duties, teachers in the College also taught abbreviated and accelerated math and science classes for the 3,500 army trainees that began to arrive on the campus in 1943. Home economics, vocational agriculture and agricultural engineering teachers took on the extra duty of instructing teachers in "Rural War Production Training Courses" as well as drawing up lessons for the courses. The College trained 3,236 local vocational teachers and master farmers in all phases of agricultural production, farm machinery repair, agricultural engineering and electrification. In addition to these extra teaching duties, members of the College of Agriculture faculty served on the Arkansas State Planning Board and the State Land Use Committee, helping to plan the most efficient use of agricultural resources in war production and the transition to a peace-time agricultural economy.[48]

The students of the College contributed to the war effort through sacrificing their educations to serve in the armed forces. Some made the ultimate sacrifice and paid with their lives. By the end of the war 20 former agriculture students had been killed in action. They were:

Major Clyde R. Benbrook, B.S.A. '30, M.S. '31
Lt. Edward H. Bostic, B.S.A. '38
Pvt. Doyle Morgan, '41-43
S/Sgt. James Roger Moneyhun, '41-43
Capt. Maurice L. Stephens, '40
Lt. Hugh L. Williams, Jr., '36-37
Lt. William S. Young, '37-38
S/Sgt. Jack Lincoln, '35
Lt. Edward P. Bauer, '42-43
Lt. Jack Gore McKenzie, '41-43
Lt. Lowell A. Goforth, '36
Lt. Herschel T. Hardin, '40
Lt. Edgar H. Lloyd, '43
Lt. (J.G.) Van E. Hamilton, '42
S/Sgt. Maurice C. Mills, '39-40
Cpl. Edward M. Nations, '39-40
T/Sgt. Joe Scalet, '39-40
Pvt. Thomas E. Stubblefield, '41-42
T/Sgt. James Walls Thomas, '41-42
A.O.M. 3/c Harold Ellen

Lt. Edgar Harold "Bud" Lloyd, a 1943 graduate, was awarded the Congressional Medal of Honor posthumously for his outstanding heroism in operations near Pompey, France, in the fall of 1944.[49]

The number who served is reflected in the enrollment figures for the war years.

Enrollment 1941-1945.

Year	1941-2	1942-3	1943-44	1944-45
Men	282	171	72	83
Women	200	168	170	161
Total	482	339	242	241

While the drop in enrollment was most dramatic for the men, women from the College also served in the armed forces. Five were listed as serving in the WACs, WAVs or Marine Corps. After the precipitous drop in the numbers of men in the College, the women took up the slack. Headquartered in the new Home Economics Building that was completed in 1940, the women took up many of the responsibilities in directing student activities in the College.

The new, three-story Home Economics Building, located west of, and adjacent to, the Agriculture Building, was approximately two-thirds the size of the Agriculture building and mirrored its architectural style. The ground floor housed various laboratories: bacteriology, nutrition, food research and child development. Also on the ground floor were three offices and one classroom. The first floor housed the meal planning laboratory, a food preparation lab, a lecture room, a classroom, a clerical room, a reading room and the offices of the Extension and staff members. On the second floor were laboratories for clothing construction, home management, dermatology and textile testing in addition to rooms for teacher training and other offices.

With the increased responsibilities undertaken by the Home Economics Department during the war years, the additional space was welcome. The home economics students directed some of their added responsibilities through a new organization, the Phi Upsilon Omicron home economics honor sorority, which was established in 1940 and granted a national charter in May 1943.[50]

Keeping in step with war rationing, the annual Agri Day celebrations during the war years were trimmed down affairs. The Agri Day Association's slogan, "A Bigger and Better Agri Day," took on a new meaning to reflect the war-time attitude. The parade was canceled due to rubber and materials shortages. The departmental displays were also dropped due to lack of time and a shortage of materials. Instead, the students used the occasion as an opportunity to sell war bonds and collect blood plasma. The Agri Queen selection and the annual Agri Ball continued with the usual enthusiasm, however.[51]

The new Home Economics Building, completed in 1940, was approximately two-thirds the size of the Agriculture Building to its east and mirrored its architectural style.

Class scheduling and coursework were tailored to reflect the war-time emergency as well. Much of the coursework was accelerated to provide potential officers and soldiers with as much education as possible before enlistment or induction. To accommodate those students going into the armed services, a three-year program toward a degree was introduced. In 1943 University President Arthur M. Harding switched the University into the quarterly semester system in order to pack as much instruction as possible in the limited time available. Some professors decried this lowering of standards to meet the war-time needs, saying that improperly prepared graduates were as much of a drag on the war effort as a shortage of trained leaders. The question would come up again in the context of educating returning veterans, but for now, most of the teaching staff did their best to adjust instructional goals to the war-time needs.[52]

As the war wound down and finally came to an end in August 1945, the students and faculty shared in the outpouring of emotion that swept the nation. Filled with the collective confidence that accompanied winning the war, all looked forward with enthusiasm to the future and to settling down to enjoy the fruits born of the years of sacrifice. Yet there was also uncertainty, as many changes had come over the country during the years of the Great Depression and WWII, especially in agriculture. The changes brought on by the New Deal programs had been accelerated by the fury of war-time production imperatives. The small family farms, traditionally regarded as the bedrock of the American Republic, were pushed to the margins as commercial agriculture, aided by mechanization, government price supports and the latest innovations discovered in the labs and experimental fields of the colleges of agriculture, took center stage. The men and women at the University of Arkansas College of Agriculture surveyed this transformation of American agriculture and saw their opportunity to contribute to the agricultural revolution that could help reduce food costs for all Americans as well as feed millions around the world. While some lamented the passing of a mythic rural America that had been slowly retreating from the scene for decades, the push toward modernization and greater efficiency was a stronger tide whose benefits accrued to a greater number of people. It was a heady prospect that would also transform the classrooms, labs and fields of the College and the way they served the needs of the people of Arkansas as well as the world.

Endnotes

1 Donald Holley, "Arkansas in the Great Depression," *Historical Report of the Secretary of State*, Vol. 3 (Little Rock: Arkansas Secretary of State, 1978), 157-158.

2 Holley, 159-160.

3 *Agricultural Triangle* (December 1930), 3, (February 1930), 5; Holley, 160; Nan Woodruff, *As Rare as Rain: Federal Relief During the Great Southern Drought of 1930-31* (Urbana: University of Illinois Press, 1985); Pete Daniel, *Breaking the Land: The Transformation of Cotton, Tobacco and Rice Cultures Since 1880* (Urbana: University of Illinois Press, 1985), 71; Jeannie M. Whayne, *A New Plantation South: Land, Labor and Federal Favor in Twentieth Century Arkansas* (Charlottesville: University of Virginia Press, 1997), 151, 154-155; Danbom, 202-203.

4 Leflar, 94-95; Hale, 115

5 *Arkansas Agriculturist* (March 1932), 3.

6 *Agricultural Triangle* (July 1930), 5, (September 1930), 1, (September 1931), 2, (September 1932), 2, (April 1933), 1.

7 *Arkansas Agriculturist* (March 1934), 4

8 *Arkansas Agriculturist* (March 1934), 4, (April 1938), 8-11, 27-30; Leflar, 97.

9 *Arkansas Agriculturist* (November 1932), 10.

10 *Arkansas Agriculturist* (December 1932), 3.

11 *Arkansas Agriculturist* (October 1931), 9

12 *Agricultural Triangle* (October 1931), 1-2.

13 *Arkansas Agriculturist* (April 1938), 3, 28.

14 *Agricultural Triangle* (March 1930), 1.

15 *Arkansas Agriculturist* (October 1930), 11.

[16] *Dean's Reports*, Boxes 1, 2, 3 & 4, Files: Bacteriology and Veterinary Science, Entomology, Horticulture and Forestry, Plant Pathology, all 1932-33.

[17] *Arkansas Agriculturist* (March 1932), 10, 14.

[18] *Agricultural Triangle* (January-February 1932), 1, 5; *Arkansas Agriculturist* (February 1933), 4; Stephen F. Strausberg, *From Hills and Hollers: Rise of the Poultry Industry in Arkansas* (Fayetteville: University of Arkansas Agricultural Experiment Station, 1995), 42-44.

[19] *Dean's Reports*, Box 4, File: Rural Economics and Sociology, 1932-42.

[20] Holley, 162; Danbom, 202-203.

[21] McNary-Haugenism refers to a plan proposed in Congress during the 1920's to raise prices on American agricultural goods. The plan, sponsored by Sen. Charles L. McNary of Oregon and Rep. Gilbert N. Haugen of Iowa, called for a protective tariff on American agricultural goods and for the U.S. government to assist farmers in dumping suplus agricultural commodities on the world market in order to raise domestic farm prices.

[22] Danbom, 207-208.

[23] Danbom, 209-211; Daniel, 92-94; Whayne, 158-165; Holley, 165.

[24] Danbom, 217-218; Daniel, 72-73, 89-90; *Dean's Reports*, Box 4, File: Rural Economics and Sociology, 1933-34.

[25] *Arkansas Agriculturist* (February 1934), 10; Holley, 167.

[26] *Dean's Reports,* Box 4, File: Rural Economics and Sociology, 1933-34; *Agricultural Triangle* (May-June 1934), 1, 2.

[27] *Arkansas Agriculturist* (October 1934), 3.

[28] Futrall to Brannen, September 8, and September 20, 1934, Brannen to Futrall, September 14, 1934, Futrall to Brannen, August 16, and Brannen to Futrall, August 18, 1934, *President's Records*, Box 11, File: Brannen, 1934; *Agricultural Triangle* (March April 1935), 3.

[29] *Agricultural Triangle* (November December 1936), 1; (July August 1936), 1; Strausberg (*A Century of Research*), 75-76; *Arkansas Agriculturist* (March 1937), 9 (June 1938), 13; *Dean's Reports*, Box 4, File: Rural Economics and Sociology, 1934-35, 1935-36, 1936-37, 1937-38.

[30] Unpublished paper in Department of Rural Economics and Sociology Department files, probably for Resettlement Agency, a New Deal Agency that was the precursor of the Farm Security Administration.

[31] *Dean's Reports*, Box 4, File: Rural Economics and Sociology, 1933-34-1939-40; *Arkansas Agriculturist* (December 1936), 3; *Agricultural Triangle* (November December 1936), 2, (July August 1936), 1; Carter, 21-22; Daniel, 178.

[32] Danbom, 221-222; *Arkansas Agriculturist* (April 1938), 13; Carter, 22, 27.

[33] Strausberg (*A Century of Research*), 78-81; *Agricultural Triangle* (July August 1935), 8, (July August 1937), 3-4, (September October 1937), 3.

[34] *Arkansas Agriculturist* (March 1935), 11.

[35] *Agricultural Triangle* (March April 1937), 3; B. Youngblood to Gray, May 26, 1938, *President's Records*, Box 11, File: Gray, 1938.

[36] *Arkansas Agriculturist* (March 1935), 9, (February 1935), 9; *Arkansas Gazette*, January 30, 1938, 2:3; University of Arkansas Committee for the Survey and Appraisal of the National Youth Administration in Arkansas, *Preliminary Report* (Fayetteville: University of Arkansas, 1940), 1, 7-8, 50; Leflar, 97, 99.

[37] *Arkansas Agriculturist* (February 1935), 3, (May 1936), 4, (April 1937), 12, (September 1937), 3, (October 1938), 4, (March 1939), 12, 14.

[38] Gray to Futrall, March 23, 1938, July 8, 1938, November 2, 1938, February 7, 15, 21 and 22, 1939, Futrall to Gray, May 3 and May 4, 1938, November 2, 1938, February 17, 1939 *President's Records*, Box 11, File: Gray, 1938-9.

[39] Edward Danforth Eddy, "The Development of the Land Grant Colleges: Their Program and Philosophy" (Ph.D. Thesis: Cornell University, 1956), 372-379.

[40] Holley, 172; Gray to Futrall, October 29 and 31, 1938, May 20, 1939, Futrall to Gray, October 28, 1938, November 1, 1938, May 19, 1939, *President's Records*, Box 11, File: Gray, 1938-39; *Farm Bureau Press*, (May 1939), 1-4, (June 1939), 1-3 (July 1939), 1-2.

[41] *Arkansas Agriculturist* (May 1939), 4.

[42] Hale, 140.

[43] Danbom, 229-231; Daniel, 242-243.

[44] Edward Danforth Eddy, *Colleges for Our Land and Time: The Land Grant Idea in American Education* (New York: Harper, 1957), 202-203; *Arkansas Agriculturist* (October 1940), 14.

[45] Strausberg (*A Century of Research*), 92-95; *Arkansas Agriculturist* (March 1943), 10.

[46] *University of Arkansas Annual Report: 1943-44*, 17-18.

[47] *Arkansas Agriculturist* (May 1942), 6; *University of Arkansas Annual Report: 1944-45*, 16.

[48] *Arkansas Agriculturist* (March 1943), 11; *Dean's Reports*, Box 11, File: Agriculture 1942-43, 3-7, 1944-45, 1-2; *University of Arkansas Annual Report: 1943-44*, 11-12.

[49] *Arkansas Agriculturist* (October 1944), 11, (February 1945), 4.

[50] *Arkansas Agriculturist* (December 1939), 10.

[51] *Arkansas Agriculturist* (March 1943), 5, (October 1944), 9.

[52] *University of Arkansas Annual Report: 1943-44*, 5, 13-14.

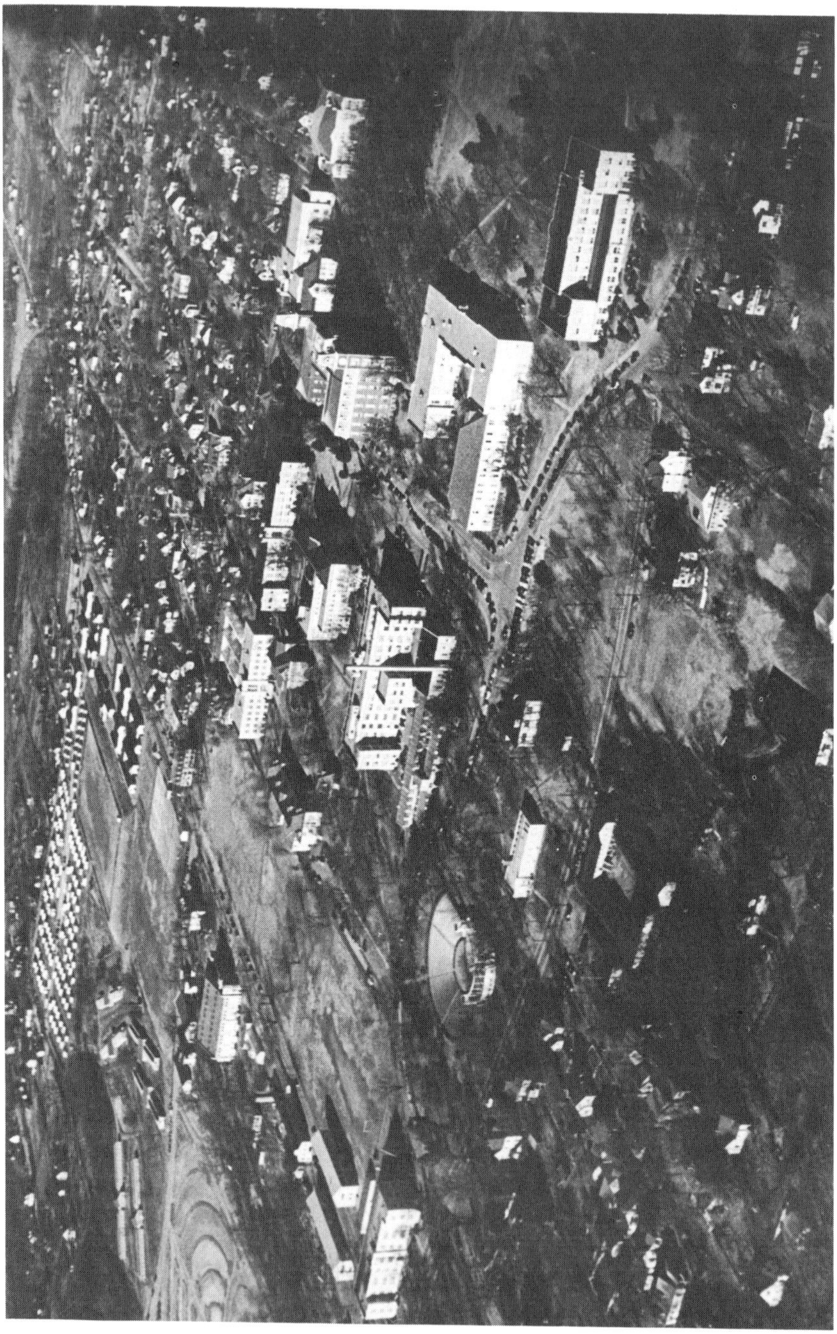

Chapter Five

The College of Agriculture and the Production Revolution: 1946-1960

Between 1946 and 1960, the College and Division of Agriculture played an active role in the revolution in agricultural production that transformed American agriculture. It was a golden age for scientific agriculture as the accumulated knowledge gained during the first 50 years of coordinated research was augmented with new discoveries. More important was the way that farmers eagerly adopted new discoveries and put them into practice to stay competitive with the new commercial agricultural operations that came to dominate the scene in the postwar years.

In Arkansas, as in other parts of the South, farm operations became increasingly mechanized, capital-intensive operations. Between 1940 and 1954 the number of tractors increased seven-fold. In 1940 there were 24 farm workers per tractor; by 1954 there were three. The number of farms with electricity in Arkansas increased from barely 5,000 in 1940 to 131,500 in 1954, or 91 percent of all farms. Capital invested in land, buildings, machinery and livestock per farm worker in Arkansas increased 247 percent between 1940 and 1954. Farmers spent more than 10 times as much on feed, gasoline and oil and fertilizer materials in 1954 as in 1940.[1]

The College's contribution to modernizing agricultural production included research into improved crop varieties and livestock breeds, the use of pesticides, herbicides, growth hormones, antibiotics and the application of anhydrous ammonia fertilizer. These new discoveries boosted yields per acre and increased livestock and poultry production. With these technical, mechanical and chemical innovations, farm sizes doubled while the number of farm workers and operators fell by half as farm operations became more efficient.[2]

Meanwhile, a debate raged in the agricultural community over the wisdom of retaining government price support programs and acreage restriction

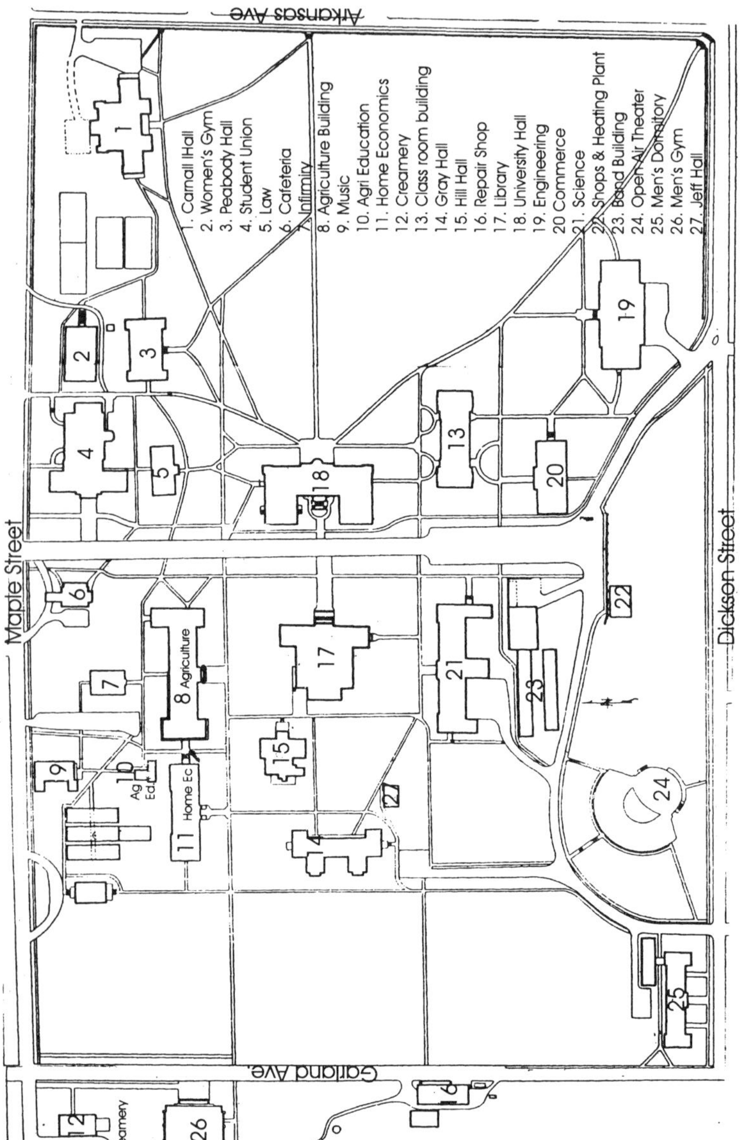

The University of Arkansas campus in 1940-1941. The Home Economics Building (11) has been added, and the 1905 Agri Building has become the Infirmiry (7). The athletic field and track have been moved down to their current location, and the "Old" Student Union (Memorial Hall) (4) was new at the time. The Dairy Building (12) is still standing on the corner of Garland and Maple, next to the Men's Gym (26).

policies in the midst of the production revolution. What to do with commodity surpluses that resulted from the stunning success of applying modern scientific farming techniques and how to retain the favorable prices enjoyed during the war were also matters of concern for farmers. The ASULGC and the Farm Bureau both favored lifting government controls over agricultural production after the war and allowing the free market to determine farm output. Other groups, including the Farmer's Union and some government policy makers, advocated continuing high price supports and production controls in order to protect operators of smaller farms.[3]

All groups agreed that agricultural policy should encourage the development of foreign markets for agricultural goods. Indeed, during the postwar years the transfer of agricultural commodities and expertise became a key component of U.S. foreign policy. The University of Arkansas College of Agriculture and other land-grant institutions sent agricultural specialists abroad and trained foreign students and experts. The College addressed one of the nation's major foreign policy concerns through participating in the U.S. government's programs to aid in rebuilding war-ravaged Europe and to assist developing countries. The College sent agricultural experts to Germany, Korea, South Vietnam, Panama, Peru, British Guyana, Bolivia, Turkey, Egypt and Senegal, to name only a few countries. The College also educated hundreds of foreign students and agricultural experts through the Foreign Visitors program sponsored jointly by the University of Arkansas and the U.S. State Department.[4]

The introduction of mechanical cotton pickers, chemical defoliation, flame weeders and new varieties revolutionized cotton production in Arkansas and the South. Increasingly, cotton production in Arkansas was confined to the richer, alluvial Delta lands more suited to the new production methods. By the end of the 1950's, soybean replaced cotton as the state's leading crop. California, Arizona and Texas became the centers of cotton production as a result of irrigation projects that turned deserts into farmland. Other production changes encouraged by discoveries at the College resulted in growth of the livestock and timber industries in Arkansas.[5]

After the war, leading figures in Arkansas industry, agriculture, education and government began a campaign to industrialize the state. One of the campaign's principal highlights was developing processing industries for agricultural commodities and streamlining the state's farm operations to reflect the postwar emphasis on commercial agricultural production. Previously farmers were forced to ship most agricultural produce out of the state for processing because of the shortage of commercial refining, freezing and preservation facilities. The College of Agriculture played a major role in helping develop these new agricultural industries through the work of the Experiment Station on quick freezing and other processing techniques. The Bureau of Research and its successor, the Institute of Science and Technology, created in 1948, coordinated research and planning in this area. Dean Lippert Ellis and University President Lewis Webster Jones served on the board of the Arkansas Economic Development Committee that spearheaded the industrial development program and encouraged developments along these lines.[6]

The College of Agriculture faced many challenges in the postwar era and addressed them in a variety of ways. The massive influx of students enrolling under the terms of the GI Bill of Rights in the years immediately after the war created a logistical problem to which there was no easy solution. Educating the returning veterans called for innovative programs and approaches that the College developed, approaches that were imitated by other institutions across the nation. The war had called many of the faculty into service in either the military or other branches of government, and restaffing the departments with competent teachers and researchers with the low salaries provided by the Legislature was a problem that continued into the 1950's. As farm operations grew more complex and new agribusiness concerns branched in many different directions, there was a need for more agricultural specialists. Consequently, the College addressed the need through developing graduate programs in most departments.

Agriculture's greater technical sophistication in the postwar years called for larger maintenance costs for research conducted in all fields. With the state of Arkansas perennially short of funds, agricultural research looked increasingly to private sources or corporate financing for specific projects or to foundation grants. Since the total sales of equipment, feed, fertilizers, fuel, pesticides and other farm supplies increased more than ten-fold in the years between 1940 and 1954, it was a logical outcome that agribusiness concerns became both sponsors and clients for all land-grant colleges of agriculture.

Lippert Ellis - Dean of the College 1946-1959.

In September 1946 Lippert S. Ellis took over the Dean's duties after Horlacher resigned to take a job at Colorado State University. Ellis remained in the Dean's chair from 1946-1959 and guided the College with an able hand through the transformations brought on by the production revolution. Lewis Webster Jones, who served as University President from 1947-51, was a tireless advocate of agricultural education in much the same way as President Futrall before him. From his position on the Agricultural Committee of the Arkansas Development Council, Jones aided Dean Ellis in many ways. The new Dean was uniquely qualified for his new post as he had directed a study on projected changes in Arkansas agriculture in the postwar period while employed as an analyst for the state's Bureau of Agricultural Economics.[7]

The first challenge that Ellis dealt with was how to adjust the College to the massive influx of students following the end of the war. A large number of these students were veterans taking advantage of the provisions of the Servicemen's Readjustment Act of 1944, otherwise known as the GI Bill of Rights, legislation that was perhaps as influential as the Morrill Act in bringing opportunities for higher education to a greater number of people. The GI Bill provided veterans with educational and housing benefits. The College enrollment climbed dramatically until the Korean War, when once again agricultural students were called away from their studies. The record enrollment in the 1949-50 year would remain the high mark for the College until the 1960's.

Enrollment: 1945-50.

Year	1945-46	1946-7	1947-8	1948-9	1949-50
Men	267	460	534	614	627
Women	153	152	155	127	146
Total	420	612	689	741	773

The College made special provisions for the veterans to take full advantage of their opportunities; this included a sub-freshmen class similar to the preparatory courses of earlier years, with a special curriculum adapted for those at lower educational levels. The College also instituted a special two-year program similar to the Short Course in earlier years. After successfully completing the requirements of the course, the students earned associate degrees, or they could then go with the regular college course toward a B.S.A. or B.S.H.E. The College also instituted an innovative program to train teachers for the veterans' vocational education program to help ease the transition into the peacetime job market for returning veterans.[8]

The overcrowded conditions in the College were aggravated by a shortage of faculty. Dean Ellis, ever the careful architect, did not act rashly by snatching up unqualified personnel to fill the gap. Working with the limited resources given him by the Legislature, he made a concerted effort to attract the best people for the job. However, the 26 teachers and researchers he hired in the 1946-47 year did little more than fill in the gaps created by the wartime emergency. By the beginning of the 1950's, Ellis had hired more than 60 new faculty and staff to buttress the research and teaching program at the College.

Ellis' selective recruitment program succeeded in bringing many notable men and women to the College whose achievements brought acclaim for the institution. However, low salaries continued to be a problem and caused a slow attrition of staff seeking salaries commensurate with their achievements. Ellis grappled with an additional problem for the staff in these overcrowded conditions. With the teaching load so heavy, Ellis feared that research work would suffer. The only solution to this problem, as Ellis saw it, was an expansion of the faculty, but under budget constraints, perhaps expanding the graduate program could help in easing the shortfall in both research and teaching.[9]

The Dean understood clearly that before any headway could be made toward a successful graduate program, a solid undergraduate curriculum had to be in place, taught by competent faculty. Consequently, he evaluated the present course work and the curricula at other universities with successful graduate programs. Coordination of instruction with other departments in the University outside the College was also deemed necessary, particularly in zoology, botany, mathematics and chemistry. An outgrowth of the efforts to coordinate instruction was the establishment of a degree program in agricultural engineering in 1948 under the joint supervision of the Agriculture and Engineering Colleges. Ellis concluded that among the departments as they existed in 1947-48, only Plant Pathology, headed by the veteran V.H. Young, and assisted by three other faculty who held Ph.D.'s, was in a position to conduct a credible graduate program.[10]

> *The University was really caught in a time warp. They were unprepared to deal with the numbers of students or the maturity and outlook of the veterans. When I enrolled, I was sent to see the Dean of the College of Agri. I was 24, married, one son, on the GI Bill, an orphan who had been making my own way since I was 17, and he asked who I wanted him to send my grades to. I said send them to me. I'm totally responsible for me. But, no, that wouldn't work– so I let him send them to my in-laws, who thought that was a blast.*
>
> *Edgar S. Milton, BSA 1949*

Ellis also made a plea for more space needed for both teaching and research. Among his suggestions were allotting more space for the Departments of Animal Industry, Plant Pathology, Horticulture and Forestry, Veterinary Science and Home Economics. In consideration of the postwar expansion of the livestock industry in Arkansas, Ellis submitted a proposal for the construction of a new Animal Science Building that would also incorporate the bacteriology and veterinary science operations. While there was ready agreement among University officials that Ellis' proposal had real merit, budgetary constraints delayed the project for many years, and the Animal Science complex was not completely finished until the early 1960's. Ellis also suggested moving the Botany Department out of the Agriculture Building and devoting the resulting space to Horticulture and Forestry. To make more room for the Department of Rural Economics and Sociology, Ellis recommended that the Agricultural Library move to the new addition being made to the campus' Main Library (Vol Walker Hall). For the Home Economics Department the

Dean proposed that the old Experiment Station Building, then used by the Music Department, could more profitably be used to house the nursery school and perhaps a home management house could be constructed nearby.[11]

In 1947, to effectively direct and counsel the larger number of students in the College, many of whom had special needs as returning or disabled veterans, Ellis secured the go-ahead to appoint G.T. Hudson as Assistant Dean in charge of resident instruction. Hudson came to Arkansas from Colorado A&M College and held degrees from the University of Illinois, University of Wisconsin and Cornell. Hudson made a special effort to make himself available to students in his official capacity as counselor. He also aided students in organizing extracurricular activities and in forming new clubs and organizations that reflected the postwar students' new orientation. In career counseling, Hudson directed students into traditional areas such as vocational and extension education and advised them of new opportunities in the burgeoning agricultural industries as researchers. In addition to counseling students on course work and careers, Hudson made a thorough evaluation of the curriculum at the College and came up with a set of recommendations. Hudson also retooled the program for reaching out to recruit students from agricultural junior colleges and vocational schools.[12]

The students in the College quickly set to work in reorganizing student life to meet the postwar conditions at the University. One important development that spoke to the needs of returning veterans, many of whom were older and married with families, was the establishment of the College of Agriculture Married Students Club in the fall of 1945. Nearly one-third of the enrollment in the College in 1946 were married students. One of the purposes of the club was to provide non-student spouses an opportunity to meet the spouses of other agriculture students and share child care responsibilities and to make spouses aware of additional educational opportunities sponsored by the University.

Other student organizations reconstituted themselves after a wartime hiatus. The Alpha Gamma Rho fraternity reorganized in the spring of 1946. Eleven men previously active in the fraternity joined 13 new members to reactivate the organization. Alpha Zeta, also inactive during the war, was reinvigorated with new membership. Fourteen Alpha Zetas were joined by 16 new members at the fall 1947 induction ceremonies. The Phi Upsilon Omicron sorority, which had carried the torch for the agricultural Greek societies during the wartime absence of men, took on a new life in the postwar years as well, growing to 19 members in 1946. In 1949 the FarmHouse club, a men's agricultural social fraternity, was organized on campus with 14 members. In addition to the Agriculture Students Association, the men's and women's 4-H Clubs, the FFA college chapter and the Home Economics Club, which changed its name to Colhecon in 1952, many new organizations came into being during the postwar period. The Animal Science, Agronomy, Agricultural Economics, Agricultural Engineering and Entomology Clubs were all geared toward promoting interest and knowledge in these various disciplines. The Agri Day Association, the organization responsible for putting together the annual Agri Day celebrations, took on new life after the subdued Agri Days during the war. The program was expanded to include a rodeo, a new event that sparked added interest in the day's events.[13]

Part of the expanded student body included a number of foreign students invited to the campus to study and conduct research as part of an outreach and assistance program initiated by the U.S. State Department. Three Chinese students, Phi Kung, Tai-tsoo and Shih-chao, arrived on campus in 1946 to study with M.C. Kik in his ground-breaking research on rice and nutrition. The College also received visiting researchers from Argentina, the Philippines and Peru during the years immediately after the war. The program for hosting foreign researchers and students expanded during the 1950's as the College took on international extension work as a part of the land-grant mission in addition to serving United States foreign policy objectives.[14]

> *I was among the hundreds of servicemen who enrolled at the UofA following WWII in 1946. As freshmen our average age was probably around 25 years, and we definitely wanted to succeed in college. Most of us had been out of high school 5 to 7 years, so it took a lot of dedication and hard work, especially during the first year or two, to progress satisfactorily. Our numbers taxed the faculty and facilities to the utmost, so I have a great deal of respect for the faculty and the administrators of that time who worked so well under difficult circumstances.*
>
> *Dr. Joseph O. Legg, BSA 1950, MS 1951*

While the College's international extension program took definite shape during the 1950's, efforts in that direction were begun in the years immediately after the war. E.M. Cralley, from the Department of Plant Pathology, spent 18 months beginning in 1946 in the American occupation zone in Korea. Cralley spent most of his time working with rice diseases and investigating new soybean varieties. In return for his service in Korea, Cralley brought back between 300 and 400 varieties of Asian rice to use in plant breeding experiments at experiment stations in the United States.[15]

Another early foreign agricultural mission sent entomology professor Dwight Isely to Peru in early 1947. The project was under the administration of the U.S. State Department's Institute of Inter-American Affairs in cooperation with the Peruvian Ministry of Agriculture and the University of Arkansas College of Agriculture. Isely, recognized as one of the outstanding entomologists in the United States for his work with cotton pests, spent three months helping to reorganize and modernize research facilities in Peru. Isely also spent time in the field collecting information that could prove useful in combating insect problems in Arkansas.

Isely's sojourn was followed up the next year by his protege, Charles Lincoln. During his three-month stay, Lincoln also concentrated on cotton pests and made general observations regarding cotton cultivation in the dry equatorial regions of the country. Lincoln noted that insecticides used to control pests in Peru were about the same as those used in Arkansas: calcium arsenate, along with recently discovered chemical controls, DDT and toxaphene. One goal of the mission was to set up a Peruvian extension service system that functioned with its own personnel after being trained by American extension and technical experts.[16]

In other developments in foreign agricultural work, C.O. Brannen took a leave of absence during 1947 in order to serve as a member of the U.S. Army research commission sent to Germany to help plan that country's rehabilitation. O.B. Brown, an Extension farm management specialist, was sent to British Guyana to study rice production. Earl Rambo, an Extension agricultural engineer, went to Panama to assist with agricultural engineering problems and began what would become one of the College's longest-running foreign agricultural missions that lasted well into the 1950's.[17]

The College's research staff was also busy at this time working with postwar adjustments for agriculture in Arkansas. The staff tackled problems involved with mechanization, herbicide and pesticide application and increasing livestock and timber production. Researchers also conducted experiments on new crops to replace acreage previously planted in cotton and rice. Keeping in step with the statewide plan to encourage the growth of industry in Arkansas, researchers also studied problems associated with starting up new industries devoted to processing agricultural products in Arkansas. Work continued in developing new crop varieties suited to conditions in Arkansas.

The introduction of the mechanical cotton picker and chemical defoliation techniques as an aid to mechanical harvesting of cotton in the years after the war were topics of great interest to planters and researchers alike. After flame and mechanical cultivators appeared on the scene, followed by herbicidal weed control, the need for information on these revolutionary methods of controlling weeds in cotton fields was markedly increased. Agricultural engineers, agronomists and horticulturists at the Cotton Branch and Main Experiment Stations all joined in the study of the efficiency of the new techniques. Their studies showed clearly that the farmer or planter with enough land and capital to invest in the new production techniques could produce more cotton on fewer acres for less money than ever before. Researchers also began experiments to develop new cotton strains that would open more uniformly and had other characteristics adaptable to machine harvesting.[18]

The research staff in the Department of Rural Economics and Sociology were also busy determining the effect mechanization had on rural labor and land tenure patterns. C.O. Brannen headed up the Southwest Land Tenure Research Committee to study the problem of displaced farmers and farm laborers that resulted from the modernized production techniques. The committee's work, and its final report issued in 1948, dealt with rural rehabilitation and the social impact of modernization on farm workers.

The study was among the last of its kind for the region as the mood of the country and agricultural planners turned more conservative in the postwar years. Accordingly, the focus for rural economists and sociologists turned to production and marketing problems. The name *Rural Sociology* was even dropped from the department for many years.[19]

The research scientists at the College began experiments on the new hydrocarbon-based pesticides that began to arrive on the market during the war years. W.J. Baerg, the veteran entomologist on the staff, began investigations on DDT in 1944. Baerg concluded that indiscriminate use of the new pesticide also destroyed beneficial predatory insects, upsetting the balance that was the key to successful biological control of crop pests. Dwight Isely and Charles Lincoln developed an insect scouting program to pinpoint the location and the level of serious infestations so that pesticides could be applied judiciously and at a time when the chemicals could do the most good. Lincoln, the self-described "clod-hopping entomologist," noted that after DDT became widely available in 1946, his scouting missions found fields "practically sterilized of insect life" because of haphazard over-applications of DDT. Lincoln observed that young, returning, veteran farmers took up the chemical pest control program with enthusiasm whereas insecticide dusting was considered, "an antisocial activity in some farm communities."

Charles Lincoln - Head of the Entomology Department 1951-1969.

Conflict also erupted between Experiment Station scientists and chemical company representatives over the timing and amount of pesticide used. By the early 1950's corporate support for investigations into pesticide use at the Experiment Stations became a major source for funding entomological research. However, Lincoln, who became head of the Entomology Department in 1951, and other scientists in the department, resisted pressure to tailor their opinions to suit the chemical companies' recommendations. Lincoln did his best to keep Extension Service personnel and farmers informed about the latest scientific findings regarding timing and application of chemical pesticides. But his was an uphill battle as chemical companies flooded farm communities and organizations with their own representatives and promotional literature, which many times offered advice that was contrary to the findings of Experiment Station scientists. Lincoln, ever an outspoken advocate of sensible use of the new chemical tools, was considered a troublemaker by many in the corporate agriculture community. He bristled at suggestions made by some farmers that the College entomologists were in league with the chemical companies in advocating wider use of pesticides. Some people in the chemical industries attempted to get Lincoln fired, but University administration stood behind Lincoln. Later, other cotton growing states adopted Lincoln's cotton insect control methods, and the environmental movement supported the Lincoln approach.[20]

Agricultural engineering research was also keyed to the modernization of farm operations in the postwar period. Agricultural engineer E.J. Matthews worked with Charles Lincoln on calibrating spray mechanisms for pesticide use to avoid damage to growing crops. As already noted, mechanization increased rapidly in the relatively flat Delta cotton areas. Engineers studied how to adapt mechanization in the upland farm areas for different farm crops as cotton production became increasingly confined to the alluvial areas of the state.

A major innovation that developed in agricultural engineering research was the invention and patenting of a rice dryer by staff engineering professors Xzin McNeal and Kyle Engler. The device, which dried the grain as it was harvested, greatly aided in mechanical rice harvesting and cut losses from mold contamination as well as the time and work involved bringing the crop from the field to the processing and storage stage.

The Home Economics and Agricultural Engineering Departments conducted joint research on improving farm home life through the use of labor-saving devices, which were now available in increasing numbers due to the rapid progress of the rural electrification program in Arkansas. Studies continued on the condition of rural housing. Agricultural engineers, assisted by home economics home planners, provided rural families with house and building plans as well as advice on safer and more efficient uses of electricity on the farm.[21]

Research in fertilizer use revealed how quickly agricultural research adapted to changing needs. During the war the research staff was busy advising farmers on the best ways to handle and preserve the nutritive value of barnyard manure. After the end of the war, researchers turned their attention to the best way to use the new concentrated and mechanically applied anhydrous ammonia nitrogen fertilizer.

Research continued in developing new crop varieties. 'Arkot 2-1' was a new variety of cotton bred for qualities making it more adaptable to mechanical harvesting. Other varieties of oats, vetch, watermelon and tomatoes were developed also, to name but a few.[22]

To help farmers diversify, the Experiment Station established fruit, vegetable and alfalfa substations near Nashville, Clarksville and Bald Knob in eastern Arkansas during 1948-49. Closely related to these developments were experiments done in quick-freezing fruits and vegetables to retain both their nutritive value and their pleasing appearance for consumers. These experiments also facilitated the push to develop new agricultural processing industries in Arkansas.

At this time Barnett Sure was working to develop a totally new food source to help supplement diets in countries dependent on grain proteins that lacked essential amino acids to build complete proteins. The new protein food could also help American consumers stretch their food budgets at a time when meat prices were high. Sure's concoction was manufactured principally from milk solids and dried yeasts fortified with calcium and iron and enriched with A, D and B complex vitamins. The University Research Bureau secured a patent for Sure's discovery, and a Canadian firm began manufacturing the food product for sale in Great Britain in 1948. While the "new tasty meat-like dairy food" never caught on with consumers in the late 1940's, Sure's efforts foreshadowed developments in the food processing industry in the 1990's with the marketing of "textured proteins."[23]

The 1950's opened with University President Lewis Webster Jones being named Arkansas's Man of the Year in Agriculture for his advocacy on behalf of the College of Agriculture. In particular he was cited for his encouragement of diversified farming through expanding the livestock industry. Jones believed strongly that replacing idled cotton lands with beef cattle, dairy herds and poultry operations would be a key element in the development and growth of the agricultural industry in Arkansas. He spearheaded a successful drive to have the University buy a $30,000 herd of registered Aberdeen Angus cattle. A welcome addition to the animal industry program was the opening of a Beef Cattle Experimental Farm in 1952 on 1,300 acres near Savoy in western Washington County. The University bought the acreage from the USDA, which had combined nearly 4,000 acres in the area as part of the Wedington Land Use project. In 1953, F.W. Dye, a leading beef cattle breeder in northwestern Arkansas, donated seven champion Herefords to the College's beef herd, making a major contribution to the project. The work with beef cattle made major strides under the direction of the Animal Science Department head Warren Gifford, who, along with President Jones, deserved much of the credit for the expansion of the beef cattle industry in Arkansas in the postwar years.[24]

While beef production increased a respectable 75 percent between 1949 and 1952 in Arkansas, poultry broiler sales increased a whopping 1100 percent. The poultry industry took notice of the progress in Arkansas, and after a successful lobbying effort by R.M. Smith and University President Jones, the national "Chicken of Tomorrow" contest took place on the University campus in 1951. This "world series" of the national poultry industry not only

A total of 32,400 eggs were shipped from 25 different states for competition in the 1951 Chicken of Tomorrow Contest in Fayetteville. The University gained new facilities from this contest, and the poultry industry gained a better broiler.

brought attention to poultry production in Arkansas and the work at the College but provided the Poultry Experimental Farm at the Main Experiment Station with two 24 x 420-ft modern poultry buildings worth an estimated $25,000. The new buildings would enable the College to begin essential research projects that had been delayed due to lack of funds and facilities.[25]

In spite of this boon to the poultry program, Dean Ellis still made a strong appeal for the construction of a new facility for the Animal Industry and Veterinary Science Department. Animal Science classes and research were scattered across the campus, taking place in the Dairy, Agriculture, Home Economics and Chemistry Buildings, in Gray Hall and at the University farm, not to mention at the Livestock Branch Station in Batesville and the Wedington Beef Farm.

The Animal Science Building would have to wait for several more years before construction began. Nevertheless, the department took on a new name and expanded its responsibilities. Beginning in October 1950, the department was known as Animal Industry and Veterinary Science after the Veterinary Science operation was moved from the Bacteriology Department. The expansion was part of Ellis and Jones' plan to develop the livestock industry in Arkansas. Ellis suggested that only a minimum amount of teaching was needed to satisfy the needs of the general agriculture students. The program would then be able to concentrate on the research needed to assist livestock producers. Ellis also asked Jones to include money in the budget to hire someone to care for the livestock so that the staff could devote their full time to teaching and research.[26]

Ellis decried the overcrowded conditions for the staff and students in the College. The Dean told Jones that in spite of the enrollment drop that accompanied the outbreak of the Korean War and a decrease in the number of veterans taking advantage of the GI Bill, classrooms, labs and equipment were inadequate to serve the needs of the College. Faculty were stuffed three to an office, leaving inadequate room to counsel students, conduct research or answer to public service requests for information.

Enrollment 1950-1959.

Year	1950	1951	1952	1953	1954	1955	1956	1957	1958	1959
Agriculture	443	362	365	299	334	362	355	345	345	288
Home Economics	143	140	145	139	156	158	174	162	169	150
Total	586	502	510	438	490	520	529	507	514	438

Fall enrollment figures provided by Dean's Office, DBCAFLS.

Even operating under these handicaps, Dean Ellis rated the performance of the faculty and students as "excellent."

In addition to making breakthroughs in the modernization of agriculture in Arkansas, the College also made a major contribution to equal education in the South after it admitted Benjamin F. Lever as the first African-American graduate student in residence at the University of Arkansas in the fall of 1950. Lever was a graduate of Tuskegee Institute in Alabama and took up graduate work in agronomy, studying germination rates of selected grain sorghums under A.B. Burdick. Unlike the situation that prevailed when other African-American graduate students took classes in the Law School or in the College of Education previous to Lever, the College of Agriculture made no "special arrangements" to segregate Lever inside the agriculture classes or laboratories while he attended to his work. He said that his fellow agriculture students showed a good attitude toward him and that the racial atmosphere was much better than he had expected. Lever received his M.S. degree in agronomy at the June 1951 commencement ceremony with full honors.[27]

The College expanded its involvement in foreign agricultural extension at this time as well. Two programs were set up in cooperation with the United States State Department and the USDA Office of Foreign Relations. In the Foreign Visitors Program, initiated in 1950-51, the College arranged with the State Department and the USDA to bring foreign scholars, agriculturalists and students to the University. The College attended to the visitor's needs, arranging housing, research trips, course work, special lectures and entertainment. In the program's first year, George Steinbruegge, from Agricultural Engineering, and Marie Lavallard, the associate agricultural editor, were in charge of making arrangements for the 20 visitors who came from seven different countries.

The program expanded rapidly with 97 visitors from 30 countries arriving during 1951-52. The principal area of interest was with the work at the Rice Branch Experiment Station at Stuttgart and M.C. Kik's research on rice and nutrition. The Visitors Program continued into the early 1960's with the majority of the guests from developing nations in Asia and Latin America where the United States had national security interests. The State Department, the USDA and the United Nations agencies involved in the program praised the College for its efforts; however, the College was short of personnel, and the program took up valuable time and resources for which Dean Ellis felt the College was not adequately compensated.[28]

In 1951 the College became the first land-grant institution in the nation to assemble an agricultural foreign mission. The object of the mission was to assist Panama in establishing an agricultural teaching, research and extension program similar to the one that had been so successful in modernizing agriculture in the United States. In dispatching an agricultural mission to Panama, the College not only helped the people of Panama, but also assisted the United States in carrying out foreign policy objectives. The State Department developed The Point IV program in order to stabilize economic, political and social life in countries considered vital to the national security of the United States. In January 1951 a special committee that included Dean Ellis, Experiment Station Associate Director R.P. Bartholomew and Extension Service Director L.A. Dhonau went to Panama. The committee members, U.S. government and USDA officials met with Panamanian officials, businessmen and farmers and toured the chief agricultural regions in the country. Following the consultations, the College, the U.S. government and Panama signed a contract in May 1951 pledging the College to aid Panama in setting up an agricultural program.[29]

> *The instructors and professors were the highlights. I can still see most of them at their podium. Their ethics, character and devotion to us were memorable. [They] took an individual interest in our development, gave good tests [and] ensured that we learned and could deliver the details.*
>
> *Maj. Gen. USAF (Ret.) William P. Bowden, BSA 1954*

The agricultural mission to Panama originally was contracted for one year, but subsequent developments extended the mission's life for six years, from 1951-1956. During those years the College sent 24 different agricultural specialists, extension workers and administrators to Panama from its staff. Bartholomew directed the mission from 1951-3, Dwight Isely served as director from 1953-56, and Carl D. Koone administrated the project through its final year from 1956-7. Roy Roberts, head of extension education at the College, also played a key role in helping at the Divisa National Vocational Agriculture School.

The mission was guided by the philosophy that the College could impose no successful program from above, but that enduring progress could be made only if an indigenous program of development, which relied on Panamanian expertise, was put in place. Taking a page from the accomplishments of the land-grant system of education in the United States, the mission set to work helping Panamanians develop and improve their resident teaching, research and extension capabilities.

The National Institute of Agriculture at Divisa became the focus of improving resident teaching. The school, established in 1941, required that students have only the equivalent of a sixth-grade education in order to enroll. The curriculum was based almost entirely on agricultural subjects with little attention paid to basic science, math and other subjects. A more solid foundation curriculum was needed to give the students an understanding of scientific agriculture. Panamanians enthusiastically adopted a program to revise the curriculum and raise academic standards. Admission requirements were raised from completion of the sixth grade to completion of the ninth grade. The Institute hired a faculty of college graduates and broadened the curriculum to include general academic subjects. Work on the school farm and in the shops underwent considerable revision to reflect a more research-oriented approach. Graduates of the school became eligible for college admission, and many went on to enroll at the University of Arkansas College of Agriculture.[30]

> *If we graduates could (dared!!) write candidly about our personal remembrances of the ag farm, your book would be a best seller! The farm was a favorite "parking spot" in the 50s! Campus police never seemed to patrol that far from campus. The square dances at the farm to welcome new and old students back were special too.*
>
> *Anonymous on request, BSHE 1957, MS 1965*

The agricultural mission staff considered its contributions to the development of a Panamanian Experiment Station program its most constructive effort in Panama. In 1951 the experimental farm at Divisa was little more than a collection of demonstration plots or a practice field for the students at the Institute. The mission staff introduced the experimental plot technique on the farm, and the Panamanians readily absorbed the idea and proved to be keen observers and recorders of results.

The Panamanian experiments produced 28 published research bulletins on a wide range of topics that included fertilizer studies, insect control, plant disease control, comparing new crop varieties, livestock feeding and management and economic studies. The experimental work was so successful that the program was introduced in outlying provinces as well.[31]

Establishing an agricultural extension program in Panama ran into political difficulties as the system in place before the agricultural mission arrived was mired in an entrenched system of political patronage. *Fomentos* (equipment and tractor stations), under the jurisdiction of the Ministry of Agriculture, were established throughout the country whose purpose was to disseminate agricultural advice and to provide farmers with access to tractors and other modern equipment. The fomentos became what the Minister of Agriculture described as "a national disgrace" as the few trained agricultural experts were replaced by political appointees who doled out advice and equipment on a highly selective basis and padded their staffs with friends and family at the expense of using funds for extension activities. The mission sought to alleviate the shortage of trained personnel through inaugurating a training program for extension workers at the National Institute. By 1953 there were 14 extension agencies with trained personnel operating in Panama. The Panamanians also adopted a girls and boys club program, and by 1956 there were 56 clubs in the country with over 2,000 members.[32]

The College's agricultural mission to Panama ended in 1957. Differences between the College's staff and the U.S. State Department's personnel over the direction of the mission were responsible for curtailing the College's participation in the program. The College's staff felt very strongly that in order for a successful agricultural development program to take hold in Panama, the Panamanians themselves must be in control of the program. The goal of the Arkansas mission all along was to train Panamanian people to effectively manage and administer an agricultural teaching, research and extension program for the benefit of Panamanians. However, the Arkansas mission felt that the United States Operations Mission's staff constantly undermined their efforts to promote indigenous responsibility and competence. Therefore, though they believed their work was far from being completed, the College's staff chose to curtail their participation rather than challenge the sponsors of the project, the U.S. State Department.

While the mission was terminated before its completion, the College's staff could be proud of their many accomplishments during their six years in Panama. As a token of their appreciation, the Republic of Panama presented Bartholomew and Dhonau with medals signifying their membership in the Order of Vasco Nunez de Balboa, the highest honor given non-citizens of Panama. The Panama mission was the first of many foreign agricultural development projects in which the College would participate over the years. As the United States expanded its international responsibilities in the postwar years, so did the nation's land-grant institutions.[33]

Meanwhile, back in Arkansas, the College struggled to develop its own teaching, research and extension program in the face of government and public disinterest in domestic agricultural studies. After the Korean War, enrollment dropped off, not only in the College but also in the number in the College as a percentage of students in the University overall. Land-grant institutions across the country experienced the same leveling off of interest in agricultural studies as the postwar student population became more absorbed in liberal arts studies and other technical and scientific fields. The fact that there were fewer farmers or people living in rural areas also affected the enrollment drop. News reports emphasized the decreasing farm population and agricultural surpluses that accumulated as a result of the production revolution and government price support and commodity credit programs. Some critics maintained that in the face of such surpluses, further expenditures for research and price supports were unnecessary and suggested that perhaps the colleges of agriculture had succeeded too well. With the drop in enrollment, the lessening influence of rural areas in the state and the overwhelming successes in agricultural production, the Legislature felt less compelled to provide financial support for the College. To counter the legislative apathy, Ellis pointed out that as agriculture had grown more complex and modernized, the need for highly trained agricultural experts and new research was greater now than ever before.[34]

> *When I was a student, very few people had cars on campus. There was a closeness in the student body because money was scarce and education was the driving force.*
> *Jack G. Justus, BSA 1954*

Ellis continued to be concerned about salary levels that were low when compared to those of other universities and industry. During 1951-52 alone 15 staff members resigned. Recruitment was especially difficult in agricultural engineering, agricultural economics, veterinary science and home economics, fields of growing interest in the 1950's. The College's salary situation was also handicapped because the pay rate was calculated on a nine-month basis when most of the resident teaching staff were on a 12-month research schedule. Under these circumstances, the College experienced a continual turnover in faculty and staff during the 1950's. The conditions were mitigated somewhat many respected members of the faculty stayed on and new recruits found other reasons besides salary issues to stay in Arkansas.[35]

The resident teaching program felt the budget constraints keenly as Dean Ellis struggled to develop a graduate training program and provide upper-level courses to answer the need for technically trained agricultural experts for the increasingly complex farm operations and agribusiness. The low point in the ongoing effort to build up the teaching program came in 1957 after the Legislature slashed the College's budget requests 68 percent. The cuts forced the Dean to cancel 26 upper-level and graduate courses in seven departments and to withhold 17 new courses that he had already approved. Courses considered of critical interest to Arkansas farmers and to the development of a graduate program, including marketing, livestock production, crop irrigation, plant diseases and home economics studies, were delayed until the 1959-1960 year.[36]

In spite of the inability of Dean Ellis to enhance the resident teaching capabilities, Associate Dean Hudson's review of the curriculum led to revisions in this area intended to address the needs of students entering into new fields opening in agribusiness. The curriculum remained basically the same for freshmen and sophomores but branched out in the junior and senior years to provide more options. The general curriculum, enhanced with more electives, remained in place for those going into vocational and agricultural education. The new curriculum provided 11 majors in specialized fields, including agronomic crops, soils, animal industry, dairy husbandry, poultry husbandry, animal nutrition, entomology, horticulture, plant pathology, rural sociology and agricultural economics. New courses in zoology, botany and mathematics were introduced to prepare students entering a specialized field of study and to give them a foundation for graduate work should they choose that option. In 1959 the College further broadened the curriculum in the Agronomy Department by adding a range management program, one of the first in the nation. Under the direction of A.M. Davis in the Agronomy Department, the program combined instruction and research in animal science, botany and agronomy to provide an integrated approach to the use of range lands for the expanding livestock industry in Arkansas.[37]

The College made some advances in improving its facilities during the last half of the 1950's, in spite of budget cuts. The Experiment Station opened a new $40,000 Soils Testing Laboratory at the Cotton Branch Experiment Station in Marianna in July 1954. The new facility was touted as the most modern of its kind in the country. The lab was made possible after the Legislature levied a tax on fertilizer sold in the state and set aside part of these funds for the construction of the lab. R.L. Beacher, associate professor of agronomy, was made head of the new facility. A $20,000 greenhouse was built in 1955 to augment the work done at the lab.

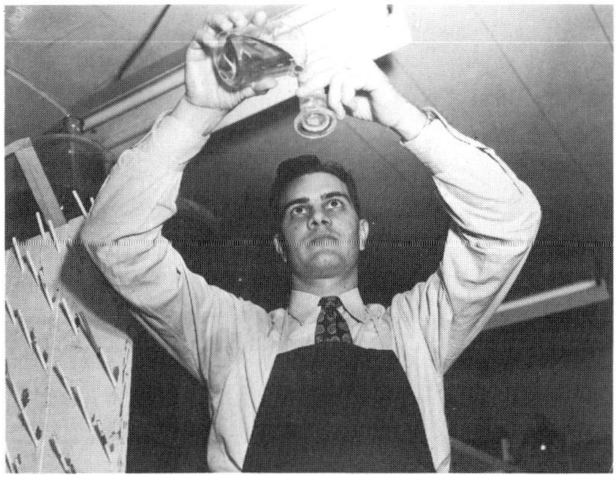

R.L. Beacher, head of the Soils Testing Laboratory.

An Animal Disease Laboratory, which had operated on a part-time basis on the campus since 1950, was upgraded to operate full-time in the summer of 1954. Using additional appropriations made available by Governor Francis A. Cherry, the College installed additional equipment and hired full-time staff to oversee operations. In addition to the full-time lab assistants, C.F. Hall and J.M. Oyler from the Animal Industry and Veterinary Science Department, served as veterinarians in the lab. The facility, located in the overcrowded labyrinth of labs in Gray Hall, dealt primarily with poultry diseases but also diagnosed and investigated diseases in larger animals.

> *I was Editor of* The Agriculturist *magazine when Perry Lee Adkinson was Business Manager. What wonderful training and associations. My agri school provided me many extracurricular activities and an association with some of the finest and most capable people found anywhere. I was fortunate to be there when so many outstanding leaders came along.*
>
> *Herschel H. McClurkin, Jr., BSA 1951, MS 1952*

Addressing the needs of the booming poultry industry, the College established a 240-acre Turkey Experimental Farm six miles west of Fayetteville in early 1958. E.L. Stephenson, an early proponent for increasing turkey production in Arkansas, was the guiding force behind the effort to expand the College's well-regarded poultry program. R.M. Smith, the College's resident poultry expert, backed the project. R.N. Gyles was put in charge of breeding and genetics work at the farm, and Stephenson conducted nutrition studies. A new 108 x 30-ft horticulture lab focusing on fresh fruit and vegetable production and processing was built on the University Main Experiment Station during 1958-9. A.A. Kattan, who worked with ripening tomatoes and quantity storage of fresh produce, supervised the lab.[38]

Dean Ellis' pleas and plans for a new Animal Science Building to better serve the needs of the growing livestock industry in Arkansas and the student interest in the subject were at least partially realized beginning in 1955. However, the construction and equipping of the new facility was hampered by budget shortfalls that plagued the agriculture program at the University. Two wings of the building opened in the fall of 1955. The first floor of the west wing housed experimental animals, feed preparation facilities, storage and large and small animal research labs. The second floor contained genetics and animal nutrition labs. The east wing was devoted to the poultry and dairy sections and various labs associated with their work. A section connecting the two wings accommodated graduate student offices. While the availability of the new facilities was welcomed, the enthusiasm was dampened somewhat because funds to equip teaching laboratories in the two wings were not available for another two years. The remaining three wings of the building were

Dean Lippert Ellis and architect Herbert Bowler with a model of Animal Science Building. The first two wings of the facility were built in the 1950s, but the last three wings were not built until 1964. When the faculty moved into the Animal Science Building in 1964, the Dairy Building (built in 1908) was torn down to make way for a swimming pool. (Photo courtesy Photograph Collection, Special Collections, University of Arkansas Libraries.)

not completed until 1964 and would contain the veterinary science facilities, the new University creamery, classrooms and office space.[39]

In 1951 the College adopted a new name, the College of Agriculture and Home Economics, to reflect the growing interest in home economics studies and research that took place in the years following the war. Increasing numbers of women enrolled in colleges and universities across the nation in the postwar years eager to follow up on opportunities opened to them during the war when women filled many positions vacated by men away at the battle front.

Beginning in 1952 and continuing through the rest of the decade, Eleanor Barnes headed the Home Economics Department. Barnes' qualifications, a B.S. in chemistry, an M.S. in foods and nutrition and a Ph.D. in biochemistry from the University of Minnesota, combined with her interest and experience in a variety of other fields, made her a perfect match for the position. Barnes presided over the expansion and modernization of the department to meet student's educational and professional needs in the 1950's. Other qualified and able faculty members assisting Barnes during the decade were Vera McNair, elevated to Professor Emeritus in 1954; Clara Ridder, a specialist in housing and home management; Isabel Irwin, a nutrition and dietetics expert; Lorene Raymond, teaching in foods and meal management; and Lily Wu, a textiles and clothing specialist.

In addition to receiving a more modern approach to the field previously known as "domestic science," home economics majors received instruction in subjects designed to prepare them for employment outside the home. Teaching and Home Demonstration Extension work were still the major fields of study, and the students choosing this career path were under the supervision of Mrs. Carl Whorley. An increasing number of students began to look into careers in nutrition, public health and dietetics. Home economics studies provided training for careers in appliance sales, which gave women an entree into the business world. Journalism, advertising and merchandising were other business areas open to home economics majors.

Fashion design was a specialty field that began to flourish as Americans began to generate more disposable income during the 1950's. Research careers in all fields related to home economics in academics, government agencies and private industry began to beckon qualified and interested graduates. Another primary concern for Americans in the 1950's, child development and child care, was included in the home economics curriculum, and the training in this field provided students with career opportunities as teachers, social workers or researchers.[40]

During the 1950's agriculture and home economics students took an active interest in the College and promoting its activities. Active departmental organizations included the Agricultural Economics, Agricultural Engineering, Agronomy, Animal Science, Colhecon, Entomology and Horticulture Clubs. There were more of these departmental clubs in the College of Agriculture than in all other divisions of the University combined. Alpha Zeta and Phi Upsilon Omicron agricultural and home economics honor fraternities continued their activities and were joined by the Alpha Tau Alpha national honor fraternity for vocational and agricultural education majors, organized in 1954. Students who applied themselves academically could look forward to an increasing number of scholarship awards during the decade. A variety of corporate and private donors awarded 62 scholarships ranging from $150 to $500 for the 1958-9 school year.[41]

The Agriculture Students Association applied themselves as usual to organizing and promoting the annual Agri Day celebration. The parades that were the highlight of earlier Agri Day activities had been abandoned during the war. While the exhibits that replaced the parade were well thought out and educational, they lacked the excitement and popular attention given the earlier tradition. A tug of war with the engineers garnered some excitement and helped tame some tempers through ritualizing the traditional agriculture-engineering rivalry.

The students came up with a new addition to Agri Day that certainly grabbed attention and also had some students grabbing for a bottle of aspirin. The rodeo that highlighted the 1950's celebrations began in 1951 with students building a makeshift chute from a set of barn doors "borrowed" from the University Farm. Henry Walker, a local livestock dealer, generously lent the intrepid bronco busters a herd of unbroken mules for the occasion. Agriculture students Myron Cassidy and O.P. Hilliard turned in the most outstanding rides of the day. The rodeo took on a more polished air in 1959 after students built a 240 x 120-foot arena and began using regular rodeo livestock for the event.[42]

The annual Agri Day picnic found a new home in 1951 following improvements to a five-acre plot of land near the University Farm donated by the University in 1949 for use as a park. Agri Park became a gathering place for students and faculty for a host of social events. The park was gradually improved over the years, with the crowning event being the construction of the Barton Pavilion, completed in 1957. Col. T.H. Barton, Chairman of the Board of the Lion Oil Company, noted for his support of numerous research projects at the College and his generous student scholarship program, do-

Construction of the "fireplace room," Barton Pavilion, Agri Park. In 1951 a five-acre "abandoned" wedge of property near the farm was made into Agri Park. Col T.H. Barton, Chairman of the Board of Lion Oil Company, donated $20,000 for the project. The park still functions as a gathering place for the College and community.

Barton Pavilion, Agri Park, University of Arkansas. Mr. Antonio Alcides Jimenez-Escobar, Republic of Panama, is observing the Lignum Vitae beam engraved with the Coat of Arms of the Republic of Panama, a gift from the Republic of Panama to the College for its assistance in the Panama Mission. The six stones immediately below the beam are from Old Panama City, which was sacked by the pirate Morgan in 1670.

nated $20,000 for the project. Numerous other organizations and individuals also provided gifts. The Republic of Panama presented a lignum vitae[43] beam engraved with the country's coat of arms and six stones from the old Panama City fort as a token of their appreciation for the work done by the agricultural mission. Students and faculty alike donated their time and labor in improvement projects, including pulling stumps, leveling the grounds, seeding grass and drawing plans. Agri Park and Barton Pavilion brought an even greater sense of social cohesion to the College. The effort to transform the "abandoned" wedge of property into an area where people could gather in a comfortable and refreshing outdoor setting combined the skills of members of the College for the benefit of all in the community.[44]

As the College teaching, research and extension programs expanded and grew more complex to fit the needs of modern agriculture, there also grew a need to reorder the administration of the College's tripartite mission. The Experiment Station program had grown to include 12 stations and substations that were conducting over 200 research projects by the end of the decade. The

Extension Service's role also changed along with the transformation of agriculture during the 1950's. While the number of farms and farmers dropped dramatically in the postwar period, the technical complexity of commercial farm operations increased. The need for more technical advice elevated the importance of Extension specialists. Marketing agricultural products also became a leading concern as farmers searched for additional outlets for the bounty produced by modern methods. With more rural people employed in non-farm occupations by the end of the decade, the Extension Service had to widen the scope of its services in order to stay relevant to the needs of the population.[45]

The Board of Trustees created the Division of Agriculture in June 1959 to oversee the expanded mission. The Board appointed John W. White as Vice President to direct the operations of the Division, which consisted of the Agricultural Experiment Station and the Cooperative Extension Service. It would have been hard to find someone better qualified for the position. White had considerable experience in all three phases of the agricultural program over the years. After he graduated from the College in 1935, White took a job as a field agent for the Department of Rural Economics and Sociology. In 1936-1937 he worked as an Extension agent in Lawrence and Craighead Counties. From there he went on to serve as a research supervisor at the Cotton Branch Experiment Station in Marianna from 1938-1940. After receiving his master's degree in agricultural economics from the University of Minnesota in 1940, he became an instructor in the Agricultural Economics and Rural Sociology Department in 1940. He earned his Ph.D. in agricultural economics from Minnesota in 1943 and worked as an Extension specialist in marketing for several years while continuing to teach at that university. In 1944 he was appointed assistant director of the Rice Branch Experiment Station in Stuttgart. In 1947 he returned to the University campus as professor and head of the Department of Agricultural Economics and Rural Sociology. He taught classes in marketing, farm management and agricultural policy and conducted research on land and labor problems, farm management and marketing. In 1953 White was appointed Associate Director of the Experiment Station before his appointment as Vice President.[46]

John W. White - Vice President for Agriculture 1959-1974.

The administrative reorganization called for the College to operate as a separate unit responsible for resident teaching and academics. The Dean of the College no longer had administrative responsibility for the Experiment

Station or the Extension Service. Both of these organizations would now have their own directors who would answer to the Vice President. E.M. Cralley was appointed Experiment Station Director, and C.A. Vines, the Associate Director of the Extension Service, became Director. Dean Ellis, after 13 years of faithful service, stepped aside during the reshuffling to make room for a new Dean. Ellis took a leave of absence, after which he intended to return to teaching. However, he changed his plans and resigned from the College and joined the International Development Agency as an Agricultural Economics advisor for the United States Agricultural Mission to Turkey. Gerald T. Hudson, the Associate Dean since 1954, took over the Dean's duties. The new arrangement allowed Dean Hudson, with his experience in the academic role of the College, to concentrate his energies on the teaching program.[47]

G.T. Hudson - Dean of the College 1959-1965.

The call for trained research scientists and extension specialists to aid farmers in developing modern farm operations would take on an even greater importance in the coming years. The burgeoning number of agricultural industries known collectively as agribusiness also needed agricultural experts and formed an increasingly important clientele. American consumers, growing accustomed to a wider and more plentiful variety of food and agricultural products during the 1950's, began to make their voices heard in the marketplace and in public forums. As a result of this widening clientele, it was necessary for the College to adapt its program to meet not only the needs of the farmer, but also the needs of the farmer's customers and suppliers as well.

The College's vital connection with research continued. As they do today, most teachers of college courses also conducted Experiment Station research programs; the faculty depended on research in order to teach students approaches to solving agricultural problems. The College's connection with Extension work was less important than in earlier days when the Extension staff had little formal training. But, with the rapid development of new technologies, techniques and varieties, there was a constant need for workshops and training sessions that brought Extension agents in touch with the latest developments.

The new administrative organization was keyed to coordinating an integrated agricultural program to better serve the needs of Arkansas and its people. The College's contribution to the great production revolution in the postwar era made more food and fiber available in greater varieties to more people for less money than ever before, not only in Arkansas, but around the world. With the tremendous changes that had taken place in agriculture in the years since World War II, new approaches were needed in solving problems that affected agriculture and the faculty and students at the College were ready to face the challenge.

Endnotes

1 John A. White, et al., "The Arkansas Agribusiness Story" (Fayetteville: University Of Arkansas College of Agriculture, 1959), Arkansas Collection, University of Arkansas Mullins Library Special Collections, Fayetteville.

2 Danbom, 234-235; Boyce Drummond, "Arkansas, 1940-1954," in *Historical Report of the Secretary of State of Arkansas* (Little Rock: Arkansas Secretary of State, 1978), 178-179.

3 R. Douglas Hurt, *American Agriculture: A Brief History* (Ames: Iowa State University Press, 1994), 322-325.

4 John L. Shover, *First Majority-Last Minority: The Transformation of Rural Life in America* (De Kalb: Northern Illinois University Press, 1976), 194-198,

5 Pete Daniel, *Breaking the Land: The Transformation of Cotton, Tobacco and Rice Culture Since 1880* (Urbana: University of Illinois Press, 1985), 178.

6 *President's Annual Report, 1945-46*, 31.

7 United States Department of Agriculture, Bureau of Agricultural Economics, *Farm Production Adjustments in Arkansas Under Post-War Conditions* (Little Rock: USDA, Bureau of Agricultural Economics, 1944), 1-2, in Arkansas Collection, Mullins Library Special Collections, University of Arkansas, Fayetteville; *Farm Bureau Press* (September 1946), 1.

8 *President's Annual Report, 1945-46* (33-34); *Arkansas Agriculturist* (January 1945), 9, (March 1946), 15.

9 *Dean's Report,* 1947-48, 3, 1948-49 3, 1949-50, 5, 1950-51, 2.

10 *Dean's Report*, 1947-48, 5, 1948-49, 7.

11 *Dean's Report*, 1947-48, 2-4;

12 *Dean's Report*, 1947-48, 5; *Arkansas Agriculturist* (September 1947), 11.

13 *Arkansas Agriculturist* (February 1946), 14, (April 1946), 14, 21, (January 1947), 5, (April 1947), 4, (January 1950), 8, 16, (November 1949), 13.

14 *Arkansas Agriculturist* (April 1946), 16.

15 *Arkansas Agriculturist* (January 1947), 14, (February 1948), 7.

16 *Arkansas Agriculturist* (February 1947), 8, (April 1950), 15, 21.

17 *Arkansas Alumnus* (October 1948), 3; *Dean's Report*, 1947-48, 3.

18 *Arkansas Agriculturist* (February 1946), 4, (December 1948), 14, *Dean's Report*, 1948-49, 9.

19 *Arkansas Agriculturist* (February 1946), 4, (December 1948), 8, 16; *Dean's Reports*, Box 4, File: Rural Economics and Sociology, 1948.

20 Charles A. Lincoln, *The Personal Journal of a Clod-Kicking Entomologist* (Fayetteville: 1983), 2-1, 2-5, 2-10, 3-2, 3-17, 4-12, 4-14 (numbers indicate chapter number followed by page number). Unpublished manuscript in Mullins Library Special Collections, University of Arkansas-Fayetteville.

21 *Arkansas Agriculturist* (December 1947), 13, (March 1949), 7.

[22] *Dean's Report*, 1947-48, 9-13, 1948-49, 8-9; *Arkansas Agriculturist*, (December 1947), 13, (March 1949), 7.
[23] *Arkansas Agriculturist* (January 1948), 7, (February 1948), 14.
[24] *Arkansas Agriculturist* (February 1953), 5, 9-10, (February 1960), 6.
[25] *Arkansas Agriculturist* (February 1950), 3 (March 1951), 4-5; White, 20-21.
[26] Jones to Ellis, October 27, 1950 and Ellis to Jones, August 9, 1950, *President's Records*, Box 13, File: Ellis.
[27] *Arkansas Agriculturist* (November 1950), 10; Leflar, 287; Benjamin F. Lever, "A Study of Germination in Six Varieties of Grain Sorghum" (Master's Thesis: University of Arkansas, Fayetteville, 1951).
[28] *Dean's Report*, 1950-51, 16-17, 1951-52, 35-36, 1956-57, 30-31, 1957-58, 33.
[29] University of Arkansas, *The Arkansas Agricultural Mission to Panama: A Report* [Fayetteville, Arkansas], 1957, 1-2. In Mullins Library, Special Collections, University of Arkansas-Fayetteville.
[30] Ibid, 19-20.
[31] Ibid, 20-21.
[32] Ibid, 7-8, 21-22.
[33] Ibid, 15-17.
[34] *Dean's Report*, 1956-57, 8-9.
[35] *Dean's Report*, 1951-52, 31-32.
[36] *Dean's Report*, 1956-57, 12.
[37] *Arkansas Agriculturist* (Autumn 1953), 9, 11.
[38] *Arkansas Agriculturist* (April 1954), 5, (November 1955), 7,
[39] *Arkansas Agriculturist* (April 1955), 13-18, (December 1955),7.
[40] *Arkansas Agriculturist* (April 1957), 10-11.
[41] *Arkansas Agriculturist* (October 1958), 10-11.
[42] *Arkansas Agriculturist* (May 1951), 5-6, (March 1952), 14, (February 1957), 10, (February 1958), 8, (April 1959), 8.
[43] Lignum vitae is a tropical American tree having very heavy, durable wood.
[44] *Arkansas Agriculturist* (December 1951), 3, (October 1956), 7, (April 1957), 10-11).
[45] *Dean's Report*, 1957-58, 9-10, 16-28, 30-32;
[46] *Arkansas Alumnus* (October 1974), 8-9.
[47] *Arkansas Alumnus* (January 1960), 5, *Arkansas Agriculturist* (April 1962), 9.

Chapter Six

Challenge and Change, the College Adopts New Roles: 1960-1979

As the 1960's dawned, the changes taking place in American agriculture called for a reevaluation of the land-grant mission. The application of the scientific discoveries from agricultural colleges made a more abundant and varied supply of food and fiber available to consumers at a lower cost than ever before. In addition to the progress in agricultural production, there were also revolutionary developments in the technology of food handling that made the delivery of food products to consumers more efficient. No longer was agriculture confined to farm operations. As the number of farmers declined due to more efficient production methods and a drift to urban areas by the new generations, the number of non-farm people involved in the processing of agricultural goods increased dramatically. It was estimated in 1962 that for every farm worker on the farm, there were three employed in agricultural processing, distribution and marketing industries in towns or cities. During the 1960's, integrated agribusiness concerns were involved in all phases of production, processing, marketing and distribution of agricultural products; consequently, agriculture came to play a more important role in all sectors of the national and international economy and life than ever before.

A report funded by the Carnegie Foundation to mark the centennial of the Morrill Act addressed the problems confronting land-grant colleges of agriculture in adapting their programs to meet the challenges ahead. Charles E. Kellogg and David C. Knapp, authors of the report, pointed out that the rapid pace and the results of mechanization and modern technological farming methods had "exceeded even the wildest dreams of the agricultural leaders of 1862."[1] In redefining the mission of the land-grant college of agriculture, Kellogg and Knapp highlighted the growing complexity of agriculture and its widening clientele. They delineated five areas of concern in their report.

1. **The Science and Technology of Modern Agriculture.** The science and technology of agriculture reached into all areas of scientific inquiry and

called for an interdisciplinary approach that encompassed and affected developments in other fields of research. This was particularly true in the emerging field of food science and technology as well as in weed management and pest control. By the end of the 1970's, nearly all areas of agricultural inquiry encompassed achievements in other scientific fields and vice versa.

2. **Rural People and Their Institutions.** "Indirectly the colleges of agriculture have helped create rural social problems and rural-urban imbalance by promoting technological changes." The authors urged the colleges of agriculture to join with social scientists and extension workers in helping evaluate the social effects of technological change in rural areas. Out of this effort could emerge plans to address the problems of rural poverty and rural education and how to foster economic growth in rural areas with declining populations.

3. **Environmental Science.** With the publication of Rachel Carson's *Silent Spring* in 1962, the public became more concerned with the use of chemical pesticides and their cumulative effects on the environment and human health. Science and technology had made possible higher crop yields and lower production costs, but in the rush toward modernization, some had lost sight of the possible ill effects cited earlier by University of Arkansas College of Agriculture scientists W.J. Baerg, Dwight Isely and Charles Lincoln. Colleges of agriculture would need to address these concerns in a meaningful way in the decades ahead in order to maintain credibility with the public on such issues.

4. **Urban and Non-farm Resource Use**. In the years ahead, the colleges of agriculture would have to pay more attention to their urban and non-farm clientele. Suburban and urban populations could profit greatly from principles and techniques developed by experiment stations and taught in agriculture classes. Homeowners had a need for updated information on lawn and garden care as well as instruction on efficient use of plants, soils, water and lawn and garden chemicals. The increasing urban population was also making greater use of rural areas for recreation. An increasing amount of land formerly used in farming or forestry was adopted for recreational use by urban dwellers.

5. **International Agriculture.** The University of Arkansas College of Agriculture and Home Economics had already become intimately involved in educating foreign students and training agricultural experts. That role would continue in the next two decades, though further expansion of agricultural missions abroad would wait until the 1980's.

Kellogg and Knapp also mention forging good relations with private business and the corporate world as an outstanding concern for expanding the mission of the land-grant college of agriculture. This field was especially critical as agribusiness concerns came to dominate agriculture and the need for trained specialists in corporate positions became a major employment source for graduates of agricultural colleges. With increased corporate funding for agricultural research during this period also came a rising awareness of the need to protect the public interest.[2]

The University of Arkansas College of Agriculture and Home Economics did an exceptional job of expanding and balancing its mission during the 1960's and 1970's. New courses, programs and curricula were developed to meet the needs of the expanded mission. Vice Presidents John W. White and James Martin; Experiment Station Directors E.M. Cralley and Lloyd Warren; Dean Hudson and Hudson's successor, Dean Glenn Hardy, all worked tirelessly to provide students, resident faculty and researchers with programs and equipment necessary to keep the College in the forefront of the latest developments in agricultural education and research. They paid special attention to increasing interdisciplinary approaches to research and teaching, recruiting outstanding faculty and students and developing the graduate programs at the College. During this period the College's enrollment tripled, expanding from 401 in 1960 to over 1,300 in 1979. Major additions to the College's facilities included completion of the Animal Science Building in 1964 and opening the first phase of a new Plant Science Building in 1977. It was an exciting era for students and faculty alike as they became involved in adapting the College to the challenges of the 1960's and 1970's.

The April 1960 Agri Day activities reflected the changing face of agriculture in this period. Except for the Alpha Zeta initiates, still clad in overalls, bandanas and hats, the Agri Dance that evening saw coats and ties and evening dresses replace the traditional overalls and gingham dresses. Men no longer risked having their ties snipped off for violating the traditional Agri Day dress code. While agriculture had always been considered a business proposition for many farmers, modern production techniques and the attendant changes in rural home life had eliminated much of the cultural context from farming as a "way of life." On another level, the mission of the Agriculture College mission had turned toward providing more in-depth scientifically trained agriculturalists for both the farm and the business world. The rise of agribusiness industries during this period meant even more graduates joined the corporate world. Coats and ties became a truer representation than overalls and bandanas of the role that graduates would play in American and international agriculture during the 1960's and 1970's.[3]

Home economics study at the College revealed its changing face at the 1960 Agri Day as well. Agri Day breakfast had traditionally been hosted and prepared by the home economics students. However, for the 1960 celebrations, a banquet, held at the ballroom in the Student Union Building, was prepared and served by the Student Union Food Service. Home economics studies no longer emphasized domestic science; instead they began to stress career and professional development for women. Courses in home meal management gave way to classes in quantity food production to prepare students for careers outside the home. Nutrition research classes replaced basic food preparation and preservation courses as more and more people began to buy ready-made foods and became concerned about eating habits. Fashion merchandising and design classes replaced the sewing and clothing courses of earlier periods.[4]

The departmental and club displays, a feature of Agri Days since World War II, did not grace the Agriculture Building during the 1960 celebrations. Instead, the Agriculture Students Association integrated the displays with the University-wide presentations on GAEBALE Day the following week. This development was in line with programs that emerged in the College during the following two decades that stressed an interdisciplinary approach to agricultural education. Interdepartmental studies within the College and programs that reached out to include more course work in the other colleges at the University, particularly in sciences, liberal arts, business and law, became a focus for the College's curriculum development.[5]

Beginning in 1960 Dean Gerald Hudson and the faculty took the first steps toward retooling the College for its expanded mission. The University Graduate School approved Ph.D. programs in animal science and agronomy developed by the College's curriculum committee, although implementing the programs would have to wait for funds in the 1962 year. The committee also revised the master's degree programs in general agriculture and home economics in order to better equip graduates for the emerging job market in these fields. More three-week graduate-level courses were developed and offered at both on-campus and off-campus locations to address the needs of Extension Service personnel and other agricultural professionals requesting more information on the latest technological and scientific advances.[6]

Enrollment 1960-69.

Year	1960	1961	1962	1963	1964	1965	1966	1967	1968	1969
Agriculture	251	254	268	272	291	306	357	402	432	451
Home Economics	150	167	160	168	145	201	261	270	281	283
Total	401	421	428	440	436	507	618	672	713	734

Fall enrollment figures provided by Dean's Office, DBCAFLS.

After years of declining numbers of students, Dean Hudson and the faculty also worked on ways to increase the enrollment in the College. The College's drop in enrollment in the years between 1955 and 1960 mirrored developments in other land-grant institutions across the nation as more students became interested in liberal arts studies. The student scarcity ran counter to the widening career opportunities available to agriculture and home economics graduates and caused some alarm in agribusiness circles that there would not be enough trained agricultural experts to meet the needs of the industry. In response, Dean Hudson and the faculty created a recruitment committee in the College and set to work designing plans to attract students to agriculture. The College prepared departmental supplements to include in the

University Redbook sent to Arkansas high schools. Dean Hudson and Evelyn Cox, the new head of the Home Economics Department, and other faculty, opened direct lines of communication by visiting high schools around the state during the winter months.

In the fall of 1962, Billy B. Bryan of the Agricultural Engineering Department began an enthusiastic recruiting campaign for his department. Bryan contacted every county agent and vocational agriculture teacher in the state asking for the names of students interested in an engineering career. His contacts sent him a sizable list of names, and he sent out letters to all the students and followed up with personal visits to quite a few of his correspondents. Bryan succeeded in raising the enrollment in agricultural engineering and perhaps saved the department from folding because of low enrollment. Bryan's efforts were successful and would be applied in later years to College-wide recruitment with similar results.[7]

A committee of freshmen advisors, along with the Dean's Office and the head of the Home Economics Department, worked together to provide an orientation program for incoming freshmen in order to familiarize new students with opportunities in agriculture and home economics. The faculty also pushed forward plans to improve instruction in the College. Three faculty members participated in a workshop on improving instruction held at Athens, Georgia, during the 1960-61 year. The committee on instruction also recommended developing a series of seminars and forums on improving classroom teaching in the College.[8]

Paul Noland, from the Animal Science Department, provided one example of teaching excellence in the College. Noland received Alpha Zeta's outstanding teacher award four times, in 1955, 1957, 1963 and again in 1983. In 1963 the Alumni Association honored him with a university-wide award for outstanding achievement in classroom teaching and awarded him membership in the Blue Key National Leadership fraternity as well as a $500 check

Evelyn Cox, Billy Bryan, Paul Noland.

from the Alumni Association Endowment fund. Since his arrival at the College in 1951, he had built his reputation for working with students as an instructor and researcher in swine production and feeds and feeding. Noland was responsible for some of the pioneering research in the use of alternative feeds in livestock nutrition.

Noland's teaching and research reputation had international dimensions as well. He served two years on the agricultural mission to Panama from 1955-57 and spent six weeks in Costa Rica during 1961 advising Ministry of Agriculture employees in animal research methods. In 1965 he was off to Guatemala for the summer, working for the Agency for International Development. During his stay in Guatemala, Noland advised the Ministry of Agriculture there on setting up an Extension program in animal science and helped them develop breeding, feeding and livestock production programs. He also spent considerable time working directly with farmers in the countryside, advising them on livestock production and using surplus agricultural byproducts as inexpensive, available feed sources.[9]

Revision of the curriculum to conform to the enlarged mission for the College was implemented by Dean Hudson and the faculty in the early 1960's. The faculty curriculum committee, chaired by E.L. Stephenson, worked steadily on the issue for more than two years before the results were approved in the spring of 1964. As in other agriculture colleges across the nation, the new curriculum required more work in English, speech and journalism in order to improve agricultural graduates' communication skills. As people began to see that agriculture was integrated in the workings of society and the economy at large, rather than being strictly confined to farm production, there was a growing feeling among agricultural education experts that agricultural education should broaden its scope. Consequently, more liberal arts classes in the humanities and social sciences were deemed necessary in order that agriculture students might receive a broader education to better serve society.

> *I was always interested in livestock, so I began an animal husbandry major. Then I made the livestock judging team, which I credit with a large part of my success. I learned to think "under pressure," giving reasons for the way I placed the class to a professional judge. I'm not afraid to appear before any crowd, large or small. To me this is very important to one's success.*
>
> *Sherman D. Cullum, BSA 1960*

The faculty adopted the curriculum committee's recommendation of five study plans. All plans were enriched with added liberal arts requirements. One plan was designed to give the student a broad background in agriculture by requiring courses in several departments. Another was designed for the preparation of secondary school teachers of agriculture. Options in ag-business and journalism were developed. The ag-business option became very popular since it fit the needs of much of the job market emerging in the 1960's. Students in many departments chose this plan, which gave them the opportunity to develop their major and gain some exposure to the business world by taking courses in agricultural economics and in the Business College.[10]

Throughout the early 1960's, there was an increasing effort to strengthen and modernize the faculty and curriculum in the Home Economics Department. The course on school lunch preparation was replaced with a more advanced course that emphasized nutrition. As people became more accustomed to buying readily prepared foods at the supermarket, the department dropped food preservation classes as part of the regular home economics curriculum in favor of food demonstration classes. These classes prepared students for careers in presenting prepared foods to consumers in an appealing way in order to increase sales.

Research activities in nutrition took on an added emphasis at this time as well. Isabel Irwin and Catherine Carroll carried on research in nutrition, studying vitamin deficiencies and lipid metabolism in new labs in the Home Economics Building built and equipped for that purpose. The graduate program in the department, which for many years had focused on nutrition research, continued in that direction as Carroll supervised the work of graduate nutrition students in her lab. Child development courses took on added significance at this time as well with Mary Burton supervising the work in this area. The University Nursery School, located in the basement of the Home Economics Building, served as the setting for both instruction and research in this area of vital concern. Sewing and tailoring classes gave way to studies that emphasized evaluating ready-made clothing and textiles. Fashion design and merchandising also held an interest for home economics students, although it would not be until after 1965 that actual courses were developed in this area.[11]

With developing enthusiasm for new directions in home economics studies also came an increased demand for qualified professors in the field. There had been an historic shortage of home economics professors and researchers with advanced degrees, and the new demands made in the 1960's for more qualified personnel made the need even more critical. Unfortunately, the chronic problem with low salaries at the University made the situation worse. However, the College identified the problem and made some strides in this area, and the development of a strong Home Economics Department in the College made real progress in following years.[12]

The seminal event in the physical development of the College in the early 1960's was the completion of additional wings on the Animal Science Building in 1964. Under various stages of construction since 1955, the new facility was a welcome addition to the College and Division of Agriculture. The main wing housed the meats and dairy products technology laboratories as well as offices and an auditorium. The second wing, projecting from the south part of the main building, was home to offices for department heads and other faculty as well as additional office space for secretaries. This wing also contained a conference room and two classrooms. The third wing was the veterinary science section of the new building. This section included research and student labs for ongoing studies in physiology, reproduction, histology, virology and parasitology. In conjunction with the new building, the University also built a large animal laboratory on the University farm for convenience in handling and transportation of the animals kept on the farm. As reported at the time, the new facilities opened a new era in teaching animal science at the College.[13]

Coinciding with the opening of the new building there was also "a change of command" in the Animal Science Department. Warren Gifford, head of the department since 1939, decided that the moment had come to retire. In his long career Gifford received many honors for his research and teaching activities in animal breeding and his promotion of the Southern livestock industry. He had taught the undergraduate course in animal breeding required for all animal, dairy and poultry husbandry majors since first coming to the College in 1939. One of his many notable achievements included organizing and directing the Animal Industry Club, one of the earliest departmental organizations in the College.

Gifford's successor, E.L. Stephenson, came to the College in 1949. Stephenson was well known for his research in poultry nutrition and swine feeding as well as his courses in feeds and feeding and animal nutrition. At the time of his appointment, Stephenson counted 22 of his former graduate students that had gone on to earn their Ph.D.'s, and 30 other students had their master's degrees.[14]

As a new era was dawning in animal science at the College, a long-lived symbol of the first beginnings of the College made its exit. After the new Animal Science Complex opened in the summer of 1964, the Dairy Building, which had been in continuous use since it first opened in 1908, was razed to make way for an olympic-sized swimming pool. Sixteen animal science faculty vacated the building in August for their new quarters across the street. The venerable old stone building did not surrender without a fight, however. It was

Warren Gifford, E.L. Stephenson.

estimated that it cost more than the original price of the building, $5,000, to tear it down.[15]

The new Animal Science Building symbolized the beginning of an era that would see tremendous changes in animal agriculture over the next several decades. Dairy production would diminish, at least in terms of the number of dairy farms. Poultry production would increase beyond the wildest dreams of most people. Beef cattle production would become more entwined with the poultry business, and swine production would become an increasingly integrated operation. E.L. Stephenson served as department head from 1964 to 1981. He was followed by Wayne Kellogg, Park Waldroup and Paul Noland during the 1980s. During the 1960s and 80s the following professors were located on the campus and were prominently involved in teaching the classes, carrying out research that was vitally needed by the dynamic animal industries and extending it to the industry: O.T. Stallcup, Bernie Daniels, Wayne Kellogg, Jerry Rakes in dairy research with Ralph Williams in teaching and extension; Maurice Ray in beef cattle production with some focus on forage utilization; Connell Brown in beef cattle breeding followed by Hayden Brown in the same field; Ed Piper in reproductive physiology; A.L.Goetsch, David Kreider, Lloyd Patterson, Jack Perkins, Rick Rorie and Thomas Yazwinski in teaching and/or research; veterinarians Joe Beasley, John Brown, Howard Peterson and Kirk Skeeles in animal diseases; Chester Meinecke in Extension and teaching; and Paul Lewis and Marshall Heck in the meat processing area. Those working in poultry included David Andrews, Nicholas Anthony, Walter Bottje, Richard Forsythe, Tommy Goodwin, Roy Gyles, Grover Harris, Talmadge Nelson, Michael Slavik, Amy Waldroup, Park Waldroup and Lionel Barton, Extension Specialist. Paul Noland worked in swine nutrition and production practices.[16]

One other symbol of the passing of an era was the retirement of M.C. Kik in 1964. Kik contributed enormously to the College's reputation as a first-class research institution during his 42 years of service. His studies on rice and nutrition and his most recent research on formulating "fish flour" to mix with rice had vital implications for improving the diets of people around the world. After Kik's retirement, the College discontinued the Agricultural Chemistry Department and merged the faculty with that of Home Economics, yet another symbol of the passing of an era. The growing complexity of the agricultural industry and agricultural studies that branched into all scientific fields meant that agriculture students' knowledge of chemistry required several courses of chemistry.[17]

While student activities in the College during the early 1960's took on a new look during this period, much remained the same. The changes in the 1960 Agri Day did not signal an "all business" approach to social life in the College by any means. College social affairs were known for their friendly and informal air, whether the people attending were dressed in blue jeans or evening gowns. There was one additional change in the Agri Day schedule. In 1964 the banquet was held on Friday night in order to avoid crowding too many events into one day. An additional advantage was that the Queen selected on Friday evening would have the entire Saturday to reign over the day's activities. Saturday's schedule still included the Queen's Milking Contest, a barbeque lunch at the Agri Park, a tug-of-war with the engineers, a rodeo, and finally, to cap off the day, the Agri Day Dance.[18]

New departmental clubs formed during the early 1960's and the established organizations carried out new activities that enhanced the image of the College. The Agri Economics club was reformed in the fall of 1962, and the Horticulture Club was officially organized. Other established clubs, the Agronomy and Entomology Clubs in particular, continued holding meetings, hosting guest speakers and planning various projects. Home economics students were busy with their organization, which was renamed the AHEA in 1964 after becoming a part of the American Home Economics Association as an affiliate chapter. The Animal Industry Club, the oldest continuing departmental club in the College, was active at this time planning and sponsoring new activities. Members of the Animal Industry Club organized a new event in the College with a barbeque held at Agri Park in November 1962. Kirk Hale, the resident barbeque master, concocted his famous sauce for the event and supervised the cooking of the chicken.[19]

> *The college experience, in general, gave me the tools that I need to succeed. Most of the academic work is geared toward enabling the individual to "learn how to learn." My participation in the activities of organizations like Alpha Zeta, AHESA and the Ag Mech Club enhanced my ability to interact with others while in the pursuit of a common goal.*
>
> *Dennis Lichti, BSA 1987*

Members of the Alpha Zeta honors fraternity were also active in College activities at this time. In addition to running a non-commission used textbook service for the students in the College, sponsoring a low-cost loan program and providing scholarships, the members also began a recruiting program for the College in April 1963. In order to boost lagging enrollment in the College, Alpha Zeta members fanned out to high schools across the state, spreading the word about the new directions in agricultural education and the widening opportunities for careers in agricultural industries. Their message was clear enough: agriculture was no longer confined to farm production but involved the whole process of food and fiber production from field to table. There was a niche in agricultural industry for every interest. Advanced education in everything from business to biology and from economics to engineering (with the humanities fitted in for good measure) was needed in the emerging fields of agriculture.[20]

The program at the College was expanding in the early 1960's to accommodate the needs of the new agriculture. Through the combined efforts of Dean Hudson, the faculty and students, significant inroads were made. However, a period of sustained and, some would say, explosive growth was still on the horizon as the College branched out in many directions with new programs and a more aggressive approach to student recruitment and retention in late 1960's and 1970's.

Glenn Hardy - Dean of the College 1965-1987, longest tenure of any Dean to date. He is shown here at the dedication of the Infant Development Center, April 1987.

The chief architect of this period of growth and change, Glenn W. Hardy, was appointed Dean of the College in August 1965 and would serve in the post until 1987, the longest tenure of any Dean of the College to date. Hardy had come to Arkansas in 1956 as an assistant professor of agronomy. Originally from Baxter Springs, Kansas, Hardy earned his undergraduate A.B. degree in chemistry from Pittsburg State University in Pittsburg, Kansas. After serving in the Navy during WWII, he went on to get his M.S. in secondary school administration from Pittsburg and his Ph.D. from Kansas State University in Manhattan, where he majored in soils. After coming to the the University of Arkansas in 1956, Hardy began research in fertility problems with cotton and soybean, taught the beginning soils class and sponsored and revitalized the Agronomy Club. In 1962 he was tapped to supervise the State Soils Testing Program, which he coordinated with the help of Leslie Hileman in Fayetteville and Richard Maples and Joe Keogh at the eastern regional soils testing laboratory on the Marianna Branch Experiment Station. Before his appointment as Dean, Hardy was involved with relocating the western regional soils testing laboratory at the University farm. The laboratory had been housed in Gray Hall until it was razed to make way for Mullins Library.[21]

Dean Hardy set to work immediately after his appointment developing approaches that would both increase student enrollment and keep students in the College once they were there. The recruitment program was strengthened in a number of ways. Dean Hardy and Billy B. Bryan worked together with other faculty members in devising an intensive recruiting program comparable to the one that Bryan had used with such success previously. The departmental secretaries typed and mailed 400 individual letters to prospective students, not an easy task in years before word processors, photocopiers and desktop computers. Letters were also sent seeking invitations for speaking engagements at high schools, junior colleges, agricultural organizations and civic groups. Hardy, Bryan and other faculty members spoke to science classes, vocational education classes, FFA and FHA groups, commencement exercises, career days and banquets across the state. In the spring of 1966, Hardy and Bryan organized 13 regional meetings with school administrators, teachers of vocational agriculture and vocational home economics, counselors and science and math teachers.[22]

In the meetings, attended by over 800 people, discussion centered around the nature of the changes taking place in agricultural industries and why there was a critical need for more agricultural experts trained in a variety of fields that most people had previously thought were not related to agriculture. For the 1966 Agri Day celebration, 800 FFA members and their teachers were invited to share in the festivities. Private donors provided the extra food and refreshments and promotional literature needed for the day. Alpha Zeta members and faculty volunteered to guide the visitors through the day and provide information about the College's program. As an adjunct to the recruiting efforts, the Dean's office put together a student guide that combined all the information needed to apply for admission, seek financial aid and learn about the academic opportunities available at the College.[23]

> *In comparing my training and course work at UofA with other universities (I was Department Head and hired many teachers over the years), the University of Arkansas was far more demanding in requirements....Sciences required at UofA far outnumbered those at Florida State, University of Georgia, Nebraska, etc.*
>
> *Dr. Patricia M. Wallace, BSHE 1961*

The College's recruitment activities stepped up even more in following years. In 1968-69, 43 faculty members visited 131 high schools around the state. The Dean arranged seven workshops with Extension Service personnel across Arkansas and provided them with information on admission requirements, housing, financial aid and the College's curriculum that would allow them to assist interested students. In addition to entertaining FFA members, the College also played host to over 100 visiting high school seniors and juniors in the spring of 1969. The recruiting efforts were continually upgraded to take advantage of new communications technologies during the 1970's as the College reached out to an increasingly diverse clientele.[24]

Combined with the recruitment program, efforts were made to strengthen the student advising system and freshmen orientation procedures already in place. In 1966 the freshmen orientation class that had previously been voluntary became a requirement for all incoming freshmen. Since this was Dean Hardy's suggestion to the faculty, he taught the class from 1965 to 1976.

Other measures undertaken to reduce attrition included assigning some of the best teachers to teach the freshman classes, a measure that had appeal for both students and professors. Remedial classes were set up for students with deficiencies in key areas, such as English, math or chemistry. While there were objections to remedial classes among some of the faculty and a few students, most people agreed that the courses were helpful to many students.[25]

The College's Scholarship Committee worked tirelessly during this period to find new sources for funding agricultural education in all fields and to make students aware of the possibilities available. Seventy-one students shared $20,000 worth of scholarship money in the 1965-66 school year alone, and the total amount of funds available continued to increase throughout the decade, into the 1970's and beyond. Besides scholarships, the various departments also provided part-time work-study jobs. It was estimated that one-third of the students in the College held work-study jobs, ranging from bailing hay to working in sophisticated laboratories to duties in various offices. To further student involvement in the workings of the College, in 1969 the faculty voted to include student representatives on all of the faculty committees except the scholarship committee. Through these efforts of the Dean and faculty, the University of Arkansas College of Agriculture and Home Economics soon became known as a place where students mattered and where the faculty cared about the student's welfare. That this was true was evident in the drop in the attrition rate of incoming freshmen and the dramatic and sustained increase in enrollment during the last half of the 1960's and into the 1970's. In the 1964-65 year, the College lost 54 percent of its freshmen by the end of their sophomore year; by the 1968-9 year the figure dropped to 34 percent. In the 1965 term the total enrollment in the College stood at 507; by 1969 the numbers increased to 734.[26]

With an increasing number of students needing course work in a widening number of fields, the Dean, the department heads and the faculty curriculum committee came together to draw up more interdepartmental courses and programs and plans that shared work in other colleges of the University.

In home economics several new programs emerged in the late 1960's. A new curriculum was adopted in vocational home economics so that students could specialize in two teaching areas through a careful use of electives. There was also a new program leading to a major in food and nutrition that followed the guidelines set down by the American Dietetics Association. The program was designed so that a graduate could enter directly into a dietetics internship without taking additional courses. Plans were laid during this period to develop majors in fashion design and fashion merchandising to replace the clothing and textiles majors. Consideration of a joint program with the College of Education in child development was also in the planning stage. However, these programs would not come to fruition until the 1970's.[27]

Expanding the mission of the land-grant college of agriculture called for publicizing the changes taking place in agriculture and home economics and what the changes meant for the general public. To answer this need, the 1968 College curriculum committee formulated a major in agricultural journalism. The program drew on course work in many of the departments in the College as well as classes in the College of Arts and Sciences.

The College initiated a major development in 1967 that addressed the needs of the changing agricultural industry. An interdepartmental food science major was created that embraced studies in the Departments of Horticulture and Forestry, Animal Science, Agricultural Engineering, Agricultural Economics and Rural Sociology and, later, Home Economics. The following year the College created the Department of Horticultural Food Science, headed by A.A. Kattan, a nationally recognized leader in the food science field. Horticulture and Forestry remained as a separate department under leadership of George Bradley and concentrated on plant breeding and cultural practices, areas in which the faculty was noted for its many accomplishments. The new Department of Horticultural Food Science was more strictly confined to pre- and post-harvest physiology, harvest technology and preservation and packaging of fruits and vegetables. The department was formed around the staff at the Main Experiment Station Physiology and Processing Laboratory, which had worked in the area of raw product physiology since the late 1950's. Establishing processing plants for agricultural goods had been a major part of the efforts to industrialize Arkansas since that time as well.[28]

The Experiment Station's research scientists were busy during this time breeding plants with characteristics adaptive to mechanical harvesting, devising machinery for harvesting and packaging foods and experimenting with methods to preserve the quality and taste of processed and preserved foods. The work of A.A. Kattan, Bill Sistrunk, Bob Davis, Roy Sharp, Justin Morris, Daniel Tompkins, Ron Buescher, Michael Johnson, Harry Snyder, Alfredo Gonzalez and Phil Crandall (who served as Department Head after Kattan retired), to name only a few, was particularly important in these developments. These professors worked closely with faculty members in the Department of Horticulture and Forestry where plant breeding and cultural practices were the work of several renowned scientists, including Joe McFerran, George Bradley, Roy Rom, John Bowers and Jim N. Moore. Later Teddy Morelock and F.S. Huang added critical expertise to the program.[29]

A.A. Kattan

These efforts all contributed to the growth of the food processing industry. Consequently, there was a growing need for trained experts in all

Horticultural Food Science Building.

phases of the food industry, and industry groups gave substantial support to the program at the College. With a grant from the Ozark Freezers and Canners Association, a building at the University farm was enlarged to 14,000 square feet and equipped with laboratories, classrooms and office space to house the Department of Horticultural Food Science. The industry also made scholarships available for students and liberal grants to fund research in the area. Food science soon became one of the most popular majors in the College.[30]

Under the leadership of Billy Bryan, the faculty in Agricultural Engineering developed a new major in agricultural mechanization in 1965. The major was intended to prepare students for careers in sales, management and service of the wide variety of increasingly sophisticated agricultural technology. In addition, the curriculum offered many other options that included work in other departments of the College and other divisions in the University. Combining the agricultural mechanization major with an agricultural business option was popular and led to careers in marketing and sales of agricultural technology.

Agricultural Engineering also formed an integral part of the Food Science program, especially as it applied to harvesting technology, processing and packaging. While work in farm structures and field terracing and drainage still continued, the agricultural engineer in this era was just as likely to be working on setting up an analog computer, designing a sophisticated mechanical harvesting system or helping with interior design problems in home economics. To provide space for the new programs, a new 11,000-square-foot building was built at the University farm. While some of the more sophisticated instrumentation and laboratory equipment remained on campus, the bulk of the research was conducted in the new $90,000 facility. The department was further aided in improving its program with two National Science Foundation grants totaling $20,000 to buy undergraduate teaching equipment.[31]

Another important interdepartmental program initiated in 1967 included a major in agricultural pesticides. The name of this major was later changed to plant protection and still later to plant protection/pest management to better reflect the complex nature of this major. The Departments of Agricultural Engineering, Agronomy, Horticulture and Forestry, Entomology and Plant Pathology were all involved in this effort. After completing the requirements of the program, a student was qualified to enter graduate-level studies in any of the last four departments. The chemical pest control industry generously supported the new program and offered scholarships for students and funding for research as well as jobs for graduates from the program.[32]

While chemical pest control had become a dominant feature in agriculture since the end of WWII, the traditional focus for pest control in the Entomology Department emphasized biological and ecological methods. This was true even before scientists and consumer groups became concerned about the environmental impact of pesticide use. Public concern raised the profile of the work in biological control, and grants became more plentiful, helping accelerate work in some areas. By the end of the 1960's, George Templeton in plant pathology and Roy J. Smith, a USDA research agronomist, were also doing pioneering research in biological control, but their work concerned the control of weeds rather than insects. Their work and the recognition paid the entomology program ensured that the University of Arkansas College of Agriculture and Home Economics was in the forefront in developing alternative pest control methods. Templeton was invited to lecture in many foreign countries as a result of his work.[33]

In 1967, under the leadership of the Director of the Experiment Station, E.M. Cralley, and department heads Charles Lincoln (Entomology) and Derald Slack (Plant Pathology), a Virology and Biological Control Laboratory was built on the Main Experiment Station at Fayetteville. This lab was to become of tremendous importance to the efforts of entomologists and plant pathologists in the field. The lab was equipped with an electron microscope, an expensive but necessary tool for viral research. K.S. Kim was in charge of the electron microscope, but scientists throughout the university used this laboratory since it was the only electron microscope on campus. This facility and Kim became world renowned for their work in identification of plant diseases. Agricultural scientists who made important use of the microscope facility from the beginning included Bill Yearian, J.R. Phillips and Seth Young from entomology and Howard Scott, Joe Fulton and, later, Rose Gergerich from plant pathology.[34]

K.S. Kim and the electron microscope, obtained by the Department of Plant Pathology in 1967. The facilty became world renowned for work in identification of plant diseases.

Research on parasites, pathogens, predators and methods using plant substances to control weeds and harmful insect populations was carried out at the lab. J.R. Phillips in entomology developed an interesting approach to controlling the boll weevil. Phillips was searching for a component in the weevil's diet that could prevent the onset of the insect's hibernation cycle. Without the physiological protection gained from hibernation, Phillips reasoned that weevils would die from the cold or starvation during the winter, thus obviating the need for chemical controls. Scientists at the lab even called up the services of a parasitic variety of the nefarious nematode to battle the peach tree borer, a serious pest in orchards. Through working at the lab as assistants, doing graduate research, or merely observing the processes at work there, the students at the College gained valuable insight in the many dimensions of agriculture and were able to see first-hand, "cutting edge" research in a field that would gain greater prominence in the years ahead.[35]

In addition to drawing up new curricula and interdepartmental approaches to instruction, the Dean and faculty also created pre-professional programs in veterinary medicine, forestry and law. The pre-veterinary medicine program took a minimum of two years to complete. Veterinarians Joe Beasley, Howard Peterson, John Brown and Kirk Skeeles often served as advisers for these students. After completing the course of study, if the student was accepted by certain out-of-state veterinary schools, the state of Arkansas paid the non-resident part of the tuition for the student. If for some reason the student was not accepted into veterinary school, the course work taken could be counted toward a degree in animal science. Under the plan of subsidized tuition, the spaces for entering Arkansas freshmen in colleges of veterinary in other states grew from 6 to 27. In time a shortage of veterinarians in Arkansas was overcome.[36]

A pre-forestry curriculum was also offered as an option at this time. Like the veterinary medicine program, the pre-forestry plan called for a minimum of two years of study. College advisors took special care to tailor the course work to the requirements of particular forestry schools. In this way, the student could enter directly into professional studies. H.A. Holt, a forestry ecologist, was brought into the Horticulture and Forestry Department to coordinate the program in 1970. Laurin Wheeler, forest ecologist, later coordinated the program. Brad Murphy, plant physiologist, and Fred Stephen, entomologist, contributed to this program with research headquartered at the Forest Resource Center at Monticello. The program was discontinued in 1997.[37]

Though no specific program existed for agricultural law in the College or the Law School at this time, agricultural economics was particularly suited for providing a background for legal studies. The growing sophistication of commercialized agriculture meant that farmers and agribusiness professionals needed legal advice in a number of areas, including contracts, property issues, landlord-tenant relations, corporate law and labor relations. By using the business option in the College curriculum and concentrating in agricultural economics, a student could be well prepared to enter legal studies.[38]

Other innovative approaches to undergraduate studies at the College included a program in the Plant Pathology Department funded by the National Science Foundation that allowed upper-level undergraduates to participate in the department's research projects. George Templeton coordinated the efforts in this area. Students were recruited from around the state, especially from schools that lacked advanced research facilities, and spent 12 weeks in the summer working with a plant pathologist on a specific project. Many of the students went on to do graduate work at the College, and 18 of the 25 students went on to work on their doctorates. National Science Foundation grants aided in improving undergraduate programs in other departments as well. The Departments of Agricultural Engineering, Agronomy, Horticulture and Forestry and Entomology all benefitted in some way from the grants.[39]

As the 1960's ended, the people in the College had reason to be proud of their many accomplishments. The enrollment had nearly doubled, student attrition rates were cut, all departments had active graduate studies, and Ph.D. programs were initiated in animal science and agronomy. Many innovative programs were operating at the interdepartmental and inter-university levels that answered the needs of the new agricultural industries and the public at large. New facilities equipped with the latest technology and staffed by recognized experts in a variety of fields put the College in the forefront of creative research and teaching efforts in the land-grant system. Yet there was still more work to be done as the pace of technological change in agriculture and food science proceeded at a dizzying rate in the years ahead.

Enrollment 1970-79.

Year	1970	1971	1972	1973	1974	1975	1976	1977	1978	1979
Agriculture	449	482	488	497	526	711	777	829	878	995
Home Economics	311	333	315	305	328	344	343	389	382	383
Total	760	815	803	802	854	1055	1120	1218	1260	1378

Fall enrollment figures provided by Dean's Office, DBCAFLS.

As the 1970's opened, the academic programs in the College continued to expand, both in the variety and number of courses offered and in student enrollment. Master's degree programs in all departments and new doctoral programs initiated during this time increased the graduate enrollment in the College, which rose dramatically between 1970 and 1979. The College developed more innovative interdisciplinary programs, pressed ahead with student recruiting and strove for teaching and research excellence, all with a great deal of success. Building on the foundations laid down in the 1960's, Dean Hardy and the faculty and students continued to adapt the College to the expanded land-grant mission.

Just as the land-grant mission continued to expand in the 1970's, so did the traditional Agri Day celebrations. In 1970 the celebration became known as Agri Week and began on Thursday night with the awards banquet and the crowning of the Agri Queen. On Saturday the departmental exhibits returned in the form of an open house at the various departments. A new event was a touch football tournament pitting the departmental clubs against each other. The traditional picnic followed the tournament, and the tug-of-war with the engineers and the rodeo filled the rest of the afternoon.[40]

A visitor to the open house displays in the Entomology Department would have been treated to one of the finest arthropod collections in the South. The University of Arkansas Arthropod Museum included 331,157 specimens in 1970. The collection began as far back as 1887 and included 170,000 specimens from the United States and 143,000 specimens from around the world, mostly from Central and South America, no doubt collected by W.J. Baerg, Dwight Isely and Charles Lincoln during their many trips in that part of the world. R.T. Allen, an insect taxonomist from the University of Illinois, came to Arkansas in 1967 to oversee the collection, which attracted professional and scholarly attention across the nation. Such an assortment, perhaps a curiosity for the casual observer, was a valuable educational tool for advanced entomological studies.[41]

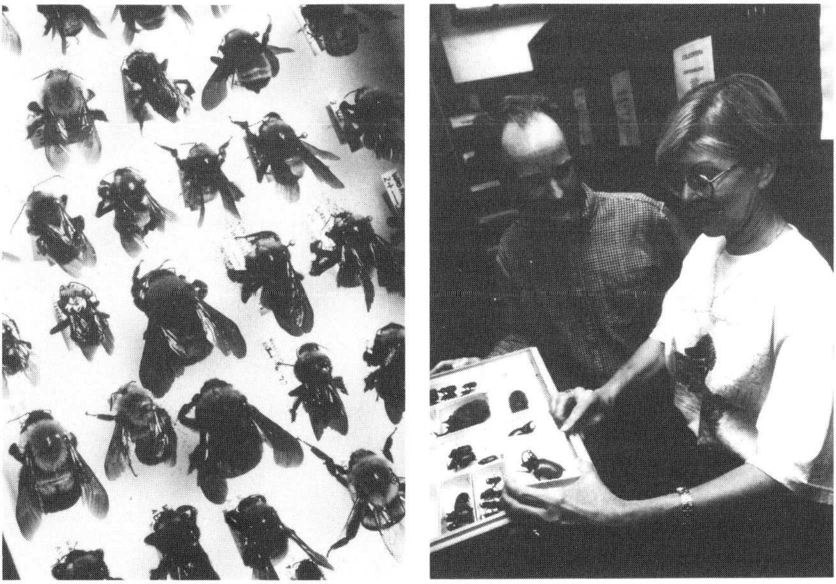

Arthropod museum - On the left, specimens from the collection. On the right, Jim Whitfield, curator of the University of Arkansas Arthropod Museum, and graduate student Carolyn Lewis examine beetle specimens in the museum's collection. The museum has grown from about 38,000 specimens representing about 4,000 species in 1964 to 750,000 specimens and 25,000 species. (Photos by Fred Miller)

The increasing complexity of agricultural production and the industries involved in all facets of production, processing and marketing of food and fiber goods made the need for highly trained agricultural scientists and professionals more critical than ever. Consequently, developing doctoral programs became a major focus in fulfilling the College's expanded mission during the 1970's. Drawing on the strength of the faculty and the availability of such research tools as were found in the Arthropod Museum, two new doctoral programs appeared in 1970, one in entomology and another in plant pathology. While the faculty in these departments had held the professional credentials necessary to supervise doctoral work for many years, the accreditation of the programs capped two years of planning, curriculum adjustment and hiring. Entomology had 10 Ph.D.'s on the faculty in 1970. Their average tenure was 11.2 years with six of them at the rank of professor. Four of the professors in the Entomology Department, F.D. Miner, the head of the department, W.D. Wylie, Charles Lincoln and J.L. Lancaster, all had 20 or more years of experience in research and teaching.

> *...the teacher training and lab instructional methods I learned in Home Ec made me a great science and technology teacher! I always tell people that Home Ec teachers make the best lab instructors because they can easily manage to do 10 things at once!*
>
> *Teresa J. Franklin, BSHE 1977*

While the department had historically emphasized biological and environmental pest control in Arkansas, with the increased interest in reducing pesticide use nationwide, more national and international attention was paid to the department's research. The doctoral program grew slowly, as the faculty intended it to, so as to not overreach their resources and weaken the master's program already in place. By 1975 there were seven Ph.D. candidates in the program as well as one more associate professor and one more assistant professor. After getting the doctoral program approved, the department was then well situated to contribute to other interdisciplinary programs in the College and at the University level.[42]

The Plant Pathology Department had also shown a high level of professional and scholarly competence over the years. Of the 14 faculty members on the staff in 1970, 11 ranked as full professors with an average of 11.8 years of teaching and research experience. In addition to Templeton's work with biological herbicides, M.J. Goode was widely recognized for his work in developing disease-resistant vegetable varieties. D.A. Slack, the head of the department, was known for his studies involving fungicidal control of plant diseases. J.P. Fulton, R.D. Riggs, J.L. Dale and J.M. McGuire all contributed to research confirming nematodal transmission of viral plant diseases. The plant pathology program, while not large, was an important one because its research and teaching were relevant in many other departments in the College and the University and, therefore, a key component in the College's expanding mission.[43]

Taking the movement toward more interdisciplinary programs seriously, the College's curriculum committee, the Dean's office and the various departments involved began planning two more doctoral programs in the mid-1970's that could successfully serve multiple departments. The food science Ph.D. program brought horticultural food science studies together with the animal science studies in meat, dairy and poultry processing. Home economics joined in the program, drawing on their foods and nutrition curriculum. The resulting food science Ph.D. program was implemented in 1978.

A plant science doctoral program joined the recently developed plant pathology doctoral studies with horticulture and forestry to produce a hybrid Ph.D. in which the student could choose to emphasize one field or the other. The graduate advisory committees for the plant science program included faculty from both departments. The committee required a minimum of six hours of course work in each department along with six hours in botany in addition to course work recommended by each student's committee. Students were required to enroll in an interdepartmental colloquium each semester. The chairperson of the colloquium rotated between departments each semester and within the departments every year in order to give the students exposure to different faculty areas of expertise. The plan at this time was to phase out the plant pathology Ph.D. program by 1980 and graduate future doctoral students in plant science. The undergraduates from the plant pathology program would be in the plant protection/pest management program.[44]

Undergraduate curriculum development in the College also benefitted from interdisciplinary approaches. An interdisciplinary program that had particular relevance with the rising concern for environmental issues in the 1970's was the environmental science program. This undergraduate program was an intercollegiate effort involving the College of Agriculture and Home Economics, the College of Arts and Sciences and the College of Engineering. Some of the courses in the program were team-taught by professors from different areas, and some courses rotated among colleges in alternate years. Each college contributed course work, with the Departments of Agronomy, Entomology and Plant Pathology contributing the principal course work for the College of Agriculture and Home Economics. Students could take the courses in, and graduate from, either the College of Agriculture and Home Economics or the College of Arts and Sciences with a B.S. in environmental science. However, the Engineering College already had its own environmental science program, and though they provided some classes for the interdisciplinary program, a degree in engineering environmental science had requirements separate from those for the interdisciplinary program.[45]

South entrance, Agriculture Building.

The Departments of Entomology and Plant Pathology, along with the weed science faculty in the Agronomy Department, put together an undergraduate program in the early 1970's that replaced the pesticides major. The program was first named plant protection in the early 1970's and later renamed plant protection/pest management in 1977. A student could choose to concentrate on any of the three areas that supported the core curriculum of the program. Because the College was not generating enough graduates from any one of the three areas to meet the requirements of the State Board of Education, the plan was to combine the course work from the three departments into a common curriculum and thereby increase class sizes as well as provide more interaction among the three disciplines. Combining similar studies that had emerged from the traditionally separate disciplines followed a trend that was evident in other land-grant institutions around the country. Most of the disciplines were categorized and separated around the turn of the century as agricultural education experts were trying to bring a degree of sophistication and professionalism to the field. However, much had changed since that time, and those categories had little relevance in the "systems" approach to agricultural production, processing and marketing that gained prominence during this period.[46]

Rapid growth of the urban areas in Arkansas after 1960 created a need for more horticulturalists trained in the production, merchandising and use of ornamental trees, shrubs, lawn grasses and flowers. More city dwellers also meant more golf courses, parks, ballfields and rural areas converted to recreational uses. The Department of Horticulture and Forestry answered the need through developing courses in ornamental horticulture and floriculture in 1970-71 at both the undergraduate and master's degree levels. The Agronomy Department also developed course work in turfgrass culture under J.W. King in response to the call for more experts in lawn, ballfield and golf course care. A.E. Einert was brought in to coordinate the ornamental horticulture and floriculture program. Besides teaching the basic principles and practices in producing commercial flower crops and ornamental shrubs and trees, Einert also instructed students in basic landscaping techniques, as well as nursery and greenhouse management. By 1975, Einert had developed, and the College curriculum committee had approved, a new undergraduate major in landscape design and urban horticulture. The new major became very popular, and many of its graduates went into landscaping and nursery businesses themselves or found jobs with municipal parks and recreation departments. Extension agents stationed in urban areas found the program valuable as well. Gerald Klingman received a half-time teaching/extension appointment in order to provide this valuable service in northwestern Arkansas. The successful expansion of programs in ornamental horticulture and home landscaping was further proof of the College's ability to adapt to the new mission of the land-grant college of agriculture.[47]

During the 1970's the Home Economics faculty and curriculum continued to be revised and strengthened. In this department, perhaps more than any other in the College, one could see the transformation of the mission of the College in expanding to include a wider clientele. The traditional focus on home-based production and domestic enterprise was almost completely replaced with a program that emphasized career development for women outside the home and consumer education. The child development and family relations program was renamed human development and family studies to reflect broader concepts being taught in these classes and the research-oriented work being done in the field of family relations and child development. After years of frustration in being unable to recruit Ph.D.'s for the department, four were appointed in the early 1970's, raising the number to six, and by the end of the decade there were nine Ph.D.'s on a staff of 17. One of these new Ph.D.'s was Dorothy A. Larery, who replaced Evelyn Cox as the department head in 1971.[48]

Under the leadership of Larery in the early 1970s, the department developed several new curricula that included more classes in fashion merchandising and design, interior design and housing, and child development and family relations. The fashion merchandising major taken in conjunction with the business option soon became the most popular program in the department. By the end of the decade it overtook home economics education, traditionally the major with the most students, as the program with the highest enrollment. Mary Cotton was the architect of this program. Her efforts were augmented later by the addition of Mary Warnock and Charlene Mooty to the faculty. Interior design and housing became the second most popular major during the same period, followed closely by child development and family relations. Cora McKown took the lead in putting together the first steps toward developing

Dorothy Larery

a major in interior design and taught the first courses. There were further changes in the curriculum in 1977 and 1978. The foods and nutrition major in home economics was revised so home economics students could participate in the interdepartmental food science program at both the undergraduate and graduate levels. This involved the faculty in foods and nutrition, including Bethel Cunningham, Lorene Raymond, Ruth Griffith and Catherine Carroll, Harriett McCoy and Mary Alice Kenney.[49]

There was a concerted effort during this time to encourage more research in all fields in the department. In 1975 several of the nine-month teaching appointments were changed to twelve-month positions in order that more research could be done in child development, interior design, housing, textiles and clothing. Carroll continued with her nutrition studies begun in the early 1960's on the correlation between fat and carbohydrate metabolism. Mitsuko Inano, brought in to teach and conduct research in nutrition and institutional management in 1971, studied the growth and food preferences of children of low-income mothers. Harriett McCoy conducted research in nutrition as related to pre-adolescent and adolescent girls that was keyed to economic level and ethnic background. Sue Martin and Mary Burton investigated the development and improvement of programs for day care facilities. Burton had been instrumental in developing the Headstart Program in Washington County during the early 1970's. Harkening back to the housing studies done in the 1930's by Deane Carter and others, Janet Navin researched the design and construction of low-income housing. Navin also looked into the attitudes and economic constraints that influenced low-income families to either accept or reject alternative housing options. In textiles and clothing, Mary Warnock studied the durability and wear characteristics of many new fabric blends. As in other departments, the research done in the home economics labs and field studies had a direct effect on classroom instruction. The typical teacher/researcher shared the results of the latest research with students in the classroom and laboratories, and in this way instruction was kept current and relevant.[50]

> *I'll never forget Dr. Sistrunk cooking possum for one of our get-togethers. I still use that example when I talk to my students about cultural diversity.*
>
> *Cynthia Wilks, MSRD, CFCS, BSA 1970, MS 1972*

In addition to the changes taking place in home economics studies during the 1970's that reflected the changing role of women in society, such changes were also seen in the College at large. Many more women began enrolling in course work in the College outside of the Home Economics Department. An estimated 60 women worked on majors in pre-veterinary medicine, animal science, ornamentals and floriculture and food science in 1973, and the numbers increased throughout the decade. In 1975 the Experiment Station hired

Susan Murphy as a forestry researcher in the Horticulture and Forestry Department. Murphy was the first woman on the faculty in any department other than Home Economics.[51]

In 1977 the learning resources unit of the Division of Student Affairs offered to provide a student mentor to each undergraduate college to enhance the advising system of the college. The Dean of each college was responsible for recommending a person for this position and, to a large extent, defining the role of the mentor in the college. Dr. Elizabeth 'Libby' Wheeler was hired in the College of Agriculture and Home Economics. Her role was to work directly with problems encountered by students, to review class schedules, to inform students of opportunities for remedial assistance and, in general, to look after the welfare of the student. She worked in the summer with freshman orientation programs and eventually taught the freshman orientation course for the College.[52]

Under the leadership of Henry F. Meenen, head of the Department of Agricultural Economics and Rural Sociology, the department worked steadily toward fulfilling the needs of the expanded mission of the college. The curriculum and course work in the department evolved during the 1970's to fit the needs of students seeking jobs in commercial agriculture and agribusiness industries. Research and course work also addressed both short-term and long-term economic and social issues emerging in the 1970's. Reflecting the tendency to view farming as a business and the robust nature of the agricultural economy during the 1970's, agribusiness became a very popular program in the College during this time. The program grew from 80 students in the early years of the decade to 240 by 1979. In addition, there were 15 agribusiness graduate students in 1975, and the numbers continued to grow throughout the decade. Students in the program also took a number of courses in other colleges and departments, particularly in the College of Business Administration. Faculty who helped develop the agricultural business major included Calvin Berry, J.L. Charlton, Mark Cochran (who would become department head in the 1990's), Waymon Halbrook, Hillard Jackson,

> *I am very grateful for the opportunities I had at the UofA. Having spent 20 years as a faculty member at another university, I have called many times on my experiences as a student. Some of the activities enabled me to be a better faculty member and to help many students in the manner I was treated at the University of Arkansas....When I was fortunate enough to receive the most prestigious teaching award at the University of Missouri, I reflected that I was motived by the excellent teaching and advising I had received as a student.*
>
> *Dr. Ron Morrow, BSA 1968, MS 1972*

Bill Morrison, Carter Price, D.G. Lafferty, William Windham and James White (who later became Associate Dean of the Graduate School). Later this group was supplemented with additional faculty, including Gail Cramer (who was appointed to the L.C. Carter Chair for Marketing and Policy), Martin Redfern and Robert Shulstad (who became department head in 1982 following Henry Meenan). J.C. Headley became department head in 1989 after Shulstad left for another university. Others who joined the department in the 1970s and 1980s included Eric Wailes, Bruce Dixon, John Goodwin (after he left the Vice President's position), Ed Fryar and Luke Parsch. In rural sociology, Frank Farmer and Michael Miller joined Don Voth.[53]

Since there was no Ph.D. program in the department at this time, an agreement was worked out with the College of Business Administration to permit an agricultural economics professor to serve as the major professor and direct the dissertation of Ph.D. students in economics who had a particular interest in agriculture. Professors from both colleges would be on the graduate student's committee. Since the business option was a popular program in the College, quite a number of agriculture and home economics students from other departments were also in agribusiness and agricultural economics classes. With exports of American agricultural commodities during the 1970's at an all-time high, there was also a great deal of interest in international marketing and commodity and futures trading. When the department offered classes in these areas, the classes quickly filled. A special night class offered in commodity marketing attracted 67 students from all walks of life. A farm management class doubled its enrollment from 40 to 80 students in the years between 1977-79. There were plans to develop a domestic and international agricultural marketing major, but the idea did not come to fruition until the 1980's.[54]

The rural sociology section of the department was busy at this time in answering the needs of rural people who were rapidly being pushed to the margins of contemporary society. Michael Miller and Don Voth researched and taught courses in development of rural areas, rural health care and how to assist rural people displaced by technical farming innovations.[55]

One other major curriculum and program innovation in the 1970's involved retooling the master's degree program in agricultural science, a program that had fallen into disuse. Dean Hardy and the faculty curriculum committee considered dropping the program altogether. However, further study of the issue revealed that there were many extension workers and possibly teachers of vocational agriculture out in the state who wanted to obtain a master's degree. Others were willing to take courses that would update their education. Most of those who were interested could not leave their jobs or their homes for long periods of time. So a decision was made to reinvent the program to better serve extension personnel, vocational agriculture teachers and others who needed or desired graduate courses. A faculty committee chaired by Ron Talbert administered the program, which was renamed General Agriculture. Master's degree-level classes were scheduled at locations across the state. The program proved to be very successful in reaching out to non-traditional student populations around the state and even appealed to other clientele such as bank employees, soil conservation personnel and farmers. This was a forerunner of distance learning that was to come later in the century.[56]

The highlight of the 1970's campaign to improve the physical plant at the College came in August 1975 when construction on the new Plant Science Building began. The first phase of the projected two-phase project was completed in June 1977. However, the second phase of the project never got off the ground as other priorities continued to take precedence over the years. The new building greatly relieved the overcrowded conditions in most departments and housed the Department of Horticulture and Forestry and portions of the Departments of Plant Pathology and Agronomy, including the offices of the department heads. Teaching facilities in the building included a 144-seat lecture hall equipped for multi-media and closed circuit television presentations. A teaching laboratory used primarily for horticultural instruction was well equipped and ready for use. There was a drafting room to be shared jointly by landscape design, interior design and housing and agricul-

The Plant Sciences Building, completed in 1977, was the "first phase" of a two-phase project, the second phase of which is still pending. The building houses the Departments of Agronomy, Plant Pathology and Horticulture. (Photo: Fred Miller)

tural engineering classes. An audio-visual tutorial laboratory was outfitted for the crops section of the Agronomy Department. The AV tutorial lab was based on a concept developed by Sam Offutt, of the Agronomy Department, in the early 1970's and for which he received many awards for teaching excellence. There was also a 30-seat conference room for the use of all departments in the College. The rest of the building devoted ample space to laboratories and offices for approximately 40 faculty and staff.[57]

There were a number of administrative changes affecting the College during the 1970's that were keyed to the College's expanding role. In 1974, the Vice President of the Division of Agriculture, John W. White, stepped down because of ill health after 43 years of service to the College and University. C.A. Vines, the veteran Director of the Extension Service, was named interim Vice-President until a suitable sucessor for White could be found. The Board of Trustees appointed James E. Martin, the Dean of Agriculture and Life Sciences at Virginia Polytechnical Institute and State University, as Vice President.[58]

James E. Martin, Vice President for Agriculture, 1975-1980.

In the fall of 1975 there was an adminstrative change in the College. Nolan Arthur was appointed Assistant Dean of the College to help with student affairs. Arthur came from Oklahoma State University where he received his doctorate in animal science and vocational agriculture.

By 1975 College enrollment had pushed past 1,000. Surpassed in numbers by only Home Economics, Animal Science was the second largest department in the College, both in terms of students and faculty. By 1979 the department had over 300 undergraduate and graduate students and 29 faculty. The department was a collection of many diverse disciplines covering breeding, production, nutrition, diseases, and other areas that were then divided into animal groups.[59]

In 1978, in an attempt to bring order to the department, E.L. Stephenson, the department head, reorganized the department into four sections: poultry, animal sciences, dairy and animal diseases. Section leaders reported to Stephenson, who was also the head of the diseases section. Stephenson initiated the change in hopes of achieving greater productivity and better communication between sections, with other departments, and with industry constituents.[60]

Concurrent with the Animal Science Department's reorganization, Charles Bishop, the President of the University, designated certain areas of the University for concentrated development. One of the areas of excellence selected was the poultry program. The College funneled more resources for both teaching and research to the department in anticipation of developing the poultry program. An additional $52,800 was requested in 1979 for construction of a poultry teaching laboratory to complement the teaching and research facilities already approved by the Legislature but awaiting construction. While the College had long played a key role in the development of the poultry industry in northwestern Arkansas, the expansion of the poultry program at the end of the 1970's and the designation of the program as "an area of excellence" anticipated even greater developments in the future.[61]

> *My original interest in research began at the UofA when I worked in Dr. Catherine Carroll's lab comparing the effect of various types and amounts of fat in the diet of rats on the body content of cholesterol and triglycerides. This also stimulated my interest in cardiology and how lipids and atherosclerotic disease are related. I feel my training and degree in nutrition have been an advantage in pursuing a career in medicine.*
>
> *Dr. Rita Frances Marr, BSHE 1978*

In preparing for the future, in the fall of 1979 the College also established a Media Instructional and Statistics Laboratory equipped with computer terminals. David Marx supervised the lab and taught statistics classes in order to improve the student's numeric and statistics capabilities. Several departments in the College, especially the Department of Agricultural Economics and Rural Sociology, revised their curricula and strengthened the statistics and econometric requirements at this time. In 1979 Dean Hardy cited computer literacy as a goal to work toward for all graduates from the College during the 1980's. Computers were also installed in the Animal Science Building for student use, and requests from many other departments were coming in for the same equipment. In an interesting experiment with the possibilities of multi-media learning, a series of special one hour "mini-courses" in plant pathology made their appearance in the media lab in 1979. The special courses, which relied heavily on the new audiovisual tutorial equipment, were intended to replace two three-hour courses that had been taught in the conventional manner.[62]

As the 1970's drew to a close, there were abundant signs that the University of Arkansas College of Agriculture and Home Economics had succeeded quite well in adapting to the challenge of its expanded mission. Enrollment had tripled over the course of two decades, and new facilities supported the latest developments in research and teaching. Interdisciplinary programs that drew on strengths of other departments in the College and throughout the University brought a new dimension to agricultural education. The College's programs, led by an excellent teaching and reseach faculty, in turn reached out and enriched the University and the larger community as well. New programs developed during this period addressed the needs of the College's expanded clientele and found new ways to speak to the requirements of the traditional constituents. If the period was one of challenge and change, the administrators, faculty and students of the College rose to the test and expanded the College's mission; in the process they laid a foundation for future progress.

Endnotes

1 Charles E. Kellogg and David C. Knapp, *The College of Agriculture: Science in the Public Interest* (New York: McGraw Hill, 1966), ix-x.

2 Ibid, 42-56.

3 *Arkansas Agriculturist* (May 1960), 6, 9.

4 Glenn W. Hardy interview (written response) with the author, August 8, 1998, Fayetteville, 22.

5 *Dean's Reports, College of Agriculture, 1960-64*, passim, University of Arkansas Archives, Records of the Chancellor' Library Special Collections, Fayetteville Arkansas. Hereafter cited as *Dean's Report* followed by year and page number.

6 *Dean's Report, 1960-61*, 2-3, 13-14.

7 Billy B. Bryan, "Most Influential People In the History of the BAEG Department," unpublished manuscript in the possession of Glenn W. Hardy, 1.

8 *Dean's Report, 1960-61*, 2-3, 13-14.

9 *Arkansas Agriculturist* (December 1960), 5, (October 1961), 8; *Arkansas Alumnus* (June 1963), 14-15.

10 *Arkansas Agriculturist* (April 1964), 4; *President's Report, 1963-64*, 13.

11 *Arkansas Agriculturist* (February 1969), 5, 10, (October 1961), 14, (June 1963), 6, (April 1964), 9, 18; *Hardy Interview*, 29-30.

12 *Dean's Report, 1961-62*, 17-18, *Hardy Interview*, 30.

13 *Arkansas Agriculturist* (April 1964), 13.

14 Ibid, 12, 18.

15 *Northwest Arkansas Times*, October 31, 1964, 12.

16 Glenn Hardy, written contribution to this manuscript

17 *Arkansas Agriculturist* (November 1963), 15, 17.

18 *Arkansas Agriculturist* (April 1964), 5.

19 *Arkansas Agriculturist* (December 1962), 10-11.

20 *Arkansas Agriculturist* (April 1963), 13.

21 *Hardy Interview*, 5.

22 Glenn Hardy, written contribution to this manuscript.

23 Ibid.

24 *Arkansas Agriculturist* (April 1966), 4, (January 1967), 4, 10; *Hardy Interview*, 3-4.

25 Glenn Hardy, written contribution to this manuscript.

26 *Arkansas Agriculturist* (January 1966), 9; *Hardy Interview*, 5-7.

27 *Dean's Report, 1966-67*, 5-6.

28 Glenn Hardy, written contribution to this manuscript.

[29] Ibid.
[30] *Hardy Interview*, 22-23; *Arkansas Farm Research* (September-October 1974), 10-11.
[31] University of Arkansas, *Institutional Self-Study, 1969-70* (Fayetteville: University of Arkansas, 1970), 179-180; *Arkansas Agriculturist* (January 1968), 6; Bryan, 2.
[32] *Institutional Self-Study, 1969-70,* 181.
[33] Glenn Hardy, written contribution to this manuscript.
[34] Ibid.
[35] *Arkansas Agriculturist* (April 1967), 4; *Self-Study 1969-70*, 186; *Arkansas Alumnus* (April 1970), 2-3.
[36] *Hardy Interview*, 7-8.
[37] Glenn Hardy, written contribution to this manuscript.
[38] *Arkansas Agriculturist* (April 1967), 4, (Fall 1968), 16, 18.
[39] *Arkansas Agriculturist* (May 1970), 7; *Self-Study 1969-70*, 196; D*ean's Report, 1968-69*, 6.
[40] *Arkansas Agriculturist* (May 1970), 13, 16.
[41] *Arkansas Agriculturist* (Jan. 1970), 7.
[42] University of Arkansas, *Institutional Self-Study, 1975*, (Fayetteville: University of Arkansas, 1975), 199-200.
[43] *Arkansas Agriculturist* (November 1963), 12-13; *Institutional Self-Study, 1975*, 211-213.
[44] *Dean's Report, 1977-78*, "Academic Programs," 2-3; *Hardy Interview*, 10-11.
[45] *Hardy Interview*, 14; *Institutional Self-Study, 1975*, 152.
[46] *Hardy Interview*, 13.
[47] *Arkansas Alumnus* (April 1987), 6-7; *Arkansas Agriculturist* (May 1970), 6, (Fall 1970), 8; Glenn W. Hardy, "Historical Information About the Activities of the Office of the Dean of Agriculture and Home Economics from August 1, 1965 to July 1, 1987," unpublished manuscript, 10.
[48] *Dean's Report, 1976-77*, 1-2.
[49] *Dean's Report, 1971*, 3, *1972-73*, 4-5, *1975-76,* 1-2, *1977-78*, "Academic Programs" 2-3; College of Agriculture and Home Economics, *Five Year Projections, 1975-80* (Fayetteville: College of Agriculture, 1975), 75-84.
[50] Dorothy A. Larery, *Special Report: Department of Home Economics* (Fayetteville: College of Agriculture, 1979), 9-10.
[51] *Dean's Report, 1972-73*, 2, *1975-76*, 10.
[52] Glenn Hardy, written contribution to this manuscript.
[53] Ibid.
[54] *Dean's Report, 1977-78*, 2, *1978-79*, 2,
[55] *Five Year Projections, 1975-80*, 43-46.
[56] *Hardy Interview,* 11; *Dean's Report, 1977-78*, 1.
[57] *Dean's Report, 1976-78*, 1.
[58] *Arkansas Alumnus* (April 1975), 8-9.
[59] *Arkansas Agriculturist* (Fall 1975), 16; *Hardy Interview*, 7.
[60] *Dean's Report,1978-79*, 1.
[61] Ibid.
[62] *Dean's Report,1979-80*, 1-2.

Chapter Seven:

The Crisis Years: The 1980's

The College moved into the 1980's with a higher enrollment than could have been imagined in 1960. The recruiting program that Dean Hardy, Bryan and others developed had succeeded in getting the message out about the exciting developments and opportunities available at the College. Interdepartmental and interdisciplinary programs initiated during the previous decade drew teachers and researchers together from many different fields to work together on a myriad of projects designed to address the needs of the variety of clienteles that emerged as agriculture came to mean more than production.

The College's strong position as the decade began was a great advantage because the 1980's turned out to be years of crisis in agriculture. The farm crisis of the 1980's affected the College of Agriculture and Home Economics in a number of ways. Arkansas' economy was already ranked 49th in the nation before farm incomes in the state fell nearly 60 percent between 1978 and 1983 before recovering slightly and then declining once again in 1986-87. Farm values dropped from $972 per acre in 1982 to $761 in 1987 for the state as a whole. By this time manufacturing and service industries contributed more to the state's economy than the sale of agricultural products. Many of the new industries that took root in the state after the industrialization campaigns of the 1960's and 1970's processed agricultural commodities or built farm equipment and contributed an estimated 25 percent of the state's gross income. In any case, considerably less money came into the state treasury.

Revenue shortfalls were translated as reduced appropriations for the University and other state programs. Federal funds were cut back in the 1980's as well, with an increasing amount of the available research funding coming in the form of competitive grants that routinely went to high-profile institutions. The Dean, department heads and faculty did their best to deal with budget cuts during the 1980's. One avenue to remedy the budget situation was private funding through corporations, foundations or individuals, with which some departments had more success than others. Grantsmanship became critical as College faculty and administrators targeted key corporate and government sponsors in their fields. The situation was partially remedied during the 1980's through the efforts of Sen. Dale Bumpers, Sen. David Pryor and Rep. John Paul Hammerschmidt. After Sen. Bumpers secured the chairmanship of the Senate Agricultural Appropriations Subcommittee in 1988, funding levels for programs of the Division of Agriculture increased, with the College benefitting from the new funds. Bumpers' efforts directed funds to key areas of research and teaching where the College already had a proven track record of accomplishment. Poultry science, alternative pest control, food science and rural development were specific projects targeted for assistance; these areas, with the input of the College's expertise and the backing of agricultural industry, would grow into centers of national and international reputation in the 1990's.[1]

Enrollment 1980-89.

Year	1980	1981	1982	1983	1984	1985	1986	1987	1988	1989
Agriculture	900	846	797	828	764	686	558	560	527	537
Home Economics	399	385	364	361	336	281	317	300	278	277
Graduate									258	269
Total	1299	1231	1161	1189	1100	967	875	860	1063	1083

Fall enrollment figures provided by Dean's Office, DBCAFLS.

The changing times affected the College in another way that impacted funding. After reaching a high in 1980 of almost 1,300 students, the College's enrollment decreased through the rest of the decade. The College's portion of the total student population in the University steadily declined. When University budget considerations came up for review, the areas of greatest growth received the most attention. News reports on a "farm crisis" made students reluctant to enroll in the college. However, agricultural industries in Arkansas, particularly the food processing and poultry sectors, continued to expand, as did their need for trained agricultural experts.

In spite of budget crunches and the shifting priorities of administrators, the people in the Division of Agriculture and College continued on the path of retooling their programs to meet the needs of agriculture as it approached the late twentieth century. Computer literacy, statistical expertise, interdisciplinary course work and research, alternative pest control, excellence in teaching and research and beginning biotechnological research were all parts of the evolving program. Faculty excellence reinforced exceptional work on the other side of the lectern. Recruiting took on new dimensions, as faculty and staff took advantage of new communication technology, tapped private resources for scholarships, fine-tuned advising procedures and offered internships and cooperative education plans.[2]

Innovative programs and unceasing efforts undertaken by administrators, faculty, staff and students carried the College through the budget shortfalls and funding difficulties of the 1980's to emerge at the end of the decade ready to face the challenges of the 21st century. The element that held the whole enterprise together was the tradition of excellence that had sustained the College through difficult times in the past. As John Goodwin, the Vice President for Agriculture, said in 1983:

> Establishing excellence in the University's Division of Agriculture is not a challenge, the challenge is securing the resources which will allow us to retain and enhance the high degree of excellence which existed when I assumed the vice presidency two years ago.[3]

The excellence indeed continued, and when the resources finally did arrive near the end of the decade, the Division of Agriculture and the College of Agriculture and Home Economics were poised for an era of excellence.

John Goodwin - Vice President for Agriculture, 1981-1986.

The 1980's brought changes in the Agri Week celebrations. The Showdeo, devised as a kinder, gentler and more frivolous display of the prowess of agriculture and home economics students, replaced the rodeo in the 1970's. The program was filled with events such as cow chip throwing, greased pig catching, sheep dressing, pig kissing and hay stacking competitions. During the 1980 celebrations, a large crowd turned out for the various events, sponsored by the AHESA. After the mid-1980's, the Showdeo was discontinued, as were the departmental exhibits. The week's celebrations became known as Agriculture and Home Economics Week instead of "Agri Day," which some thought was too rustic a name for pursuits so urbane as computerized commodity trading and fashion design. More attention was paid to recognizing academic achievements and leadership as the agribusiness side of the College became more involved in the events.

Other clubs organized in the 1980's reflected the new directions in the College curriculum. The Fashion Merchandising Club was formed around the most popular major in the Home Economics Department. The Food Science Club, although small in number like the department it represented, received national recognition as the most outstanding Food Science Club chapter in the country for its many activities in rounding up scholarships and efforts to prepare its members for future employment opportunities. The club and the department were poised for dynamic growth as the decade rolled on and food processing became one of the key growth areas in an economy dominated by slow- or no-growth industries.[4]

The College initiated a dynamic and fresh approach to recruiting in 1980 called the Beginning Scholars Program. The plan had a dual purpose, to recruit outstanding students and to foster student excellence. The Beginning Scholars Program involved high school students who had just completed their junior year and who expressed an interest in attending college in either science or social studies and who had high ACT scores. If they chose the College of Agriculture and Home Economics, which many of the beginning scholars eventually did, they were prepared with a familiarity of the College's programs and goals that went beyond any information they could glean from brochures or recruiting lectures. All students presented an oral report on their research project in an informal setting, which prepared them for the concept of peer review. Their research projects provided an uncommon insight for a young scholar into the research process. The students were housed in dorms where an upper-class undergraduate lived in the building to provide them with planned social activities as well as counseling and advice. The Beginning Scholars Program was an innovative approach to recruiting and preparing students for the College that succeeded in bringing a number of high-caliber students into agriculture and home economics.[5]

> *The single most important event was the Beginning Scholars Program, which was created in the early 80's through the efforts of then Dean Glenn Hardy. It was a brilliant mixture of a class for college credit, work experience in a research project, social interaction, and college life in general. This program gave me the assurance that I could excel in the college environment. I owe a tremendous debt of gratitude to the people, who are too numerous to mention, instrumental to this program.*
>
> *Dennis Lichti, BSA 1987*

Dean Hardy took a keen interest in the Beginning Scholars program from its inception. The plan had a dual purpose, to recruit outstanding students and to foster student excellence. Hardy indicated that there were a number of dedicated faculty members who went the extra mile to make this program work. A donor became interested in the program and gave the College a farm with the understanding that the proceeds from the sale of the farm would be used to establish an endowment fund to perpetuate this program. The resulting endowment fund in excess of $80,000 provided continuing support for this program.

A new Associate Dean appointed during this time, Tom Westing, was hired to develop and direct new international programs. It was anticipated that this would require about half his time, and he was to assume some other administrative duties. The international programs developed more rapidly than anticipated and tended to take all of Westing's time. He was able to take care of a few administrative tasks the first year but was full time in international work after that.[6]

The most conspicuous event of 1980 that highlighted the achievements of the College and Division of Agriculture was the dedication of the Altheimer Laboratory in October. John W. White, the first Vice President for Agriculture and former Director of the Experiment Station; E.M. Cralley, Director of the Agricultural Experiment Station from 1959 to 1973; and Lloyd Warren, Director from 1973 to 1986, all played a large part in getting funding for and development of this laboratory.[7]

Located on the Main Experiment Station Farm, the lab had been taking shape in stages since the late 1950's, built with funds provided by the Ben J. Altheimer Foundation. The most recent improvements in the facility included enlarging the laboratory to include more than 19,000 square feet of space and over $1 million worth of scientific equipment. The newest improvements provided the space and technology to conduct multi-disciplinary research in crop physiology, pest residues and weed science led by internationally recognized scientists who were part of the Agronomy Department faculty.[8]

One of the scientists who worked at the lab was Terry L. Lavy. Lavy was a leader in establishing testing standards and assessing the environmental risks from pesticide residues. Lavy coordinated the pesticide residue lab located at the Altheimer facilities that provided a testing service for Arkansas farmers, businesses and government agencies. Over the course of the decade, Lavy and his associates tested thousands of water and soil samples each year and developed procedures to contain the risks involved in the application of agricultural chemicals.[9]

The Altheimer Laboratory took shape over a span of 30 years, leading to its dedication in 1980, and has continued to grow and improve since then. It is shown here in a 1999 photo. (Photo Fred Miller)

Another outstanding and internationally recognized scientist from the Agronomy Department who used the lab to facilitate his work was Robert Frans. Frans received many honors for his teaching and research work in weed science at the College and Experiment Station. In 1980 he not only received the Arkansas Alumni Association's Outstanding Teacher Award but was appointed to fill the Elms Farming-Richard S. Barnett, Jr., Chair for Weed Science endowed by the Altheimer Foundation. Other endowed chairs in cotton and soybean provided by the Foundation were awarded to Brad Waddle for his research in cotton wilt disease and Charles A. Stutte for his work in soybean physiology and pioneering studies in understanding the photosynthetic process.[10]

Also working at the Altheimer facility were Ronald E. "Ron" Talbert, Lawrence R. "Dick" Oliver and Derrick M. Oosterhuis. All the scientists working at this laboratory were successful, productive scientists and were as effective in carrying out their responsibilities in the teaching arena as they were in research. They each advised graduate students who went on to successful careers, some in high-profile positions. Most of them taught at both the graduate and undergraduate level and were well received by the students. Some participated in team-taught courses with professors from other departments. Oliver coached a team of students in weed science that won contests at the national level more than once.

Other scientists in the Department of Agronomy included Charles Caviness. Caviness first joined the faculty of the Division and College in 1951 as a member of the agricultural mission to Panama and was appointed an assistant professor of agronomy in 1956. Caviness' specialty was in the genetics, breeding and management of soybean, and his teaching and Experiment Station research efforts were prodigious. It was estimated that over half of the soybean planted in Arkansas in 1980 were varieties bred by Caviness and his colleagues. Caviness won many Division, College, University and professional awards over the years for his teaching and research. Jack Walters, a plant pathologist, frequently collaborated with Caviness as they developed disease-resistant soybean varieties.[11]

The Agronomy Department was one of the larger departments in the College. The department included not only the weed science, physiology, and pesticide residue group at the Altheimer Laboratory but crop scientists, soil scientists, the Soil Testing Laboratories and the Statistical Laboratory. In addition, the department had oversight for agronomic personnel at several branch experiment stations.[12]

D.A. Hinkle became head of the Department of Agronomy in 1949 and served as head until 1975. He was very effective in hiring people and assembled a productive group of scientists. Wally Moline, Fred Miller, John Gilmour, Bobby Wells and Jim Barrentine followed Hinkle as heads of the department.[13]

Other members of the faculty were appreciated for their work by members of the agricultural community and attained national prominence (and not a few gained international recognition) for their work: R.L. Thurman, Fred Collins and Robert Bacon in breeding and in developing important cultural practices for small grains; J.O. York with corn and sorghum; A.E. Spooner in pasture management in cooperation with Maurice Ray of the Animal Science Department (Spooner's work has been followed by work done by Charles "Chuck" West); Bobby Wells in rice fertility; and Sam Offutt in legumes.[14]

The work of the soils faculty was also impressive. D.A. Brown's work on root growth netted him international speaking engagements, including one at Oxford University. His work has been carried on by Craig Beyrouty. John Gilmour, soil chemistry; Moye Rutledge, soil classification and genesis; Don Scott, soil physics; and Duane Wolf, soil microbiology, all have made important contributions to the state and to their profession. These professors have influenced irrigation practices and fertility practices and developed different types of septic disposal systems that can be used in rural areas where soil type interferes with drainage and have taken a leading roll in helping to alleviate environmental problems.[15]

Fertility experiments were established across the state to gather data that could be used to refine soil test recommendations. Lyell Thompson followed R.L. Beacher as leader of the soil test group in 1958. Glenn Hardy, who was heavily involved in fertility work related to soil testing, lead the laboratory from 1962 to 1965, at which time he became Dean of the College. Wayne Sabbe then became the leader, and many important advances have been made in the years he has led the group. With the development of technology, many operations have been automated and new services introduced.[16]

David Marx was the first leader of the statistical laboratory established in 1979. Up to this time, faculty members, whose main source of expertise was in another field, taught beginning courses in statistics. While they did a very commendable job, they needed to be doing research in their first field of interest. Brad Waddle and Fred Collins from agronomy as well as C.J. Brown and David Kreider from animal sciences were most helpful in this regard. The Stat Lab, as it was known, also provided opportunities for advanced courses that fit the needs of students in the College. In addition, the laboratory was available to research personnel who wished to have project plans reviewed or who needed assistance in interpreting data.[17]

The 1981-82 year saw some changes in the Animal Science Department that had far-reaching implications for the development of the department in general and the poultry science program in particular. Wayne Kellogg began work as the new head of the Animal Science Department in August 1981. Under his leadership the department was reorganized completely. The section leaders system was replaced with a committee program to provide more individual faculty input in departmental decision making. Kellogg also revamped the research program through upgrading facilities in physiology, parasitology and animal health for sheep, cattle and rabbits. He computerized financial records for improving accountability and gave more management responsibilities to graduate assistants so that faculty researchers could devote more time to their projects. Coinciding with the departmental reorganization was the construction of a new $1.5 million poultry research farm. As construction of the new facilities neared completion, Kellogg and others on the faculty made a special effort to elicit grant and contract support from the poultry industry to equip and maintain the new facilities. The Legislature had provided funds for construction of the new poultry facility but no additional operating funds. The Vice-President for Agriculture gave the Dean the budgeted research monies for the 10 departments for allocation for the next fiscal year. He was told that the Animal Sciences Department needed more maintenance money to operate the new research facilities. He was further told that the department heads were to operate within those budgets since the outlook for discretionary money that year was bleak. However, this increase was not adequate to cover the maintenance requirements for the new facilities, making it necessary to seek funds from private sources. The Dean asked to be relieved of any further research budget duties and the request was granted.[18]

Even in the face of these budget problems, the department began to make plans to further enlarge the program in poultry disease studies and to renovate the teaching laboratory in order to answer the call for more trained graduates in this area. The teaching laboratory was renovated the next year with teaching funds. These efforts began a series of developments that, by the end of the decade, had the poultry science program prepared to enter the field as a center of excellence.[19]

Another area where planning by Dean Hardy and others made significant changes during this time was in creating the Department of Agricultural and Extension Education in 1982. The vocational agriculture education program was housed in the College of Education for many years. With only 12 students in the program in 1980-81, the College of Education made plans to reduce the teaching faculty in the section. Through an agreement with the College of Education, Dean Hardy arranged to have three of the vocational agriculture teachers moved from the College of Education to the College of Agriculture and Home Economics in order for the program to continue. Lamar Love, James Scanlon and John Rolloff transferred to the College on July 1, 1982, and the administration agreed to add a half-time person from the Cooperative Extension Service to teach courses dealing with that important arm of the agricultural triangle. In 1985 Donna Graham joined the department in a half-time teaching and half-time Extension assignment. Clifton "Pete" Braker, lead advisor in the agricultural mechanization program, was brought over from Agricultural Engineering to add what would be a very successful dimension to the department. Love served as the interim department head dur-

Aerial view of the University farm, 1988. North is to the right in this view. The "bypass" is visible in the upper right hand corner. The triangular shape and trees of Agri Park are in the upper left hand corner.

ing the first year. Nolan Arthur, the Assistant Dean, was appointed the interim head the following year so Love could give more time to teaching. After Dean Hardy announced there would be a search for a permanent head of the program, Arthur asked to be considered for the position. After resigning from the Dean's office, he was appointed to the post in 1984. The new department rapidly took shape and grew from 12 students in 1982 to nearly 100 three years later.[20]

New courses added in the College in 1981-82 included a mini-series of one-hour courses in the Entomology Department similar to those previously offered in plant pathology. The self-paced classes were keyed to investigating pest management programs for specific commodity crops such as rice, soybean and cotton. The Plant Pathology Department began a new course in clinical plant pathology in the summer of 1982. A new plant disease clinic operated by the Experiment Station at this time offered clinical diagnosis of diseased plant specimens sent in to the clinic by farmers, county agents and householders. Students in the class got hands-on experience diagnosing plant diseases while under the supervision of faculty in the department. The course performed a triple function: it served a public need and provided an instructional and research role at the same time.[21]

Interdepartmental and intercollegiate programs continued to thrive in the College, making the best use of available resources at a time when budgets were being scaled back. Three horticulture classes designed for students in the landscape architecture program continued to be taught. Faculty from the Home Economics Department served on master's and Ph.D. committees in horticultural food science, elementary education, vocational education and other higher education programs. The nursery school run by the Home Economics Department also served as an instructional and research center for students in psychology, and their graduate courses in human development also filled the requirements for students in education and psychology.

In addition to fulfilling the social science and business course requirements for students in the College, the Department of Agricultural Economics and Rural Sociology also hosted Ph.D. candidates in economics, marketing, management and finance. Undergraduates from nearly all colleges took course work in the department, and the offerings were well suited to the needs of the Law School students also. As the 1980's progressed, the College continued its outreach and integrated its programs in the context of the whole University. The outreach process involved institutions across the state as the Division of Agriculture and the College made reciprocal arrangements with the University of Arkansas-Monticello to share instructional and research duties in forestry and with the University of Arkansas at Pine Bluff in aquaculture.[22]

The College also served the state in outreach. Besides the courses being taught in the new Department of Agricultural and Extension Education, classes were arranged around the state at various locations in order to provide instruction in everything from managing pastures to computer literacy. College staff and faculty taught 11 off-campus classes in 1982-83 with 188 students. James Ferguson taught a class in irrigation in Forrest City. Robert Frans drove to Pine Bluff one night a week to teach an upper-level undergraduate class on the principles of weed control. Faculty from entomology, agronomy and plant pathology also instructed Extension workers in the integrated pest management concept intended to cut down on the amount of agricultural chemicals that farmers used. The innovative approaches had both an environmental and economic component. The College scientists pointed out that by reducing the amount of pesticides or herbicides farmers needed for effective pest control, not only would there be less environmental damage, but farmers would save money, especially on oil-based chemicals.[23]

A major step forward in the internationally known alternative pest control program in the Division of Agriculture and the College was the patenting and manufacture of George Templeton's biological herbicide. Marketed under the trade name "COLLEGO" by the Upjohn Company beginning in 1982, the new biological agent was a derivative of a fungus, *Colletotrichum gloeosporiodes*, which attacked northern jointvetch, a major weed pest in rice and soybean fields. Templeton initiated his research in 1969 and was joined in the project by Roy Smith, D.O. TeBeest and a host of graduate students in the 13-year span between the initiation of the project and bringing the final product to market. Ever aware that this new tool could also pose risks to the environment, Division of Agriculture scientists TeBeest, Templeton and G.J. Weidemann followed up the development phase of the research with a body of work intended to assess and address the possible risks involved in releasing a biological control agent into the environment. Using new techniques developed in biotechnology, TeBeest took the research a step further and induced a genetic change in the fungus that provided tolerance to a fungicide frequently used in rice fields.[24]

Throughout their research, the scientists were also teachers. With the support of the Division of Agriculture and the College, government funding and private donors, Templeton and his colleagues were able not only to develop COLLEGO but also to train a whole new generation of alternative pest control scientists and attract further funding for more research in the field at the College. Templeton's work attracted international attention. In the following years he received invitations from universities around the world to either lecture or start other research projects in locations as far-flung as South Africa and Japan. The University granted Templeton frequent sabbatical support in order that he might carry the important research to the world community.[25]

While Templeton and others were giving their individual attention to international research cooperation, Dean Hardy worked during the 1980's to increase the College's participation in International Programs. In 1980, a University Task Force on International Programs and Services recommended that the University increase its participation in the international sphere, particularly in agricultural development. The University was awarded a USAID grant in 1980 to help it prepare for a leading role in international projects. The "strengthening grant" was designed to give funding support for such things as administrative training in international programs, language training, background studies in international agricultural problems as well as studies in the history, politics and social/cultural development of different countries. In 1981, USAID requested the assistance of the Division of Agriculture and the College in helping Egypt develop a Rice Research and Training Project. The Egypt project was a cooperative effort among the University of California at Davis, the International Rice Research Institute in the Phillippines and the College. College personnel involved in the project were John Paul Jones, a plant pathologist; Carter Price, an agricultural economist; and Ron Talbert, who did weed control studies on several short-term stays. Eric Wailes also contributed with brief visits for economic evaluations. The goal of the program, to help Egyptians increase their rice production through weed and pest control strategies and aid them in improving the quality of the rice produced, was fairly successful. The other part of the project, aimed at helping the Egyptians make their domestic rice marketing system more efficient, was an important contribution, but it was difficult to measure the lasting impact.[26]

After Tom Westing was appointed Associate Dean in 1983, the Division of Agriculture and the College took a more active role in international agricultural programs. In the following years the Division and College became involved in four major projects in Jamaica, Haiti, Rwanda and Burundi and, to a limited degree, in other developing countries. Using the training provided through the strengthening grant, in 1983 the College assigned James Scanlon to head up a two-year project with USAID called the Jamaica Basic Skills Training Project. Scanlon helped the Jamaicans institute a program for vocational and technical education for the country as well as develop an instructional curriculum for the program.[27]

The Division and College also led a cooperative project with the University of Minnesota, the University of Illinois, Lincoln University (Missouri) and the University of Puerto Rico called the Rwanda Farming Systems Improvement Project from 1984-89. The purpose of this five-year project was to strengthen Rwandan agricultural research and extension institutions and improve the farm productivity of small landholders. Based on an understanding of the unique character of Rwandan agriculture and knowledge gained from working with the South Central Family Farms Research Center at Booneville, Don Voth drew up what USAID officials termed "a model project design." The project leaders emphasized growing food crops rather than commodities and worked at helping to build up the capabilities of the Rwandan extension program. Lucas D. Parsch, Eric J. Wailes and Diana Danforth also assisted with the Rwanda project.[28]

The Haiti Agricultural Development Support Project, 1984-88, in cooperation with the Winrock International Foundation, was developed to improve Haiti's agricultural research and extension capabilities and work with Haitians to formulate a realistic agricultural development plan. Other objectives included providing owners of small farms with appropriate technologies to enable them to improve crop production and arrest environmental degradation that was endemic in much of the countryside. The project was hampered by political instability that had plagued the country for many years.[29]

The fourth major international project was the Burundi Small Farming Systems Research Project. Arkansas led this program and was joined by the University of Minnesota, Lincoln University and Winrock International. Similar in some ways to the Rwanda assignment, this project targeted owners of small farms and strove to develop indigenous agricultural research and extension institutions. One unique feature of this project was the work done with two farming communes located in Burundi's central plateau region. The goal, to improve nutrition among small farming families through the distribution of high-quality seeds and improved farming practices, read like a page out of a typical U.S. Extension Service handbook. Voth, Wailes, Parsch and Danforth also helped in the design and evaluation of the Burundi project.[30]

The College cooperated with USAID and Winrock International in sponsoring graduate students from developing countries in agricultural studies. Between 30 and 40 students from 20 different countries were brought to the College each year to study in agricultural economics and rural sociology, food science, agricultural engineering, animal science, plant pathology, agronomy and entomology.[31]

A six-week summer short course for international students in intensive poultry production systems was offered beginning in the summer of 1985. Coordinated by E.L. Stephenson, the course gave technical officers, Extension specialists and other agriculturalists from developing countries the opportunity to catch up on the latest knowledge in poultry breeding, nutrition and disease prevention and control.[32]

The value of the international programs cannot be calculated in dollars and cents alone. In addition to the value added in instruction and participation at the College by including diversity in its programs, the humanistic impulse was a guiding principle in developing these programs overseas. The international programs conducted through the Division and College are nothing less than the land-grant mission writ large, albeit in a number of different languages—"science in the public interest" elevated to a worldwide level, a concept with which Sen. Justin Morrill or Seaman Knapp would indeed be proud to be associated.

While Dean Hardy and Tom Westing succeeded in securing funding for international programs, they were less fortunate with the state College money. The Agricultural Experiment Station Director also suffered a shortfall in research funds that was so severe it became necessary to leave 11 sorely needed research positions unfilled and to reduce research maintenance in order to give a long-promised salary increase. The academic side also had to make some changes. For example, courses in the Department of Agricultural Engineering were rescheduled to be taught in alternate years, and the teaching load in the department was reduced. The call went out for extramural monies to sustain critical programs. However, even within these constraints, some significant gains were made.[33]

In agronomy a course in seed production technology was offered for the first time. Because Arkansas produced more certified seed than any other state, program planners in the department felt it was time that such a course was offered, regardless of budget constraints. Two new courses appeared in horticulture and forestry. "Modern concepts in pomology" addressed the newest research in the field, or rather, orchard. "Tissue culture and biochemical genetics" was keyed to the newest developments in biotechnology, where scientists had only recently made tremendous strides. In a realm less arcane, the Department of Agricultural Economics and Rural Sociology offered a class in agricultural law for the first time. The class was another interdisciplinary effort being taught by two professors from the Law School. Because the Department of Horticultural Food Science had turned to research in cereals and also studied poultry and meat processing, the name was changed in 1982 to reflect this expanded focus—the Department of Food Science.[34]

The Infant Development Center was a facility used for day care and as an observation/research center for human development studies. The center was a joint venture with the College, the Fayetteville Christian Church (Disciples), who owned the building and the surrounding property, and their volunteers, who assisted the students and faculty in developing programs at the facility. In 1982 the church informed Dorothy Larery, head of the Home Economics Department, Mary Burton, the director of the human development program, Dean Hardy and the Infant Development Center Board of Directors that they wanted to turn the entire operation over to the University and offered to sell the building and property. In this time of tight budgets it seemed that the College would have to pass on the opportunity to take over the facility, which had become a vital part of the program in home economics. However, Ellen Meenen, the wife of Henry Meenen in the Department of Agricultural Economics and Rural Sociology, and a member of the church, was able to convince the church to hold the property off the

The "in-house" preschool lab and Infant Development Center gave me true experience in the field I was entering. Becoming a preschool teacher was a very smooth transition after having the hands-on education I received as opposed to textbook only.
Mary Sugg Bassett, BSHE 1979

market until Dean Hardy had an opportunity to try to secure private funds to buy the property since the University did not have funding available for this purpose. Six prospective donors were approached, and five agreed to provide the needed funds over a three-year period. A contract was negotiated, and the final payment was made in 1987.[35]

The 1983-84 year was the year of the computer in the Division of Agriculture and the College. Following Dean Hardy's injunction that all students in the College should work toward becoming computer literate, both the Animal Science and Agricultural Engineering Departments were equipped with computers, which the students "received eagerly and readily," according to one report. The Department of Agricultural Economics and Rural Sociology developed a computer-simulated commodities trading course, and Joel T. Walker in agricultural engineering developed an interactive computer program for high school vocational education. The following year, as part of a University-wide effort, the Vice Chancellor for Academic Affairs funded a new microcomputer lab for the College with 40 terminals installed by the Texas Instruments Corporation. Further upgrading of the College's computer resources became a prime consideration, which superceded many other needs during the rest of the decade, even in the midst of budget shortfalls.[36]

The College also continued to find resources to establish new courses during the year. The Department of Agricultural and Extension Education continued to expand, created four new courses, held a cooperating teachers workshop and developed a K-6 curriculum for agricultural studies with 64 lesson plans. Following the success of the agricultural law class, the Department of Agricultural Economics and Rural Sociology ran an "agricultural law II" class, once again with the assistance of professors from the Law School. The College received an endowed scholarship from the Katharina Bollenbacher Memorial Scholarship Fund for the first time in 1984. The endowed memorial was designed to give advanced undergraduate students an opportunity to participate in the summer research program that had been going on in the Department of Plant Pathology for many years. Qualified students from any Arkansas institutions of higher learning were eligible, and Clell Ford, an undergraduate biology major from Hendrix College, was the first recipient. Further movement toward fostering excellence in undergraduate studies was made during the year also. A budget and a curriculum for a new undergraduate Honors Program were developed during the year and awaited implementation as soon as funds were available.[37]

Division of Agriculture researchers in the Department of Horticulture and Forestry made advances in establishing biotechnology studies in 1983-84. F.S. Huang's work to increase the desirable germplasm in agricultural and forestry species through the use of tissue culture showed great promise. Entomology made advances in their work in alternative pest control, concentrating in three areas. Max Meisch reported that a bacterium released in ultra-low-volume sprays by air or titration into irrigation water was effective against the anopheles mosquito, which was always a pest in the rice growing regions of the state. Research was ongoing with a fungus for the same purpose. Researchers also worked with three new viruses to be used against the soybean looper and developed a non-chemical cattle tick control.

The Department of Animal Sciences reported progress in developing a method for early detection of a disease in poultry. Nutrition studies emphasized the use of forages by beef cattle with attention given to fescue toxicity. There were other nutritional studies with swine, dairy cattle and poultry. All departments had active research projects that were adding to the knowledge base. The Department of Agronomy, in cooperation with Plant Pathology, released three new rice varieties and one soybean variety for seed increase or to growers. Over 9000 copies of three Experiment Station publications by Sue Martin in the area of child development were requested and distributed in less than a year. Studies on breakage of rice kernels during drying were carried out in the Agricultural Engineering Department and in Food Science. A set of models for the rice marketing and distribution system were developed in Agricultural Economics. These models form the basis for the most comprehensive management models available for the rice industry at this time.[38]

These are only a few examples of the work by faculty members who taught the classes in the College as well as developed curricula and other policy matters under which the College functioned. Faculty members had appointments split between the College and the Experiment Station. Funding for the teaching program came from the central administration of the Fayetteville campus. Funding for the Experiment Station budget came from a state appropriation earmarked for that purpose.

Grantsmanship through the Division of Agriculture was beginning to pay off in some areas. Over $2 million was received in the Departments of Agricultural Economics and Rural Sociology, Agronomy, Entomology, Food Science and Home Economics. The Food Science grant came in the form of a $115,000 contribution from private industry to serve as matching money for a special $365,000 appropriation from the Legislature. The money went toward adding space to the food science lab at the University farm and housing a food microbiologist. The Legislature was responding to public concern over food safety, and the food processors saw an opportunity to get potentially important research in food safety.

Relying on grants still did not solve the problem of funding shortfalls. In order to operate as an integrated learning and research institution, the Division of Agriculture and the College needed a dependable source of long-term funding that covered maintenance, salaries and overall operations. Grants for specific projects were certainly welcome, but they did little to alleviate day-to-day operational problems. Buildings, equipment and even scientists get worn out.[39]

In 1984-85 the University recognized the contributions of six of the professors in the Division and College for their outstanding research, teaching and public service by elevating them to the new rank of University Professor: C.E. Caviness from agronomy, O.T. Stallcup, from animal science, J.R. Phillips from entomology, Justin Morris from food science, George Templeton from plant pathology and James N. Moore from horticulture and forestry.[40]

A new major in international and domestic agricultural marketing, which was several years in the works, was available for the first time in the Department of Agricultural Economics and Rural Sociology. The major was developed in part as a response to the "farm crisis." Marketing was often considered the weakest link in a farming operation. It was the goal of this program to improve that situation by raising the profile of this problem and providing educational experiences to future Extension personnel and other farm advisers as well as some future farmers. This major was especially helpful to students who were interested in working in businesses that used the commodity market to control price risk, e.g., work in the grain merchandising business. A centerpiece of the program was course work in commodity marketing with emphasis on how the markets could be used in a non-speculative way to manage risk.[41]

The College suffered an unfortunate loss of additional funding in 1985 when $140,000 expected from military land funds failed to materialize. The funds were reallocated without notification to the College. Consequently, anticipated repairs to the electric wiring in the Home Economics Building and the renovation of the building's auditorium had to be postponed. Repairs on the Animal Sciences auditorium also had to be delayed. Three laboratories for soils research and teaching were renovated and air conditioned with a combination of College and Experiment Station monies and a small contribution from previously allocated military land funds providing the resources for the projects. The renovation resulted in two excellent teaching laboratories and two small research laboratories.[42]

Preston LaFerney - Vice President for Agriculture 1987-1992. Director of the Station 1983-1987.

The 1985-86 year opened on a hopeful note as a much-needed 10 percent salary increase was made available by the Central Administration for merit salary increases in the College. The Vice President for Agriculture, Preston LaFerney, who had replaced John Goodwin, did not feel comfortable with 10 percent, so he allocated 8 percent for use by the Experiment Station for research faculty.

Since most faculty members had split appointments, this gave rise to administrative difficulties each time an appointment was changed. The significant salary increases did address what was always indicated as the highest priority for funding: salaries were low in comparison with those in other states. However, four different budget cuts by the state during the year, due to less-than-anticipated tax collections, strained already tight maintenance budgets. The expectation of another upward adjustment of salaries in the second year of the biennium evaporated as much of the available funding had to be used to restore maintenance funds. The result was a 2.5 percent salary adjustment on a merit basis in the second year for both teaching and research. Some of the gain realized the previous year was lost. However, some small progress was made in upward salary adjustments in the two-year appropriation cycle, although inflation significantly diminished any real gain.[43]

While much of the economy was in a slump during the late 1980's, the poultry and food processing industries in Arkansas and the nation enjoyed a boom time. However, the rest of the agribusiness sector of the economy, particularly in commodities, showed no such vigor. Those involved in agriculture pinned their hopes on the 1985 Farm Bill working through Congress at the time, which promised to do something about sagging American commodity prices in the international markets. Such hopes proved illusory as prices remained depressed. Media coverage of the farm crisis tended to leave students with the idea that agriculture would not be the best area in which to seek a career. This was at a time when there were more jobs than graduates in agriculture.[44]

Declining enrollment in the College became a major issue in 1985. Not only did this affect funding levels, but it also had an impact on College morale. The Dean's office and individual departments began an intensive recruiting drive during the year to help reverse the course of a steady drop that resulted in over 400 fewer students in the College by 1987 compared to 1980. The Animal Sciences Department and the agribusiness section of the Department of Agricultural Economics and Rural Sociology were the departments hardest hit, but there was a general decline across the College, except in the Department of Food Science, where there was a slight overall increase.

> *I started out in the College of Arts and Sciences and transferred to Ag/Home Ec. It was such a friendly atmosphere and the level of commitment from faculty and personal attention from the college was irreplaceable.*
>
> *Jawanda Barnett Must, BSHE 1983, MS 1989*

Before he left the Dean's office in 1987, Dean Hardy suggested that the College should try to organize a group of interested laymen into a "Friends of

Agriculture and Home Economics" type of booster organization that had been effective in other universities. While some progress was made in that direction, such a group did not coalesce at the time. In 1985 Hardy targeted the College's recruitment campaign in high-interest areas such as food science, fashion merchandising, poultry science, agricultural education and foods and nutrition, where student levels were increasing and employment opportunities were good, although all areas were covered in the presentations. Hardy, Mooty and other faculty members fanned out across the state visiting high school career days, FFA and FHA meetings and 4-H functions. They even visited schools in Oklahoma and Missouri on occasion. With the help of Billy Bryan, they developed a slide show presentation and several videos that highlighted the advantages of attending the College and pointed to the opportunities available for graduates in agriculture and home economics. High school students were invited for on-campus weekend visits so they could meet the students, participate in activities and familiarize themselves with the College in a direct way. Mooty set up communications with the various community colleges around the state, examined their offerings and made determinations on which courses the College would accept for credit in order to ease the transition for students wanting to continue studies in Fayetteville. To cap it all off, the Dean's office sent over 3,300 personalized letters and colorful, new informational brochures to prospective students during the campaign.[45]

Various departments also developed recruiting programs of their own. The Departments of Agricultural and Extension Education, Food Science and Home Economics were the most active and, significantly, attracted the highest enrollments in the coming years. Recruiting was intensified over the next several years. While it is hard to determine how effective the recruiting program actually was, because student numbers in the College continued to decline into the 1990's, in the long run, the campaigns probably kept the decline in enrollment from becoming even worse and at the least kept the College in the public eye.[46]

A development closely connected with the recruiting efforts was the implementation of an undergraduate Honors Program in the College in the fall of 1985. John Gilmour in Agronomy helped develop the program and directed it until 1987. A.E. Einert took over the program in 1987. The program had approximately 40 students that first year and consisted of three courses in addition to work in other classes. Students took an honors orientation class, which was devoted to problem solving. Faculty from each academic department came to the class and gave a description of their programs and then presented the students a problem dealing with some aspect of their field. In the 1986 Honors Colloquium, the students investigated different features of agriculture in various countries, gave oral reports and prepared a written report of their findings. Two additional courses were added to the program in 1987: "introduction to applied entomology" and "introduction to applied animal science."

Each student was required to write an honors thesis on a topic of his or her choice in the form of a research project, special project, internship or other "enriching experience." The program benefitted both the students and the College. Since most Honors graduates went on to graduate school, it improved their chances for getting into the school of their choice and for obtaining assistantships once they were there. An Honors degree also made them more marketable in their career choices. The College gained from being able to attract more of the brightest students.[47]

Another important program that allowed the College to recruit, retain and then be able to place more students in desirable jobs was the Internship and Cooperative Education Program implemented in 1986. Under the direction of Martin Redfern, this program was designed to help students finance their education and give them valuable on-site training in their field of study with the possibility of employment after graduation. Companies and employers benefitted from the program through being able to recruit high-quality, long-term employees with pre-professional backgrounds. Internships were one-time paid or unpaid positions, either part time during the school year or full time in the summer, that placed student "pre-professional" workers with selected employers for a specific length of time. The unpaid positions were intended to give students on-the-job training that earned college credit, in lieu of class or lab work. A cooperative was a paid work assignment with a company while the student was in college. In its first year of operation, the program placed 43 students in internships/cooperatives. These positions were with the poultry industry, agricultural chemical companies, feed and grain companies, brokerage houses and banks and other businesses. The program continued to grow through the rest of the decade, with generous grants from the U.S. Department of Education, and benefitted all involved, including the College.[48]

> *I already knew that I wanted to come back and farm. However, the professors in the College of Agri and Home Ec made sure I came home with more "tools" to work with and the ability to step back and take a second look before getting in....I try to farm as an example of what a good education can do in farming. I encourage young men in the area to go to college and mature. Education is never wasted if we use it wisely.*
> *Clifford Bruce Jacobs, BSA 1981*

Under the leadership of A.A. Kattan, the Food Science Department had an enviable record of attracting industry support for its program, especially in regard to its support for students in the department. During the 1986-87 year the department received 27 grants that generated over 60% of its operating funds, including maintenance, salaries and the purchase of new equipment and supplies. The students also benefitted from industry largesse. Twenty-one undergraduates shared $26,580 in scholarships, and there were seven additional scholarships provided by the Institute for Food Technologists during the year. Since its inception, the department had routinely employed a large number of the undergraduates as lab assistants and technical assistants and in other positions, and the tradition continued as the number of grant projects increased. The department was also an eager participant in the Beginning Scholars Program and had developed its own summer internship program for interested high school students. Public service programs conducted by the department included courses offered on Saturdays to provide local/regional USDA meat and poultry inspectors with additional training. The department also operated a certification school for food processing industry workers that had certified over 1,100 food technologists by 1987.[49]

There were further expansions in the Poultry Science program during the year that helped prepare the department for future growth. A new experimental poultry processing facility was built for the program using funds raised by the faculty through grants and foundations. The construction of the plant greatly enhanced the research and teaching potential in poultry science at the University. As with the development of other projects of this kind, building new state-of-the-art facilities had a domino effect in attracting further funding for the program. Foundations, government projects and private industry were more likely to fund projects at locations best equipped to bring them to successful conclusion. In the next several years the poultry science program received tremendous benefits from these initial investments that would match the explosive growth of the poultry industry in Arkansas.[50]

All departments experienced the same pattern of increasing dependence on private support for new projects as well as ongoing programs. But, as mentioned before, some departments had more success in this area than others. Derald Slack, the head of the Department of Plant Pathology, reported at this time that his department had received $400,000 in grants for 1986-87 and that state and federal funds contributed less than 20 percent of the maintenance monies for the world-class program underway there. The grant monies came nowhere near meeting the maintenance needs of the department, however. Appropriated funds for faculty salaries continued to be scarce at this time. In addition, many of the greenhouses needed to carry on teaching and research obligations were torn down during the year because of their deteriorated conditions. Money was simply not available to replace these essential components. All the plant sciences, agronomy, horticulture and forestry, as well as plant pathology, were affected by the loss of the greenhouses.[51]

In 1986-87, a request was made to move the Department of Agricultural Engineering to the second floor of the Engineering Building after the completion of the new Bell Engineering Center. This request was later approved, and the department made the move later in the decade. At about the time the department moved, the name of the department was changed to the Department of Biological and Agricultural Engineering. This more accurately reflected the nature of the research and teaching program in the department, which had grown to include quite an important food science function.

Gerald Musick - Dean of the College and Director of the Station 1987-1993.

Perhaps the most important change during 1986-87 was the announcement made in early 1987 that Dean Hardy would be stepping down from his post on July 1, 1987, after 22 years of service as Dean. Gerald Musick, from the Department of Entomology, and most recently, the Interim Director of the Experiment Station, was chosen to succeed Dean Hardy. An administrative reordering accompanied Musick's appointment. As Musick assumed the Dean's post, he was also named Director of the Experiment Station, Division of Agriculture. This was the first time the positions had been combined since 1959.[52]

The change was made in hopes of improving coordination and communication between the two functions, teaching in the College and research in the Agricultural Experiment Station. As Director of the Experiment Station, Dr. Musick reported to Preston LaFerney, Vice President for Agriculture. As Dean he reported to the Vice Chancellor for Academic Affairs and to the Chancellor, Dan Ferritor. This was especially critical in a time of budget shortfalls and declining resources available for agricultural research and education. The Chancellor wanted a direct line to the Director of the Experiment Station, who previously reported only to the Vice President for Agriculture.[53] Hardy's departure was not voluntary; he was asked by the administration to step aside so that the changes described above could be made. At his request, Hardy was reassigned to a teaching position in the Department of Agricultural and Extension Education, where he served until his retirement in 1994, after 38 years of service to the College.[54]

Dean Musick faced the same dilemmas that had dogged Dean Hardy: declining enrollment and shrinking budgets. However, he would soon be getting some extra help at the federal level as the Arkansas Congressional delegation kicked into gear. A key player in the campaign for funding was Sen. Dale Bumpers, who had taken over the chairmanship of the Senate Subcommittee on Agricultural Appropriations in 1987.[55]

Some innovative courses added during the year included a videotaped course in biological and agricultural engineering, developed by James Ferguson for an off-campus graduate class. Ferguson produced the video in the College of Engineering studio. The course was unique for the time and was quite successful in reaching engineers in outlying areas who were interested in working toward advanced degrees but had no access to a college or university nearby. The Department of Agricultural and Extension Education revamped its curriculum to include more humanities courses, reduced the number of education courses and revised the technical agriculture core courses to include new offerings in farm power and machinery. The Department of Entomology, in an attempt to draw more undergraduates into the program and to provide course work for potential majors who had biology deficiencies, developed two new courses that gave an "appreciative" view of insects and their interaction with humankind.

> *The day I graduated from high school Dean Hardy showed up at my school and talked with me. At that time I was planning on majoring in Biology. Dean Hardy offered me an OFPA scholarship and asked me to consider changing my major to Food Science. That one visit made a big difference in my life. It really impressed me that he would travel all the way to Harrisburg, Arkansas, and take his time and talk with me. He will probably never know the impact that visit made.*
>
> *Andrea Dunigan, BSA 1990, MS 1991*

In the Department of Home Economics, Dennis Papini designed a fully equipped audiovisual laboratory intended to facilitate research and teaching in human development and family relations. The new facilities included an observation booth equipped with one-way mirrors and video cameras so researchers could both observe behavior and interactions and record them for later study. A new computer system was also installed to facilitate recording of data. Additional computer facilities for research in food and nutrition were also installed.

Events moved quickly toward establishing the commitment of the College and the Division of Agriculture to a larger research and teaching role in poultry science during 1989. Indeed, 1989 was a banner year for the Division and College in other fields as well. Through the efforts of Sen. Bumpers and others in the Arkansas Congressional delegation, $3 million in federal grants were secured: $1.4 million for alternative pest control, $1.4 million for food safety and $200,000 in additional funds for a poultry disease isolation unit. In other developments, arrangements were made to transfer the Appropriate Technology Transfer for Rural Areas research unit from Memphis to Fayetteville, and the National Center for Agricultural Law Research Information was established at the Law School in 1988.[56]

An indication of the increasing level of support given the Division and College by the poultry industry came toward the end of the year when it was announced that the Arkansas Poultry Foundation would provide funds to endow a chair in poultry science.[57] The Division and College completed plans for establishing a Center of Excellence for Poultry Science in 1989. In keeping with the trend toward interdepartmental cooperation that had been evident throughout the decade, the plan envisioned interdisciplinary support for the program from the Departments of Biological and Agricultural Engineering, Agricultural Economics and Rural Sociology, Entomology, Food Science and Home Economics. The funding obtained for a new state-of-the-art poultry disease isolation unit and a broiler production demonstration unit assured that the program would attract yet further support. To reflect the new focus in teaching and research, the name of the Department of Animal Sciences was changed to the Department of Animal and Poultry Sciences. Richard Forsythe, retired vice-president of the Campbell's Soup Company, was named to fill the newly endowed Arkansas Poultry Federation Chair for Poultry Science, giving the poultry industry needed visibility in the Center.[58]

Because of my understanding of basic science and my background in the cattle industry, I am able to implement technology relatively easily. My education at the University of Arkansas laid the groundwork for my career in biotechnology.

Dr. Charles Looney, BSA 1978, MS 1979

Funding for the Food Safety Consortium was also tied into poultry research. The University of Arkansas, through its Division of Agriculture, was the lead institution in the effort to identify and control contaminants at processing plants for poultry and other meat. Kansas State University and Iowa State University, along with the University of Arkansas for Medical Sciences, were the other members of the Consortium. The funds for a poultry disease isolation unit would be added to earlier federal appropriations of $365,750 and a state appropriation of $405,618 for the construction of a state-of-the-art technology system. Another $474,000 federal appropriation was also made for the purchase of equipment and the operation of the system. Amy Izat (Waldroup) was able to expand her research in treating poultry in processing lines at plants to reduce bacteriological contamination. The research of Park Waldroup on reducing infections in live broiler chickens that can cause food-borne illness in humans would also take on a new emphasis.[59]

The increased appropriations to the Division of Agriculture for alternative pest control provided the impetus to set up the Center for Alternative Pest Control on the campus. In 1989 The Center was not confined to one building but rather was a collection of labs across the campus and at the Main Experiment Station and the University farm. The goal of the Center, as stated in 1989, was to:

> ...establish and maintain a multi-disciplinary research center for the collection, evaluation and development of biological alternatives for control of agricultural and urban pests including weeds, insects, plant diseases and nematodes.

The work of the Center was the combined effort of the Departments of Agronomy, Entomology, Horticulture and Forestry and Plant Pathology. While the focus of the Center was research through the Division of Agriculture, there was also a strong instructional component of the research, whether through student assistants working in the labs or the fact that most faculty researchers also held teaching appointments.[60]

The funding bonanza overshadowed other developments in the College that year, but there were some of significance. The Beginning Scholars Program, now under the direction of Donna Graham, was revamped. Instead of the student "shadowing" a professor involved in one particular field, now the scholars received a broad-based introduction to all fields studied at the College. One part of the program was devoted to a topic relevant to all the departments. In 1989 the topic was "Contemporary Issues in Agriculture and Home Economics." Graham said that it was good way to get the kids

> ...thinking about a lot of societal concerns, and there are many that relate to agriculture and home economics, such as the environment and safety in the food supply, changes in population, animal rights and genetics. I wanted them to open their eyes and think about the broader picture.

Twenty students participated in the three-week program that summer, learning about genetics and crop improvements, chemistry and the analysis of soils and foods and a host of other topics that applied to the study of agriculture and home economics.[61]

The College's Internship program continued to grow. Fifty-two interns were placed during the 1988-89 year. Employers were especially pleased with the program, hiring many of the intern "pre-professionals" once they graduated. The students were quite satisfied with the results of the program as well. Many of them got multiple job offers before graduation. Not only that, but the majority received much higher starting salaries than those who had not held internships.[62]

In other developments, the Department of Agronomy became heavily involved in International Programs during the year. The department hosted two visiting scholars from the People's Republic of China. John Gilmour visited Mexico, Colombia and Honduras to acquaint the faculties there with the College's graduate program. The department also sponsored graduate students from Mali, Tunisia, Rwanda, Indonesia, Haiti and Malawi through agreements made with USAID.[63]

The decade had indeed ended with a bang. Although there was still concern over the availability of basic maintenance funds and the deteriorating state of equipment and facilities, the poultry center, the food safety consortium and the alternative pest control center were projects that augured well for the ability of the Division of Agriculture and the College to attract funding for high-profile programs and further expand the mission of the Division and College. As the decade ended and the "farm crisis" faded into memory, the College was well situated to approach the challenges ahead in the 1990's and the 21st century.

Endnotes

1 Danbom, 253-255, 262-263; R. Douglas Hurt, *American Agriculture: A Brief History* (Ames: Iowa State University Press, 1994), 331-34; Bureau of Econmic Analysis, *Regional Economic Data for Arkansas: 1969-1996* (Washington, D.C.: GPO, 1996), Available online: http://govinfo.library.orst.edu/stateis.html (January 10, 1999); *Arkansas Alumnus* (Winter 1992-93), 8-9; *Arkansas Land and Life* (Fall/Winter 1995), 5-6.

2 University of Arkansas-Fayetteville, College of Agriculture and Home Economics, *Annual Report: 1985-86*, Manuscript copy in possession of Glenn W. Hardy, 1. Hereafter cited as *Annual Report COAHE,* followed by year and page number.

3 *Arkansas Alumnus* (October 1983), 15.

4 Melissa Lester, "Agri Day History Slide Show Presentation"; University of Arkansas; *Razorback, 1980, 1982* (Fayetteville: University of Arkansas).

5 *Annual Report COAHE, 1980-81*, 2, *1981-82*, 2, *1982-83*, 44.

6 *Annual Report COAHE, 1979-80*, "Item 1," 2-3.

7 Glenn Hardy, written contribution to this manuscript.

8 Ibid.

9 Ibid.

10 *Arkansas Alumnus* (December 1980), 2, 4-5, (February 1980), 3-4; *Arkansas Farm Research* (November-December 1980), 9.

11 Glenn Hardy, written contribution to this manuscript.

12 Ibid.

13 Ibid.

14 Ibid.

15 Ibid.

16 Ibid.

17 Ibid.

18 Ibid.

19 *Annual Report COAHE 1981-82*, 2, 17-18.

20 *Hardy Interview*, 26, 39; *Annual Report COAHE, 1981-82*, 2; Historical Information, 18.

21 *Annual Report COAHE, 1981-82,* 2.

22 Ibid, 2, 8-9.

23 Ibid, 14, 23.

[24] Glenn Hardy, written contribution to this manuscript.

[25] G.J. Weidemann and W.C. Yearian, eds., *Alternatives: Accomplishments of the University of Arkansas Alternative Pest Control Center1989-1995* (Fayetteville: Arkansas Agricultural Experiment Station, Spec. Rep. 180, 1995), 27, 36.

[26] *Hardy Interview*, 18; *Annual Report COAHE, 1985-86*, "International Programs," 5-6.

[27] Glenn Hardy, written contribution to this manuscript.

[28] Ibid.

[29] *Annual Report COAHE, 1985-86*, "International Programs," 3-4, *1986-87*, "International Programs, 2-3.

[30] Ibid.

[31] Ibid.

[32] Ibid.

[33] *Annual Report COAHE, 1982-83*, 5, 11, 21.

[34] Ibid, 7, 14.

[35] *Hardy Interview*, 14.

[36] *Annual Report COAHE, 1983-84*, 9-10.

[37] Ibid.

[38] Ibid, 14-15.

[39] Ibid, 17-20, 27.

[40] *Annual Report COAHE, 1984-85*, 1.

[41] Ibid, 2.

[42] Ibid.

[43] Ibid.

[44] *Annual Report COAHE, 1985-86*, 1; Danbom, 265-266.

[45] Glenn Hardy, written contribution to this manuscript.

[46] *Annual Report COAHE, 1985-86*, 3, 40.

[47] *Annual Report COAHE, 1986-87*, "Honors Program"; *Arkansas Alumnus* (November 1989), 5.

[48] *Annual Report COAHE, 1986-87*, "Internships," 1-4.

[49] *Annual Report COAHE, 1986-87*, "Food Science,"1-6.

[50] *Annual Report COAHE, 1986-87*, 24, "Animal Science," 3.

[51] *Annual Report COAHE, 1986-87*, 8, 12, "Plant Pathology,"1, 10.

[52] Glenn Hardy, written contribution to this manuscript.

[53] *Annual Report COAHE, 1986-87*, 35-36.

[54] Glenn Hardy, written contribution to this manuscript.

[55] *Arkansas Alumnus* (April 1986), 4-5.

[56] *Arkansas Alumnus* (December 1988), 18; *Arkansas Farm Research* (Winter 1992-3), 8-9.

[57] *Annual Report UAF*, 1987-88, 37-38, 41; *Arkansas Alumnus* (May 1989), 7.

[58] *Annual Report UAF*, 1988-89, 9.

[59] *Arkansas Alumnus* (December 1988), 18.

[60] Weidemann and Yearian, iii; *Arkansas Farm Research* (May-June), 2 .

[61] *Arkansas Alumnus* (September 1989), 6-7.

[62] *Annual Report UAF, 1988-89*, 11.

[53] Ibid, 36.

Chapter Eight:

Toward the 21st Century: The 1990's

The 1990's were years of momentous change in the College of Agriculture. The scope of the change is perhaps best illustrated by the name change approved by the Board of Trustees in July 1995. The new name, the Dale Bumpers College of Agricultural, Food and Life Sciences, and the elevation and renaming of the Home Economics Department to the School of Human Environmental Sciences were both parts of a larger effort to redirect the College's mission to prepare for the 21st century. Advances in biotechnology, genetic engineering and the processing and marketing of agricultural products changed the way that agricultural education approached the scientific study of agriculture. The name changes reflected more accurately what the teachers and students did at the College, but the changes were also intended as a kind of public relations announcement. The College was saying that its programs addressed public concerns with the environment, food safety and other broader societal issues. While still involved in teaching of production agriculture as a fundamental concern, the College had steadily expanded its mission over the years. Barely pausing to shake the chalk dust from their hands or to change their lab coats, the faculty and administration were well positioned to contribute toward yet another revolutionary transformation of American agriculture.

The decade of the 1990's opened with a year of planning for the future. Dean Musick directed that each department should make a thorough review of their curricula and bring them in line with the new University core curriculum. Adapting the course work to keep ahead of the times enabled the College to offer majors that prepared agriculture students for the 21st century. Internationalism and writing across the curriculum were encouraged as part of any revisions. A.E. Einert and the father-son team of R.C. and C.R. Rom combined their efforts to develop a "freshmen experience" program in the Department of Horticulture and Forestry. The Roms would serve as instructors in the unique course that also served as a model for the development of the Freshman Scholars Seminar. Other curriculum innovations included changing the food science option in the Department of Home Economics to hospitality and restaurant management to more accurately reflect the focus of the program.[1]

Battling the onus of declining enrollment, recruitment efforts continued to be a focus for the College throughout the 1990's. The Agronomy Department took the lead in this area. Martha Davis directed recruitment activities that included visits to high schools around the state, visibility at meetings of agricultural organizations, building displays stationed at key locations across the campus, mail outs and the development of a recruiting video. Faculty members also established a fund used to bring interested high school students to the department for on-campus visits. Eighteen Beginning Scholars were brought to study in the department during June-July 1990, and the visits were followed up with phone calls and letters to students who expressed an interest in continuing agronomy studies at the University. Unfortunately, the Honor's program was put on hold in 1990 due to lack of funds.[2]

Enrollment 1990-99.

Year	1990	1991	1992	1993	1994	1995	1996	1997	1998
Agriculture	532	574	587	560	588	608	595	591	629
Home Economics	296	231	281	260	249	275	260	245	274
Graduate	258	267	265	253	217	224	201	225	246
Total	1086	1072	1133	1073	1054	1107	1056	1061	1149

Fall enrollment figures provided by Dean's Office, DBCAFLS.

Once the students were at the College, measures to retain them included an expanded scholarship program. During 1990-91, the program awarded scholarships worth $166,500 to 182 students. The Internship/Cooperative program continued to enlarge its scope as well, providing students with internships in agricultural economics and rural sociology, agronomy, animal and poultry science, food science and home economics. The program also took on international dimensions with internships in the United Kingdom and Europe.

The various departmental clubs also did their best to reach out to new students. The Agronomy Club sponsored a "get acquainted picnic" at Agri Park at the start of the fall semester and followed up with other events throughout the year. Nearly every department had its own student organization at this time. Home Economics sponsored no less than five, a Phi Upsilon Omicron chapter, a chapter of the National Association for the Education of Young Children, an American Home Economics Association (AHEA) chapter, a Fashion Merchandising Club and an Interior Design Club. A newly organized Poultry Science Club joined the pantheon of established departmental clubs that included those in animal science, agribusiness, entomology, food science, horticulture, plant pathology, agronomy and home economics.[3]

Administrators and faculty of the Division of Agriculture and the College began planning capital improvements that would transform the research and teaching capabilities in key areas. Plans for the Center of Excellence for

Poultry Science were developed in 1990. A site immediately west of the Animal Science Building was selected for the projected 112,000-square-foot structure. Faculty members assisted architects in planning the building, visiting other recently completed poultry science facilities at North Carolina State and Texas A&M and submitting outlines of anticipated needs. The faculty also made recommendations concerning the recently funded Pilot Poultry Processing facility, and it was decided that the new facility should be located at the University Farm.[4]

To address the needs of the internationally recognized pest management program at the University, plans were also developed at this time for the Alternative Pest Control Center and a building to house the Center. The first stages of plans for the design and construction of a greenhouse containment facility were made. The new facility was to be used for interdisciplinary research and teaching being conducted by the Departments of Plant Pathology, Entomology and Horticulture and Forestry. The planned facility, with 36,000 square feet, would come to occupy the site next to the Plant Science Building where the old greenhouses were razed in 1988-89. The design of the new building was intended to compliment the architecture of the Agriculture and Plant Science Buildings. These plans would go a long way toward providing necessary experimental and teaching space in this critical field.[5]

> *...I found my agri background prepared me to do anything. I spent 20 years in the Air Force in a highly technical area (avionics) which was possible due to my science courses in college.*
>
> *Claude R. Jones, BSA 1955*

In a related development, the Entomology Department was pleased to have a new 3,400-square-foot insectary completed and their insect host interaction lab renovated and equipped with the latest state-of-the-art technology. With these new facilities, the researchers and teachers would be able to expand their interdisciplinary work on host plant insect resistance studies with plant pathology and continue with the pathbreaking integrated pest management program inaugurated in the department during the 1970's, as well as expand their biological insect control work.[6]

"Interdisciplinary" continued to be the word to describe most of the new programs developed in the Division and College at this time. Not only did the interdisciplinary focus meet the complex needs of much of the research and teaching in agriculture, but the combining of resources to meet experimental and instructional goals made sense in a time of limited maintenance funding. While the sharing of resources was not always possible or desirable in all situations, it did have the advantage of bringing together different approaches to a particular problem.

One example in the instructional program was that of teaching the principles and techniques of biotechnology. The rapid changes in this newly emerging field of agricultural and biological research required five advanced courses to be team taught. Close coordination of lectures and sharing of lab time called for cooperation and sharing of facilities. Such cross-fertilization in the disciplines of plant pathology, horticulture, agronomy and entomology had a positive result in both experimental results achieved and maintaining student interest. The internationally recognized fruit and vegetable breeding program in the Department of Horticulture and Forestry, led by James Moore, George Bradley, R.C. Rom and others, was credited in part to interdisciplinary cooperation among the Departments of Horticulture and Forestry, Plant Pathology, Entomology and Food Science.

The Department of Agricultural Economics and Rural Sociology was also well versed in cooperation in research and teaching. Course work and research were shared among departments, colleges, universities and government agencies at the state, regional, national and international levels. As had been the case for many years, students throughout the university took course work in the department to satisfy requirements in nearly all the colleges. The Rural Policy Research Institute, established at the University through the help of Sen. Dale Bumpers, shared research responsibilities with both the University of Nebraska and the University of Missouri. Department personnel were also involved in projects in Japan, Rwanda, Burundi and Bolivia.[8]

Yet, in spite of all these advances in curriculum development, teaching improvement, student recruitment and retention and capital improvements, there was widespread concern over the deterioration of basic equipment and facilities due to a shortage in basic maintenance funds. Extramural fund raising had reached new levels, and grantsmanship had become an expectation in most departments. But the dependence on so-called "soft money" sources had drawbacks. There was increased competition for available monies that created rivalries inside departments and within the College, not to mention among disciplines nationwide. The process of grantsmanship also took valuable time away from research and teaching duties. The fundamental problem with relying on soft money was that the grants rarely made allowance for overhead costs involved in carrying out long-range projects. Buildings deteriorated, and equipment and machinery broke down, but these costs had to be borne by maintenance budgets, which continued to shrink as costs grew for conducting high-profile research that required increasingly sophisticated equipment. There was no clear solution to the dilemma other than appealing to legislators for higher maintenance budgets or pleading with the University system administrators for a larger slice of appropriated monies.

A new core course program for all undergraduate students in the College began in the fall of 1993. The new core curriculum included six courses. First, the new curriculum required a freshman orientation course patterned after those developed earlier in agronomy and horticulture and forestry. Students would also be required to take specially designed introductory courses in animal and poultry science, plant science, agricultural economics and plant protection and pest management. The highlight of the core curriculum would be a "capstone" course in agricultural issues. The capstone course provided students with an overview of all agricultural disciplines and the attendant critical issues involved with them. Critical thinking, problem solving and written and oral communication skills were emphasized in the class. In a related development, C.R. Rom helped organize a university-wide course, a Freshmen Scholars Seminar, to encourage excellence among incoming freshmen.[9]

To assist faculty members in implementing the new curriculum, the College sponsored a two-day College Teaching Retreat. The meeting brought together 70 teachers from the College where they discussed instructional techniques, possible problems arising from teaching the new curriculum and how teachers' different personalities affected teaching styles. The workshop proved a valuable addition to ongoing efforts to improve teaching and academic standards in the College, and those who participated recommended that such meetings be held on an annual basis in the future.[10]

More major changes were underway in the Poultry Science Department. Poultry Science was designated as a separate department within the College in July 1992. J.H. Denton from Texas A&M University was appointed to head the new department and was made Director of the Center of Excellence for Poultry Science. In a reversal of the trend toward reduced state funding, the Arkansas Legislature earmarked $2.2 million to new hire faculty and increase support for existing poultry science researchers and teachers in order to further the design for the Poultry Center. Private industry also supplied considerable support for the new department. The Tyson Foundation established the Randal Tyson Memorial Scholarship fund endowed with $500,000 for scholarships in the department. Under these circumstances, undergraduate enrollment in poultry science jumped from 30 to 50 in the space of one year. In the most exciting news of the year, the Board of Trustees approved the design plans for a four-story building to house the Center of Excellence for Poultry Science. Construction on the $17.5 million facility and the Pilot Processing Plant was slated to begin in 1993 and be completed by the summer of 1995.[11]

> *The Home Management House experience was something more young people today need, even though I still hate ironing tablecloths to Miss Fergusons' perfection.*
>
> *Mabel Rogers Coleman, BSHE 1964, MS 1966*

All the emphasis on the poultry science program posed a potential morale problem in animal science. With the large influx of funds into poultry science, other livestock commodity producers in Arkansas, especially the beef, dairy and swine industries, expected treatment commensurate to their contributions to the agricultural economy of the state. Consequently, the department developed a strategic plan that anticipated expansion in key areas of concern. Over the next several years the department would hold many meetings with industry and government officials and concerned citizens in search of extramural support. An additional problem with declining student enrollment was brought home after two majors in animal science, the dairy science and the domestic animal biology programs, were eliminated during the year because of low enrollment. The department and the Animal Science Club immediately launched an intensive recruitment and retention program that had some success in arresting the decline and that also boosted morale in the department.[12]

The Department of Agricultural Economics and Rural Sociology received a boost in their research and teaching program during 1991-92. In addition to work begun with the Rural Policy Research Institute under the supervision of Don Voth and Frank Farmer, funding shepherded through Congress by Sen. Bumpers and others in the Arkansas delegation designated the department as the primary location for the new Center for Farm and Rural Business Finance. The department undertook the project in cooperation with the University of Illinois. Bruce Ahrendsen and Bruce Dixon directed the research at the center for the Division of Agriculture. Congress also approved a rice marketing modeling project in conjunction with the University of Missouri at this time. Gail Cramer and Eric Wailes provided their expertise in this project, which was designed to give Arkansas rice producers easier access to international markets. Enrollment increased substantially in the department, from 113 in 1990-91 to 157 in the 1991-92 year. Departmental support for students at this time included 35 internships as well as numerous scholarships. The Agricultural Business Club continued to be very active in departmental affairs. In addition to fund raising activities for student scholarships, the club sponsored a symposium on Agricultural Issues.[13]

Developments in the Departments of Horticulture and Forestry and Agronomy during the year foreshadowed the beginnings of future directions in those department's programs. A.E. Einert developed an upper-level course in landscape and turfgrass management that was team-taught in conjunction with agronomy. In the Agronomy Department the program in environmental science was strengthened by cross listing courses taught in the Fulbright College of Arts and Sciences to ensure that such classes would be taught on a regular basis and to improve accessibility of the courses to a wider range of students.[14]

Another major change in 1992 was the hiring of Milo Shult as the fifth Vice President for Agriculture. Shult came from Texas A&M to replace Preston LaFerney, who returned to the Department of Agricultural Economics and

Milo Shult - Vice President for Agriculture, Director of the Arkansas Agricultural Experiment Station and Director of the Arkansas Cooperative Extension Service, 1992 to present.

Rural Sociology. For the first time in history, the Vice President for Agriculture's primary office was in the University of Arkansas System Offices in Little Rock rather than in Fayetteville. Vice President Shult also assumed the titles of Director, Arkansas Agricultural Experiment Station, and Director, Cooperative Extension Service, and the Division assumed a role equivalent to that of the other "campuses" in the University of Arkansas system. Musick's title became Dean and Associate Vice President for Agriculture-Research, and the two administrative lines, one to the Division of Agriculture and the other to the UAF, were sharpened.

In early 1993 Dean Musick announced his plans to return to teaching and research in the Entomology Department. L.B. "Bernie" Daniels, the Associate Director of the Experiment Station, was named Interim Dean while a national search was conducted for Musick's successor. Dean Musick had directed efforts to improve the academic programs in the College through mandating the development of strategic plans for the departments as well as guiding the development of the new core curriculum. Musick also presided over the new initiatives to enlarge the scope of the poultry science, food science and alternative pest control programs. Important core research programs related to commodities such as beef cattle, cotton, rice, soybean and wheat were also given increased support through the Dean's efforts with commodity organizations to provide check-off funds for basic research in these areas. Leaving these many accomplishments behind was a difficult decision for Dean Musick. He said:

> This is, in part, a selfish act, as I trade the complexity of making the decisions required to support diverse research and educational programs for the luxury of focusing on specific scientific objectives. I am using this forum to express my appreciation to my University colleagues and to the many people who have worked to keep our programs relevant and who use the information from our research and the academic training we provide to make our world a better place.[15]

Keeping the programs at the College relevant was the key purpose of developing the new agriculture core curriculum, which was implemented in the fall of 1993. The departments in the College had revised their curricula and course work as well.

The Department of Agricultural Economics and Rural Sociology revamped their program to correspond to the new University and College curricula. A major in agricultural business became the primary focus of the department. Within this major students could chose four optional plans of study: farm management, agricultural marketing, agricultural business or science. The science plan was intended for those students preparing for graduate study. To increase academic standards, the new program also required a senior seminar course for each student. Two majors, agricultural economics and domestic and international marketing, were eliminated and integrated into the new curriculum.[16]

In agronomy, the new curriculum featured minors in agronomy, environmental science and plant protection/pest management that were available to students in other fields of study. A new major in environmental soil and water science was developed by the department to address the current concern with the impact of agricultural production practices on the environment.[17]

In animal science a five-year strategic plan was developed to expand research, teaching and extension activities in beef and dairy cattle and swine studies. In the wake of the establishment of the Poultry Science Department, the animal science major was redefined, and the new curriculum encouraged students to enroll in a broader spectrum of course work. Responding to a request by the swine industry to provide 100 trained graduates for jobs emerging in the field in Arkansas, Kansas and Tennessee, the department made special efforts to develop an internship program during the year.[18]

The passage of the 1992 Food Nutrition and Labeling Act introduced changes in the food industry nationwide and heightened the need for professionally trained scientific personnel in the industry. The Food Science Department strove during the year to incorporate the new College core curriculum into the Institute of Food Technologists' minimum standards. The department strengthened science training, increased the math load requirements and emphasized oral and written communications skills for its students in order to better prepare them for future job prospects.[19]

The Department of Biological and Agricultural Engineering's strategic plan called for improving contacts with the food industry and a shift in their program toward emphasizing teaching and research in food science. The department also began a drive to recruit more faculty in the area.

The Entomology Department continued to work on their new curriculum throughout the year for implementation in 1994 and scheduled a two-day retreat in January 1993 devoted exclusively to revising the curriculum. People in the department contributed in a significant way to the development of the new College core curriculum as well. M.V. Meisch, with cooperation from faculty members in agronomy, animal science and plant pathology, drew up the core course in plant protection/pest management. The renovation of one of the entomology labs into a molecular biology lab was a major improvement not only for the department but also for programs across the campus.[20]

The Department of Home Economics established departmental minors in human development and family living and in foods and nutrition as a part of their curriculum revision. The department's nursery school, which had been located in the Home Economics Building basement since it was founded in the 1950's, was moved to a newly renovated location on Duncan Street. The move freed up space in the building for the expansion of the interior design program. The space was used to provide a studio, a lighting lab, a work room and an enlarged resource room for the ever-expanding program. In addition to improvements in the curriculum and facilities, another great boon for undergraduates in home economics was a $250,000 endowed scholarship fund established by alumnus Ethel O. Adair.[21]

> *I think the graduate assistantship that I had as a lead teacher in the university nursery school was a great importance in that it not only provided me with tuition for graduate school, but...the experience of working with preschool children and their families [and] supervising staff helped prepare me for my work as a Child Guidance Specialist and Guidance Clinic Director.*
> *Janice Curry Hubbard, BSHE 1989, MS 1992*

While still maintaining its research and teaching commitment to a world-class fruit and vegetable breeding program, the Horticulture and Forestry Department expanded into the area of propagation, raising and care of ornamentals and nursery stock in 1993. The Division of Agriculture was given 27 acres of land adjacent to the University farm to be developed into the Arkansas Landscape Plant Evaluation Center. The McClelland family of Fayetteville donated the land in memory of C.K. McClelland, the well-known Experiment Station agronomist who was with the station from the 1920's to the 1940's.[22]

International Agricultural Programs at the College continued their outreach overseas with new initiatives developed in 1992-93. The Global Marketing Support Services (GMSS) project, headed by Preston LaFerney, brought the talents and technology of the Division of Agriculture and the College in service to Arkansas producers to enable them to compete more effectively in international markets. Using electronic databases, international publications and computer network services, GMSS provided agricultural industries and private individuals with the latest information on the state of international markets.[23]

Robert Frans from the Department of Agronomy and Laurin Wheeler from the Department of Horticulture and Forestry, with the aid of International programs and USAID, put together a university linkage project between the UAF College of Agriculture and Home Economics and the Universidad Privada de Santa Cruz de la Sierra in Santa Cruz, Bolivia. In the project, Frans and Wheeler assisted the Bolivians in setting up a Department of Ecology at their university to study environmental issues in Bolivia, including reforestation, natural resource management and water and soil engineering and conservation. Tom Westing from the University and Andy Martinez from Winrock International developed a long-term collaboration with the former Soviet republic of Kyrgyz that would include two-way exchanges of faculty and students. The College's involvement in Burundi was extended for two more years in 1991, and the last phase of the project was being evaluated in 1993. A new involvement in Haiti was also initiated during the year under the direction of Richard Swanson. James Ferguson and Hani Muhtar from the Department of Biological and Agricultural Engineering taught the Experiment Station Operations Management class in the summer of 1993 for 18 participants from Egypt, Burundi and Rwanda. A special five-week course in station budgeting, taught by Phil Serafini, was held in the West African country of Mali during the year. Many international students from a wide range of developing nations were brought to the College for graduate studies as the USAID-funded project continued to expand throughout the 1990's.[24]

1994 was a year of major administrative changes that would influence the direction of the College's development in meeting the challenges of the 21st century. After a year-long, exhaustive search, the Board of Trustees appointed Charles J. Scifres as Dean of the College and Associate Vice President for Agriculture-Research. As Dean of the College, Scifes reported to the Chancellor of the Fayetteville campus. As Associate Vice President for Agriculture and Chief Operations Officer for the Arkansas Agricultural Experiment Station, he reported to the Vice President for Agriculture, headquartered in Little Rock.

Charles J. Scifres - Dean of the College and Associate Vice President - Research, 1994 to present.

Scifres' background in teaching, research and administration made him ideally suited for the new position. Scifres earned his B.S. and M.S. at Oklahoma State University and his Ph.D. at the University of Nebraska, all in agronomy. He began his teaching and research career at Texas A&M on the range science faculty in 1968 where he served with distinction until 1987. During his career at Texas A&M, Scifres developed the idea of integrated brush management and studied the ecological impact of various rangeland management practices. He held the Thomas M. O'Connor Endowed Chair as professor of Weed Science at Texas A&M from 1982-87 and then went on to become professor and head of the Agronomy Department at Oklahoma State University in 1987 and Associate Director of the Oklahoma Agricultural Experiment Station in 1990. In 1992 he was appointed to serve also as Director of the Center for Agriculture and the Environment at Oklahoma State where his duties included "developing opportunities for the advancement of agriculture consistent with the protection and enhancement of the environment."[25]

Dean Scifres began his duties at the University of Arkansas in March 1994 and immediately outlined the challenge of managing change in such a way as to accomplish positive outcomes within the highly constrained state maintenance budgets. According to Dean Scifres, the need to relate to a broader set of publics was a key component in facing the challenges ahead and called for developing a vision that "embraces change appropriately and anticipates the needs and desires of all our publics." Accordingly, Dean Scifres appointed a blue-ribbon panel, headed by James Moore from horticulture, to study the appropriateness of the College's name and to make recommendations for a name change that would more accurately describe the College's mission. Scifres also formed a Dean's Faculty Advisory Council, which included department heads, Experiment Station administrators and selected faculty members, to move forward and develop a strategic plan that would include a modernized mission statement, set priorities for programs and resource allocation and identify areas where resources need to be redirected.[26]

To help manage the projected changes in the College, the administrative and support staff in the Dean's Office were reviewed. Faculty members were appointed to serve internships in the College and Experiment Station. Libby Wheeler continued and expanded her mentoring activities through the Dean's Office, which she had begun in the 1970's under Dean Hardy. Melissa Lester was appointed as Coordinator of Student Affairs in the Dean's Office to help manage recruitment efforts, serve as a liaison with student organizations and oversee the new Student Ambassador Program.[27]

The Student Ambassador Program put students in the forefront of the public relations efforts. As emissaries for the College, the ambassadors reached out to prospective and newly enrolled students in the College and made them feel welcome, comfortable and involved in College activities. In a similar development in the Home Economics Department, 12 upper-class home economics students were selected for their academic and leadership abilities to serve as peer advisors—counselors, tutors and mentors to incoming students.[28]

Departmental clubs also increased their involvement in student outreach. The Entomology Club operated a concession stand at football and basketball games to fund scholarships and provide grants for students to attend professional meetings. The Food Science Club's graduate students operated a tutoring service for undergraduates taking agriculture prerequisites or food science classes. In order to raise money for an undergraduate scholarship fund, the club also processed two tons of apples into apple butter. Money from the sales was also used to send students to professional meetings. The Horticulture Club was awarded the University's Outstanding Club Award for their ongoing efforts to increase student involvement in the College's activities during the year.[29]

In addition to the new planning initiatives, new courses and majors were developed. The minor in agricultural mechanization was moved from the Department of Biological and Agricultural Engineering to the Department of Agricultural and Extension Education. In conjunction with the move, James A. Ferguson, who had developed and directed the program for many years, was transferred to Agricultural and Extension Education to direct the program in its new location and to teach the agricultural mechanics classes. In agronomy the new major in environmental soil and water science had attracted 40 students. The department initiated a new minor in turfgrass management to satisfy the call for more trained professionals in the field. Food Science established a senior capstone course that emphasized critical thinking and writing skills and made the course a requirement. Other new courses included a class in food law, one in the sensory evaluation of food and a food science orientation class for incoming freshmen. The facilities in the department were upgraded during the year, and a new Food Sensory Lab was completed. The lab included six totally computerized sensory booths, fully automated data collection software and adjustable lighting. In poultry science M.S. and Ph.D. programs were established along with an M.S. program developed in conjunction with the College of Business Administration. All these new programs prepared the way for momentous developments in store for the College in 1995, as the College made giant leaps toward preparing for the 21st century.[30]

> *Being a College Ambassador gave me confidence when speaking one on one as well as to a group. I developed many skills such as working with computers and writing proposals...the best advice I received was from the Ambassador Advisor when she told me to always listen to myself and make myself happy.*
>
> *Patti Priest, BSA 1995*

1995 was indeed a year of "firsts and changes" for the College. The name of the College was officially changed to the College of Agricultural, Food and Life Sciences. The Board of Trustees approved the name change at their November 11, 1994, meeting. At a special July 20, 1995, meeting the Board dedicated the "new" College to Sen. Dale Bumpers in honor of his unceasing efforts on the College's behalf, as well as that of the entire University. The Department of Home Economics also received a new name and was elevated to the School of Human Environmental Sciences. Dean Scifres commented on the name changes in this way:

> The old names reflect the strong traditions upon which our programs have been built, and we will not abandon those traditions. We are proud of our history. The new names more accurately reflect the relevance of our programs to everyday life and to societal concerns that transcend production agriculture. Agricultural and Food Sciences commit us to continued involvement in the food and fiber system from "field to fork." Life Sciences encompass the complex biological processes of plants and animals and related human biological and societal issues that are critical to our mission.[31]

Scifres went on to speak about the concepts that provided the foundation for the School of Human Environmental Sciences:

> Human Environmental Sciences accurately describes the set of principles encompassed by the vintage term *home economics*. What humans eat and wear, where and how they live, how they develop individually and collectively—the human environment—are addressed by HES programs.[32]

Further name changes were also in store during the year. The name of the Department of Horticulture and Forestry was changed to simply the Department of Horticulture. Preparations were made to headquarter forestry teaching and research at the Forestry School at the University of Arkansas-Monticello. It was felt that the change more accurately reflected the major thrust of the program in the department.[33]

One of the major changes of the year, which dramatically expanded the Poultry Science Department's research and teaching capabilities, was the opening and dedication of the Center of Excellence for Poultry Science on September 10, 1995. The John W. Tyson Building was named to honor one of the pioneers of the poultry industry in Arkansas and the founder of Tyson Foods. On hand for the dedication by the Division of Agriculture were nearly 1,000 people, including Governor Jim Guy Tucker, Sen. Dale Bumpers and Sen. David Pryor and Congressmen Ray Thornton, Tim Hutchinson and Jay Dickey. Sen. Bumpers once again was the force behind securing the $10 million in USDA special grants as federal matching funds for the construction of the building, the research feed mill and a pilot processing plant and was the keynote speaker at the dedication ceremony. A $5 million state bond issue and $5 million raised from poultry and utility companies matched the federal funds. An Allied Industries Capital Equipment Campaign solicited continuing sup-

The John W. Tyson Building is the physical home of the Center of Excellence for Poultry Science. (Photo by Shawnna Brannon)

port from the poultry industry in an ongoing effort to raise money to equip the Center's laboratories.

The four-story building was divided into three sections. The laboratory section contained 100 laboratory modules, environmental chambers, walk-in freezers and coolers, teaching labs and a computer lab, a kitchen and tasting booths for sensory evaluation and other research facilities. The office section housed 35 faculty offices with additional space for graduate students and staff. These two wings were connected through an atrium that accommodated the Pioneer Room highlighting the beginnings of the poultry industry in Arkansas, a conference room, a resource room and a lounge. The atrium also connected to the adjoining 172-seat Leland Tollett Auditorium, named to honor the chairman and CEO of Tyson Foods, whose contributions did so much to make the center a reality.[34]

The 19 faculty members of the Department of Poultry Science, headed by James H. Denton, and faculty from other departments and colleges involved in interdisciplinary projects carried out the research, teaching and extension responsibilities of the Center. They were joined by four adjunct faculty in the new Poultry Production and Product Safety Research Unit of the USDA's Agricultural Research Service under the direction of William Huff. Concurrent with the increased potential for expanded research capabilities, the department also raised academic standards and revised their curriculum in order to make the most of the new facilities. The department enrolled 77 undergraduates, 16 M.S. students and 17 doctoral students during 1995, doubling the enrollment figures from 1993. Thirty-one of the students were provided scholarships, and the department arranged for 14 internships during the summer of 1995. A new course added to the curriculum, "careers in poultry science," was a great help to students in identifying their particular interest in the rapidly expanding field and helped students hone skills needed in their career choices.[35]

In addition to the accomplishments in poultry science, the Division of Agriculture also developed plans for the Institute of Food Science and Engineering, to be headquartered on the Fayetteville campus. Three interactive centers would be part of the Institute without walls: Food Processing and Engineering, Food Safety and Quality and Human Nutrition. Justin Morris was appointed to direct the Institute and the Center for Food Processing and Engineering, and Terry Siebenmorgen assisted in coordinating the program. The institute would be a multi-disciplinary unit involving the College, the Experiment Station and the Extension Service to deliver scientific information and educational programs related to the processing, storage and marketing of food products. The food processing industry pledged considerable support for the new effort in providing funding to update equipment and facilities for the center.[36]

Other changes occurred in the administration. Bernie Daniels decided to return to teaching and research in the Department of Animal Sciences. Dr. Greg Weidemann, plant pathology faculty member, accepted the position of Associate Director, Agricultural Experiment Station, and Associate Dean for Research in the College.

In other developments during the busy year, the Strategic Planning Committee continued to work on a new strategic plan for the College and Agricultural Experiment Station. The strategic plan addressed research and extension needs of the Division of Agriculture and worked out a program for implementing their recommendations. *Advantage Arkansas*, as the plan for the College and Experiment Station was called, was the result the committee's year-long efforts. The goal of the plan was to set priorities and identify areas in which the redirection of resources could best accomplish preparing students, faculty and the public for the future of agriculture in the 21st century. In the words of Dean Scifres, the goal of the administration, faculty and staff of the College is "not just to respond to change, but to be proactive agents of positive change."[37]

Changing trends in funding for agricultural research and education presented a challenge. With continual shortfalls experienced in the amount of real dollars provided from state and federal sources for maintenance and equipment upkeep at the College and Experiment Station, difficult decisions had to be made regarding resource allocation. Essential base programs were identified and maintained, and investments were made in new programs through redirecting resources and targeting priority needs. The whole process was dependent on keeping lines of communication open among administrators, faculty, students and public and private constituents

A number of initiatives were undertaken during the year to open and maintain communication channels. The Dean periodically updated the faculty, staff, students and alumni relative to important issues with Kiplinger-syle "Dean's Notes." In addition to the Dean's Faculty Advisory Council, a

Student Forum was initiated. Designed to foster communication between the administration and the student body, the Forum met once a month with Dean Scifres, giving students the opportunity to voice concerns and discuss ideas for improving the learning environment in the College. College alumni were also active in the campaign to bring the needs of the College to public attention. The Dale Bumpers College of Agricultural, Food and Life Sciences Alumni Society was officially chartered by the Arkansas Alumni Association in the summer of 1995. James T. Batchelor was elected the first President of the new organization with Dale Reed as Vice President and M. Portia Short as Secretary-Treasurer. The new alumni society, eager to support the College, quickly established a scholarship. They then established the first recognition of faculty for advising and mentoring students, the Alumni Advising Award.[38]

Getting the word out on a regular basis regarding the accomplishments and the needs of the College was taken care of by the able and dedicated staff at Agricultural Publications. Ray Barclay, Howell Medders, Elaine Williams and Nancy Wyatt combined their talents and energy in providing timely and interesting information about the College to the public. A new magazine, *Arkansas Land and Life*, devoted to the accomplishments of the Division of Agriculture and the connection between the College and Arkansas agriculture, was first published in February 1995. The new magazine joined other new or updated publications, including *Vision*, the bimonthly newsletter of the Experiment Station; *The Graduate*, the College alumni newsletter; and *Partners*, newsletter of the Center of Excellence for Poultry Science, in an ongoing effort to keep communication channels open. Showing that excellence was not confined to teaching and research activities in the College and Experiment Station, in 1996 *Arkansas Land and Life* placed second in the National Agricultural Alumni and Development Association Contest for the best alumni magazine in the country, and *Partners* placed first in the alumni/donor newsletter category.[39]

Agricultural and Extension Education initiated three programs in 1995 that spoke directly to reaching out to a wider clientele. One project, undertaken in conjunction with the Agricultural Experiment Station, was a study to identify and overcome obstacles for minorities in professional roles in agricultural sciences and technologies. George Wardlow, Freddie Scott and Donna Graham led the research team that interviewed and observed minority professionals and high school students in Lee and Columbia Counties to gather data for the first phase of the project. Wardlow and Donald M. Johnson also began a four-year project that developed a model agriscience program for at-risk students. One other program, headed by Wardlow and Scott, involved curriculum development through the use of on-line data provided through a $5,000 grant from the Arkansas Farm Bureau.[40]

Under the leadership of their new department head, Stephen Myers, the Horticulture Department implemented several innovations in their instructional program that indicated an expansion in their program in new directions. An internship program in horticulture and landscape design and urban horticulture was also developed at this time to provide pre-professional training and linkages with private business groups and individuals in this increasingly important area of the "new" agriculture. C.R. Rom took the instructional program in horticulture in a new direction through his development of an on-line course for introduction to plant science. He also set up a class-wide e-mail discussion group that was well received by the class and led to increased participation in "virtual" classroom discussions via electronic means for students who otherwise would not have joined class debates.[41]

The year of 1996 was another year of dedication for the people at, and involved with, the College and the Division of Agriculture. Early in the year, new Animal Science head Keith S. Lusby announced the outlines of the "Animal Science 2000 Capital Improvement Campaign." The goal of the plan was nothing less than to elevate the animal science program to a level commensurate with the importance of the livestock industry in Arkansas and to develop a program second to none in the nation. The initial campaign was funded by USDA special grants of $1.87 million and a state bond issue for $930,000, to be buttressed by $1 million in matching contributions. Over the next four years, an intensive campaign was waged to elicit federal, state and private support for the construction of new facilities and the renovation of existing facilities. Seventy-two livestock industry leaders volunteered to lead the campaign in the private sector to solicit funding and support for the total bill of the projected improvements, which was expected to reach $6.4 million after plans for the renovation of the Animal Science Building were announced later in the year.[42]

Plans were drawn up for renovation and construction of animal science facilities at the Batesville Livestock and Forestry Substation, the Savoy Research Farm and the "University Farm," which had been newly redesignated the Arkansas Agricultural Research and Extension Center. The area began to take on the characteristics of a campus rather than a farm. An agreement was reached with athletics to establish a cross-country track around the emerging campus. The Feed Mill was repaired, old structures were razed, and new fences were built. Also the pasture on the west side of the farm was designated as a research site for ongoing endophyte fescue experiments. The focus of the renovations was a new multi-purpose livestock arena and activities center, an equine facility and a new meats laboratory. Tom Westing, recently reassigned to the Animal Science Department from International Programs, directed the efforts toward developing an Equine Center, holding five seminars for local and regional equine enthusiasts to drum up support for the project. The Batesville Livestock and Forestry Branch Station was put under the direction of the head of the Animal Science Department, Keith Lusby. The obsolete dairy operation at the facility was eliminated, and research concentrating on beef and forage systems was led by Hayden Brown, Bernie Daniels (fresh from the Director's Office) and Ken Coffey. A new office building, feed handling facilities and other research/teaching/extension facilities were constructed at the station. Construction of new facilities at the Savoy Beef Research Farm and long-range plans for the development of swine and cattle research at the farm were also underway at this time.[43]

Enthusiasm generated by the new initiatives spilled over in dedication ceremonies held in the spring of 1996 for the new Alternative Pest Control Center and the (re)dedication of the College. Ceremonies for the Harry R. Rosen Center for Alternative Pest Control were held on April 14, 1996. Named for the noted University of Arkansas plant pathologist, Harry R. Rosen, whose career at the University spanned the years from 1918-1958, the building was made possible with $5.5 million in USDA special grants secured through the efforts of Sen. Bumpers, state funds and private contributions. A substantial gift toward making the Rosen Center a reality was made by Rosen's daughter Miriam and her husband, Judge William Enfield. The new 36,000-square-foot facility provided state-of-the-art controlled environment greenhouse facilities, labs and classrooms for teaching and research. A conference room in the building was named in honor of Distinguished Professor of Plant Pathology George H. Templeton for outstanding and innovative contributions to the field of alternative pest control. In the words of the keynote speaker at the

The Harry R. Rosen Center for Alternative Pest Control, dedicated in the spring of 1996, was named for noted University of Arkansas plant pathologist Harry R. Rosen.

dedication, David McKenzie, the Rosen Center would help Experiment Station and College scientists, teachers and students "harness the power of living organisms...to make agriculture more environmentally friendly."[44]

The environment at the College was indeed friendly on the next day. April 15, 1996, as the College officially dedicated the Dale Bumpers College of Agricultural, Food and Life Sciences. The day's festivities began at 11:30 with a picnic held in the courtyard of the Animal Science Building as the dedication procession gathered before moving to the Bud Walton Arena where the dedication ceremony commenced at 2:00. The Keynote speaker, Secretary of Agriculture Dan Glickman, spoke to 1,200 people gathered at the arena. He mentioned Sen. Bumpers' outstanding achievements in bringing support to the Division of Agriculture and the College for the Alternative Pest Control Center, the Center of Excellence for Poultry Science, the Agricultural Law Research Center, the Food Safety Consortium, the Center for Farm and Rural Business and Finance, the Rural Policy Research Institute and other programs. With these programs the College could stay on the cutting edge of teaching and research to serve the needs of the people into the 21st century. As Glickman said, "Dale doesn't just deserve a building; with all he has done for agriculture, he's justifiably earned a whole college." Bumpers recalled the day he learned that the College was to be named for him and said,

Senator Dale Bumpers

> In my many years of public service, I have been honored many times and in many ways. Other than the honor of being elected governor and senator by the people of Arkansas, this is easily the highest honor ever bestowed on me.[45]

About the future of the College and of agriculture, Bumpers remained confident. Through the College's teaching and the Experiment Station's research programs, further advances would be made in protecting the environment, assuring a plentiful and safe food supply, and improving people's lives. He went on to say:

> We can take great pride in the knowledge that future advances in the production of food and fiber will be developed right here on this campus and in the facilities you've heard about today. We are here to celebrate the lives of those dedicated researchers and scientists

Dedication of the Dale Bumpers College of Agricultural, Food and Life Sciences. Top (l to r) Senator Dale Bumpers, Dean Charles Scifres, Secretary of Agriculture Dan Glickman and James Batchelor, president of the College Alumni Society, at the presentation of the College plaque, which has its permanent home beside the south entrance to the Agriculture Building. Center, the plaque. Bottom, Chancellor Dan Ferritor (center) and University of Arkansas President B. Alan Sugg (right) congratulate Senator Bumpers.

without whom the buildings we have built would be mere bricks and mortar.[46]

Reflecting on the meaning of the dedication, Dean Scifres had this to say: "Arkansans and many other Americans know Dale Bumpers as a public servant devoted to helping individuals and communities improve the quality of their lives. This is precisely how we wish our College to be known."[47]

As part of the strategic plan adopted by the College and Experiment Station, further changes in the program were initiated in the fall of 1996 to consolidate the limited resources available in other areas. The rural sociology program was moved from the Department of Agricultural Economics and Rural Sociology to the School of Human Environmental Sciences. By joining the rural sociology program with the human environmental sciences community development and family services program, it was hoped that there would be closer coordination and expanded development on quality of life issues in the College. Mark Cochran, who had served as interim head for the previous year, was named department head in 1997. Under his leadership, the Department of Agricultural Economics and Rural Sociology was renamed the Department of Agricultural Economics and Agribusiness to reflect the new focus of its program. The undergraduate curriculum was revamped to focus on agribusiness management and environmental issues. Courses in agricultural marketing, which for many years had been a key strength in the department, were also revised, and a capstone course in marketing was developed for implementation in the spring of 1997. In a further consolidation of resources, the turfgrass management program was moved from the Agronomy Department to the Horticulture Department where it would be linked with the ongoing efforts, led by the new head, Stephen Myers, to expand and develop its urban horticulture and landscape design program. The department initiated a turfgrass minor in conjunction with the relocation. Plans were made to develop a major program in this growing field as soon as additional faculty could be hired.[48]

To improve the quality of the student experience at the College, a new freshman orientation program began in the fall of 1996. The Freshmen Academic Support and Tracking (FAST) Program was modeled after a similar program initiated in the Fulbright College of Arts and Sciences in 1994. Groups of up to 20 students could voluntarily register for the program, which allowed the student groups to enroll in nine hours of the same classes, attend the classes together and attend group mentoring sessions and campus events as a unit. The program was designed as a way to encourage student bonding and the development of student social support groups to ease the transition of the students into the College environment. In the Fulbright College the program had succeeded in boosting student retention and academic performance as well as making the college experience more enjoyable for those students involved.[49]

A successor to Agri Day/Agriculture-Home Economics Week was inaugurated in October 1996. The new event, Land Grant Days, was held October 2-5 and revealed that not only was the College a place where serious scholarly work was taking place, but that College administrators, faculty, students, alumni and other supporters could have fun doing it. The events included a "family photo" taken at the beginning of the week of events similar to those taken in earlier years. "Dogs with the Dean" featured Dean Scifres as the master chef serving the public in the finest land-grant college mission tradition. The week's activities also included an Alumni Society meeting where recruiting and College support strategies were discussed, as well as departmental open houses and information sessions for prospective students. The week's events were wrapped up with a brunch at the Student Union, followed by a silent auction to raise money for scholarships for the College. The event was a resounding success in putting across the message that the land-grant mission was still alive at the University of Arkansas and that some of the most exciting and innovative teaching and research programs were centered in the Dale Bumpers College of Agricultural, Food and Life Sciences and the Arkansas Agricultural Experiment Station.[50]

"Dogs with the Dean" chefs Charles Scifres (left) and Jason Apple.

The year of 1997 opened with new initiatives in the Animal Science 2000 Campaign. The Arkansas Cattlemen's Association pledged $100,000 to get the campaign off on the right hoof, and the Arkansas Farm Bureau pledged to match individual contribution from country Farm Bureau organizations. The Arkansas Pork Producers Association also pledged their support for rennovation and constructions iniatives at the Savoy Research Farm. In the wake of this outpouring of public support for the campaign, Keith Lusby announced that bids would be opened and construction would begin on the Animal Science Activities Center and the renovation of the Animal Science Building during the year.

The Institute of Food Science and Engineering expanded into its second phase of development during the year. The Center for Food Safety and Quality opened in January 1997. Michael Johnson was appointed as chairman of the committee to oversee the center's operations. The center's mission was to assess the risks of food-borne illnesses and harmful chemical residues and to develop technology to improve food safety and quality. Interdisciplinary research and links with private industry would be key elements in the successful operation of the new center.[51]

1997 was a year of change for the campus as well as for the College. A year earlier, Dan Ferritor announced that he was stepping out of his role as Chancellor, a position he had held for 10 years, and would return to the faculty. President B. Alan Sugg formed a search committee of faculty, administrators, students and stakeholders to find a new Chancellor capable of leading the University into the new millenium. After an exhaustive search, John Austin White, Dean of Engineering at Georgia Tech was persuaded to take on the challenges of Chancellor of the University of Arkansas-Fayetteville. White is an alumni and came to campus with a vision for the future and a deep and abiding fondness for his alma mater. Beginning his new job on July 1, 1997, he said "...I've been preparing for this job all my life...."

Chancellor John White.

Chancellor White also brought a fresh respect and support for programs in agricultural, food and life sciences. He made it clear that success of the College was critical to achieving his vision for the University.

> Our vision for the University of Arkansas to be the model state and land-grant university in the 21st Century depends critically on the Bumpers College continuing to grow and excel in its teaching, research, and outreach programs. We expect the faculty and staff of the Bumpers College to continue their tradition of leadership in contributing to a healthier, wealthier, cleaner, greener, wiser, smarter, and safer state, nation, and world.

Later, he indicated his confidence in Dean Scifres by appointing him to chair the search for a new football coach. After a highly visible and exciting search engaging some of the top talent in the nation, the committee recommended Houston Nutt, an Arkansan then coaching at Boise State, to return home to lead the Razorbacks. The first season was a success with the Razorbacks posting a winning season and a trip to face Michigan in the Citrus Bowl.

Changes and innovations in the College curriculum developed during the year that were a part of the overall strategic plan for the College. International Programs, under the direction of Preston LaFerney, began a European studies program during the fall of 1997. Two students, Sylvia Morrison, an environmental soil and water science major, and Joseph Ham, an agricultural and extension education major, spent a semester abroad studying in Scotland at the Scottish Agricultural College in Edinburgh and Auchincraive. Raymond Barclay was sent to Scotland as resident director of the program to coordinate activities on site and develop further exchanges in the future that would include bringing Scottish students to the University of Arkansas. The following year, Carole Locke endowed the Richard D. Locke International Scholarship, in honor of her husband, a noted international agribusiness consultant, to facilitate further studies abroad for students from the College.[52]

The College's outreach to the wider academic world via the Internet also made important advances during the year. David Barnes from poultry science and Charles Rosenkrans from animal science developed an "introduction to animal and poultry science" course for use over the Internet. The course would serve as a model for development and delivery of distance education resources via the Internet in the coming years. Agronomy also developed plans at this time to offer several courses via the Internet for Northwest Arkansas Community College in Rogers, Arkansas, in order to smooth the transfer of students into the department's program. The Internet was also used as an extension education tool as the Agricultural and Extension Education Department posted lesson plans on the Internet for use by Arkansas public school teachers as a public service. A new web page for the College was also developed that included important enrollment and recruitment information for prospective students.[53]

The School of Human Environmental Sciences made considerable progress in developing and implementing their strategic plan during the year, including an expansion of interdisciplinary studies as well as a consolidation of existing instructional resources. A joint program between the interior design program and the School of Architecture provided an experience in interdisciplinary instruction for freshmen in interior design. The degree, Bachelor in Interior Design, was approved to receive its own diploma. Students in the course worked together with architecture students on several residential design plan projects. Students from both schools also took a joint tour of Great Britain to study interior design and architecture there. In order to make the most of scarce teaching resources, the School consolidated two languishing minors in clothing and textiles with the minor in fashion merchandising to create a major in apparel studies.[54]

The next year, 1998, began with as much change as the previous one, including a new face in the Dean's office. Randall G. Luttrell arrived to replace Charlene Mooty, who had retired at the end of 1997 after serving 13 years as Associate Dean. For Luttrell it was a return to his alma mater as he had earned both his M.S. and Ph.D. degrees in entomology at the College before going on to a teaching and research career at Mississippi State Univer-

Randy Luttrell - Associate Dean of the College 1998 to present.

sity. While at Mississippi State he had served as interim head of the Entomology Department and Director of Undergraduate Studies and developed the program there in pest management. Luttrell's experience in the development of undergraduate and graduate academic studies, his interaction with students over the years and his familiarity with the research side of the land-grant college equation made him a good match for his new position.[55]

The progress of the Animal Science 2000 initiative continued at a brisk pace throughout 1998. A new office and conference building at the Branch Station in Batesville was dedicated in May 1998. New beef cattle research and teaching facilities at the Savoy farm, including a cow/calf component for studies in reproduction and nutrition, a stressed cattle facility and an anticipated forage studies building, were scheduled for completion in the near future. A new Swine Center for the farm was in the design stages while the existing unit was refurbished, and an off-site nursery was added.[56]

The 45,000-square-foot animal science activities center, dedicated in April 1999, included a 25,000-square-foot arena, three classrooms, a 9,600-square-foot animal holding barn and a parking lot to accommodate the many public

Pauline Whitaker Animal Science Center.

activities anticipated to be held at the center. The new center was named the Pauline Whitaker Animal Science Center in honor of Mrs. Whitaker and her family, who donated $500,000 toward the completion of the project.[517]

Renovation of the original Animal Science Building got underway in March 1999. The plans included moving the entrance from the south side of the building to the east side courtyard, renovation of classrooms, labs and offices in the building and the construction of a covered walkway leading to the adjoining parking lot. After the renovations were complete, the administrative offices for both the College and the Experiment Station were to be moved into the refurbished building.[58]

Further progress toward addressing public concern with food safety was made during the year at both the national and international levels, and the University of Arkansas was in the forefront of the advances in the field. In the summer of 1998 it was announced that Dean Scifres had taken a leading role in creating the National Alliance for Food Safety (NAFS). Ultimately, 26 universities and the USDA's Agricultural Research Service agreed to coordinate their efforts to address food safety issues. Dean Scifres developed the guidelines for the operation of the NAFS, which worked to bring government agencies, food industry experts, university and college scientists and consumer groups together in a common effort to address public concerns regarding food safety and to increase consumer confidence in the safety and quality of the nation's food supply. An agreement with the United Nations Food and Agriculture Organization allowed the UN agency to establish a secretariat within the Institute of Food Science and Engineering. Justin Morris was appointed as a member of the secretariat's coordinating committee. The purpose of the international initiative was to coordinate efforts to assure that foods imported into the United States from abroad met the same quality and safety standards as those produced in the U.S.[59]

International Agricultural Programs at the College also expanded its mission during the year. The Scottish Agricultural College in Aberdeen and Bumpers College cooperated in establishing a transatlantic master's program in agribusiness. As part of the arrangement, the colleges exchanged faculty as well as students. Eric Wailes, with many years of international work under his belt already, was selected to teach and direct research in Aberdeen, and Garth Entwistle came from Aberdeen to Arkansas for a semester. Four American students and three Scottish students participated in the program that required them to either develop a thesis project or begin an internship in their host countries.[60]

Liaison among academic institutions in Arkansas also made advances during the year. Dean Scifres was instrumental in developing a coordinated academic program among Arkansas universities and colleges and the two-year colleges in the state. The Arkansas Consortium for Teaching Agriculture (ACTA) initially included the Bumpers College, the University of Arkansas at Pine Bluff's Agriculture College and six community colleges across the state. As Dean Scifres noted:

> Our purpose is to promote quality education and to improve access for all Arkansans to the bachelor's degree program in our college....We want to do all we can to help students with an interest in agricultural, food and life sciences make as seamless a transition as is possible to the University of Arkansas.[61]

Courses in introductory animal and poultry science and environmental soil and water science meeting the requirements previously taught via the Internet now would be offered in classes at community colleges in Batesville, Hope, Harrison, Mena, Rogers and West Memphis. Animal scientists at the Batesville Branch Station and the Southwest Research and Extension Center at Hope also taught classes in community colleges that were accepted by the College. Additional courses in agronomy, horticulture, animal science, agricultural education and other subjects were reviewed and adapted to better meet the standards at the Bumpers College. The program was designed to benefit all involved. The recruitment program in the Bumpers College was able to draw on the fastest-growing higher education student population in the state, those students in Arkansas' community colleges. When the students arrived at the College, they were better prepared to begin advanced studies. The community colleges benefitted from higher academic standards and the dissemination of the most current research and teaching information. Arkansas industry also gained through having a better-trained pool of graduates from both the two-year institutions and the College.[62]

Students arriving at the Bumpers College in the fall of 1998 found a new curriculum and several new programs awaiting them. The number of credit hours required for graduation with a Bachelor's degree from the Bumpers College was reduced from 132 to 124. The core agriculture courses initiated in 1993 were abandoned in favor of a more flexible program. The 16-hour core was replaced with nine hours of electives that could be selected from a broad field of subjects outside the major field but still within the College.[63]

The Department of Agricultural and Extension Education was restructured with the merging of the department with Agricultural Publications, which became the Communications Services unit within the department. Communication Services was headed by Howell Medders, who had served as interim director of Agricultural Publications since 1997. Under the leadership of Don Herring, the Department of Agricultural and Extension Education also developed a program of four areas of concentration: agricultural education, extension and industry education, agricultural systems technology management and agricultural communications. Students in all four options complete the program with a degree in agricultural education, communication and technology. In addition, George Wardlow and Don Johnson obtained a USDA grant in 1998 to open a College Teaching Resource Center. They serve as co-directors of the center.[64]

Expanding the horticulture program further in the direction of the "green industry," one of the rapidly developing areas in the horticulture industry in Arkansas and the nation, was accomplished by department head Stephen Myers during the year through establishment of a turfgrass major in the department. Michael Richardson, a specialist in golf and sports turf management, was appointed earlier in the year to direct the new program along with John King, who had brought the turf program from the Agronomy Department. The four-year program was the only one of its kind in Arkansas and required a broad background in agronomy, horticulture and landscape design as well as management techniques and soil and water engineering. In addition to Richardson, the department also brought in Jon T. Lindstrum, a specialist in woody ornamental and native plants for landscape design to further buttress the green industry component of the program. In line with these new initiatives, the department also developed a new specialization in horticulture merchandising. The program combined traditional horticultural studies with business/management techniques in order to prepare graduates to start their own nursery or landscaping businesses, one of the fastest growing business areas in the nation.[65]

One other program developed that fall was particularly keyed to developing research programs in the Experiment Station and training the College's students in the revolutionary transformation of agriculture taking place through the use of biotechnology and molecular-cellular biology. Plant pathology brought in Yinong Yang to develop research and courses in the field.[66]

The commercial development of transgenic crops, genetically engineered livestock varieties and other developments in the "new biology" during the 1990's has been touted as the third great transformation of American agriculture. These new developments will undoubtedly have a greater impact on agricultural production, research and education than either of the two "great transformations": the revolution in agricultural mechanization and education that followed the Civil War and the introduction of agricultural chemicals and mechanization that followed World War II. The College and Experiment Station responded to the transformation through its strategic plan, which called for enhancing existing programs and creating new ones, as Dean Scifres noted, to "position our research and education efforts to directly support the needs of Arkansas and its citizens." A proposed doctoral program in molecular and cellular biology, Scifres said,

> ...would draw on the campus-wide capacity to coalesce faculty expertise across programs in biology and agriculture. By drawing on existing strengths and progressively adding expertise, we believe that a high-quality graduate program will be the result.[67]

Dean Charles J. Scifres congratulates 1999 Senior Scholars William and Gwenith Gordon. The husband and wife spoke at commencement. They had the same grade point average of 4.0 and the same major: Agricultural Education, Communication and Technology. (Photo by Fred Miller)

As a part of Chancellor White's redirection of campus resources to high priority opportunities, the proposed degree in cell and molecular biology received special funding for development in 1999. The proposal, championed by Dean Scifres, emphasized a campus-wide collaborative degree program with Bumpers College and Fulbright College taking the lead. Administration of the program was initially headquartered in Bumpers College with John Kirby, the new interim associate director and associate dean, accepting responsibility as the first faculty director for the program. Such programs assured that the College would stay in the forefront of agricultural teaching as the pace of change in modern agriculture accelerated and the 21st century approached.

The combination of improved facilities in key areas, innovative programs that spoke to the needs of Arkansans and a resourceful recruitment program, as well as an excellent student support and advisement system, resulted in the first increase in enrollment for the College in 12 years. The total fall 1998 enrollment stood at 1,149: 903 undergraduate and 246 graduate students. The largest expansions took place in the Departments of Horticulture and Animal Science, with increases of 30 percent and 28 percent, respectively. The interior design and poultry science programs showed increases of over 20 percent. Associate Dean Luttrell said he expected the enrollment to continue to grow because the College's degree programs, ranging from interior design to biotechnology, appealed to a wide range of students. Lutrell reflected further on the reasons for the enrollment increase:

> We will assure that our academic programs continue to reflect the needs of society. We will always embrace our land-grant heritage of teaching, research and service to meet those needs....Our unique ability to place students in learning environments with innovative research and extension scientists will remain our strongest asset.[68]

In the precedent-setting spirit of recognizing excellence in teaching, friends of Jack Justus, 1997 Outstanding Alumnus of the College and former Executive Vice President of the Arkansas Farm Bureau, established the Jack G. Justus Award for Teaching Excellence. The first recipient (1999) was Craig Beyrouty, Department of Crop, Soil, and Environmental Sciences (formerly Department of Agronomy).

As always, at the heart of the College's program were the needs of the people, whether students, consumers, producers or processors of agricultural products. The administrators, scientists and teachers at the College recognized that the public interest remained their special imperative.

Much had changed in agriculture and agricultural education in the years since Thurston delivered his first lectures on agriculture at Arkansas Industrial University in 1872 and Purinton and Menke began their first agricultural experiments in 1888-89 at the Arkansas Agricultural Experiment Station. Yet, while the mission had expanded over the years to include a broader public than "the industrial classes" targeted in the Morrill Act, the emphasis on science in the public interest had remained constant. Undoubtedly, the earlier promoters of agricultural education and research at the University over the years would have approved of the mission statement contained in the 1996 strategic plan:

> The mission of the Division of Agriculture is to improve the quality of life for all Arkansans through excellence in education, research, service and the transfer of knowledge. This will be accomplished through superior programs that prepare the next generation of leadership; research that reduces barriers to social and economic well being of the state and its people; providing research-based information through informal education to help Arkansans improve their economic well being and the quality of their lives; and providing services and leadership within the University of Arkansas System and to the State essential to achieving our potential.[69]

With the development of innovative programs on the horizon of agricultural research and education during the 1990's, the people in the College and Experiment Station, administrators, faculty, staff and students were well prepared to accept the challenges of the 21st century. As Sen. Bumpers has said, all the ingenious programs and modern facilities in the world mean nothing without a corps of dedicated people working together to bring the results of their combined efforts to a wider audience. Since the early struggles to establish agricultural education and research at the University during the nineteenth century, through the trials of war and depression of the twentieth century, the College and Experiment Station, collectively, have emerged as one of the premier research and teaching institutions in the land-grant system. As the new century approaches, both the people who have made the history of the Dale Bumpers College of Agricultural, Food and Life Sciences and those who will contribute to its future can be justifiably proud of their accomplishments.

Endnotes

1 *Annual Report COAHE, 1990-91,* 2-3.
2 Ibid, "Agronomy," 7.
3 Ibid, passim.
4 Ibid, "Animal and Poultry Science," 1.
5 Ibid, "Plant Pathology," 2-3; *Arkansas Farm Research* (May-June 1990), 2.
6 Ibid, "Entomology," 4.
7 Ibid, "Horticulture and Forestry," 14.
8 Ibid, "Agricultural Economics and Rural Sociology," 1-2, 7, 14.
9 *Annual Report COAHE, 1991-92,* 1.
10 Ibid.
11 *Annual Report COAHE, 1991-92* ,"Animal and Poultry Science," 10-11; *Arkansas Alumnus* (Spring 1992), 4.
12 Ibid.
13 Ibid, "Agricultural Economics and Rural Sociology," 1-2, 5-7.
14 Ibid, "Horticulture and Forestry," 18, "Agronomy," 4.
15 *Arkansas Farm Research* (July/August 1993), 2.
16 *Annual Report COAHE, 1992-93,* "Agricultural Economics and Rural Sociology," 1.
17 Ibid, "Agronomy," 2, 7, 14.
18 Ibid, "Animal Science," 1-3.
19 Ibid, "Food Science," 3-4.
20 Ibid, "Entomology,"1-2.
21 Ibid, "Home Economics," 2-4, 23.
22 Ibid, "Horticulture and Forestry," 1-2.
23 Ibid, "International Agricultural Programs," 1-3.
24 Ibid.
25 Charles J. Scifres, "Professional Brief" and "Curriculum Vitae," 1-5.
26 *Station News* (March-April 1994), 1, 5, (May-June), 1, 6.
27 *Annual Report COAHE, 1993-94,* 2.
28 Ibid, "Home Economics," 3, 45.
29 Ibid, "Entomology," 7, "Food Science," 8, "Horticulture and Forestry," 6.
30 Ibid, 5, "Agricultural and Extension Education," 1-2, "Agronomy," 1, "Food Science," 2, 8, 11 "Poultry Science," 1-2.
31 *Vision* (November 1994), 1.

[32] Ibid.
[33] *Annual Report COAHE, 1994-95,* "Horticulture and Forestry," 9.
[34] *Arkansas Land and Life* (Fall/Winter 1995), 14-17.
[35] *Arkansas Land and Life* (Fall/Winter 1995), 17; *Annual Report COAHE, 1994-95,* "Poultry Science," 12; *Partners* (Spring-Summer 1996), 1-2.
[36] *Annual Report COAHE, 1994-95*, 1-2, "Food Science," 17.
[37] University of Arkansas College of Agricultural, Food and Life Sciences, *Advantage Arkansas* (Fayetteville: University of Arkansas College of Agricultural, Food and Life Sciences, 1996), 11.
[38] *Annual Report COAHE, 1994-95*, 1-2; *Vision* (January 1995), 1-2, (July 1995), 1, 4.
[39] *Vision* (July 1995), 1; *The Graduate* (September 1996).
[40] *Annual Report COAHE, 1994-95*, "Agricultural and Extension Education," 7-8.
[41] *Vision* (March-April 1996), np, online version at *www.uark.edu/agripub/Publications/.* All online versions at same site unless otherwise noted.
[42] *Vision* (January-February 1996), np, online version; *Annual Report COAHE, 1995-96*, "Animal Science," 1; *Arkansas Land and Life* (Winter 1996), np, online version.
[43] Ibid.
[44] Ibid (May-June 1996), 3.
[45] *Arkansas Land and Life* (Fall/Winter 1995), 5.
[46] *Vision* (May-June), 3.
[47] *Arkansas Land and Life* (Fall/Winter 1995), 5.
[48] *Vision* (September-October 1996), np, online version.
[49] Ibid (July-August 1996).
[50] Ibid (September-October 1996).
[51] *Vision* (January-February 1997).
[52] *Vision* (September-October 1997); *The Graduate* (Fall 1998), 3.
[53] *Annual Report COAHE, 1996-97*, 1-2.
[54] *Annual Report COAHE, 1996-97*, 3.
[55] *Vision* (March-April 1998), 1.
[56] *Vision* (January-February 1998), 5.
[57] *Vision* (November-December 1997).
[58] *Vision* (November-December 1997), *Dean's Notes*, November 18, 1998.
[59] *Vision* (July-August 1998), 5; (September-October 1998), 6.
[60] *Vision* (July-August 1998), 1.
[61] *Arkansas Land and Life* (Spring/Summer 1998), 7.
[62] Ibid.
[63] *The Graduate* (Fall 1998), 3-4.
[64] *Vision* (March-April 1998), 2, (November-December 1998), 6.
[65] *Vision* (March-April 1998), 5; *News Release*, July 23, 1998, Communication Services, online Version; *Arkansas Traveler*, August 24, 1998, 4.
[66] *Vision* (May June 1998), 5, (November-December 1998), 1, 6.
[67] *Vision* (November-December 1998), 1, 6.
[68] *Arkansas Land and Life* (Fall/Winter 1998), 25; *The Graduate* (Fall 1998), 3.
[69] *Advantage Arkansas, the Arkansas Connection: Strategic Plans for the Dale Bumpers College of Agricultural, Food and Life Sciences, the Arkansas Experiment Station and the Cooperative Extension Service, University of Arkansas* (Fayetteville: Dale Bumpers College of Agricultural, Food and Life Sciences, 1996), 3.

APPENDIX I

John W. White Awards

Each year, awards honoring the late John W. White are given to outstanding faculty and students. Three of the four awards go to outstanding faculty for their contributions in teaching, or research, or extension. All professional employees in the Dale Bumpers College of Agricultural, Food and Life Sciences, Fayetteville, may be considered for the teaching award. All professorial ranks in the Division of Agriculture may be considered for the research awards. All non-classified personnel with the Cooperative Extension service may be considered for the extension awards. The student award goes to the student selected as the outstanding all-around student in the Dale Bumpers College of Agricultural, Food and Life Sciences, including the School of Human Environmental Sciences. Recipients of these awards include the following:

1976	
Glenn Nelson, Professor, Agricultural Engineering	Research
1977	
Mark Waldrip	Student
1978	
Nicholas R. Gyles, Professor, Animal Sciences	Research
Max Meisch, Professor, Entomology	Teaching
Mescal Johnston, Extension Food Marketing Specialist	Extension
Rick Short	Student
1979	
George Templeton, Professor, Plant Pathology	Research
Sam Offutt, Professor, Agronomy	Teaching
Ruel Nester, Extension Agronomist	Extension
Gordon Askew	Student
1980	
Charles Caviness, Professor, Agronomy	Research
Lorene Raymond, Professor, Home Economics	Teaching
Dorris Vick, District Program Leader	Extension
Richard Drewry	Student
1981	
Connell J. Brown, Professor, Animal Sciences	Research

Jack L. Perkins, Professor, Animal Sciences	Teaching
Raymond C. Benz, Extension Agricultural Engineer	Extension
No student designated this year	

1982

Odie T. Stallcup, Professor, Animal Sciences	Research
Robert N. Shulstad, Professor, Agricultural Economics	Teaching
Leo Rainey, District Community Development Specialist	Extension
Kevin Chris Threlkeld	Student

1983

Justin Morris, Professor, Food Science	Research
Joe McFerran, Professor, Horticulture	Teaching
Lowell T. Lankford, Extension Poultryman	Extension
James Alan Hopkins	Student

1984

James N. Moore, Professor, Horticulture	Research
Waymon Halbrook, Professor, Agricultural Economics	Teaching
John D. Shults, SE District Director	Extension
Miriam Flippo	Student

1985

Terry Lee Lavy, Professor, Agronomy	Research
Paul R. Noland, Professor, Animal Sciences	Teaching
Marjorie M. Phillips, Extension Foods & Nutrition Specialist	Extension
Deborah Lee Capps	Student

1986

Park W. Waldroup, Professor, Animal Sciences	Research
John T. Gilmour, Professor, Agronomy	Teaching
Carl Farler, State Specialist-Agricultural Economics	Extension
Dana Renea Brown	Student

1987

Roy J. Smith, Rice Research & Extension Center	Research
Carl Griffis, Professor, Agricultural Engineering	Teaching
Mrs. Jean C. Frisby, County Extension Agent-Staff Chairman	Extension
Ingrid Rathburn	Student

1988	
Seth Y. Young III, Professor, Entomology	Research
Alfred E. Einert, Professor, Horticulture & Forestry	Teaching
Robert Pierce, State Specialist Natural Resources	Extension
Dana Spigner	Student
1989	
Robert D. Riggs, Professor, Plant Pathology	Research
Lawrence R. Oliver, Professor, Agronomy	Teaching
Bobby A. Huey, Extension Agronomist-Rice	Extension
Cecilia Belle Ferguson	Student
1990	
Frederick M. Stephen, Professor, Entomology	Research
Gregory J. Wiedemann, Asso. Professor, Plant Pathology	Teaching
Dorothy Rodgers, SE District Director	Extension
Tracy C. Baltz	Student
1991	
Howard A. Scott, University Prof., Plant Pathology	Research
Craig Beyrouty, Associate Professor, Agronomy	Teaching
Joel N. Ferguson, County Extension Agent-Staff Chairman	Extension
H. Dwight Wynn	Student
1992	
H. Don Scott, Professor, Agronomy	Research
Clifton "Pete" Braker, Professor, Agricultural & Extension Education	Teaching
James McCloud, Assistant to Director	Extension
Jim W. Frisby	Student
1993	
Michael G. Johnson, Professor, Food Science	Research
Donna L. Graham, Associate Professor, Agricultural & Extension Education	Teaching
Alva Ray Siler, County Extension Agent-Staff Chairman	Extension
Jeff J. Simpson	Student
1994	
David O. TeBeest, Professor, Plant Pathology	Research
Duane C. Wolf, Professor, Agronomy	Teaching

Claude M. Bonner, Extension Cotton Specialist	Extension
Jennifer L. Hare	Student

1995

Derrick Oosterhuis, Professor, Agronomy	Research
Luke Parsch, Assoc. Professor, Agricultural Economics	Teaching
Quinton Hornsby, County Extension Agent-Staff Chairman	Extension
Dennis Howell	Student

1996

Walter G. Bottje, Professor, Poultry Science	Research
James A. Ferguson, Professor, Agricultural & Extension Education	Teaching
Lynn R. Russell, Family Resource Management Specialist	Extension
Rebecca Kettle-Gay	Student

1997

Nicholas Anthony, Poultry Science	Teaching
Terry J. Siebenmorgen, Professor, Agricultural Engineering	Research
Lawrence R. Oliver, Professor, Agronomy	Research
Kent Rorie, County Extension Agent-Staff Chairman	Extension
Lanny O. Ashlock, Extension Agronomist-Soybean	Extension
Tracy Leonard	Student

1998

Donald M. Johnson, Assoc. Professor, Agricultural & Extension Education	Teaching
Ford Baldwin, Extension Weed Scientist	Extension
Betty L. Archer, County Extension Agent-Staff Chairman	Extension
Joseph Ham	Student

No research award this year.

1999

Gisela Erf, Assistant Professor, Poultry Science	Teaching
John Kirby, Associate Professor, Poultry Science	Research
John Langston, Ext. Agricultural Engineer, Section Leader	Extension
Betty Oliver, County Extension Agent-Staff Chairman	Extension
Lance Keeling	Student

APPENDIX II

Spitze Land-Grant University Faculty Award for Excellence

1996	George Templeton, Plant Pathology
1997	James N. Moore, Horticulture
1998	Ronald E. Talbert, Agronomy
1999	Fred Stephen, Entomology

APPENDIX III

Outstanding Alumnus Award

1989	Francille M. Firebaugh
1990	Perry Lee Adkisson
1991	Leland E. Tollett
1992	Irma Fitch Giffels
1993	L.D. Williams
1994	Robert G. F. Spitze
	Hazel Taylor Spitze
1995	Sherman D. Cullum
1996	Steven A. Slack
1997	Jack Justus
1998	Mark Bryles
1999	Howard Young

APPENDIX IV

Jack G. Justus Award for Teaching Excellence

1999	Craig Beyrouty, Crop, Soil, and Environmental Sciences

APPENDIX V

College Alumni Society Outstanding Advising Award

1997	Mary Warnock, Human Environmental Sciences
1998	Ed Piper, Animal Sciences
1999	Nick Anthony, Poultry Science

APPENDIX VI

The Honor Society of Agriculture, Gamma Sigma Delta
Senior and Sophomore Awards

The senior award goes to the graduating senior with the highest grade point average. They are given a certificate and a cash award of $100.00. In case of ties, multiple awards are given. The sophomore award goes to the highest g.p.a. student who has completed at least 60 hours credit and is classified as a junior for the next academic year. The sophomore award consists of a certificate, a cash award of $100.00, and frequently they are also awarded a scholarship for the next academic year.

Date	Senior Awardees	Sophomore Awardees
1954-55	Leo Rainey	Johnnie Jenkins
1955-56	Margaret J. Lowe	Laura Hemby
1956-57	JoAnne Rife	Virginia Tucker
1957-58	Hazel Newsome	Elbert Van Wann
1958-59	Martin S. Stauber	William Morton
1959-60	Elbert Van Wann	James V. Gramlich
1960-61	Hollis H. Bowen	Joe Lee Gaston
1961-62	William H. Morton	Sandra Sue Pate
1962-63	Joe Lee Gaston	Judith L. Taylor
1963-64	Fredrick Collins	Virginia M. Ford
1964-65	Joan Dickerman Pharr	John R. Bridges
1965-66	Joyce R. Carter	William J. Fambough
1966-67	John R. Bridges	Shelby C. Bryant
1967-68	Carole B. Bunge	Diana McKinney
1968-69	Hollis T. Reed	Patricia Sue Keeling
1969-70	Diana McKinney	Kathryn Counts
1970-71	Patricia S. Keeling	Ann Stillman
1971-72	Willie Sue Tucker	Vikki Whitlock
1972-73	Mary Ellen Stephenson	Helen K. Crawford
1973-74	Vikki W. Pitts	Cindy C. Carrington
1974-75	Lewis D. Mayfield	William J. Ormsbee
1975-76	Glenda J. Christian	Susan K. Tiemann
1976-77	David L. Griffith	Nancy Jo Oden
1977-78	Carol L. Eisenbauer	Pamela R. McCutcheon
1978-79	Sheryl Satterfield	Kevin Bost
1979-80	Joe Fagan	Andrew T. Bain
1980-81	Dickie Petry	Nicole Nguyen
	David Swayne	
	Kevin Bost	

1981-82	Beverly Baughman	Laura Jacimore
	Sandra Barclay	Brad Taylor
		Lisa Rice
1982-83	William P. Alexander	Anna Sue Westfall
	Tracy R. Archer	
1983-84	Justin Garrett III	Carla Jane Sinor
1984-85	Anna B. Westfall	Debbie Nida
1985-86	Carla J. Sinor	Dina Grammar
		Michael Wood
1986-87	Debbie L. Nida	Doyle J. Nichols
1987-88	Dina G. Alden	Stephanie Hardy
1988-89	Janice C. Hubbard	Kimberly D. Carter
	Elizabeth A. Law	Michael A. Pennington
	Betty S. Loewer	Casey A. Sams
	Brian J. Teppen	
1989-90	Doyle J. Nichols	Megan Atkinson
1990-91	Pamela L. Olienyk	Tricia L. Rozypal
1991-92	Jeff Jaret Simpson	Kristin Lea Cozort
		Keith D. Martin
1992-93	Glenda F. Samuels	Timothy A. Roberts
1993-94	Timothy Simon	Rebecca Kettle
1994-95	Joffrey Mendoza	Maya Scott
	Dennis Howell	
1995-96	Rebecca Kettle-Gay	Lance T. Keeling
		Lori M. Newton
1996-97	No awards this year	
1997-98	Jennifer Attak	Reema Persad
		Lance Keeling
1998-99	William R. Gordon	Paul C. Bryant

APPENDIX VII

Alumni Association Distinguished Faculty Achievement Awards

The Arkansas Alumni Association Distinguished Achievement Awards Program, inaugurated in 1961, recognizes, rewards and encourages superior classroom teaching, and particularly outstanding research or creative activity. Earlier awards were given in two categories—teaching or research. Since 1969, a third category evolved to recognize faculty members known to be both strong researchers and excellent teachers. Currently, recipients of the awards are presented a crystal apple and $1,500.

Past recipients of these awards from the Dale Bumpers College of Agricultural, Food and Life Sciences, the recipient's departments, and type of award for each person include the following:

1961 - Derald A. Slack, Plant Pathology - Research
1962 - Joseph P. Fulton, Plant Pathology - Research
1963 - Paul R. Noland, Animal Sciences - Teaching
1965 - Bradford A. Waddle, Agronomy - Research
1966 - Odie T. Stallcup, Animal Sciences - Research
1968 - Charles L. Wilson, Plant Pathology - Research
1969 - Grover Harris, Jr., Animal Sciences - Teaching/Research
1971 - Joe McFerran, Horticulture & Forestry - Community and University Service
1973 - Charles Lincoln, Entomology - Research
1974 - James N. Moore, Horticulture & Forestry - Research
1975 - Charles E. Caviness, Agronomy - Research
1979 - Max Meisch, Entomology - Teaching
1980 - Jacob R. Phillips, Entomology - Teaching/Research
1983 - Robert E. Frans, Agronomy - Research/Creativity
1987 - George E. Templeton, Plant Pathology - Research
Alfred E. Einert, Horticulture & Forestry - Teaching
1988 - Terry L. Lavy, Agronomy - Research
1989 - K. S. Kim, Plant Pathology - Research
1992 - Gail Cramer, Agricultural Economics & Rural Sociology - Research & Public Service
1993 - Robert Riggs, Plant Pathology - Research
Justin Morris, Food Science - Research & Public Service
1995 - Park W. Waldroup, Poultry Science - Research
1997 - Lawrence R. Oliver, Agronomy - Teaching/Research
1999 - Derrick Oosterhuis, Crop, Soil, and Environmental Sciences - Teaching/Research

APPENDIX VIII

Chancellors, Vice Presidents for Agriculture, Deans and Department Heads

CHANCELLORS

B.A. Nugent	1981-1983
Willard Gatewood, Jr.	1984-1985
Daniel E. Ferritor	1986-1997
John A. White	1997

VICE PRESIDENTS FOR AGRICULTURE

John W. White	1959-1974
James Martin	1975-1980
John Goodwin	1981-1986
Preston La Ferney	1987-1992
Milo Shult	1992-Present

AGRICULTURE DEANS

Albert Menke (First Experiment Station Director)	1888-1891
Robert L. Bennett	1891-1903
William G. Vincenheller (Title of Dean added)	1903-1908
Charles F. Adams	1908-1913
Martin Nelson	1913-1920
Bradford Knapp	1920-1923
Dan T. Gray	1924-1939
Walter R. Horlacher	1939-1946
Lippert Spring Ellis	1946-1959
Gerald T. Hudson	1959-1965
Glenn W. Hardy	1965-1987
Gerald Musick	1987-1993
Charles Scifres	1994

DEPT. HEADS FOR AGRICULTURAL ECONOMICS AND AGRIBUSINESS

Claude O'Brannen	1927-1945
Trimble R. Hedges	1945-1947
John W. White	1947-1953
Henry J. Meenan	1953-1982
Robert N. Shulstad	1982-1987
J. C. Headley	1989-1996
Mark J. Cochran	1996

DEPT. HEADS FOR AGRICULTURAL STATISTICS LAB

David Marx	1979-1989
Ronald McNew	1989

DEPT. HEADS FOR AGRONOMY

Martin Nelson	1908-1940
R.P. Bartholomew	1940-1947
William H. Garman	1947-1949
Dale A. Hinkle	1949-1975
Wally J. Moline	1976-1981
Fred P. Miller	1982-1986
John T. Gilmour	1986-1993
Bobby R. Wells	1994-1996
James L. Barrentine	1998

DEPT. HEADS AGRICULTURE AND EXTENSION EDUCATION

Nolan Arthur	1984-1995
Don Herring	1997-Present

DEPT. HEADS FOR ANIMAL SCIENCE

R.W. Wade	1905-1907
R.R. Dinwiddie	1907-1909
A.K. Short	1909-1915
H.E. Dvorachek	1915-1932
None**	1932-1936
W.L. Horlacher	1936-1939
Warren Gifford	1939-1964
E.L. Stephenson	1964-1981
Wayne Kellogg	1981-1986
Park Waldroup	1986-1988
Paul Noland	1988-1994
Keith Lusby	1995

DEPT. HEADS FOR BIOLOGICAL AND AGRICULTURAL ENGINEERING

F.W. Wirt	1920-1921
Deane G. Carter	1921-1941
Kyle Engler	1941-1964
Billy B. Bryan	1964-1985
Otto J. Loewer	1985-1992
Carl Griffis	1992-1997

DEPT. HEADS FOR ENTOMOLOGY

Charles F. Adams	1905-1914
George G. Becker	1914-1918
William J. Baerg	1918-1951
Charles G. Lincoln	1951-1969
Floyd D. Miner	1969-1979
Gerald J. Musick	1979-1986
Jacob R. Phillips	1986-1987
David E. Foster	1988-1988
William C. Yearian	1989

DEPT. HEADS FOR FOOD SCIENCE

Kattan, Ahmed A.	1968-1988
Crandall, Phil	1989-1996
Siebenmorgen, Terry	1999

DEPT. HEADS FOR HORTICULTURE

John Ralph Cooper	1918-1949
Victor M. Watts	1949-1968
George Bradley	1968-1991
Lloyd Martin	1992-1996
Stephen Myers	1996

DEPT. HEADS FOR HUMAN ENVIRONMENTAL SCIENCES

Sarah Petit	1913-1915
Ruth Peck	1915-1917
Stella Palmer	1918-1924
Helen Goodspeed	1925-1927
Henrietta Burton	1927-1930
Isabella C. Wilson	1936-1942
Zilpha Battey	1943-1944
Grace M. Henderson	1945-1946
Beulah V. Gillespe	1947-1951
Eleanor O. Barnes	1952-1960

Evelyn M. Cox	1960-1971
Dorothy Larery	1971-1985
Sarah R. Jordon	1985-1988
Mary M. Whan	1988

DEPT. HEADS FOR PLANT PATHOLOGY

J. Lee Hewitt	1909-1917
John A. Elliott	1917-1923
Vive H. Young	1923-1953
E. M. "Monty" Cralley	1953-1959
Joseph "Joe" P. Fulton	1959-1964
Derald A. Slack	1964-1990
Sung M. Lim	1991

DEPT. HEAD FOR POULTRY SCIENCE

James Denton	1992-Present

APPENDIX IX
Who's Who in the History of the College

L. David Andrews, 1957-1994. He coached the very successful poultry judging team from 1958 to 1994, which included students who became leaders of the poultry industry. He taught and conducted research and service programs in the area of poultry production and management systems and environment.

Nolan Arthur, 1975. He served as assistant dean (1975-81) and associate dean (1981-85) of the College and was the first head of the Department of Agricultural and Extension Education. In 1995, he stepped down as department head to serve as coordinator for recruiting and the student teaching program.

William J. Baerg, 1918-1951. As a researcher, teacher and publisher of books on entomology, ornithology and arachnology, he was an inspirational teacher and mentor of students. He was known for a description of symptoms he suffered from the bite of the black widow spider and his work with tarantulas.

Robert P. Bartholomew, 1926-1957. He was head of the Agronomy Department in the 1940's and 1950s, and he was director of the Arkansas Agricultural Mission to Panama in 1955 and 1956. As a soils and crop management teacher and scientist, he strengthened the University's soil testing and research program, leading to increased use of soil tests by farmers.

T. Lionel Barton, 1967-1995. As a Cooperative Extension Service poultry specialist on the College faculty, he maintained strong ties between the College and the poultry industry as it developed in Arkansas. He developed a computerized feed formulation program and provided leadership for water quality and animal waste educational programs involving the poultry industry.

Robert L. Beacher, 1949-1957. He directed expansion of soil testing services to meet high demand from Arkansas farmers.

Joseph N. Beasley, 1949-1995. A veterinarian and poultry pathologist, he taught and conducted research on poultry and animal health issues. He conducted pioneering research on Marek's disease, including identifying how the disease was spread in poultry flocks.

John L. Bowers, 1940-1987. He taught and conducted research primarily in breeding and culture of vegetables, especially spinach, pickling cucumbers and southern peas, of which he released a number of varieties. He was the first Arkansan elected a Fellow of the American Society for Horticultural Science.

George A. Bradley, 1955-1991. His teaching and research concentrated on the physiology and culture, especially mineral nutrition, of vegetable crops. He was head of the Horticulture Department from 1968 to 1991. He was presented the first Leadership and Administration Award given by the Southern Region-American Society for Horticultural Science.

Clifton R. Braker, 1977-1997. As an original member of the Agricultural and Extension Education faculty, he helped Arkansas schools develop agricultural mechanization programs.

Connell J. Brown, 1950-1990. He developed a nationally recognized Animal Science teaching and research program in quantitative genetics in cattle. He obtained the first digital computer at the U of A in 1957, used to calculate observations of hereditary traits in cattle. He developed a statewide bull performance testing program.

Donald A. Brown, 1950-1986. As a leading soil scientist, he directed studies of many agronomy graduate students who became prominent in the profession. He pioneered use of the video camera scope and other new methods of observing growth of plant roots in the soil.

Billy B. Bryan, 1953-1991. He served as acting head of the Department of Agricultural Engineering from 1969 to 1985 and head until 1991. His recruiting efforts helped establish a viable agricultural engineering teaching program in the 1960s. He instigated the department's move to the Engineering Building in 1985. A cotton management and irrigation researcher, he employed students for field work and lectured on water management in agriculture courses.

Mary Burton, 1948-1982. She taught a wide range of courses in Home Economics, now the School of Human Environmental Sciences. Her main area was family issues, aging and retirement.

Deane G. Carter, 1922-1941. As the first professor and head of the Agricultural Engineering Department, he taught and conducted research on structures, machinery, electricity, irrigation and waste management. He wrote the first college textbook on farm housing. He helped establish the first rice irrigation project.

Charles E. Caviness, 1949-1991. As the Arkansas soybean breeder, he led a team effort to develop seven UA variety releases and collaborated on releases from other states that were widely planted in the South and in other countries. His description of soybean growth stages was widely used by researchers. He taught and supervised research projects by graduate students who became prominent in the profession.

J.R. Cooper, 1918-1949. He was head of the Horticulture Department during his entire tenure. In 1947, substations were established for work with fruit, vegetable and ornamental crops: Strawberry Substation at Bald Knob, Peach Substations at Lamar and Nashville, and the Vegetable Substation at Van Buren. Current locations are the Vegetable Substation at Kibler and Fruit Substation at Clarksville.

Mary Cotton, 1965-1994. She was a leader in developing curriculum and teaching fashion design and marketing, which included collaboration with art and business faculty and the fashion and retail clothing industries. She was instrumental in developing a clothing and textiles master's degree program and research program.

Elza Monroe "Monty" Cralley, 1931-1973. He was head of the Department of Plant Pathology, 1953-1959, and director of the Arkansas Agricultural Experiment Station, 1959-1973. As a preeminent plant pathologist in his early career, he established credibility for University programs to identify and provide solutions to major crop disease problems. He was internationally known for work in rice and cotton diseases.

L. B. Daniels, 1969. He is a popular teacher in the Department of Animal Science and conducted pioneering research in fescue toxicity. As Associate Director of the Experiment Station and Interim Associate Vice President for agriculture and interim dean (1987-1996), he was instrumental in obtaining funding for the John W. Tyson Poultry

Science Center, the Rosen Center for Alternative Pest Control and the Pauline Whitaker Animal Science Center.

James H. Denton, 1992. He was the first director of the Center of Excellence for Poultry Science and head of the Poultry Science Department formed from the Animal Science Department faculty who specialized in poultry. He presided over a $20-million construction program for new poultry teaching, research and extension facilities, a doubling of the size of the poultry science faculty, and a significant increase in student enrollment.

Kyle Engler, 1937-1964. As head of the Agricultural Engineering Department from 1941 to 1964, he assembled faculty that established a curriculum that was accredited in 1955. He taught and conducted research on rice irrigation.

Richard H. Forsythe, 1989-1995. After a distinguished career in poultry industry research and development, he joined the College faculty in 1989 as the first holder of the Arkansas Poultry Federation Endowed Chair. He played a leading role is the creation and expansion of a separate Poultry Science Department from the Animal Science Department. He provided leadership for poultry food safety research and educational programs.

Robert E. Frans, 1955-1994. He developed a leading weed science research program and graduate degree program. He obtained substantial funding for research and for construction of the Altheimer Laboratory complex for agronomic crops research.

Joseph P. Fulton, 1947-1987. He is known internationally for virology research on vectors, transmission and characterization of plant viruses. He mentored many graduate students who became leaders in the profession. He developed methods for producing disease-free propagating stock of strawberries and vegetables.

M.J. Goode, 1957-1988. He taught advanced plant pathology for 20 years, among other courses. He was internationally known for research in vegetable diseases, particularly for developing disease-resistant cultivars in cooperation with horticulture professors Joe McFerran and John Bowers.

Donna Graham, 1985. She provided leadership in developing a major in extension education and a master's degree program in agricultural and extension education.

Carl Griffis, 1973. He was head of the Biological and Agricultural Engineering Department from 1992 to 1997. He taught in the food processing area and was active in recruiting students through the Arkansas Science Teachers Association. His research included mathematical modeling of aquifers in the Grand Prairie.

N. Roy Gyles, 1952-1985. As a poultry breeder, he worked closely with the poultry industry to bring about steady improvement in growth rate and feed efficiency of broilers. He taught genetics and breeding courses, and he was the first to show that a genetic mechanism acting through the immune system was involved in spontaneous regression of a malignant tumor in chickens.

Vernon L. Hall, 1956-1966. As a teacher and researcher in the Agronomy Department on rice and cotton physiology, he reported observations of rice plant response to nitrogen fertilizer that had a major impact on fertilization practices.

Glenn W. Hardy, 1956-1994. He was Dean of the College from 1965 to 1987. Before accepting the Dean's position, he taught and conducted research in the Agronomy Department and directed the statewide soil testing services of the Arkansas Agricultural Experiment Station. Following his tenure as Dean, he joined the Agricultural and Extension Education faculty, where he helped develop a major in extension education and a master's degree program.

Warren Harris, 1959-1994. He taught a popular house planning course and provided engineering services for many College and Experiment Station building and renovation projects. His agricultural engineering research was in sub-surface drainage and irrigation systems.

Dale Albert Hinkle, 1948-1977. He was head of the Agronomy Department during a period of major growth from 1950 to 1975. The first M.S. degree was awarded in 1949, and the Ph.D. degree program was started in 1963. His teaching and research were in soil science and weed science. Faculty were assigned to outlying branch experiment stations, and a second soil testing laboratory was built at Marianna.

Dwight Isely, 1921-1956. As a pioneer in establishing the modern crop pest management strategy known as IPM (integrated pest management), he inspired and trained a generation of prominent applied entomologists.

Marinus C. Kik, 1927-1967. He was internationally known for developing a process, later patented, for parboiling rice to increase retention of water-soluble nutrients during cooking. His research in agricultural chemistry, later merged with home economics, on nutritive values of proteins documented benefits of adding fish and chicken to supplement diets composed mostly of rice or other grains.

Marie Lavallard, 1946-1982. As director of the Agricultural Publications Department, she taught writing courses and managed public information services and publication of faculty research results. She started the *Arkansas Farm Research* journal (1952-1994). She was a voluntary mentor to international students.

Charles G. Lincoln, 1951-1981. An innovative teacher, he also was the state's first extension entomologist. He developed integrated pest management principles into applied methods of judicious pesticide use based on scouting fields to estimate actual threat of insect pest damage to crops.

Edwin Lamar Love, 1965-1992. He was instrumental in moving the agricultural education program from the College of Education to the College of Agriculture and Home Economics to establish the new Department of Agricultural and Extension Education in 1983.

Otto Loewer, 1985-1992. The Agricultural Engineering Department was changed to Biological and Agricultural Engineering during his tenure as department head. Additional biological science courses were added to the curriculum, and enrollment increased. He was a leader in computer modeling of biological systems as a management tool.

Vera MacNair, 1929-1954. Her research in the Home Economics Department in the 1950's showed that sweet potatoes stored over winter retained their value as a source of vitamins A and C.

C.K. McClelland, 1920-1943. He started soybean breeding and management experiments that helped Arkansas farmers produce soybeans as a new crop in the 1920's and 1930's. He taught a variety of agronomy courses. The 'Mack' soybean variety released by the UofA in 1971 was named for him.

Joe McFerran, 1946-1987. As a teacher and scientist he contributed much to the vegetable industry by developing adapted disease-resistant varieties of tomato, okra, and turnip greens. Four of his tomato varieties were the mainstays of the southeastern Arkansas fresh tomato industry from 1961 until 1986. He was honored regionally and nationally for distinguished teaching and research.

Xzin McNeal, 1943-1972. His "agricultural structures" course was required for several majors for three decades and included results of his research on tornado damage and other issues. He was a pioneering researcher on rice irrigation, harvesting and drying, and he designed and built the first pallet system for storing cotton in the field. He provided a $100,000 trust for scholarships.

Max V. Meisch, 1970. A charter member of the Chancellor's Teaching Academy, he has received many teaching awards. The "insects and people" course he developed is taken by students from all colleges, particularly the College of Education.

Floyd D. Miner, 1942-1982. As Entomology Department head, 1969-1979, he instituted the Ph.D. degree program (1971).

James N. Moore, 1964-1997. His teaching and research in genetics and breeding of fruit crops received national and international acclaim. Fruit cultivars he developed include three strawberries, eight blackberries, five grapes and six peaches. The blackberry and grape varieties are currently the most widely planted in Arkansas and several other states. His blueberry research led to production of blueberries as a new Arkansas fruit crop.

Martin Nelson, 1908-1940. The number of agriculture graduates increased from about six to more than 100 a year during his 32 years as professor and head of the Agronomy Department. He was Director of the Agricultural Experiment Station from 1913 to 1920 and then became vice-dean of the College and vice-director of the Station.

Talmadge S. Nelson, 1968-1994. As a poultry nutrition teacher and researcher he had a significant impact on the Arkansas poultry industry. His research focus was on dietary minerals, especially calcium and phosphorus, and their effect on bone development.

Paul Noland, 1951-1996. He was a leading researcher and teacher in swine nutrition and management and served as head of the Animal Science Department from 1987 to 1993. His work as a consultant in Central America, particularly Panama, was recognized by presentation of that country's highest award for a non-Panamanian, the Manuel Amador Geurero award.

Eula Noyce, 1953-1980. A teacher of health, safety and childbirth preparation in Home Economics, she was known for public education on prepared childbirth. She provided health services in the College's Nursery School program.

E.H. Peterson, 1957-1984. His research on use of antibiotics to prevent chronic respiratory diseases in poultry had a major impact in the industry. His book, *Poultry Diseases*, was used as a reference in the poultry industry worldwide.

Jacob R. Phillips, 1964-1991. A disciple of UA entomologists Lincoln and Isely, he founded the area-wide approach to pest management in which farmers cooperate in the scheduling of insecticide applications based on area-wide scouting reports.

Robert D. Riggs, 1958. He is nationally and internationally recognized as a teacher and researcher in plant pathology. His work on the biology of soybean cyst nematode provides much of the basis for strategies used to control this pest.

John A. Rolloff, 1967-1993. He was a member of the original faculty of the Department of Agricultural and Extension Education, established in 1982. As a veteran professor in the former vocational education program in the College of Education, he helped establish the credibility of the new department.

Roy C. Rom, 1958-1991. He taught and conducted extensive research in tree fruit physiology, nutrition, and orchard herbicides, as well as breeding and development of peach and apple varieties and rootstocks.

Harry R. Rosen, 1918-1958. He was a leader in developing disease-resistant crop varieties as a plant disease management strategy. He developed improved varieties of wheat, oats and roses. The Rosen Center for Alternative Pest Control was dedicated in his honor in 1996.

Howard A. Scott, 1967-1991. A leader in developing biotechnology teaching and research programs, he was a pioneer in virus serology, purification methodology, and beetle transmission of plant viruses. He helped establish the Arkansas Biotechnology Center, which included a hybridoma unit for producing monoclonal antibodies.

J. Kirk Skeeles, Sr., 1978-1999. He taught and conducted research in poultry health and veterinary science. He held diplomate status in the American College of Veterinary Microbiologists and was a leader in several industry-wide programs to manage poultry disease problems.

Derald A. Slack, 1952-1990. He was head of the Department of Plant Pathology from 1964 to 1990. He was noted for his research and teaching on development of crop disease control strategies.

Arthur E. Spooner, 1955-1983. His teaching and research were in forage crop management to provide year-round pastures for livestock. He was instrumental in making tall fescue the main winter forage crop in Arkansas. He identified plants and practices best suited for highway right-of-ways and reclaiming mined land.

Odie T. Stallcup, 1950-1988. He was nationally recognized as a leading animal scientist and teacher in the areas of ruminant nutrition, digestive physiology and reproductive physiology.

E.L. Stephenson, 1949-1987. As head of the Animal Science Department from 1964 to 1982, he influenced many students who became leaders in the poultry and allied industries. His research and teaching areas were poultry nutrition and management.

Barnett Sure, 1920-1951. He received international attention for research that helped document the importance of vitamin E in the diet. He and Marinus Kik collaborated in the early Agricultural Chemistry Department, which merged in 1964 with Home Economics.

George E. Templeton, 1958-1996. He was internationally known as a leading scientist and teacher in the area of biological control of weeds. He developed COLLEGO, the first biological herbicide to use an indigenous fungus to kill weeds in a field crop. He was instrumental in securing funds for the Rosen Center for Alternative Pest Control.

Lyell F. Thompson, 1958-1991. A soil science teacher and researcher, he encouraged the University to welcome minority students in the 1960s and was a mentor to minority students and international students. He directed the soil testing program for a period and identified micronutrient requirements of certain crops.

J.E. Vaile, 1936-1969. As a horticulture professor, he taught in the area of fruit production and published extensive research results, primarily on grapes, strawberries and other small fruits.

Bradford A. Waddle, 1951-1985. He helped lay the foundation for advances in cotton genetics and management as a teacher and researcher. He also taught a statistical analysis course and was instrumental in establishing an agricultural statistics program for teaching and research. His breeding program emphasized traits for mechanical harvesting and early crop maturity.

Park Waldroup, 1966. He is an internationally prominent poultry nutrition scientist and has directed M.S. and Ph.D. degree programs for more than 50 students. He is a founding member and first president of the Southern Poultry Science Society and has served since 1981 on the Poultry Nutrition Subcommittee of the National Research Council, which evaluates research on nutritional standards for poultry.

Hubert J. Walters, Jr., 1954-1986. His work on soybean diseases helped establish Arkansas as the leading soybean producing state in the South. He worked with soybean breeder C.E. Caviness to develop disease-resistant varieties. He identified Phytophthora root rot, Cercospora leaf blight and sudden death syndrome as important soybean disease problems in Arkansas.

Lloyd O. Warren, 1947-1983. He was director of the Arkansas Agricultural Experiment Station, 1973-1983. As the state's survey entomologist, he expanded the insect scouting program in the 1950's, which stimulated interest in many young people to study entomology and provided income to help pay their tuition. He taught a popular apiculture course.

Victor M. Watts, 1927-1968. The Horticulture Department grew significantly during his tenure as head (1949-1968). He started breeding programs for vegetables, with a focus on tomato and watermelon, and crape myrtle. He taught basic and applied plant physiology and breeding.

Bobby R. Wells, 1966-1996. He designed and taught a rice culture course and was head of the Agronomy Department from 1994 to 1996. As rice soil fertility researcher at the Rice Research and Extension Center from 1966 to 1982, he was a leader in developing the DD-50 computer program widely used for timing of rice management practices.

V.H. Young, 1923-1958. As head of the Department of Plant Pathology from 1923 to 1953, he recruited an outstanding faculty and coordinated research and teaching programs that had a major impact on Arkansas agriculture.

APPENDIX X
Graduates of the College[1]

de Yampert/William Benjamin BSA 40
Aarant/Hugh Augustus BSA 40
Abbott/Frances Meriam BSHE 41
Abbott/Mary Rowe BSHE 51
Abbott/William W. ANNU BSA 67
Abbott/William W. POSC MS 69
Abboud/Martha Lenore BSHE 34
Abboud/Thomas H. BSA 29
Abdelkarim/Mohamed Rajeb ANSC PHD 86
Abdelshife/Musa A. PLPA PHD 83
Abdelshife/Susan H. PLPA MS 86
Abdelshife/Susan H. PPPM BSA 82
Abdul/Mohd Nazari AGBS BSA 85
Abdullah/Shaikh Awadz Bin Shaikh AGRN PHD 86
Abernathy/Chad W. AGBS 98
Abiraad/Mounir FDSC MS 86
Abiraad/Mounir FDSC BSA 83
Abraham/Lucien BSA 50
Abraham/Lucy L. BSHE 52
Abram/Julian Lee ANSC BSA 75
Abram/Julian Lee POSC MS 77
Abram/Susan M. Story HEED BSHE 86
Abram/Susan M. Story GHEC BSHE 82
Abston/Barbara White BSHE 60
Abston/James W. GNAG BSA 59
Abu-Oun/Mahmud ANSC PHD 85
Abu-Oun/Mahmud ANSC MS 81
Ackerman/Caroline J. Clardy BSHE 55
Ackerman/Eugene Balliet AGRN MS 83
Ackerman/Jean Zenz CLTX BSHE 86
Acord/Brenda Mae Erke CLTX BSHE 79
Acord/Kristina Louise CLTX BSHE 91
Acosta/Leigh Ann CATX BSHES 97
Adair/Ethel Owen HOEC BSHE 24
Adair/Marinell BSHE 65
Adams/Brian Wayne AGBU BSA 97
Adams/Carl J. BSA 48
Adams/Charles T. AGRI BSA 57
Adams/Charlotte Gee BSHE 40
Adams/Christopher L. GNAG MS 90
Adams/Dana Leigh HOEC MS 81
Adams/Donald BSA 53
Adams/Forrest L. HORT MS 86
Adams/Gail Anne CDFR BSHE 76
Adams/Gail Anne HOEC MS 81
Adams/Georgetta Turney BSHE 43
Adams/Hugh G. AGRI BSA 62
Adams/Kaylan L. FDSC MS 81
Adams/Kaylan L. FDSC BSA 79
Adams/Kenneth W. CRSC BSA 72
Adams/Louan VanDover HEED BSHE 55
Adams/Melanie K. BSA 82
Adams/Micah Henry DANB BSA 87
Adams/Michael Keith HORT BSA 85
Adams/Oliver L. BSA 29
Adams/Patrick Leon AGRN MS 93
Adams/Sarah Hughes BSHE 50
Adams/Sherry CLTX BSHE 86
Adams/Sheryn Lee CLTX BSHE 92
Adams/Wilburn C. BSA 49
Adarve/Rodrigo ENTO PHD 90
Adarve/Rodrigo ENTO MS 62
Addington/Jo Merida CLTX BSHE 86
Aderholt/Penny Carolyn CLTX BSHE 85
Adhikari/Chaitali FDSC PHD 95
Adhikari/Chaitali FDSC MS 91
Adkins/James BSA 58
Adkison/Lynette Kemp CRSC BSA 77
Adkisson/G. W. AGSC BSA 68
Adkisson/Perry L. MS 54
Adkisson/Perry L. BSA 50
Adkisson/Sam E. BSA 57
Aeny/Titik Nur PLPA MS 91
Agburu/Ogbu AGEC MS 84
Aggus/Larry R. ENTO MS 65
Aggus/Lynda Wilson HEED BSHE 64
Ahmad/Ismail B. PTSC PHD 83
Ahmad/Yusof H. ANSC MS 80
Ahmad/Zahoor GNAG MS 94
Ahmed/Abdaiia Aboubaker ANSC PHD 81
Ahmed/Alauddin ENTO MS 67
Ahmed/Rhonda E. J. HOEC MS 77
Ahn/Byung Joon AGRN PHD 86
Ahrens/Arnold M. ANSC BSA 74
Ahrens/Elmer H. ENTO MS 70
Aikman/Bruce Lynn BSA 52
Aikman/James E. BSA 68
Aikman/James E. ENTO MS 73
Aikman/Patricia Moore HEED BSHE 70
Aist/Eugene H. AGED BSA 60
Aist/James R. PLPA MS 68
Aist/Shelia B. PLPA MS 68
Ajmera/Sukumar AGEC MS 87
Akbar/Mohammad Karim ANSC PHD 81
Akbar/Wanda Jean POSC MS 82
Akbari/Ahmad AGEC MS 81
Akers/William George BSA 54
Akhter/Muhammad AGRN PHD 95
Akin/Kristy Donette IDHS BSHE 94
Akinola/Comfort Yetune HOEC MS 81
Akinola/Comfort Yetune HOFS BSHE 80
Akinsanmi/Josiah Akinlayo AGEC MS 86
Akkari/Khalid H. AGRN PHD 83
Al-Kahf/Duduha Chehadeh CHEG BSHE 88
Al-Khayri/Jameel M. AGRN MS 87
Al-Khayri/Jameel M. PTSC PHD 91
Al-Molla/Raad Mushin AGRN PHD 85
Al-Salman/Mudhafar H. ANSC PHD 83
Alam/Md Zinnatul ENTO PHD 83
Alberini/Joao Luiz AGRN MS 77
Albers/David Warren AGRN MS 83
Albers/David Warren AGRN PHD 87
Albritton/Charles William ANNU MS 64
Albritton/Charles William BSA 56
Albritton/Ronald L. AGRN BSA 80
Alden/Dina Gay FDNU BSHE 88
Alderman/Liz HEED BSHE 67
Alderson/Renata Ray HDFS BSHE 92
Aleman/Rudolfo E. SOSC MS 63
Aleman/Rudolfo E. BSA 60
Alexander/Bill AGBS BSA 83
Alexander/Donita Ann BSHE 65
Alexander/Gerald Wayne ANSC MS 79
Alexander/Gerald Wayne ANSC BSA 70
Alexander/J. S. HOFS BSHE 82
Alexander/Lillie Bray BSHE 57
Alexander/M. Kim HDID BSHE 80
Alexander/Margaret BSHE 41
Alexander/Mary Parker BSHE 29
Alexander/Neal S. AGEC MS 83
Alexander/Neal S. HORT BSA 81
Alexander/Robert C. BSA 52
Alexander/Susan Rachell CLTX BSHE 88
Alexander/Timothy C. ENTO MS 82
Alexander/William P. AGEC BSA 82
Alguire/Deanna Smith CLTX BSHE 88
Ali/Md Ibrahim ENTO PHD 84
Alisch-Altom/Laura Marie IDHS BSHE 89
Allain/Robert E. AGEC MS 78
Allain/Robert E. AGBS BSA 75
Allen/Cindy BSHE 75
Allen/Ernest J. BSA 49
Allen/Genie Marie CATX BSHES 98
Allen/Gladys Jane Plunkett GHEC BSHE 79
Allen/Kathryn D. HEED BSHE 67
Allen/Larry D. DASC MS 74
Allen/Mildred Bullington BSHE 53
Allen/Rodney Dwayne AGRN BSA 96
Allen/Sam AGEC BSA 79
Allen/Sandra Hartsook BSA 79
Allen/Verlie BSHE 39
Allgeyer/Judy Lynn HORT BSA 77
Allison/Bann Thompson BSHE 47
Allison/Woodrow Houston BSA 40
Allman/Sara Aldridge HEED BSHE 46

[1]Information on graduates of the College provided by the University of Arkansas Alumni Association. Graduates as of May 1999.

Allmendinger/Krista L. HOUS BSHES 98
Allred/Freeman Wesley BSA 52
Allred/Owen E. BSA 38
Allured/Robert A. AGRN MS 76
Allured/Robert A. ANSC BSA 68
Almeida/Francisco C. AGRN PHD 84
Alsup/Eileene CLTX BSHE 87
Alsup/Monica Hendrix AGED BSA 91
Alsup/Tim POSC BSA 86
Altom/Stephanie Kay ENSC BSA 96
Alvarez/Tracy Michelle IDHS BSHE 91
Alvisyahrin/Teuku AGRN MS 91
Aman/Ahmad ANSC PHD 79
Amarin/Mukhles ANSC MS 85
Amatya/Purushottam M. PLPA MS 66
Ames/Carolyn HORT MS 90
Ames/Carolyn LDUH BSA 78
Ames/Guy HORT MS 83
Amis/Mildred M. FDNU BSHE 68
Ammons/Dennis A. POSC BSA 68
Ammons/Marian Janis Walls BSHE 60
Amonsilpa/Sommai AGRN PHD 86
Amyx/James Dale AGED BSA 93
Anderson-Wyatt/Stacye ANSC BSA 86
Anderson/Alfred FDSC BSA 90
Anderson/Angela F. HEED BSHE 83
Anderson/Carlton E. AHUS BSA 62
Anderson/Dick Henry BSA 42
Anderson/Elbert H. BSA 53
Anderson/Elizabeth Kay HEED BSHE 88
Anderson/Erie Glae Dickerson BSHE 43
Anderson/George E. POSC BSA 90
Anderson/Gibson F. BSA 39
Anderson/James N. AGBS 98
Anderson/James Tyrel POSC BSA 96
Anderson/Jeannie Ellen HOEC MS 89
Anderson/Jeannie Ellen GHEC BSHE 82
Anderson/Jessie BSHE 20
Anderson/Joice Marie HOEC MS 81
Anderson/Joseph M. ANNU BSA 73
Anderson/Lisa Michelle CLTX BSHE 91
Anderson/Mary Cay HEED BSHE 63
Anderson/Naomi Rose GNAG MS 91
Anderson/Nola BSHE 39
Anderson/Patricia Berryhill LDUH BSA 82
Anderson/Phillip Sidney BSA 29
Anderson/R. Bruce AGEC BSA 78
Anderson/Robert W. BSA 42
Anderson/Tyrrell C. BSA 56
Anderson/William A. BSA 33
Anderson/William Eric POSC BSA 88
Anderton/Nan HDFS BSHE 86
Andrews/David POSC BSA 52
Andrews/Joseph K. BSA 53
Anglea/Laura Leigh CATX BSHES 98
Annis/David PPPM BSA 87
Annis/Patricia HOEC MS 80
Anthony/J. Wayne BSA 52
Anwar/Ul Haq AGRN PHD 85
Apostol/Porfirio Sarangay BSA 41
Apple/Glendon Donnie POSC BSA 70
Apple/Robert E. BSA 49
Apple/Robert O'Neal POSC PHD 88
Appleberry/Minter F. CRSC MS 68
Appleberry/Minter F. BSA 56
Appleberry/Minter F. AGBS BSA 79
Arbabi/Ali Reza ANSC MS 84
Arbabi/Ali Reza POSC BSA 73
Archer/Douglas Carl POSC BSA 90
Archer/John Thomas ANSC BSA 92
Archer/Mark A. AGBS 99
Archer/Mitchell L. AGME BSA 85
Arcos/Maria Cristina RSOC MS 94
Arif/Anwar PLPA MS 70
Armani/Dayna HDFS BSHES 98
Armbruster/Pamela L. HEED BSHE 78
Armitage/Norman L. BSA 53
Armstrong/Elizabeth Ann HORT BSA 89
Armstrong/George A. AGEC BSA 75
Armstrong/Margaret Strayer BSHE 32
Arnhart/A. D. BSA 41
Arnold/Adlai F. AGEC MS 51
Arnold/Bobbye M. BSHE 54
Arnold/Carol Ann FDNU BSHE 87
Arnold/Glenn BSA 51
Arnold/James T. GNAG MS 86
Arnold/James T. AGRN MS 84
Arnold/James T. AGRN BSA 82
Arnold/Kirby Edmond BSA 31
Arnold/Mary Frances Harrison BSHE 63
Arnold/Nan Faith LDUH BSA 77
Arnold/Rebecca Sue AGEC BSA 81
Arnold/Ruth Anne LDUH BSA 80
Arnold/Willard Ray RSOC MS 90
Arosemena/Jaime A. ANSC BSA 60
Arosemena/Jaime A. AHUS MS 62
Arrington/Dennis Ray HORT BSA 86
Arrington/Gene BSA 42
Arthur/Teresa Lynn ENSC BSA 97
Asbell/Maeda BSHE 38
Ash/Stephen Boyd AGBS BSA 82
Ashburn/Velma Crow BSHE 49
Ashcraft/Ashley Ann CLTX BSHES 97
Ashcraft/Ronald B. AGME BSA 84
Asheim/Betsy Lynn Jaicks HDID BSHE 80
Ashley/Irvin Ester BSA 60
Ashley/Thomas J. AGRI BSA 51
Ashlock/Lanny BSA 65
Ashlock/Lanny SOSC MS 67
Ashlock/Lloyd Anthony GNAG MS 88
Ashmore/Lance E. AGME BSA 81
Askew/Gordon R. HORT BSA 79
Askins/Cynthia N. BSA 78
Askins/Robert L. MKTG BSA 77
Atchley/Brian Keith PPPM BSA 81
Atchley/Susan Pharr GHEC BSHE 85
Atchley/Susan Pharr HOEC MS 86
Atha/Alma J. HEED BSHE 74
Atkins/Charles D. BSA 53
Atkins/Edward Carl BSA 24
Atkins/Laura IDHS BSHES 98
Atkins/W. Roger AGEC BSA 71
Atkins/Wayne L. BSA 50
Atkinson/Chester Barton BSA 35
Atkinson/Etna McGaugh BSHE 26
Atkinson/James V. SOSC BSA 54
Atkinson/LaDonna Malone ANSC BSA 88
Atkinson/Megan M. DIAM BSA 92
Atkinson/T. E. BSA 36
Attak/Jennifer Lynn ENSC BSA 98
Attebery/John T. DHUS MS 63
Attebery/John T. DHUS BSA 61
Attwood/Irene Gray ANSC BSA 80
Attwood/Victor G. ENTO MS 63
Attwood/Victor G. BSA 60
Atwood/Donald Wayne ENTO PHD 96
Atwood/James William ANNU MS 65
Atwood/Kathleen Marie CDFR BSHE 75
Auffert/Jacquelin Janene CATX BSHES 98
Auffet/Derrick Jay AGBS BSA 95
Auffet/Michael Wayne HORT BSA 78
Aufill/Esther Waits BSHE 40
Austin/Brian William DASC BSA 92
Austin/David L. POSC 99
Austin/Edwin L. BSA 34
Austin/Gary Virgil AGBS BSA 92
Austin/Helen M. BSHE 27
Austin/O. Dean AGEC BSA 54
Austin/Russell Hayden BSA 17
Autrey/Meredith Mullen IDHS BSHE 87
Avery/Barbara Heggie HORT BSA 78
Avery/Chris AGRN BSA 74
Avery/Harold AGBS BSA 82
Avery/Mark Allen GNAG MS 89
Avery/Raymond M. POSC BSA 70
Avey/Kenneth Leroy POSC BSA 86
Avillion/Deborah A. AGBS BSA 82
Ayers/Ronnie G. BSA 76
Aylon/Bessie Fitzgerald BSHE 43
Aynesworth/Maryetta BSHE 41
Ayres/Linnie L. BSHE 19
Ayres/Robert Sale BSA 50
Ba/Mahamadou Nassirou AGEC MS 92
Baber/Erin BSA 37
Baber/Harold Truman BSA 37
Baber/John C. BSA 27
Baber/L. C. BSA 25
Baber/Quin Morton BSA 25
Bachireddy/Veeraraghava Reddy AGRN PHD 78
Bachman/Alice L. HORT BSA 79
Bachman/Verman Clifford BSA 52
Bacon/Robert AGRN MS 80
Bacon/Robert AGRN BSA 78

Bacon/Sharon M. HEED BSHE 73
Bader/Monica Elaine AGBS BSA 94
Baer/Meika POSC BSA 95
Baer/Meika HORT BSA 95
Baertsch/Janice Elizabeth CLTX BSHE 90
Bagby/John BSA 56
Bagby/John BSA 27
Bagby/John R. ENTO MS 56
Bagby/Leslie Kaye CLTX BSHE 85
Bagley/Nathan AGME BSA 72
Bagwell/Dorothy Attebery BSHE 58
Bagwell/Ralph Dale ENTO PHD 94
Baharudin/Faiza AGBS BSA 86
Bailey/Anthoney Oluremi AGEC MS 87
Bailey/Charles Roland POSC BSA 71
Bailey/Clark Roland AGBS BSA 94
Bailey/Clayton R. ANSC 98
Bailey/Frank R. ANSC BSA 78
Bailey/Fredric L. HORT MS 64
Bailey/Fredric L. GNAG BSA 62
Bailey/Hollis F. BSA 50
Bailey/Hugh E. AGBS 99
Bailey/Joan E. CLTX BSHE 76
Bailey/Karen L. Ogawa DANB BSA 83
Bailey/Kenneth Allan AGBS BSA 94
Bailey/Kenneth Wayne AGBS BSA 80
Bailey/Mary Lee HOEC MS 68
Bailey/Mechelle FDNU BSHE 83
Bailey/Nikki Lyn HOUS BSHES 97
Bailey/Sally GHEC BSHE 82
Bailey/Timothy R. AGBS 99
Bain/Edith Austin BSHE 25
Bainbridge/Mark R. AGRN MS 82
Bainbridge/Mark R. AGRN BSA 79
Baines/Crystal Gwen PPPM BSA 95
Baker/Allen S. SOSC BSA 77
Baker/Ancil Wilbur BSA 49
Baker/Ancil Wilbur AGEC MS 54
Baker/Annetta J. CLTX BSHE 83
Baker/Billy W. POSC BSA 64
Baker/Charles Guy BSA 49
Baker/Cynthia Hiner HEED BSHE 57
Baker/Darlene Zeh HEED BSHE 76
Baker/Darlene Zeh HOEC MS 82
Baker/Donald S. BSAGE 63
Baker/Eddie Louise BSHE 43
Baker/Elbert Charles HORT MS 76
Baker/Elbert Charles HORT BSA 71
Baker/Frank E. BSA 51
Baker/Geneva BSHE 54
Baker/Helen M. BSHE 55
Baker/Horace Raymond SOSC MS 67
Baker/Imogene Elizabeth Bynum BSHE 50
Baker/John A. BSA 35
Baker/Larry M. VOAG BSA 79
Baker/Leah Ann CLTX BSHE 86
Baker/Martin Duane POSC BSA 94
Baker/Maybin S. BSA 14

Baker/Pamela AGBS BSA 88
Baker/Raymond Mark POSC BSA 92
Baker/Rodney AGEC MS 78
Baker/Rodney ANSC BSA 76
Baker/Stanley W. AGBS BSA 78
Baker/Thomas A. BSA 52
Baker/Timothy Philip HORT MS 88
Baker/Timothy Philip HORT BSA 86
Baker/Troy Y. BSA 59
Baker/Verna Mae BSHE 66
Baker/Verna Mae FDNU MS 69
Baker/Virginia Adelia IDHS BSHE 95
Baker/Wanette McKee HEED BSHE 70
Baker/Wanette McKee HOEC MS 72
Baker/Wendolyn Moore HORT BSA 90
Baker/William Holt AGRN MS 84
Baker/William Holt ENSC BSA 81
Balch/Gloria Lee CLTX BSHE 80
Baldridge/H. Devane BSA 51
Baldridge/Michael Gordon AGRN BSA 95
Baldwin/Calvin Park BSA 49
Baldwin/Danny H. GBUS BSA 75
Baldwin/Donald R. POSC BSA 52
Baldwin/Ford L. CRSC BSA 69
Baldwin/Ford L. CRSC MS 71
Baldwin/Grace Hodges BSHE 23
Baldwin/Joe F. AGEC MS 76
Baldwin/John BSA 53
Baldwin/John L. AGRN MSA 86
Baldwin/John L. AGRN BSA 82
Baldwin/Vernoice Cannon HOEC MS 93
Baldwin/Vernoice Cannon HEED BSHE 71
Balerdi/Fermin F. AGRN MS 64
Ball/Peter H. FDSC MS 88
Ball/Peter H. FDSC BSA 86
Ball/Rosalind Ann AGRN MS 92
Ball/Rosalind Ann AGRN PHD 94
Ballantyne/Paul HORT MS 88
Ballard/Mark D. LDUH BSA 83
Ballard/Willard H. AGRN BSA 35
Ballenger/Benjamin David PPPM BSA 80
Baltz/Marietta HOEC MS 86
Baltz/Marietta HDFS BSHE 85
Baltz/Tracy C. AGEC MS 92
Baltz/Tracy C. AGRN BSA 90
Bandusena/Weerasekera M. AGEC MS 93
Bandy/Steven Ray AGME BSA 97
Banks/Carreather F. HOEC MS 63
Banks/Sara Poynor FDSC MS 93
Banks/Sara Poynor FDNU BSHE 91
Banks/Van E. AGRN MS 85
Banning/Heath R. ESWS 98
Bannister/Julie Ann FDSC BSA 97
Banta/Len L. VOAG BSA 79
Bapst/Ryan F. ANSC 98
Baquet/Ramona Ruth FDNU BSHE 88
Bararpour/Mohammad Taghi AGRN PHD 95
Bararpour/Mohammad Taghi AGRN MS 90

Barber/David M. CRSC BSA 70
Barber/Leroy BSA 47
Barber/Lon T. AGRN 98
Barclay/Sandra GHEC BSHE 81
Bard/Joel W. ENTO MS 84
Barefield/Lewis A. BSA 49
Barenberg/Kevin Jay AGTE BSA 98
Barge/Carroll A. AGRI BSA 48
Barger/Anita N. CLTX BSHE 79
Barham/Jack BSA 51
Barham/Thomas B. AHUS BSA 62
Barham/Warren BSA 41
Bariola/Julius ENTO BSA 64
Bariola/Louis A. ENTO MS 62
Bariola/Louis A. GNAG BSA 58
Bariola/Paul A. AGEC BSA 65
Bark/Carolyn BSHE 50
Barker/Carol L. CLTX BSHE 84
Barker/Lance Alan AGBS BSA 96
Barksdale/Diane HOEC MS 70
Barksdale/Diane HEED BSHE 68
Barksdale/William Alexander AGBS BSA 92
Barksdale/William E. GNAG BSA 59
Barlow/James P. GNAG BSA 40
Barlow/James Paul BSA 65
Barman/Everette H. ENTO MS 69
Barnard/Roy Jackson POSC BSA 81
Barnes/Barbara Diane IDHS BSHE 95
Barnes/Chester Berton BSA 49
Barnes/Clyde J. AGRN PHD 90
Barnes/Glenn ANSC BSA 81
Barnes/Gordon ENTO MS 51
Barnes/Gordon BSA 50
Barnes/Tanya FDNU BSHE 85
Barnett/Bobby D. ANSC BSA 50
Barnett/Bobby D. POSC MS 54
Barnett/Don C. ANSC MS 69
Barnett/Don C. AHUS BSA 61
Barnett/Harold H. BSA 51
Barnett/Hershel O. BSA 50
Barnett/Ortus W. BSA 36
Barnett/Ortus W. PLPA MS 65
Barnett/Ortus W. PLPA BSA 61
Barnett/Pamela Powell HEED BSHE 66
Barnett/Ronald D. CRSC MS 68
Barnett/Ronald D. CRSC BSA 65
Barnhard/Pam LDUH BSA 79
Barnhart/Alice Raymond BSHE 67
Barnhart/Katherine Elizabeth LDUH BSA 91
Barnhill/Valerie BSHE 87
Baron/Raul Oscar Roca BSA 79
Barr/Glen R. ANNU BSA 73
Barr/Jewel T. DASC BSA 68
Barr/Jewel T. ANSC MS 69
Barr/Yvonne PLPA MS 86
Barrentine/Dennis W. CRSC BSA 73
Barrentine/Henry Frank AGEC BSA 79
Barrentine/James L. BSA 63

Barrentine/James L. MS 65
Barrentine/Jeffrey Allen ANSC BSA 86
Barrentine/William L. BSA 59
Barrett/Amy Caroline HDFS BSHES 97
Barrett/Lisa AGED BSA 95
Barrett/Michael R. AGRN PHD 82
Barrett/Patricia Falls HOEC MS 77
Barrett/Patricia Falls BSHE 53
Barron/Danny Curtis HORT BSA 89
Barron/Karol Ruth BSHE 61
Barron/Karol Ruth AGEC MS 66
Barrow/Lesley Marie CLTX BSHE 89
Barry/Virginia M. HEED BSHE 57
Barse/William P. LDUH BSA 80
Bartee/John D. ANSC MS 61
Bartelli/Dorothea Forrest CLTX BSHE 76
Bartelli/Dorothea Forrest HOEC MS 80
Bartlett/Margaret A. POSC BSA 82
Barton/James A. BSA 32
Barton/James T. POSC BSA 86
Barton/John E. BSA 38
Barton/Lionel POSC MS 60
Barton/Lionel ANNU BSA 58
Bartz/Robert O. BSA 58
Bartz/Robert O. AGEC MS 60
Basden/Gene B. BSA 52
Baser/Robert E. AGRN BSA 78
Baser/Robert E. AGRN MS 89
Basham/Gary W. AGRN MS 86
Bashaw/Vickie L. CLTX BSHE 82
Baskin/Roy O. BSA 60
Bassett/Edward Nash PLPA MS 83
Bassett/Mary S. CDFR BSHE 79
Bassett/Pamela J. FDNU BSHE 77
Bassham/Robert M. GNAG BSA 64
Bassi/Albert PLPA PHD 82
Bassi/Dana G. HORT BSA 76
Bassi/Dana G. ENTO MS 79
Bassols/Maria Do Carmo HORT MS 74
Bastawisi/Ali Orabi AGRN PHD 87
Batchelor/James T. AGRN PHD 80
Batchelor/James T. AGRN MS 78
Batchelor/James T. AGRN BSA 75
Bateman/David BSA 37
Bateman/Lanny AGEC MS 69
Bateman/Margaret Stokes HEED BSHE 69
Bateman/Patricia A. BSHE 60
Bateman/William Walter BSA 36
Bateman/William Walter CSCI BSA 71
Bates/Frances Clementine BSHE 26
Bates/Jennie B. HOEC MS 66
Bates/Jessica C. POSC BSA 97
Bates/Kenneth Stone GNAG MS 63
Bates/Kenneth Stone GNAG BSA 37
Bates/Zelpha Brown HOEC MS 77
Bates/Zelpha Brown HEED BSHE 72
Batten/Janet Bradley BSHE 67
Bauer/Ala Wilcox BSHE 41
Bauer/Roy G. ANNU MS 72
Bauer/Thomas R. ENSC BSA 80
Baughman/Arnold J. HOFS BSA 74
Baughman/Beverly Gayle HLED BSHE 81
Baughn/Alvyn BSHE 44
Baugus/Chester A. ANSC BSA 58
Baureis/Melinda Hampton CTM BSHE 75
Bausinger/Jennifer CLTX BSHE 92
Baxley/Dana Brooks IDHS BSHE 90
Baxley/Ron AGRN BSA 92
Baxter/Lisa Rice FDSC MS 85
Baxter/Lisa Rice FDNU BSHE 83
Baxter/Michael Aaron POSC MS 97
Baxter/William A. AGEC BSA 49
Bayley/Paul E. AGRI BSA 51
Bayley/Thelma Allison BSHE 50
Bays/David R. AGEC BSA 77
Bayyari/Geraldine Rose FDNU BSHE 83
Beacham/Stephen D. VOAG BSA 75
Beadles/Wayne AGRN BSA 85
Beall/Elstner D. BSA 34
Beall/George H. AGEC MS 66
Beam/David C. DANB BSA 82
Beam/Stacy Lee CLTX BSHE 91
Bean/Dolores Durnil BSHE 53
Bean/James W. BSA 53
Bean/Karen D. PHED BSA 82
Beard/Amalia Rose CDFR BSHE 77
Beard/Randy ENSC BSA 96
Beard/Suzanne BSHE 83
Beard/Virginia Clement BSHE 44
Bearden/Norma Jean BSHE 57
Beasley/Lida Ray BSHE 40
Beasley/Samuel D. SOSC MS 74
Beasley/Samuel D. CRSC BSA 71
Beasley/Vance E. BSA 39
Beatty/John M. CRSC BSA 70
Beatty/John M. BSA 42
Beaty/Bessie Birdsong BSHE 27
Beaty/Jackie Dwayne AGRN MS 91
Beaty/Ronald W. GNAG MS 92
Beaty/Sherry Lynn EXED BSA 96
Beaver/Barbara J. CDFR BSHE 75
Beaver/Graydon L. BSA 49
Beavers/Kathryn Kellye CATX BSHES 98
Beavers/Mary Katherine BSHE 70
Bebee/Paige Laine HDFS BSHE 91
Beck/Alton Edwin AGBS BSA 81
Beck/Amanda K. ANSC 98
Beck/Barbara A. FND BSHE 71
Beck/Charles F. ANSC BSA 67
Beck/Jerry L. AGBS BSA 91
Beckenholdt/Willard David GNAG BSA 62
Becker/Christopher Hans HORT BS 93
Becker/Gregory J. AGRN BSA 94
Beckham/William D. DASC BSA 77
Beckwith/Grant AGBS BSA 94
Becton/Carol M. AGRN MS 94
Becton/Scott M. AGRN BSA 85
Bedell/Calvin Wallis BSA 30
Bedingfield/Trevyr Reyn ANSC BSA 97
Bednarz/Craig William AGRN PHD 95
Bedwell/Clemon W. BSA 47
Beene/Clinton A. AGEC BSA 77
Beene/Hazel Snider BSHE 39
Beers/Kelly Wayne POSC MS 86
Beetz/Alice E. HORT MS 88
Behnke/Ann Dixon HOEC MS 69
Beighley/Mozelle Davis BSHE 25
Beight/Steven S. LDUH BSA 78
Bejarano/Francisco Xavier DIAM BSA 89
Belew/Johnny Glen AGRN BSA 86
Belew/Lucy Beth Ryall BSHE 86
Beling/Lucile Cobb BSHE 27
Beliz/Vilma Ester BSHE 57
Bell/Cecilia Anne ANSC BSA 85
Bell/Charles Osborn BSA 37
Bell/David Russell AGBS BSA 91
Bell/E. D. BSA 34
Bell/Earl L. GNAG BSA 70
Bell/Hilton L. BSA 56
Bell/Howard K. BSA 40
Bell/James H. POSC BSA 78
Bell/Jim M. BSA 37
Bell/John W. ANNU MS 63
Bell/John W. DHUS BSA 58
Bell/Joyce Sharp BSHE 27
Bell/Judy Douglas HEED BSHE 70
Bell/Kermit Otto ENTO BSA 61
Bell/Kermit Otto ENTO MS 65
Bell/Lloyd R. AGBS BSA 72
Bell/Mark A. DANB BSA 91
Bell/Mary G. HOEC MS 71
Bell/Mary G. BSHE 69
Bell/Mary Sellick CDFR BSHE 76
Bell/Melissa A. CTM BSHE 75
Bell/Pat A. AGED BSA 86
Bell/Pauline Gray BSHE 34
Bell/Richard Lewis DHUS BSA 62
Bell/Richard Lewis AGEC MS 66
Bell/Robert Allen ENTO MS 60
Bell/Sara L. CHEM BSHE 44
Bell/Valton D. AGRN MS 77
Bell/Wayne T. AGEC BSA 69
Bell/William Howard VOAG BSA 69
Bellingrath/Gail W. HEED BSHE 79
Bellinski/Susan Rogers CDFR BSHE 77
Belton/Staci Hufhines EXED BSA 96
Benavides/Luis Manuel GNAG MS 88
Benbrook/Clyde Raymond AGEC MS 31
Benbrook/Clyde Raymond BSA 30
Benbrook/Olive Kerr BSHE 24
Bender/Lloyd D. AGEC MS 56
Bender/Lloyd D. BSA 55
Bender/Ronda Renee FDNU BSHE 90
Benefiel/Fred Eugene BSA 52

Benefiel/Julie HDFS BSHE 92
Bennett/Audrey Jane FDSC BSA 88
Bennett/Billy BSA 77
Bennett/Clinton R. AGBS 98
Bennett/Dorothy Mae BSHE 57
Bennett/Ernest Milan BSA 50
Bennett/Franklin D. BSA 57
Bennett/Louise A. BSHE 38
Bennett/Ronald A. BSA 76
Bennett/Sandra B. LDUH BSA 93
Bennett/Wayne BSA 42
Benson/James A. POSC MS 80
Benson/James A. BSA 77
Benson/Nelson Ray AGRN MS 94
Benson/Nelson Ray AGRN BSA 91
Benson/Randy R. ANSC BSA 79
Bentley/Mae Cates BSHE 53
Benton/Ann Rodgers BSHE 57
Benton/Bradley AGBS BSA 97
Benton/Dorothy Mae BSHE 57
Benton/Elizabeth LaNin IDHS BSHE 94
Benton/Lenny L. AGEC BSA 72
Benton/Lenny L. AGEC MS 71
Benton/Sidney Wright BSA 17
Benton/Thomas H. DANB BSA 74
Bequette/Curtis J. POSC BSA 64
Bequette/Nan BSHE 60
Bercher/Hope Allison ANSC BSA 94
Bercher/Steven Glen AGBS BSA 96
Berger/George A. CRSC MS 64
Berger/George A. BSA 60
Berger/Jane Crawford HORT MS 83
Bergthold/Catherine Couch HEED BSHE 47
Berhanu/Abeje RSOC MS 96
Berner/Arnold BSA 49
Berner/Melvin Ray ANSC BSA 70
Berquist/Lia Dawn IDHS BSHES 96
Berrey/Carol Marie PPPM BSA 86
Berry/Benjamin Marvin BSA 14
Berry/Blake BSA 41
Berry/Calvin R. AGEC MS 53
Berry/Calvin R. GNAG BSA 52
Berry/Charles T. GNAG BSA 58
Berry/Cynthia Lee DANB BSA 85
Berry/David Harold AGME BSA 74
Berry/Everett Pershing BSA 40
Berry/Mark Alan ANSC BSA 86
Berry/Robert N. BSA 39
Berry/Thomas Henry BSA 74
Berry/Vernell Lumpkin HEED BSHE 70
Berryman/Lisa G. AGEX BSA 97
Bert/Shari Stewart HIED BSHE 81
Bertalotto/Mary Amanda FDNU BSHES 98
Bertschy/Cheryl Renee IDHS BSHE 90
Bertschy/Leon Ray AGBS BSA 86
Bess/Claude AGSC MS 79
Besser/Jean Keller BSHE 43
Bessman/Cecil Wayne AGED BSA 86
Best/Gary Max ENTO BSAGE 63
Bettison/C. Richard PPPM BSA 83
Betton/Harold B. PLPA BSA 69
Beuke/Martha Lu CDFR BSHE 77
Beuttel/Carolyn S. FND BSHE 74
Bevill/Tracy Samuel AGBS BSA 94
Bezanson/John M. BSA 58
Bhatti/Mohammad Suleman ANSC MS 90
Bi/Jianlong ENTO PHD 97
Bi/Jianlong ENTO MS 93
Bian/Yilin FDSC MS 93
Bianco/Cyril P. BSA 41
Bibb/Elinor Hale BSHE 35
Bickel/Jimmy AGRN MS 79
Bickel/Jimmy AGRN BSA 75
Bickford/Fay G. HOEC MS 71
Biddle/Marion Reed BSHE 41
Bidlack/James Enderby AGRN MS 86
Bieber/George Peter HORT BSA 85
Biermann/Janet L. LDUH BSA 80
Bierwirth/Kenneth BSA 57
Biggs/Bobbie T. HOEC MS 63
Biggs/Bobbie T. HEED BSHE 62
Biggs/Carolyn BSHE 52
Bigham/Donald G. AGBS BSA 82
Bigham/Emil Seidel BSA 32
Bilbrey/Keith J. GNAG BSA 37
Biles/Owen H. GNAG BSA 41
Bilgischer/Julie Ann CLTX BSHE 93
Billings/Billy G. HORT BSA 84
Billingsley/Faye BSHE 49
Billingsley/Ralph L. BSA 41
Billingsley/Woodrow Pleas BSA 35
Billington/Cynthia Maria CLTX BSHES 96
Bingham/Billy INEG PHD 76
Binkley/Iva Sue Thompson BSHE 38
Binkley/Virginia Randall CLTX BSHE 81
Binns/Evelyn H. BSHE 31
Binns/Gary Andrew IDHS BSHE 91
Birang/Madong A. AGRN MS 94
Birkner/Carl AGED BSA 84
Birkner/Linda ANSC BSA 82
Bisbee/Diana Chambers POSC BSA 90
Biship/Helen Young BSHE 34
Bishop/A. Wade BSA 43
Bishop/Brian William ANSC BSA 80
Bishop/Elizabeth BSHE 42
Bishop/Howard BSA 42
Bishop/Joe Allen BSA 76
Biskup/Deborah G. HEED BSHE 85
Biswell/Emmitt W. ANSC BSA 71
Bittle/Cecil M. BSA 40
Biven/Melanie K. HEED BSHE 72
Bivens/Eudell F. BSA 52
Biyer/Piper Susan CLTX BSHE 83
Bizzul/Keshav R. FDSC MS 84
Black/Bradford T. ANNU BSA 79
Black/Calvin Adam BSA 51
Black/J. Hodge ENTO MS 61
Black/J. Hodge BSA 58
Black/Jimmy D. BSA 68
Black/Kwin R. FDSC MS 79
Black/Kwin R. FDNU BSHE 76
Black/Lori Lynn HEED BSHE 85
Black/Lowell L. PLPA MS 62
Black/Lowell L. BSA 60
Black/Mark C. PLPA MS 79
Black/Mark C. PLPA BSA 75
Black/Robert E. BSA 50
Black/Russell O. AGEC BSA 67
Blackburn/Bryan Keith HORT BSA 89
Blackburn/Dean W. BSA 30
Blackburn/Lois BSHE 38
Blackerby/Donnie BSA 61
Blackman/James Ronald POSC MS 80
Blackman/Robert E. AGEC MS 67
Blackman/Robert E. AGEC BSA 64
Blackmon/Alva Askew HOEC BSHE 40
Blackmon/David W. DIAM BSA 88
Blackmon/David W. AGEC MS 91
Blackmon/Jane Puryear BSHE 47
Blackston/Mitchell F. POSC BSA 87
Blackwell/Debbie BSHE 86
Blackwell/Floyd A. BSA 83
Blackwood/William Thomas BSA 48
Blagg/Willine Rumsick BSHE 54
Blair/Bill D. BSA 51
Blair/Bill D. ENTO PHD 65
Blair/Franklin M. BSA 51
Blair/J. S. AGEC MS 88
Blake/James Howard PLPA MS 85
Blakemore/Robbie Gene BSHE 48
Blanchard/C. V. BSA 42
Blanchard/D. C. GNAG MS 66
Blanchard/D. C. BSA 52
Blanchard/Hubert AGRI BSA 43
Blanchard/Janive BSHE 44
Blanco/Pedro H. AGRN MS 91
Bland/Elizabeth Lynette CLTX BSHE 93
Bland/Wilma Carnahan BSHE 47
Blankenbaker/Lori Ann ANSC BSA 85
Blankenship/August ENTO BSA 61
Blankenship/Christina Lorrain AGBS BSA 93
Blankenship/Donald G. GNAG BSA 66
Blasdel/William E. ANSC BSA 83
Blaylock/Garland D. BSA 38
Blem/Allen AGRN PHD 78
Blevins/Ellen M. FDNU BSHE 76
Blevins/Jimmy Wade AGED BSA 94
Blevins/Wilma Lea BSHE 63
Blew/Connie Swafford HDFS BSHE 91
Blocker/John W. BSA 63
Blocker/Kristine E. CLTX BSHE 84
Blocker/William H. ANSC BSA 82
Bloodworth/Walter Allen AGBS BSA 87
Blore/Pamela J. POSC MS 82

Bluhm/Karen G. HDID BSHE 78
Blythe/Marilyn Jo ANSC BSA 70
Blythe/Michael Keith AGBS BSA 86
Blythe/Michael Keith AGEC MS 90
Blythe/Michael Keith DASC BSA 87
Blythe/Thomas O'Neal BSA 71
Boatenhammer/Thelma Perrin HOEC MS 74
Boatman/Molly Penix BSHE 63
Bodie/Teresa Arnold CLTX BSHE 78
Boettinger/Charles F. HORT MS 78
Bogan/Tamara Allen BSA 76
Bogart/James A. AGEC MS 76
Boger/Charles B. AGBS 98
Boger/Christopher S. ANSC 99
Bogle/Margaret L. BSHE 55
Bogoslavsky/Bruce A. ANSC BSA 86
Bogoslavsky/Lance M. ANSC BSA 82
Bohanan/Monty Edward GNAG MS 94
Bohannon/Gary AGBS BSA 82
Bolding/Susan D. LDUH BSA 89
Boldon/Bruce Byron BSA 53
Boles/Flossie W. BSE 41
Bolinger/Kimberly J. HEED BSHE 89
Bolland/Donna R. HOEC MS 87
Bollinger/Byron L. BSA 77
Bollinger/Fred N. AGBS BSA 72
Bolstad/Daniel Paul FDSC BSA 96
Bolstad/Jerri Lynn FDSC BSA 96
Bomblat/Christian AGRN MS 93
Bomblot/Cristina Arcos RSOC MS 94
Bona/Robert PLPA MS 83
Bonacker/Paula Denise CLTX BSHE 88
Bond/Ann Ratcliffe BSHE 41
Bond/Donna K. HEED BSHE 81
Bond/Gayle M. HORT BSA 86
Bond/Louise S. HOEC BSHE 52
Bond/Paul Dean VOAG BSAGE 63
Bond/Paul L. POSC BSA 83
Bond/Paul L. ANSC MS 87
Bond/Paul L. POSC MS 86
Bond/Richard Douglas POSC BSA 92
Bondada/Bhaskar Rao AGRN PHD 94
Bonds/David H. AGEC BSA 82
Bone/Beverly Kelley FDNU BSHE 62
Bone/Mike AGEC MS 73
Bonebrake-Medrano/Lisa CLTX BSHE 83
Boney/James Edward ENSC BSA 98
Bongo-McCain/Christina FDSC BSA 90
Bonifaz-Paz/Oscar A. AGEC BSA 81
Bonifaz-Paz/Oscar A. AGEC MS 84
Bonifaz/Oscar A. AGEC MS 90
Bonner/Beth FDNU BSHE 82
Bonner/Claude M. CRSC MS 67
Bonner/Claude M. RSOC BSA 62
Bonner/John E. VOAG BSA 73
Bonner/Margaret R. BSHE 47
Bonner/Russell Alan AGRN BSA 92
Booi/Sabina W. BSHE 45
Boon/Bobby D. PLPA BSA 81
Boon/Leesa L. FDNU BSHE 92
Boone/Judith Green ANSC BSA 68
Booth/Veronica Ann BSHE 88
Boothe/Danny Lester AGRC BSA 97
Boozman/Cathy HEED BSHE 72
Boozman/Debra Sian HEED BSHE 75
Borchers/Cammie Slutter CLTX BSHE 79
Bordelon/Bruce P. PTSC PHD 91
Bordelon/Eugene P. HORT 99
Borders/John M. BSA 71
Boren/Stephen G. DASC BSA 70
Born/Mark Robert ENSC BSA 95
Borthick/Mary E. Nunnally ANSC MS 81
Bosch/Juan R. FDSC MS 92
Bosley/Brennan Eugene AGME BSA 97
Bossemeyer/James Lee BSA 23
Bost/Kevin W. AGEC BSA 80
Bostain/David Bruce BSA 71
Bostian/Charles K. GBUS BSA 77
Bostian/David V. ANSC BSA 50
Bostian/Judy Tackett HEED BSHE 76
Bostian/Keith Lyn ANSC BSA 88
Bostian/Nancy L. GHEC BSHE 78
Bostian/Stephen F. AGEC BSA 77
Bostic/Donna M. HOEC MS 86
Bostic/Donna M. FDNU BSHE 83
Boswell/Garland Garrett BSA 35
Boswell/John C. DANB BSA 82
Bottoms/Larry K. AGEG BSA 71
Bouchard/Dermont C. AGRN PHD 83
Bouchard/Dermont C. AGRN MS 82
Boudreaux/Carolyn LDUH BSA 91
Boudrey/Kelly L. FDNU BSHES 98
Bourland/Danny R. SOSC BSA 61
Bourland/Freddie M. CRSC MS 74
Bourland/Freddie M. CRSC BSA 70
Bourne/Tim PTSC PHD 89
Bovay/Shirley HEED BSHE 73
Bowden/Henry L. BSA 57
Bowden/Lugene BSHE 47
Bowden/Robert Alton BSA 59
Bowden/William A. BSA 47
Bowden/William P. BSA 54
Bowen/Brian Franklin POSC BSA 85
Bowen/Dawnna June FDSC BSA 96
Bowen/Hollis H. BSA 60
Bowen/Hollis H. HORT MS 61
Bowen/Ken BSA 58
Bowen/Mary FND BSHE 66
Bowen/Nancy A. GFNU BSHE 67
Bowen/Rebecca Poe FND BSHE 69
Bowen/Thomas E. ANNU BSA 66
Bowen/Thomas E. AGEC MS 68
Bowen/Thomas E. POSC MS 68
Bowen/Thomas Eugene RSOC MS 89
Bowen/Tracey Lynne AGBU BSA 97
Bowerman/Earl H. BSA 65
Bowers/Nan HEED BSHE 80
Bowie/Carrie Beth HEED BSHE 89
Bowles/David E. ENTO PHD 89
Bowlin/Frankie A. BSA 52
Bowling/Clarence C. ENTO MS 55
Bowling/Clarence C. BSA 54
Bowling/David M. HORT BSA 68
Bowling/Jimmie L. CRSC MS 68
Bowling/Jimmie L. CRSC BSA 61
Bowman/George F. AGEG BSA 26
Bowman/Jennifer Lynn HDFS BSHES 96
Bowman/Roger Lyn HORT BSA 93
Bowman/Talbert Fanning BSA 39
Bowyer/Barry L. ANNU BSA 84
Bowyer/Barry L. POSC MS 87
Box/Robert L. BSA 50
Boxtic/Edward Harold BSA 38
Boyce/Ben F. HORT BSA 81
Boyce/Lillian Jane HDFS BSHE 95
Boyd/Aileen J. BSHE 38
Boyd/Bradley Gene AGME BSA 76
Boyd/Burl BSA 53
Boyd/George W. BSA 41
Boyd/Glenn T. BSA 32
Boyd/James A. ANSC BSA 73
Boyd/James E. BSA 43
Boyd/James W. ANNU BSA 73
Boyd/Jim Don GBUS BSA 71
Boyd/Larry Gene HORT BSA 77
Boyd/Leigh A. ANSC 99
Boyd/Linda Suzanne IDHS BSHES 97
Boyd/Margaret Elise HEED BSHE 77
Boyd/Martha Lou HEED BSHE 70
Boyd/Mary Turley BSHE 26
Boyd/Molly Denise IDHS BSHE 87
Boyd/Randy ANSC BSA 85
Boyd/Robert E. BSA 53
Boyd/Thurman C. BSA 53
Boyer/Elsie McNabb BSHE 48
Boyer/William P. HORT BSA 62
Boyer/William P. BSA 37
Boyette/Doug PLPA PHD 83
Boyette/Doug PLPA MS 78
Boyle/Blaise Denis HORT BSA 83
Boyle/Erin Elizabeth CLTX BSHES 97
Bozsa/Robert C. AGRN PHD 88
Bracey/Anna Belle White BSHE 53
Brack/Ellen T. HORT MS 78
Brackett/Shirley BSHE 56
Brackman/Adrian BSA 52
Bradberry/Bobby L. ENTO MS 81
Bradberry/Bobby L. ENTO BSA 77
Bradberry/Susan Y. BSHE 77
Braddy/Leon D. POSC BSA 58
Bradford/Betty BSHE 54
Bradford/Dale T. BSA 50
Bradford/Gordon HORT BSA 74
Bradford/H. C. PLPR BSA 74

Bradford/Irene Davis BSHE 57
Bradford/Kathleen Housley HOEC MS 84
Bradford/Kathleen Housley HOFS BSHE 82
Bradford/Thomas Brick AGBS BSA 86
Bradley/A. B. AGED MS 48
Bradley/A. B. AGRI BSA 43
Bradley/Jack D. AGBS BSA 82
Bradley/James E. BSA 51
Bradley/James L. BSA 47
Bradley/Janet Lea BSHE 67
Bradley/Jimmy J. GNAG MS 64
Bradley/Jimmy J. BSA 53
Bradley/John W. BSA 51
Bradley/Stephainie FDNU BSHE 84
Bradney/Marvin F. BSA 60
Bradshaw/Herbert J. BSA 57
Bradshaw/Martha Mann BSHE 58
Bradshear/Virginia Clement HEED BSHE 73
Bradsher/Bridget HDID BSHE 82
Brady/Arthur H. BSA 51
Brady/Guy AGBS MS 74
Brady/Guy AGEC BSA 71
Brady/Homer A. BSA 57
Brady/Merle G. BSHE 54
Brady/Pamela L. HOFS MS 76
Brady/Pamela L. HOFS BSA 74
Bragg/Darrell B. POSC PHD 66
Bragg/John S. AGRI BSA 43
Bragg/Lindsey A. IDHS 99
Brahmakasikara/Sirapom AGEC MS 73
Brain/Sidney George BSA 35
Brake/Andrea C. ANSC MS 88
Brake/Kelly ANSC BSA 88
Bramall/Gary Robert GNAG MS 92
Bramall/Gary Robert AGED BSA 88
Bramall/Paula Pearson ANSC BSA 90
Brammer/Aleena Kay ANSC BSA 96
Brammer/Harlan H. BSA 58
Brandenberger/Lynn P. PTSC PHD 92
Brandon/Melinda HEED BSHE 71
Brannan/Linda HEED BSHE 77
Brannen/Cecil G. BSA 40
Brannon/Annetta FDNU BSHE 62
Branscum/David L. AGEC BSA 82
Bransford/Richard Barton VOAG BSA 80
Bransford/Roger Singleton AGBS BSA 80
Brantley/Laudies D. III AGBS 98
Brantley/Mark Raymond AGRN BSA 81
Brantley/Ralph G. BSA 50
Brantley/Thomas F. AGRI BSA 41
Brar/Mohanjeet HORT MS 95
Brar/Mohanjeet AGRN BSA 93
Brashears/Jack BSA 54
Braswell/Doug K. AGME BSA 84
Bratcher/Kenneth Charles BSA 39
Bratton/Barbara Daniel HEED BSHE 79
Bratton/Nancy Carol HEED BSHE 67
Bratton/Roy Lee BSA 76
Bratton/Terri Gayle ENTO MS 87
Bratton/Terri Gayle GNAG MS 87
Braughton/Melanie Nichols CLTX BSHE 91
Braverman/Michael Paul AGRN MS 85
Brawner/Cecily E. HDID BSHE 78
Brawner/Irma Jones HEED BSHE 74
Brax/Allan W. PLPA MS 77
Braxtan/Lorraine Thorn HORF BSHE 45
Bray/Judith B. BSHE 82
Bray/Judith B. HOEC MS 85
Braymen/Edward M. LDUH BSA 82
Braziel/Jason Brandon POUL BSA 97
Brazil/Velma Alford BSHE 34
Braznell/Kathryn Virginia BSHE 87
Breashears/Joe K. BSA 57
Breaux/Janice Kay ANSC MS 81
Breckenridge/Allan ANSC BSA 88
Breckenridge/Linda CLTX BSHE 93
Breckenridge/Thomas B. BSA 50
Breeden/Katrina Michele ANSC BSA 93
Breeden/Phil ANSC BSA 79
Breeding/Barclay Elizabeth CLTX BSHES 96
Breedlove/Alan L. FDSC MS 82
Breedlove/Alan L. HORT BSA 80
Breedlove/James Dee AGBS BSA 86
Breithaupt/James AGRN MS 91
Brendel/Randall A. POSC BSA 86
Brenner/Sandra J. Hilgenberg HOEC MS 76
Breshears/David E. BSA 57
Breshears/Stanley Brock ANSC BSA 89
Bretz/Maryalice CLTX BSHE 81
Brewer/Fayte AGRN PHD 81
Brewer/I. Aaron BSA 56
Brewster/Elizabeth Ferguson BSHE 48
Brewster/Marcie Lyn AGEC MS 88
Breyel/Marilyn Snyder HDID BSHE 84
Brian/Wanda HOEC BSHE 40
Brian/William Ross BSA 40
Bricker/Lou Alice Tyree BSHE 55
Bricker/Lou Alice Tyree HOEC MS 75
Bridenthal/Hazel Sweetser BSHE 34
Bridges/John R. ANNU BSA 66
Bridges/John R. ENTO MS 68
Bridges/Marilyn A. HEED BSHE 72
Brigance/Alan R. FDSC MS 93
Brigance/Alan R. FDSC BSA 91
Briggs/Deborah Jane ANSC MS 83
Briggs/Debra Lynn Houston CLTX BSHE 82
Briggs/E. J. BSA 42
Briggs/Jay Buford POSC BSA 69
Briggs/John A. BSA 57
Briggs/Rebecca Probert BSHE 74
Briggs/Saundra Aikman HEED BSHE 69
Bright/Cindy IDHS BSHE 89
Bright/Jimmy N. BSA 65
Brigman/Margaret Stuart BSHE 44
Brim/Roland Keith FDNU BSHE 87
Brinks/Renee Lee CLTX BSHE 94
Brinneman/Gerald Q. BSA 54
Brinson/Paul H. BSA 49
Brints/Norman W. AGEC MS 64
Brister/Michael Sean AGRC BSA 98
Brister/Paula FDNU BSHES 96
Brister/Stephen M. AGME BSA 72
Britt/Jeffrey Lee AGED BSA 94
Brittain/Greg AGEC BSA 79
Brittenum/Dempsey E. BSA 51
Britton/Diane B. CLTX BSHE 82
Brock/George Foley BSA 53
Brock/Nelson B. AGED BSA 87
Brockwell/James F. BSA 41
Brockwell/Roberta Carpenter BSHE 39
Brodie/Frank T. AGRN BSA 38
Brodie/Marilou DIET BSHE 68
Bromstad/Karen L. IDHS BSHE 88
Bronaugh/Lawrence A. BSA 38
Brooker/Nancy Louise PLPA MS 88
Brooks/Andrea Anthony CLTX BSHE 90
Brooks/Bernes B. BSA 39
Brooks/Donald E. BSA 59
Brooks/Howard Leroy BSA 60
Brooks/Howard Leroy ENTO MS 63
Brooks/J. Douglas AGEC BSA 82
Brooks/Joe G. FDSC MS 79
Brooks/Joe G. HORT BSA 78
Brooks/Julia Morris FDNU BSHE 68
Brooks/Larry G. BSA 70
Brooks/Robert Bradley AGRN BSA 81
Brooks/Stephen W. AGBS BSA 79
Brooks/Sylvia Jean HORT MS 91
Brophy/Karyn HDFS BSHE 93
Brosh/Donald L. HORT BSA 82
Broussard/Dana Knoll CLTX BSHE 78
Browers/Wilma A. ANSC BSA 87
Brown/Aliza Tzila ANSC PHD 93
Brown/Amanda Suzette CLTX BSHE 93
Brown/Carolyn POSC BSA 86
Brown/Carrie W. BSHE 22
Brown/Charles L. BSA 51
Brown/Charlie L. GNAG BSA 64
Brown/Colleen Black HORT BSA 92
Brown/Colleen Black AGRN MS 97
Brown/Connell J. ANSC BSA 48
Brown/Craig David AGEC MS 94
Brown/Deborah S. FDNU BSHE 75
Brown/Don POSC BSA 50
Brown/Edward Alan AGRN MS 82
Brown/Edward Alan AGRN BSA 80
Brown/Elvin D. HORT MS 69
Brown/Elvin D. BSA 63
Brown/Emory BSA 50
Brown/George Franklin BSA 40
Brown/George G. BSA 57
Brown/George W. BSA 36
Brown/Gerald R. BSA 59
Brown/Gerald R. GNAG MS 64

Brown/Helen V. FDSC PHD 85
Brown/Hunter L. DANB BSA 81
Brown/Imon L. ENTO MS 69
Brown/Imon L. ENTO BSA 66
Brown/Jackie R. BSA 60
Brown/James E. PHUS MS 62
Brown/James E. POSC BSA 60
Brown/James L. BSA 39
Brown/James R. ENTO PHD 86
Brown/James R. ENTO MS 83
Brown/John Hardy BSA 38
Brown/John R. AGRN MS 62
Brown/John Stephen DASC BSA 74
Brown/Joyce Megan IFDS BSHES 97
Brown/Karen CLTX BSHE 85
Brown/Karen S. HEED BSHE 75
Brown/Kathleen AGRI BSA 48
Brown/Kaye HEED BSHE 68
Brown/Lans O. BSA 52
Brown/Larkin GNAG MS 66
Brown/Larkin BSA 56
Brown/Leon L. ENTO MS 74
Brown/Maurus V. AGRN MS 91
Brown/Maurus V. PTSC PHD 97
Brown/Melissa Battershell CLTX BSHE 93
Brown/Mitchell R. AGEC BSA 82
Brown/R. Stephen ANSC BSA 76
Brown/Raymond W. POSC MS 73
Brown/Richard F. ANSC BSA 77
Brown/Richard L. ENTO MS 61
Brown/Robert Kelly ANSC BSA 87
Brown/Russell D. AGBS 98
Brown/Ruth Etheridge BSHE 31
Brown/Thomas D. ANSC BSA 86
Brown/Tom A. BSA 56
Brown/Tom O. HORT MS 65
Brown/Tom O. HORT BSA 62
Brown/William G. ANSC PHD 73
Browne/Kay Meinert BSHE 60
Browne/Lloyd E. ENTO MS 61
Browne/Lloyd E. BSA 59
Brownfield/Margaret BSHE 39
Browning/M. Lucille BSHE 27
Browning/Paul D. BSA 51
Browning/Robert E. PLPA MS 71
Brownmiller/Cindi R. FDSC 98
Bruce/Mae E. Dvorachek HOEC BSHE 39
Bruce/Robby Bryan AGBS BSA 85
Bruckner/Laurin H. POSC BSA 51
Bruehl/George W. BSA 41
Brumbelow/Terry Catton ANSC BSA 74
Brumett/Jerry Frederic LDUH BSA 87
Brumley/Joel M. PLPA MS 79
Brumley/Joel M. PLPA BSA 78
Brumter/Rita J. HDID BSHE 82
Bruner/Paula Kaye Porter HOFS BSHE 83
Brunk/Feryl R. BSA 49
Brunk/Leon C. BSA 52
Brunson/J. F. BSA 48
Brush-Martin/Esther Adams BSHE 44
Bryan-Durphy/Janet Maria FDNU BSHE 86
Bryan/Jason Adam POSC BSA 95
Bryan/Kathleen S. HORT BSA 91
Bryant/Brian Wayde AGED BSA 93
Bryant/Calvin B. BSA 49
Bryant/Chad Neil AGED BSA 89
Bryant/Galen BSA 68
Bryant/Gary L. SOSC BSA 78
Bryant/Gus BSA 49
Bryant/Harry Warrington ANSC BSA 81
Bryant/James R. BSA 50
Bryant/Jayne K. HDID BSHE 83
Bryant/John Shelby AGBU BSA 98
Bryant/Kelly Jay AGEC MS 86
Bryant/Kelly Jay AGBS BSA 85
Bryant/Larry L. PLPA BSA 68
Bryant/Lois Ann FND BSHE 69
Bryant/Malinda FDNU BSHE 89
Bryant/Ralph ANSC BSA 72
Bryant/Royce S. PLPA MS 80
Bryant/Royce S. PLPA BSA 77
Bryant/Susan E. FDNU BSHE 79
Bryles/Mark B. GNAG MS 67
Bryles/Mark B. GNAG BSA 54
Bryles/William Thomas GNAG MS 92
Bryles/William Thomas AGME BSA 84
Bryson/Beatrice BSHE 30
Bryson/Clayton AGEC BSA 77
Bryson/Sophia Britt BSHE 35
Buchanan/Amy Woolwine BSHE 36
Buchanan/Arbor Jean PPPM BSA 97
Buchanan/Carolyn HEED BSHE 69
Buchanan/Susan M. FND BSHE 73
Buchanan/Thomas G. ANSC BSA 96
Buchanan/William A. BSA 33
Buchberger/Eta AGEC MS 91
Buchta/Leslie Ryburn BSHE 85
Buck/Cecelia B. Dumeny BSHE 76
Buck/Chawnerhea AGED BSA 92
Buck/Lois Jeanne VOED BSHE 55
Buck/Troy W. BSA 61
Buckley/Bob A. AGME BSA 85
Buckley/J. Blair HORT MS 85
Buckley/J. Blair BSA 78
Buckley/Marjorie J. FND BSHE 71
Buckley/Mary H. HOEC MS 75
Buckley/Ulis Martin AGEC MS 49
Buckmaster/Anna Kathleen Miller ANSC BSA 82
Buckner/Martha Lynn CLTX BSHE 91
Buckner/Nita Jo HDFS BSHE 95
Buechley/Jay F. BSA 30
Buechley/Vera BSHE 30
Bueker/Hank AGBS BSA 85
Buescher/Wayne Arnold FDSC BSA 96
Buffington/Gene H. ANSC BSA 87
Bull/Carrol E. BSA 51
Bull/Estle Burk BSA 60
Bull/Harvey P. AGRI BSA 50
Bull/Joe D. BSA 50
Bull/Marvin Clifford BSA 28
Bullington/Danny M. ANSC BSA 71
Bullington/Ethel L. Hays HOEC BSHE 64
Bullington/Leigh Ann GHEC BSHE 82
Bumgarner/Don W. DANB BSA 79
Bumpers/Sherrill H. BSA 50
Bunch/Alexander AGBS BSA 93
Bunch/Charles S. BSA 25
Bunch/Grace Lincoln BSHE 39
Bunch/Hildred G. BSA 42
Bunch/Lori Huskey FDSC MS 89
Bunch/Lori Huskey FDSC BSA 88
Bunton/Christopher Drew FDNU BSHES 98
Bunton/Clyde J. BSA 60
Bunyarat/Mungkorn ENTO MS 76
Burasco/Richard Shannon BSA 93
Burbank/Donna K. ANNU MS 79
Burch/George V. GNAG MS 86
Burch/Harold E. BSA 65
Burch/Kara Anne HOEC MS 94
Burch/Mary Lou Cornelius BSHE 48
Burcham/Gwendolyn HOEC MS 57
Burcham/L. Neil ANSC MS 71
Burcham/Michael ANSC MS 77
Burcham/Michael POSC BSA 75
Burchfield/Travis AGBS BSA 89
Burda/Irma R. FDNU BSA 78
Burden/Brooke D. LDUH 99
Burden/Marcus Dudley AGED BSA 93
Burdette/D. Lea AGRN MS 81
Burgess/Bill F. ANSC BSA 83
Burgess/Joyce Tillman HEED BSHE 61
Burgess/Joyce Tillman FDNU MS 65
Burgess/Madge Ethel Polk HOEC MS 64
Burgess/Madge Ethel Polk BSHE 34
Burgess/Nancy Ford BSHE 42
Burgess/Thomas R. BSA 42
Burgos/Nilda R. AGRN MS 94
Burk/Nellie Berry BSHE 27
Burk/Ronald Dean AGBS BSA 84
Burke/Gary L. AGRN PHD 85
Burke/Johnny Calvin BSA 60
Burke/Russell Jay AGBS BSA 89
Burkes/Jesse S. AGBS BSA 86
Burkett/Brent M. HORT 99
Burkett/Jerry AGEC BSA 79
Burkett/Peggy Killough BSHE 59
Burkett/Thomas BSA 50
Burks/Helen Denniston DASC BSA 84
Burks/Patrick R. BSA 54
Burks/Tammy Irene Winsted AGBS BSA 85
Burkwhat/Henry Edwin ENTO MS 81
Burleson/Frances BSHE 43
Burnet/Mary S. BSHE 44

Burnett/Hazel Baker BSHE 41
Burnett/Rogna M. ANSC BSA 58
Burnett/Ronald K. AGEC MS 82
Burnett/Sally Trammell BSHE 68
Burnett/Theodore B. VOAG BSA 81
Burnett/Wesley Dwayne POSC BSA 83
Burns/Barry AGEC BSA 79
Burns/Buell E. AGRN BSA 38
Burns/Coleman Dean BSA 25
Burns/Everett H. BSA 31
Burns/Fred C. BSA 34
Burns/Georgia Perchan HEED BSHE 79
Burns/Jacqueline Kay HORT BSA 79
Burns/Jacqueline Kay HORT MS 82
Burns/James E. AGEC BSA 58
Burns/Jeania Moore HEED MS 54
Burns/Jonathan A. HORT 98
Burns/Mae Harriet HEED BSHE 71
Burns/William Oran BSA 38
Burr/James B. ANSC 99
Burris/Roy E. BSA 49
Burrow/Sallie N. FDNU BSHE 74
Burrows/Douglas R. BSA 50
Burt/Melva Lou HOEC MS 68
Burton/Alice J. HEED BSHE 72
Burton/David O. BSA 60
Burton/Erwin M. POSC MS 79
Burton/Gary C. AGEC MS 78
Burton/Gregory O. AGBS BSA 83
Burton/James Edward AGEC BSA 66
Burton/Leigh Ann CLTX BSHE 93
Burton/Richard B. FDSC BSA 79
Busbee/Jennifer Appleby AGBS BSA 93
Bush/Carol Marie AGRN MS 78
Bush/Jesse F. VOED BSA 55
Bush/Mary Ann FDNU BSHE 67
Bush/Mary Ann FDNU MS 70
Bush/R. Marlene Kerr HEED BSHE 56
Bush/William L. BSA 49
Buss-Rosso/Sharon HDFS BSHE 90
Buss-Rosso/Sharon HOEC MS 94
Busse/Linda K. HEED BSHE 82
Bussell/William Denice ANSC MS 79
Bussell/William Denice POSC BSA 73
Butler/Brent AGBS BSA 86
Butler/Dwayne E. AGBU BSA 98
Butler/Helen Pauline BSHE 56
Butler/Lee A. LDUH BSA 82
Butler/Linda Eudy HEED BSHE 74
Butler/Michael W. AGEC PHD 74
Butler/Tina Thomas IDHS BSHE 89
Butler/Walter L. BSA 50
Butrum/Fred H. GNAG BSA 68
Butterfield/Loris Steward BSHE 59
Buxton/Christina Marie LDUH BSA 95
Buzbee/Linda B. FND BSHE 72
Buzbee/Martha CLTX BSHE 87
Byers/Kevin Lewis AGED BSA 92
Byers/Patrick Lee HORT MS 86
Byler/Ben L. ANSC MS 69
Bynum/Rebecca Jane ANSC BSA 95
Bynum/W. Daniels AGBS BSA 95
Bynum/William B. ENTO MS 58
Bynum/William B. BSA 57
Byram/T. J. AGRN MS 63
Byram/T. J. CRSC BSA 61
Byrd/Barett R. CNED BSA 87
Byrd/Barry Clinton BSA 50
Byrd/Claude J. BSA 25
Byrd/David F. AGRN PHD 75
Byrd/J. Brandon AGBS 98
Byrd/Jared Duane AGBU BSA 98
Byrd/John Paul ANSC BSA 80
Byrd/Ruth Bowman BSHE 27
Byrne/Julia Wells BSHE 27
Byrum/Phyllis Bartlett LDUH BSA 90
Cabel/Mario Cruz ANSC PHD 89
Cabel/Mario Cruz POSC MS 85
Cabeldue/Lisa Lynn BSHE 87
Cackler/Nancy A. FDSC BSA 85
Cackler/Steven FDSC MS 85
Cadh/Clifton Franklin BSA 51
Cagel/Marcelyn Renee BSHE 68
Cagle/Lurline BSHE 35
Cagle/Roger E. AGBS MD 48
Cahoon/Melissa AGBS BSA 89
Cain/Blakeney Jennings ENSC BSA 98
Cain/Stanley B. AGRN BSA 90
Cain/Timothy C. AGEC BSA 68
Calaway/Will J. BSA 59
Caldwell/Alice Stanford BSHE 29
Caldwell/Billy ANSC BSA 89
Caldwell/Calvin J. BSA 50
Caldwell/Dorothy Reddell HOEC BSHE 56
Caldwell/Judith Dawn PTSC PHD 81
Caley/Daniel John AGBS BSA 93
Calhoun/Earnest Leroy ENTO MS 51
Calhoun/Eurie H. BSA 50
Calhoun/Hall M. GNAG BSA 61
Calhoun/Raymond E. BSA 77
Calhoun/Zachary Herman BSA 22
Calico/Tamala Lavy POSC BSA 81
Call/Pat Bonita FDNU BSHE 68
Callahan/Janice E. HEED BSHE 69
Callahan/Jarold E. GNAG MS 80
Callahan/Marcella A. SOSC BSA 76
Callahan/Margaret BSHE 17
Callahan/Terry Lyn AGBS BSA 95
Callaway/Scott Edward ANSC BSA 75
Callaway/Shirley RSOC MS 81
Calloway/Douglas AGEC BSA 75
Calvert/Christine CLTX MS 51
Calvert/Frederic R. DASC BSA 79
Calvert/Letitia L. FND BSHE 70
Cameron/Iverson C. BSA 40
Cameron/R. Patrick AGBS BSA 92
Camfield/Peter Kipley ANSC PHD 96
Camp/Arthur A. BSA 52
Camp/Earnestine BSHE 42
Campbell/B. L. BSA 41
Campbell/Cuba Thurman BSHE 34
Campbell/Donnie R. ANSC BSA 72
Campbell/Donnie R. ANNU MS 74
Campbell/Doyle R. GNAG BSA 61
Campbell/Elizabeth M. BSHE 52
Campbell/Eula Tetrick BSHE 32
Campbell/Glenn L. BSA 39
Campbell/Jack MacGregor AGEC BSA 96
Campbell/Kenny Ray ANSC MS 72
Campbell/Kenny Ray ANSC BSA 69
Campbell/Mary Higgins BSHE 54
Campbell/Mary M. ANSC BSA 78
Campbell/Murrey BSA 38
Campbell/Teresa S. CLTX BSHE 83
Campbell/W. Peyton BSA 17
Canant/Jerry E. ANSC MS 68
Canant/Jerry E. ANSC BSA 66
Cane/Gene Clyde AGBS BSA 86
Canessa/Carlos A. Espinoza FDSC MS 90
Canfield/Cheryl Rene LDUH BSA 92
Cannaday/Floyd BSA 41
Cannaday/Hazel Bunch BSHE 40
Cannon/Barbara Patterson FDNU BSHE 62
Cannon/Helen BSHE 32
Cannon/Robert L. BSA 47
Cantrell/Camilla Kay HEED BSHE 90
Cantrell/Jeffrey Stephen AGED BSA 95
Cantrell/Mary Sue POSC BSA 83
Cantrell/Ruth Esther BSHE 30
Cantrell/Steve DASC MS 72
Cantrell/Steve DASC BSA 70
Cantrell/Thomas R. GNAG BSA 76
Capo/Brion T. PLPA MS 82
Capps/Allyce LDUH BSA 80
Capps/Emerson S. GNAG BSA 40
Cappy/James J. AGRN PHD 79
Capstick/Daniel F. AGEC MS 80
Capstick/Daniel F. GBUS BSA 77
Caraway/Vickie L. LDUH BSA 78
Cardarelli/Anthony BAEG BSBAE 97
Cardin/David POSC MS 64
Cardin/David ANNU BSA 63
Cardin/Walton V. BSA 41
Carey/Victor Frank AGRN MS 90
Carey/Victor Frank AGRN PHD 94
Carithers/Clay AGRN MS 82
Carl Lee/Ruth Wilson BSHE 56
Carl/Beulah BSHE 21
Carlile/Lisa A. Erwin BSHE 82
Carlisle/Carolyn Jenkins FND BSHE 72
Carllee/Lewis S. BSA 56
Carlock/Valerie Gerdes HORT BSA 93
Carlson/Ann H. BSHE 58
Carlson/Joan Biles HEED BSHE 72

Carlson/Judy HEED BSHE 71
Carlson/Mary Jordon BSHE 46
Carlson/Walter Lawrence AGME BSA 72
Carlton/Christopher Eugene ENTO PHD 89
Carlton/Christopher Eugene ENTO MS 80
Carlton/Jeffrey Paul ANSC PHD 94
Carlton/Jeffrey Paul AGEC MS 91
Carlton/Jeffrey Paul POSC BSA 87
Carlton/Jimmy Shawn POSC BSA 94
Carlton/Melody Kathleen HORT BSA 80
Carlton/Sherri Griffiths FDNU BSHE 79
Carmichael/Ruth BSHE 35
Carnahan/Bill POSC BSA 87
Carnahan/Catherine Sue GHEC BSHE 95
Carnell/Marilyn L. FDSC PHD 82
Carnes/Larry W. ANSC MS 77
Carnes/Larry W. ANSC BSA 75
Carney/Faye E. BSHE 37
Carpenter/David R. DASC BSA 74
Carpenter/George W. BSA 54
Carpenter/Virginia Leeper BSHE 31
Carr/Kari Lynn FDNU BSHE 95
Carr/Richard PLPA PHD 83
Carraway/Cynthia FDNU BSHE 76
Carroll/Roberta Joan FDSC MS 88
Carroll/Tammy Johnette CLTX BSHE 86
Carruth/George A. BSA 56
Carruth/Paul F. BSA 26
Carson/Kristin Hope FDST BSA 98
Carson/Rick VOAG BSA 81
Carson/Stephen Douglas AGME BSA 93
Carsten/Cynthia M. CLTX BSHES 97
Carter/Alan BSA 48
Carter/Alvis G. BSA 40
Carter/Barbara June FND BSHE 71
Carter/Ben Copeland BSA 39
Carter/Essie Mae HOEC MS 65
Carter/Frank L. ENTO MS 68
Carter/Frank L. BSA 65
Carter/Herschel H. BSA 39
Carter/Jack B. BSA 51
Carter/James E. BSA 51
Carter/James Justin AGBU BSA 97
Carter/Jason D. AGEC MS 92
Carter/Jason D. AGBS BSA 89
Carter/John W. BSA 53
Carter/Joy Carla DIAM BSA 95
Carter/Joy Carla AGEC MS 97
Carter/Joyce R. BSHE 65
Carter/Kathy L. LDUH BSA 84
Carter/L. C. AGRN BSA 31
Carter/Leonard Winton AGRI BSA 34
Carter/Marvin R. VOAG BSA 36
Carter/Preston Calvin AGRN MS 962
Carter/Sarah L. Basham CLTX BSHE 77
Carter/Scott McKinney AGBS BSA 81
Carter/Stanley D. AGRN MS 79
Carter/Stanley D. CRSC BSA 76
Carter/Sue Hawley BSHE 48
Carter/Walter R. BSA 34
Cartmell/Jamie Lynne HEED BSHE 89
Cartwright/Donald Kelly PPPM BSA 89
Cartwright/Gerald Allen AGBS BSA 85
Cartwright/Richard D. PLPA MS 87
Cartwright/Richard D. BSA 76
Carver/Jonathon Richard AGME BSA 95
Carver/Kristi Elizabeth CLTX BSHE 93
Carwell/William B. AGME BSA 70
Case/Lindsay IDHS BSHES 98
Casey/Denise Elaine HDFS BSHE 93
Casey/Evelyn S. BSHE 41
Casey/John E. BSA 18
Cash/Brian Damon AGBS BSA 95
Cash/James V. BSA 54
Cash/Jerry N. CRSC PHD 76
Cash/Larry B. DHUS BSA 64
Cash/Lesleigh Gay ANSC MS 87
Cash/Mitchell H. DIAM BSA 89
Cashion/April Moore HEED BSHE 77
Cason/Velton Winn BSA 50
Cassady/Loy Roger ANSC BSA 69
Cassidy/Doris B. DIET BSHE 59
Cassidy/M. L. VOAG BSA 67
Castleman/Clayton BSA 68
Cate/Francis Marion BSA 52
Cate/Jimmy R. BSA 65
Cate/Larry P. AGEC BSA 76
Cates/Billy L. BSA 58
Cathey/George W. CRSC MS 66
Cathey/George W. BSA 48
Cathey/Johnnie Marie FDNU BSHE 73
Catlett/Henry BSA 54
Catlett/Paul E. BSA 53
Catlett/Vernon Ray BSA 58
Cato/Cecilia Ann Nailling FDNU BSHE 71
Cato/Ralph Daryl BSA 42
Cattaneo/Elaine IDHS BSHES 97
Cattlett/Hazel Newsome BSHE 57
Caudle/Jim AGRN BSA 67
Caudle/Pauline H. HOEC MS 72
Caughman/Neil Douglas AGBS BSA 92
Caulder/Martha Lou HEED BSHE 69
Cauthron/Mary E. IDHS BSHE 92
Cavalier/Thomas C. AGRN MS 88
Caviness/Charles E. BSA 49
Caviness/Darlene AGRN MS 84
Caviness/Johnnie Marie FDNU BSHE 73
Caviness/Linda HOEC MS 892
Caviness/Linda CLTX BSHE 85
Cavness/Jack ANSC BSA 57
Cawthon/Carrie Melissa IFDS BSHES 98
Cawthon/Donald L. HOFS BSA 73
Cawthon/Donald L. FDSC PHD 82
Cawthon/Donald L. HORT MS 77
Cearley/Lon POSC BSA 77
Cearley/Ruth Dowell BSHE 29
Cearley/Woodrow BSA 40
Cefalu/Susan J. BAED BSHE 67
Center/Arthur Wayne AGBS BSA 95
Cerday/Eddie L. POSC BSA 74
Chadick/Charles Harlston AGEC BSA 71
Chadwick/Robert R. AGEC BSA 71
Chaffee/Ann Marie CLTX BSHE 94
Chairatanayuth/Pornsri ANSC PHD 81
Chalfant/Ruth Blanshard BSHE 27
Chamberlain/Dave POSC BSA 80
Chambers/Alton BSA 47
Chambers/Barney W. BSA 39
Chambers/Elizabeth BSHE 47
Chambers/James T. BSA 46
Chambers/Joe H. BSA 59
Chambers/Lendon E. BSA 43
Chambers/Leon D. BSA 41
Chambers/Steve AGBS BSA 87
Chambers/Terri Lynn HDFS BSHE 94
Chambers/Thomas H. BSA 39
Chambers/William D. BSA 55
Chambliss/Carrol G. CRSC MS 66
Chambliss/Carrol G. CRSC BSA 64
Chambliss/Mark AGRN BSA 80
Champlin/Jimmy Cox BSA 56
Chancey/Grady ENTO MS 73
Chancey/Grady ENTO BSA 71
Chandler-Twilley/DeAnna Dawn CLTX BSHE 94
Chandler/Anna HDID BSHE 84
Chandler/Leo BSA 52
Chandler/Marjorie Barger BSHE 41
Chaney/Henry M. GNAG MS 87
Chaney/Henry M. AGRN BSA 82
Chaney/Kelly Smith HEED BSHE 87
Chaney/Phillip Lynn ANSC BSA 81
Chaney/Robert C. ENTO MS 62
Chaney/Robert Cheathum GNAG MS 87
Chaney/Steve W. PPPM BSA 80
Chaney/Stuart Lynn HORT BSA 74
Chang/Yen Ming FDSC BSA 91
Chapman/Karen Rebecca AGRN BSA 95
Chapman/Stanley L. SOSC MS 67
Chapman/Stanley L. SOSC BSA 62
Chappell/Dewayne AGEC BSA 78
Charitos/Oliver AGEC BSA 79
Charlesworth/Mary M. BSHE 47
Charlesworth/Yevonne BSHE 38
Charlton/Richard D. GNAG BSA 54
Chase/Carol C. HOEC MS 932
Chastain/Wayne Edward BSA 39
Chavez/Eddie Dela Cuesta AGEC MS 94
Chavez/Rebecca Salome AGRN PHD 96
Cheatham/Emily Anne CLTX BSHES 98
Cheatum/Steven R. AGEC MS 83
Chen/Huang Hsiuluan HORT MS 86
Chen/Ming-Dang ANSC PHD 92
Chen/Ming-Dang ANSC MS 87

Chen/Sharon Ziaoyu FDSC MS 88
Chen/Xiangli BAEG MSBAE 96
Chenault/Harry K. BSA 49
Chenowith/Gary D. LDUH BSA 81
Chenowith/Gary D. GNAG MS 83
Chenowith/Robert ENTO MS 76
Chenowith/Robert ENTO BSA 74
Cherry/Jacquelyn Fern ANSC MS 88
Cherry/James Louis BSA 13
Cherry/Lisa Marie FDNU BSHES 98
Cherry/Mary Jennings FDNU BSHE 68
Chesshir/Alicia Gail GHEC BSHE 85
Chesshir/Audie R. POSC BSA 74
Chesshir/James W. BSA 59
Chester-Davis/Leah GHEC BSHE 82
Chester/Joseph M. AGBS BSA 85
Chester/Olga Dwane BSA 51
Chevallier/John Roy ANSC BSA 87
Chewning/John Jeffrey ANSC PHD 90
Chhoun/Khan CRSC BSA 64
Chi/Hongtao AGEC MS 94
Chick/Randolph Jay ANNU BSA 78
Child/Robert D. ANNU MS 66
Child/Robert D. BSA 50
Childers/Kathy HEED BSHE 72
Childers/Robert Earl AGBS BSA 95
Childress/Gary Glen ANSC BSA 86
Childress/James E. AGBS BSA 82
Childress/Karen Deen BSHE 85
Childress/Keith Alan AGBS BSA 86
Childress/Leah C. HORT MS 88
Childress/Leslie D. POSC BSA 54
Childs/Elizabeth A. GHEC BSHE 77
Childs/Kathleen Lowe ANSC MS 78
Childs/Kathleen Lowe BSA 75
Childs/Mildred S. BSHE 43
Chinault/Karen Leigh ANSC BSA 94
Ching/Alejandro PTSC PHD 85
Chisholm/Allie Pickell BSHE 36
Chism/E. Fay Taylor BSHE 51
Chism/Johnny SOSC BSA 80
Chism/P. W. BSA 81
Chitwood/Auten BSA 43
Chitwood/Rhonda HDFS BSHE 88
Chivers/Joe J. BSA 50
Chivers/Mary Hardcastle HOEC MS 62
Chivers/Mary Hardcastle BSHE 49
Cho/Mi Jin FDSC MS 88
Cho/Myoung-Rae PTSC PHD 89
Choate/Blanche Lambert BSHE 54
Choate/Dorris Norman ANSC BSA 52
Choate/Jane HEED BSHE 47
Choate/Thomas H. AGBS BSA 80
Choe/Chang-Young SOSC MS 75
Chong/Hogun AGEC MS 96
Choo/Alina ENSC BSA 97
Chrisco/Monte AGEC BSA 78
Christian/C. D. HORT BSA 24
Christian/David E. AGRN BSA 62
Christian/Glenda CDFR BSHE 75
Christian/Thomas ANSC BSA 76
Christiansen/Meryl N. AGRN MS 55
Christiansen/Meryl N. GNAG BSA 50
Christie/Parri Rae FDNU BSHE 89
Christison/Laurel Kathleen FND BSHE 73
Christman/Joe BSA 77
Christman/Susan Jablonski HORT BSA 82
Christopherson/Amy T. FDSC BSA 86
Christy/Chester D. BSA 58
Chu/Peter FDSC BSA 81
Chun/Se-Chul PLPA MS 92
Church/Christie Angela HES BSHES 95
Churchill/Mary Cheek BSHE 40
Churchill/Randle W. ANSC BSA 74
Churchman/William R. ANNU MS 76
Ciabattari/Kristin Olstad CLTX BSHE 90
Ciarletta/John Angelo BSA 53
Cichinski/Walter C. FDNU BSHE 82
Cigainero/H. L. BSA 49
Clanton/Virginia K. HDFS BSHES 97
Clapp/William Dale BSA 65
Clardy/Chester BSA 22
Clark/Arthur Hale BSA 58
Clark/Cameron Brooks AGRN MS 94
Clark/Christine Marie FDNU BSHE 90
Clark/Deborah L. BSHE 88
Clark/Dorothy Hudson BSHE 35
Clark/Elaine M. GHEC MS 65
Clark/Elizabeth A. HDFS 99
Clark/Franklin J. ANSC BSA 70
Clark/Harry J. BSA 53
Clark/James A. BSA 18
Clark/James Vernon BSA 59
Clark/Janis Carolyn FND BSHE 69
Clark/Jerry Don AGEC MS 85
Clark/Jerry R. AGBS BSA 72
Clark/John B. RSOC MS 85
Clark/John W. BSA 43
Clark/Kenneth B. BSA 49
Clark/Kenneth L. BSA 54
Clark/Kevin Dale ANSC BSA 86
Clark/Lona L. HOEC MS 89
Clark/Mark D. AGRN PHD 86
Clark/Mark D. AGRN MS 81
Clark/Mary Ellen CLTX BSHE 47
Clark/Mary Jane Langley CLTX BSHE 82
Clark/Novelle Bond BSHE 42
Clark/Patricia FDSC MS 88
Clark/Sharon CLTX BSHE 83
Clark/Sharon Presson HEED BSHE 70
Clark/Shawn ANSC BSA 94
Clark/Steven Mark HOEC MS 88
Clark/Timothy H. AGRN MS 91
Clarke/Elizabeth Ann CLTX BSHE 81
Clarke/Mark Owen ANSC BSA 81
Clarke/Stephen R. ENTO MS 83
Clarke/Susan Dennehy FND BSHE 69
Clarke/Timothy M. AGEC MS 85
Clary/Charlotte BSHE 41
Clary/Darrell N. AGRN MS 62
Clary/Joan Turner BSHE 59
Clary/Joan Turner HOEC MS 62
Claxton/Dwain BSA 51
Clay/Howard Franklin AGRI BSA 61
Clayton/Archie BSA 61
Clayton/Billie King BSHE 58
Clayton/Billy G. BSA 59
Clayton/Chet Daniel AGBS BSA 93
Clayton/Donna Reynolds HDID BSHE 82
Clayton/Joe E. AGEG BSA 59
Clayton/Johnathan Craig AGBS BSA 87
Clayton/Memori A. AGBS BSA 93
Clayton/Sherry Belle GNAG MS 92
Clayton/William L. BSA 57
Cleary/Frederick AGBS BSA 87
Cleaver/Barry Arthur AGBS BSA 86
Clem/Chad M. POSC 99
Clem/Jennifer Buechley BSHE 86
Clement/Claude H. GNAG MS 66
Clement/Ralph H. BSA 52
Clement/Scott AGBS BSA 85
Clement/Sherrill Thompson CLTX BSHE 91
Clement/Steven L. ENTO MS 80
Clements/Helen BSHE 47
Clements/John Richard ENSC BSA 98
Clements/Leigh Ann AGRN BSA 93
Clemmons/Earl H. BSA 48
Clemmons/John H. BSA 49
Clendening/Mary Bruce BSHE 44
Clerge/Gaskov AGRN BSA 91
Cleveland/Elton R. ANNU BSA 73
Cleveland/Jason Solomon AGBS BSA 87
Cleveland/Tom C. ENTO MS 58
Cleveland/Tom C. BSA 56
Clevidence/Beverly A. HOEC MS 76
Clevidence/Beverly A. HEED BSHE 69
Click/Robert Howard BSA 31
Click/Rodger H. BSA 55
Clifton/Bobby L. BSA 59
Clifton/Elizabeth Ann CLTX BSHE 77
Clifton/Ernest Clarbourne BSA 47
Clifton/Gus R. BSA 38
Clifton/John L. BSA 65
Cline/Karen Salmon BSHE 87
Cline/Nina M. BSHE 38
Clinton/Nancy Jo Williams BSHE 57
Cloninger/Thomas Lee HORT MS 70
Cloud/Gary Leonard PTSC PHD 90
Cloud/Gary Leonard PLPA MS 86
Cloutier/Hardy BSA 54
Cloutier/Jacquelyn Sabbe CLTX BSHE 83
Cloutier/Kathryn K. CLTX BSHE 88
Cloutier/Mike AGBS BSA 87
Cloutier/Steve AGBS BSA 86

Clow/Dan C. AGEC BSA 78
Clower/Bill POSC MS 53
Clower/Bill ANSC BSA 52
Clower/John D. ANSC BSA 53
Clower/Philip Mark POSC BSA 86
Cluck/Rodney Eugene RSOC MS 94
Coan/Mary Worley BSHE 29
Coats/Yvette M. CLTX BSHE 83
Cobb/Allison CLTX BSHE 76
Cobb/Debbie Cheek DANB BSA 84
Cobb/James T. DANB BSA 82
Cobb/Jerry L. AGRN BSA 81
Cobb/Monterie W. HOEC MS 66
Cobb/Sharon Anne HDID BSHE 83
Cochran/Donald E. VOAG BSA 78
Cochran/Gary E. AGEC BSA 66
Cochran/Henry Lee BSA 26
Cochran/Paul AGEC BSA 75
Cochran/Steven Lelus ANSC BSA 85
Cochran/William W. BSA 34
Cockrill/Ellon Rogers FND BSHE 71
Cockrill/J. Mitchell POSC PHD 80
Cockrum/Jack C. BSA 58
Coday/Jamie L. CLTX BSHE 93
Coe/Hall Charles BSA 46
Coe/Herbert Eugene BSA 39
Coe/Jim A. POSC BSA 50
Cofer/Kenneth D. ANSC BSA 63
Coffee/Jeffrey Allen AGED BSA 93
Coffee/Jerry G. GNAG MS 80
Coffee/Rhonda Lea AGBS BSA 87
Coffelt/Charles Eugene AGED BSA 94
Coffield/Erin Elizabeth FDNU BSHE 94
Coffman/Bob BSA 58
Coffman/Dewey I. AGRN BSA 56
Coffman/Kimberly Ann HDFS BSHES 97
Coger/Mary Ernestine CATX BSHES 97
Coker/Alice Edith BSHE 20
Coker/Charlene Dawn CLTX BSHES 97
Coker/Clifford M. PLPA MS 85
Coker/Clifford M. PPPM BSA 82
Coker/Jay Barnett AGBS BSA 88
Coker/Jerrell Joseph BSA 55
Coker/Jesse M. BSA 52
Coker/Marnie IDHS BSHES 98
Coker/R. Kirk AGBS BSA 93
Cokes/Sara Spencer BSHE 63
Colberg/Marelisa POSC MS 88
Colbert/Mark A. GNAG MS 80
Cole/Ann McGill BSHE 25
Cole/Carolyn A. HEED BSHE 75
Cole/Cynthia Hugg CLTX BSHE 79
Cole/Edgar L. BSA 56
Cole/John D. BSA 58
Cole/Kathalee AGBS BSA 96
Cole/Kathy Jane HLED BSHE 81
Cole/Kathy Jane NURS BSHE 81
Cole/Marvin L. ANSC BSA 66
Cole/Marvin Lynn ANSC BSA 76
Cole/Patsy HEED BSHE 64
Cole/Teresa Ann DASC MS 76
Cole/Terry C. ANSC MS 77
Cole/Terry C. DASC BSA 74
Cole/William T. AGRN MS 61
Cole/William T. BSA 59
Coleman/Andrew W. POSC BSA 82
Coleman/Bruce W. ANNU MS 66
Coleman/Bruce W. ANNU BSA 64
Coleman/Burke Johnston BSHE 67
Coleman/Cynthia Ann HDFS BSHES 95
Coleman/Donald AGBS BSA 78
Coleman/Eusel M. BSA 28
Coleman/Jack Furlow BSA 38
Coleman/James Arthur BSA 54
Coleman/Kevin Clinton ENSC BSA 96
Coleman/L. Quinton BSA 28
Coleman/Leslie Swanigan HDFS BSHE 93
Coleman/Lynn D. FND BSHE 70
Coleman/Mabel Faye Rogers FDNU MS 66
Coleman/Mabel Faye Rogers BSHE 64
Coleman/Neily Bernice BSA 48
Coleman/Phyllis Williams BSHE 50
Coleman/Ryan B. ANSC BSA 88
Coleman/Tom POSC BSA 52
Colemen/Jennifer Flippin IDHS BSHES 96
Coles/Anthony Keith HORT BSA 81
Coley/Charles Elton BSA 54
Coley/John I. BSA 48
Coley/John I. GNAG MS 64
Collar/Oliver Carl BSA 51
Collier/Cleon Wallis BSA 49
Collingsworth/David L. POSC BSA 79
Collins/Brodey S. AGBS BSA 84
Collins/Bryan G. GNAG BSAGE 62
Collins/Frederick C. AGRN BSA 63
Collins/Gene GNAG MS 86
Collins/Gene VOAG BSA 80
Collins/Gloria A. CDFR BSHE 76
Collins/James H. BSA 13
Collins/Jennifer Lynn EXED BSA 95
Collins/John Eric AGED BSA 93
Collins/Malcolm Wayne ANSC BSA 86
Collins/Martha J. FND BSHE 71
Collins/Merrilou CLTX BSHE 90
Collins/Millie Jo HOEC MS 83
Collins/Robert A. BSA 39
Collum/Charles G. PPPM BSA 78
Colomina/Heather R. IDHS BSHE 88
Colten/Marci Ellen CLTX BSHE 76
Colvert/Ruth HEED BSHE 61
Colvin-Snyder/Tobi L. AGRN MS 91
Combs/Elizabeth Toone HEED BSHE 83
Combs/Karen M. FND BSHE 74
Combs/Kenny J. ENTO MS 77
Combs/Kenny J. ANSC MS 77
Combs/Kenny J. ENTO BSA 77
Combs/Kenny J. ANSC BSA 75
Combs/Mary Bethel HEED BSHE 61
Combs/Robert L. ENTO MS 63
Combs/Robert L. GNAG BSA 61
Comer/Chana Annette AGBU BSA 98
Compston/Scott R. ESWS 98
Compton/Adam Don AGRN BSA 96
Compton/Agnes BSHE 26
Compton/Catherine Jabine BSHE 27
Compton/Georgia R. FND BSHE 70
Compton/Juliet CLTX BSHE 78
Compton/Nancy Coleman BSHE 45
Conard/Michael A. AGRN BSA 83
Conde-Cabrera/Hernan A. POSC BSA 81
Conklin/Ruth Marie White BSHE 74
Conley/Herman H. BSA 53
Conley/Mary Rohm BSHE 43
Conley/Rita D. AGEC MS 92
Connaughton/Claretta Cameron BSHE 38
Connell/John DASC BSA 83
Connell/Linda Standridge HEED BSHE 61
Connell/Thomas P. AGME BSA 72
Conner/Avanell Watson BSHE 44
Conner/John D. BSA 58
Conner/Melinda Stewart LDUH BSA 86
Conner/Robert Lowell ANSC BSA 85
Connerly/Laura Jane CLTX BSHE 86
Conrad/Harold Kevin LDUH BSA 93
Conrad/Irene Pittman BSHE 28
Conrad/Michael A. GNAG MS 85
Considine/Rebecca Hart CLTX BSHE 77
Conway/Tammie Dee IDHS BSHES 97
Conway/Tammie Dee POSC BSA 97
Cook/Alice Virginia BSHE 24
Cook/Bobby Joe CRSC BSA 69
Cook/Darrel Bruce ANSC BSA 74
Cook/Doyle L. AGEC MS 63
Cook/Doyle L. BSA 60
Cook/James R. ANSC BSA 58
Cook/James Rex POSC BSA 88
Cook/James S. BSA 49
Cook/Jennifer J. AGED BSA 92
Cook/Jennifer Marie HOUS BSHES 97
Cook/Joseph Luke AGBS BSA 94
Cook/Joyce Haskew BSHE 57
Cook/Mary Browning BSHE 44
Cook/Mary Moore BSHE 43
Cook/Robert John AGRN BSA 37
Cook/Shana McArthur AGED BSA 94
Cook/Warren C. BSA 65
Cooley/Ennis M. BSA 57
Coombes/DeLana L. AGBS 99
Coombes/Larry E. ENTO MS 73
Coombes/Larry E. ENTO PHD 77
Coon/Beth Anne HORT BSA 97
Cooper/Amy Squyres CLTX BSHE 90
Cooper/Anita Willis HDFS BSHE 88
Cooper/Bonnie Gibbins BSHE 64

Cooper/Carol Jean HEED BSHE 74
Cooper/Catherine Lyn CLTX BSHE 86
Cooper/Charles Albert BSA 48
Cooper/Dollie Sistrunk FDSC BSA 72
Cooper/Gary M. ANSC BSA 75
Cooper/Harper AHUS BSA 64
Cooper/James Q. AGRI MS 69
Cooper/James Q. BSA 66
Cooper/Katherine Summers HOEC MS 85
Cooper/Louise Rye BSHE 39
Cooper/Michael E. ENTO MS 78
Cooper/Michael E. ENTO BSA 75
Cooper/Myrtle White BSHE 39
Cooper/Paul E. PTSC PHD 90
Cooper/Paul E. HOFS MS 75
Cooper/Terri Denise ENSC BSA 96
Cooper/Tim L. ANSC BSA 82
Cooper/Walter Marvin BSA 32
Cooprider/Benton J. CLTX BSHE 87
Cope/Mary M. HEED BSHE 70
Copeland/Arla Wheeler BSHE 34
Copeland/Clinton Ray AGED BSA 88
Copeland/Karen S. SOSC MS 76
Copeland/Sharon Ann ANSC BSA 97
Copelin/Billy D. ANSC BSA 83
Coppedge/Bryan Russell ENTO MS 92
Corbin/Billy R. AGRN PHD 88
Corbin/Samuel V. PLPA MS 69
Core/Michelle E. HEED BSHE 75
Core/Tad McCoy HDFS BSHE 91
Corey/Mark L. AGME BSA 82
Cornelius/Theresa M. GHEC BSHE 77
Cornett/Randall Keith POSC BSA 92
Cornish/Jill Marie HDFS BSHE 86
Correa/Luis A. V. AGRN MS 77
Correll/Marion Lee SOSC MS 71
Correll/Phillip ENSC BSA 79
Cosby/Dru Ann AGBS BSA 85
Coslett/Kim Y. LDUH BSA 84
Costaldi/Mario D. ENSC BSA 94
Costello/Richard Whitman AGRN MS 95
Coston/Ray SOSC PHD 69
Coston/Ray SOSC MS 64
Cotter/Jeannie Grimes HOEC MS 87
Cotter/Jeannie Grimes CLTX BSHE 85
Cottingham/James E. BSA 56
Cotton/John A. BSA 50
Cotton/Len POSC MS 72
Cotton/Len POSC BSA 68
Cotton/Mary Boaz BSHE 49
Cotton/Thelma HEED BSHE 70
Couch/John Wesley ENSC BSA 94
Couch/Lonnie C. POSC BSA 68
Couch/Sara Catherine HOEC MS 86
Couch/Tommy D. AGEC BSA 71
Couch/Tommy D. AGEC MSAGE 73
Coulter/Curtis Lamar AGBS BSA 90
Coulter/John G. POSC BSA 78
Counce/Kathy C. HOEC MS 83
Counce/Kathy C. FDNU BSHE 82
Council/Charles D. DASC BSA 76
Counts/George E. BSA 55
Counts/Norris Dean BSA 50
Courtney/Cynthia Ann BSHE 77
Courtney/Donna Lynne HOEC MS 86
Courtney/Donna Lynne FDNU BSHE 84
Courtney/Michael L. PPPM BSA 79
Courtney/Robin Sweet ANSC BSA 88
Cousins/Martha N. HOEC MS 91
Coutts/Rebecca E. LDUH BSA 91
Covington/Kathleen Edwards FDSC MS 80
Cowan/David Wayne AGED BSA 91
Cowan/Janice L. HOEC MS 91
Cowan/Joe D. AGSC MS 77
Cowdrey/Cam Leon BSA 49
Cowger/James I. BSA 28
Cowie/R. George FDSC BSA 87
Cowling/Vicki Linn CLTX BSHE 80
Cox/Allyn B. BSA 50
Cox/Alma W. BSHE 49
Cox/Alwin P. AGRI BSA 61
Cox/Brandi Blackmon ENSC BSA 95
Cox/Cecil L. BSA 49
Cox/Cynthia GHEC BSHE 85
Cox/Glenna Foster BSHE 51
Cox/Helen Kathleen FND BSHE 74
Cox/Jack Elwyn AGRN BSA 94
Cox/James M. AGBS BSA 80
Cox/James Robert BSA 50
Cox/Jerry Doering BSHE 53
Cox/Jo K. GHEC BSHE 77
Cox/Joe L. VOAG BSA 81
Cox/Joe Roland BSA 40
Cox/John Roger DIAM BSA 92
Cox/Johnnie Haynes BSHE 49
Cox/Mary Thomas ENGL BSHE 53
Cox/Rickey G. SOSC BSA 84
Cox/Ronald W. AGRN MS 77
Cox/Ronald W. BSA 75
Cox/Ruth Golda BSHE 60
Cox/Sarah L. HDFS 98
Cox/Shirley Barrick BSHE 60
Cox/Troy GNAG MS 63
Cox/Troy BSA 49
Cox/Virginia Rainwater HDID BSHE 82
Cox/William E. AGEC MS 86
Cox/Zanetta Y. ANSC BSA 89
Crabb/Marvin D. ENTO BSA 72
Crabtree/Barbara Heaton HEED BSHE 63
Crabtree/Marsha LDUH BSA 79
Crabtree/Nathan S. AGEC MS 92
Craft/Madeline Ligia BSHE 86
Craig/Alfred H. BSA 49
Craig/Jim AGBS BSA 74
Craig/Judy J. Shumate CDFR BSHE 78
Craig/Lisa Jean CLTX BSHE 91
Craig/Patricia Jo HDFS BSHE 92
Craig/Tom AGED BSA 67
Craig/Valerie Leigh BSHE 86
Craig/William E. BSA 50
Craigmoile/Elizabeth Smith BSHE 53
Crain/Charlean Benton HOEC MS 71
Crain/Lorraine Tweedy GHEC BSHE 40
Cramer/Diane Montgomery FDNU BSHE 91
Crandall/Jessica Jean HORT BSA 97
Crandall/Walter Morris CRSC BSA 67
Cranford/Bryant Keith AGBS BSA 86
Cranford/Eva ANSC BSA 86
Cranford/Steve AGBS BSA 85
Cranford/Steven K. BSA 58
Crass/Arlen Martin BSA 51
Craun/Jerry A. BSA 68
Cravens/John Marion BSA 38
Crawford/Judith BSHE 63
Crawford/Mark A. PLPA MS 79
Crawford/Sharon Shilling CLTX BSHES 95
Crawford/Wayne Douglas ANSC BSA 84
Crawley/Carla D. FDSC MS 88
Crawley/Cathy L. Cormack ANNU BSA 82
Crawley/Dennis D. DASC BSA 79
Crawley/Margery Faye Bullard GNAG MS 90
Crawley/Robert A. BSA 54
Crawley/Russell D. AGRN MS 89
Crawley/Tim L. DASC BSA 82
Crawley/Tim L. ANSC MS 86
Creamer/Jesse Allen BSA 53
Creecy/Douglas H. GNAG BSA 67
Creelman/Richard A. CRSC PHD 76
Crews/Laura L. IDHS BSHE 89
Crider/Patricia Gamble CLTX BSHE 88
Crigler/Theron A. BSA 32
Criner/Matthew A. ESWS 98
Cripps/Eldon Eugene AGBS BSA 85
Crnko/Gary Stephen HORT MS 85
Crocker/Beverly Kaye FND BSHE 64
Crocker/Cynthia Kaye H. CLTX BSHE 82
Crocker/Thomas Floyd FDSC BSA 75
Crocker/Vonda Jones BSHE 65
Croft/Colene Elmer BSHE 54
Croft/Kenneth A. BSA 50
Croley/Powell R. BSA 26
Crom/Jean Chipman BSHE 48
Crosby/Albert AGEC BSA 90
Cross/Betty Jo BSHE 54
Cross/Catherine K. HOEC MS 75
Cross/Catherine K. BSHE 54
Cross/David W. ANSC BSA 70
Cross/Eddie AGEC MS 78
Cross/John F. BSA 57
Cross/Meta Rene CLTX BSHE 81
Cross/Opal Davis BSHE 22
Crossett/David H. AGBS BSA 84
Crossett/Jon Michael AGBS BSA 88
Crossley/Leslie Ray BSA 56

Crouch/Grace P. BSHE 43
Crouch/Hugh A. BSA 52
Crouch/Joe B. ANSC BSA 49
Crouch/Linette HORT BSA 98
Crouch/Rebecca H. HOEC MS 76
Crouse/Karl Keith AGRN BSA 85
Crow/Rodney W. HORT BSA 76
Crow/William W. BSA 53
Crowder/Donald G. VOAG BSA 73
Crowder/Randy W. AGBS BSA 74
Crowder/Ruth Marie HEED BSHE 51
Crownover/Brian Douglas ANSC BSA 94
Crownover/Jules V. AGRI BSA 41
Crowson/Suzanne FND BSHE 67
Croy/Connie Carnes BSHE 66
Crudup/Barry H. AGBS BSA 82
Crudup/Cheryl L. AGBS BSA 83
Crum/Roger C. BSA 60
Crumley/Clyde BSA 50
Crump/Clay VOAG BSA 80
Crumpacker/Carol AGBS BSA 82
Crumpacker/Kevin AGBS BSA 94
Crutchfield/M. H. BSA 40
Cucchiara/Richard P. ANSC BSA 76
Culbertson/Deborah Lee AGRN PHD 88
Culler/Hazel J. BSHE 45
Cullers/Matt Wyatt ANSC BSA 96
Cullins/Bernice BSA 46
Cullins/Bernice BSHE 42
Cullins/George BSA 46
Cullins/James A. BSA 53
Cullum/Sherman D. BSA 60
Culp/Carole Helm HOEC MS 68
Culp/Carole Helm HEED BSHE 62
Culpepper/Debra H. CDFR MS 76
Culpepper/Goldia Leona IDHS BSHE 91
Cummings/Allison ESWS BSA 98
Cummings/Bethany HDFS 99
Cummings/Cora Rogers BSHE 45
Cummings/Lula Mae BSHE 42
Cummings/Mary Jane GHEC BSHE 84
Cummings/Robert Paul BSA 23
Cummings/Wallace C. AGRI MS 65
Cummings/Wallace C. AGRI BSA 51
Cumnock/Jane Stewart BSHE 41
Cunkle/Bonna VanDalsem BSHE 41
Cunkle/John A. BSA 50
Cunningham/Bethel P. BSHE 53
Cunningham/Dearl Gene BSA 80
Cunningham/Elizabeth BSHE 38
Cunningham/Josh Kinsey AGBS BSA 96
Cunningham/Kathryn Mayden BSHE 37
Cunningham/Marsha M. HEED BSHE 70
Cunningham/Nicole Schafer HOEC MS 92
Cupp/Joe W. BSA 52
Cupples/Kenneth R. AGEC BSA 70
Cupples/Kenneth R. AGEC MS 73
Cureton/D. Alan HORT BSA 72
Cureton/Janice Daniel BSHE 72
Curless/Joseph Keith AGRN MS 97
Currie/John B. AGEC BSA 69
Currie/John B. BSA 42
Curry/Marilyn Kay HORT MS 89
Curry/Sandra K. HEED BSHE 72
Curry/Teresa D. Gregory ANSC MS 84
Curry/Teresa D. Gregory ANSC BSA 82
Curtis/Charles E. ENTO MS 62
Curtis/Jean M. CLTX BSHE 84
Curtis/Sally C. HLED BSHE 82
Curtis/William Francis DIAM BSA 93
Cyphers/Pam Swick FND BSHE 72
Czarnomski/Atlantis Alexis ENSC BSA 97
Dabney/William T. AGEC MS 77
Dace/Donald AGSC MS 61
Dace/Donald BSA 61
Dacus/Jeffery Scott GNAG MS 94
Daffron/Phillip Neal ANSC MS 85
Dagenhart/Quentin S. BSA 51
Daggett/Carol A. HORT BSA 79
Daggett/Carol A. HORT MS 82
Daher/Charles Anthony LDUH BSA 94
Dahlke/George R. AGRN MS 62
Dahlke/George R. AGRN BSA 61
Dahlke/Milton M. VOAG BSA 61
Dai/Dung Giac CHEG BSHE 88
Daily/Randell Wade DANB BSA 94
Daines/Darrah Jo CLTX BSHE 93
Dalak/Lois Cooper BSHE 43
Dale/Everett HORT BSA 78
Daley/Raymond G. ANSC MS 75
Dalla Rosa/Arthur J. AGME BSA 75
Dalton/Edna Jo Patterson BSHE 34
Damann/Catherine Ann FND BSHE 70
Dameron/John Lee BSA 34
Dameron/Sam BSA 31
Dampf/Harlie J. VOAG BSA 31
Dampf/Keith J. BSA 33
Dampf/Reece J. VOAG BSA 33
Danehower/Patricia Kim AGEC MS 85
Danehower/Samuel C. AGEC MS 85
Danehower/Samuel C. AGBS BSA 82
Daniel/Alvin Atlas BSA 49
Daniel/David E. DASC BSA 73
Daniel/Garland Simon VOED BSA 42
Daniel/Jim BSA 70
Daniel/Lolah M. FDNU BSHE 73
Daniel/Louise FND BSHE 74
Daniel/Ralph A. BSA 46
Daniel/Rebecca S. BSHE 48
Daniel/W. Peyton III AGBS 99
Daniel/William H. BSA 47
Daniels/Ann Michelle HDFS BSHE 88
Daniels/Crosby E. BSA 65
Daniels/Jimmy L. CRSC BSA 61
Daniels/John Buford BSA 33
Daniels/L. Bernard ANNU BSA 62
Daniels/Michael B. AGRN MS 88
Daniels/Michael B. SOSC BSA 86
Daniels/Sandra Morgan FND BSHE 73
Daou/Hassane AGRN MS 97
Darby/Sue Cobb HOEC MS 86
Dardon/Hugo Penagos ENTO BSA 74
Dark/Cleo H. AGED MS 51
Dark/Cleo H. GNAG BSA 50
Dark/Harris V. BSA 51
Dark/Lucille S. DIET BSHE 49
Darling/Peter R. AGEC BSA 77
Dasher/James O. BSA 59
Daugherty/Paul Ray BSA 53
Davenport/Andrea Val Matty CDFR BSHE 84
Davenport/James L. BSA 50
Davenport/Ruth Charlene Waters BSHE 49
Davey/Ronald B. ENTO PHD 77
Davidson/Amber Ellene HDFS BSHE 93
Davidson/Jane BSHE 60
Davidson/Johnny L. AGED BSA 71
Davidson/Randall Allen AGED BSA 92
Davidson/Suzanne HEED BSHE 74
Davis-Hartsook/Marsha J. BSA 79
Davis/A. Jeff BSA 38
Davis/B. Hall BSA 50
Davis/Beverly A. HEED BSHE 78
Davis/Billy C. HORT MS 93
Davis/Billy C. HORT BSA 90
Davis/Billy K. GNAG BSA 62
Davis/Charley B. BSA 59
Davis/Christine CLTX BSHE 79
Davis/Clarence B. VOAG BSA 40
Davis/Dan VOED BSA 64
Davis/David Joe AGRN BSA 83
Davis/Denise L. CLTX 98
Davis/Donald D. PHUS BSA 61
Davis/Dwight Ed GNAG MS 68
Davis/Dwight Ed DHUS BSA 61
Davis/Gary D. AGRN MS 82
Davis/George V. DHUS BSA 60
Davis/George V. ANNU MS 64
Davis/George W. AGRI BSA 40
Davis/Gregory B. FDSC 98
Davis/Gregory Scott AGED BSA 96
Davis/Harriet Goodwin BSHE 60
Davis/Jackie Lynn ANSC MS 80
Davis/Jackie Lynn ANNU BSA 79
Davis/James K. BSA 56
Davis/Jason C. ANSC BSA 96
Davis/Jodie Len CLTX BSHE 84
Davis/John F. BSA 50
Davis/Jonathan Patrick AGRN MS 97
Davis/Kaaron J. ANSC BSA 91
Davis/Larry Michael POSC BSA 70
Davis/Lynda Kay BSHE 63
Davis/Marilyn Miller BSHE 50
Davis/Mark J. ANNU BSA 89
Davis/Mary Carolyn GHEC BSHE 86

Davis/Mary Clotfelter FND BSHE 69
Davis/Mildred White HOEC BSHE 40
Davis/Natalie Gaskill ANSC BSA 71
Davis/Rachel N. CLTX BSHE 92
Davis/Ralph E. BSA 51
Davis/Reba J. GHEC MS 65
Davis/Rebecca L. Moss BSHE 72
Davis/Rebecca Strong BSHE 70
Davis/Richard Kerrigan PPPM BSA 94
Davis/Sandra L. Huenefeld HOEC MS 81
Davis/Sandra L. Huenefeld CLTX BSHE 79
Davis/Skipper AGBS BSA 97
Davis/Steven B. FDSC 98
Davis/Sue Jan ANSC BSA 82
Davis/Susan Dorothy HORT BSA 78
Davis/Terry Lynn AGRN BSA 78
Davis/Terry Lynn AGRN MS 76
Davis/Trez Harley BSA 50
Davis/Willia Lovard BSA 30
Davis/William L. AGEC MS 60
Davis/William L. BSA 58
Dawson/Bruce Christopher AGBS BSA 91
Dawson/Louise P. BSHE 52
Dawson/Margaret Ann BSHE 52
Day/Laurine Taylor HDFS BSHE 88
De Leon/Manuel Horacio AGEC BSA 60
De/Silva Punya Manjula AGRN MS 96
Deacon/Carol C. DANB BSA 82
Deal/Deborah Parsons BSHE 71
Dean/Gerald L. ENTO MS 67
Dean/Gerald L. BSA 63
Dean/Iva Fagan BSHE 47
Dean/John D. AGRN BSA 85
Dean/William F. POSC BSA 56
Deangulo/Julian AGRN MS 87
Deangulo/Julian AGME BSA 85
Dearworth/Jeff Robert AGBS BSA 93
Deaton/James W. BSA 56
Deaver/Gayle S. GFNU BSHE 43
Debhavalya/Marnop AGEC MS 54
Debhavalya/Rote AGEC MS 79
Decker/Hayden Lee BSA 49
Dedman/Robb H. AGRN BSA 94
Deema/Prayoon ENTO MS 62
Deen/Christopher F. AGEC BSA 81
Deen/Lori CLTX BSHE 82
Deer/James Milton ANSC BSA 67
Deere/Runyan BSA 42
Deere/Runyan GNAG MS 63
Dees/Aleasha Lu Stormes HEED BSHE 85
Dees/Ardella B. HOEC MS 72
Deeter/Louis Paul ANSC BSA 74
Degago/Yohannes AGRN MS 83
Degges/Annie Eva BSHE 55
Deines/Charles E. AGBS BSA 80
Deines/Charles E. AGEC MS 82
Delany/Lois S. BSHE 31
Delany/Terry Stephen LDUH BSA 95
Delap/Susan Kaye ANSC MS 87
Deller/Carolyn Jones BSHE 44
Delmonego/Frankie J. AGED 98
Delozier/Bryan K. HORT BSA 86
Delozier/Donna S. ANSC 98
Delozier/Ronald K. POSC BSA 60
Delzell/Fred E. BSA 35
Demchak/Megan Susan FDNU BSHES 98
Demille/Alice Kelly FDNU BSHE 74
Dempsey/Sonya POSC BSA 95
Demuth/Donna H. HEED BSHE 51
Denard/Tamara Marie Eleanor ANSC MS 94
Denard/Tamara Marie Eleanor ANSC BSA 92
Denherder/Mark A. HORT MS 92
Denningtion/Joseph W. BSA 37
Dennington/Daniel W. BSA 38
Dennis/Carrie L. CATX BSHES 97
Dennis/Marion D. BSA 68
Dennis/Wanda Sue BSHE 59
Denton/Danny AGEC MS 69
Denton/W. B. BSA 37
Denver/Charles E. ENTO MS 70
Denver/Charles E. POSC BSA 66
Depew/Clinton G. ANSC BSA 72
Depew/Ralph ANSC BSA 70
Depriest/Richard L. CLTX BSHE 93
Deren/Christopher W. AGRN MS 83
Deren/Christopher W. AGRN BSA 80
Derickson/Thelma Kimbrell BSHE 68
Dering/Dana Rochelle CLTX BSHES 97
Derning/Patrick J. AHUS BSA 70
Desilva/Amal Padmaja HORT MS 93
Dew/Robert Edward BSA 39
Dewett/Charles Sherry BSA 51
Dewey/Myra G. HOFS BSHE 79
Dewey/Phillip C. ANSC BSA 79
Dewey/Thomas O. ANSC BSA 74
Dewitt/Dale L ANNU BSA 82
Deyoe/J. C. AGBS BSA 74
DeBusk/Sherrell BSA 49
DeFelice/Michael AGRI MS 80
DeFelice/Michael PPPM BSA 78
DeFelice/Michael AGRN BSA 78
DeHart/Brian Allen AGRN MS 91
DeLee/James Anthony POSC MS 89
DeLee/Nancy Robins BSHE 45
DeLozier/Jeffry Derrick LDUH BSA 95
DePoyster/VirMarie Suria CLTX BSHE 90
DeRamus/Alan AGRN PHD 80
DeSalvo/Henry Joseph BSA 48
DeShazo/Deborah Marshall HDID BSHE 78
DeVaney/Joyce Ann ENTO MS 67
DeWitt/Glenda Christine GHEC BSHE 82
DeWitt/Verna BSHE 28
DeYoung/Ingrid Marie FDSC MS 88
DeYoung/Ingrid Marie FDSC BSA 87
Dhimbil/Jama Mohamed GNAG MS 84
Dhonau/Lloyd Asher BSA 27
Dial/Jamie Stevens HOEC MS 62
Dial/Jamie Stevens BSHE 60
Diallo/Hortense Atta PLPA MS 94
Diallo/Modibo AGRN MS 93
Diarra/Amadou AGRN PHD 85
Diaz-Polanco/Carlos PLPA MS 64
Diaz/Nelson ANSC MS 85
Dickens/Patsy Becker HEED BSHE 73
Dickens/Ray CRSC BSA 62
Dickerson/Carolyn Bourland HEED BSHE 56
Dickerson/Harold K. AHUS BSA 61
Dickerson/Kathryn A. DANB BSA 82
Dickerson/Ottie J. PLPA MS 57
Dickerson/Ottie J. AGRI BSA 55
Dickey/Don O. BSA 65
Dickey/Kevin Lynn ANSC BSA 98
Dickinson/Carol Lynn Williams CLTX MSHE 95
Dickinson/Carol Lynn Williams CLTX BSHE 87
Dickinson/Flora BSHE 43
Dickinson/Frances W. HORT MS 74
Dickinson/Frances W. HORT BSA 72
Dickinson/Jeremy Scott AGBS BSA 97
Dickinson/Jimmy Ray AGBS BSA 82
Dickman/Anita Bea CLTX BSHE 80
Dickson/Clifford POSC BSA 54
Dickson/Lilla Maye BSHE 40
Dickson/Maxine BSA 24
Dickson/Paul Allen GNAG MS 95
Dickson/Paul R. ARCH BSA 57
Dietrich/Robbi Rice FDNU BSHE 76
Digby/Stacy Charles ANSC BSA 95
Diggs/Mary M. BSHE 52
Dilday/Robert H. SOSC MS 67
Dilday/Robert H. CRSC BSA 63
Dilgren/Sara L. ESWS 99
Dill/Gerald L. BSA 53
Dill/Jimmy N. VOAG BSA 78
Dill/Louise Davenport BSHE 49
Dill/Nancy Bunch FDNU BSHE 76
Dillahunty/Amy Lynette FDST BSA 98
Dillahunty/John H. ANSC BSA 67
Dillahunty/Peggy R. HEED BSHE 72
Dillard/Kathleen K. Gibson ANSC BSA 76
Dillion/Debbie D. HORT BSA 91
Dillion/Scotty Lynn M. ANSC BSA 88
Dillman/Bonnie AGEC MS 62
Dillman/Bonnie BSHE 60
Dillow/Calvin AGEG BSA 84
Dillport/Jeanne BSHE 53
Dillport/Mack D. BSA 52
Dills/James Charles BSA 50
Dimitt/Helen Inman BSHE 46
Dingman/Linda F. Hall FDSC MS 84
Dingman/Linda F. Hall FDNU BSHE 82
Dinkins/Reed Leon ENTO MS 66
Dirks/Susan E. HORT BSA 82
Disney/Martha B. IDHS BS 86
Ditsworth/Whitney Weathers ANSC BSA 89

Dixon/Amy kay IDHS BSHES 98
Dixon/Charles C AGRN BSA 82
Dixon/Charles R. BSA 79
Dixon/Debbie B. HEED BSHE 78
Dixon/Joel R. BSA 57
Dixon/Lee Ann BSHE 82
Dixon/Mary BSHE 37
Dixon/Melinda Gayle FDNU BSHE 95
DiPietre/Dennis D. AGEC BSA 79
DiPietre/Dennis D. AGEC MS 80
Djiba/Saliou ENTO MS 85
Djiba/Saliou ENTO BSA 84
Djunaidi/Harjanto AGEC MS 89
Doan/John Cannon GNAG BSA 49
Dobbs/Barry Gene ANSC BSA 91
Dobbs/Mike AGED BSA 84
Dobbs/Richard E. PLPA MS 69
Dobbs/Steven H. HORT MS 86
Dockery/Barbara Keil BSHE 56
Dockins/Delma D. BSA 48
Dockins/J. O. BSA 39
Dodd/Denson Doyle BSA 38
Dodds/Athenia HDFS BSHE 86
Dodds/James Shannon AGED MAT 98
Dodds/James Shannon AGTE BSA 97
Dodgen/Rachel PPPM BSA 96
Dodgen/William Henry POSC MS 73
Dodgen/William Henry ANSC BSA 70
Dodson/Ann E. LDUH BSA 79
Dodson/Darrin W. POSC 98
Dodson/John William BSA 37
Dodson/Michael Henry ENTO BSA 90
Dodson/Sheresa Ann HDFS BSHE 95
Doggett/Sherri Tomboli CLTX BSHE 90
Doherty/Patricia L. HEED BSHE 76
Dolezal/Sue Douglas CLTX MS 79
Dolezal/Sue Douglas HEED BSHE 72
Dolezal/William PLPA MS 77
Dolezal/William PLPA PHD 81
Dollins/Cara Lynette CLTX BSHE 88
Dombek/Don HORT MS 85
Dombek/Don HORT BSA 82
Domek/John Michael ENTO PHD 89
Donaubauer/Tammi M. ELED BSA 80
Donkervoet/Yevonne Starling FND BSHE 69
Donley/Marcia GNAG MS 89
Donley/Richard Avent FDSC MS 86
Donnell/Clifford Scott POSC BSA 96
Donohoe/Marjorie S. BSHE 29
Donovan/Lisa GHEC BSHE 82
Dooley/Edwin G. ANSC BSA 82
Dorman/Danny H. POSC BSA 74
Dorman/David Earl BSA 52
Dorman/Hope Laney HEED BSHE 89
Dorn/Jacqueline ANSC BSA 91
Dorn/Ted ANSC BSA 82
Dorough/William M. BSA 53
Dortch/George L. BSA 55
Dortch/James M. BSA 49
Dortch/Robert L. AGRI BSA 54
Dotson/Cecilia Ferguson HEED BSHE 89
Dotson/Marilyn Fern Hubbard HOEC MS 69
Dotson/Marilyn Fern Hubbard HOEC BSHE 63
Dotson/Robert Travis POSC BSA 94
Dotson/X AHUS BSA 60
Doty/Julia Phillips HOEC BSHE 33
Doty/Patricia HEED BSHE 79
Dougherty/Catherine D. GHEC BSHE 82
Doughty/Harland Nellson BSA 39
Douglas/Anne R. CLTX BSHE 83
Douglas/Cerelle Fowler ANSC BSA 77
Douglas/Glenda Joan BSHE 53
Douglas/Harry W. BSA 49
Douglas/Jary Lynn GNAG MS 83
Douthit/Linda Buchanan PLPA MS 74
Dowdle/Donald W. POSC BSA 79
Dowell/Allen W. BSA 30
Dowell/Gladys Maymie BSHE 18
Dowell/Grover C. BSA 53
Dowell/Kim FDNU BSHE 86
Downall/Sara Bunyard CLTX BSHE 77
Downey/Kimbra CLTX BSHE 85
Downey/Timothy John PPPM BSA 81
Downey/William R. BSA 79
Downum/Kimberly K. ANSC 99
Doyel/Lila FDNU BSHE 80
Doyle/Jeffrey Ealem AGBS BSA 93
Doyle/Margaret Davis BSHE 34
Doyle/Stephanie Ann IDHS BSHES 96
Dozier/John P. BSA 40
Dozier/Les ANSC BSA 82
Drake/Mark D. AGEC BSA 81
Drake/Thomas J. BSA 40
Dreher/Nancy Beth Shudy ANSC BSA 94
Drewry/James Richard AGRN BSA 80
Driggers/Carla Dean POSC MS 88
Driggers/Shannon V. ANSC BSA 86
Driggers/Tommy Allan HORT BS 83
Driver/Jacquelyn Edwards AGRN MS 84
Droddy/J. Darrell ENSC BSA 78
Drummonds/Melissa Ann POSC MS 96
Dryden/William G. BSA 51
Drye/Eulas Monroe Jerome BSA 50
Dubois/Shawn D. AGBS BSA 89
Duckett/Hoyt D. BSA 52
Duckett/Mary Stokenbury BSHE 52
Duclos/Alvin L. AGRN BSA 63
Duclos/Leo A. AGRI BSA 58
Duclos/Magdelen Rose BSHE 59
Duclos/Zachary M. AGRN MS 58
Duclos/Zachary M. AGRN BSA 57
Dudley/Karin HEED BSHE 79
Dudley/Kim M. HDFS BSHE 88
Duell/Charles Matthew AGEC BSA 82
Duff/Charles L. BSA 45
Duffey/Jeannine G. CLTX BSHE 76
Duffield/Valorie S. GHEC BSHE 87
Dugal/Gordon B. BSA 49
Dugan/Jodie Halk FDNU BSHE 52
Dugan/LuAnne Dillard GHEC BSHE 78
Duggan/Dennis L. VOAG BSA 67
Duggan/Shelia Lavon ANSC BSA 89
Duke/John H. POSC MS 77
Duke/John H. ANSC BSA 73
Duke/Mary BSHE 54
Duke/Willie Belle BSHE 37
Dumas/Billy Aylmer ENTO MS 70
Dunagin/Patrice L. GHEC MS 85
Duncan/Aimee Allison GNAG MS 92
Duncan/Andrew James FDNU BSHE 95
Duncan/Bernice Newsom BSHE 39
Duncan/Gerald W. POSC BSA 79
Duncan/Gwendolyn Maureen BSHE 37
Duncan/Leland D. BSA 50
Duncan/Patricia S. FDNU BSHE 62
Duncan/Raymond L. ANSC MS 83
Duncan/Ruth A. FDNU BSHE 70
Duncan/Tommy D. AGBS BSA 94
Dunham/Margo Jean IDHS BSHES 95
Dunham/S. Craig AGRN MS 93
Dunigan/Andrea Celeste FDSC MS 91
Dunigan/Andrea Celeste FDSC BSA 90
Dunk/Fred A. ANSC MS 78
Dunk/Fred A. DASC BSA 75
Dunk/Kenneth W. DASC BSA 76
Dunkel/Robin Ashley FDSC BSA 98
Dunlap/Alex DANB BSA 84
Dunlap/Renessa Marie FDNU BSHES 96
Dunlop/Billy B. BSA 42
Dunn/Delma D. GNAG BSA 77
Dunn/J. M. GNAG BSA 61
Dunn/Mary B. HOEC MS 68
Dunn/Mary B. HOEC BSHE 46
Dunn/Rae FND BSHE 64
Dunn/Rebecca LDUH BSA 79
Dunn/Telia Ann HDFS BSHE 89
Dunn/Walter S. LDUH BSA 82
Dunnuck/Noble G. AGEC MS 58
Dupont/Shawna Faye BSHE 87
Dupree/Callie Stone BSA 30
Duren/Charles I. DIAM BSA 91
Duren/Charles I. ANSC BSA 88
Dutram/Kay HOEC MS 85
Dutram/Kay FDNU BSHE 81
Dutram/Paul W. GNAG MS 84
Dutram/Paul W. SOSC BSA 82
Dwyer/Judy K. HOFS BSHE 82
Dyck/Raymond L. HORT MS 62
Dyck/Raymond L. BSA 60
Dyer/Cyrus L. BSA 14
Dyer/Daniel AGRN PHD 82
Dyer/Daniel AGRN MS 80
Dyer/Julian Masa BSA 18
Dyer/William Rudolph GNAG MS 92

Eans/Norman J. BSA 50
Earhart/Carla Crosby BSHE 80
Earhart/Lee A. AGEC MS 81
Earhart/Lee A. AGRN BSA 78
Earhart/Treva Jo HEED BSHE 73
Earle/Mary Ella CLTX BSHE 75
Earnest/Cheryl CLTX BSHE 83
Earnest/Larry Dale AGRN MS 91
Earnhart/Jo Courtney BSHE 69
Earnhart/Roscoe Franklin BSA 53
Easley/Heather Hillman POSC BSA 95
Easley/John Charles POSC BSA 92
Easley/Rodney Lee POSC BSA 93
Eason/Cleo J. BSA 48
Eason/Linda BSHE 70
Eason/Roger L. BSA 65
Eason/Ruth Louise Robinson ENTO MS 67
Easter/Janet Ann FDNU BSHE 81
Easterling/Betty Jo Struebing BSHE 42
Easterling/Howard A. BSA 43
Easterling/Tommy Shane POSC BSA 96
Eaton/Lynda Lee CATX BSHES 98
Eaton/Mark Win GNAG MS 87
Eaton/Mark Win ANSC BSA 85
Eaton/Mary Margaret Morse BSHE 82
Eaton/Mike AGBS BSA 73
Eaton/Rayetta Lyn ANSC BSA 90
Eaton/Ronald G. HORT MS 74
Eaton/Ronald G. HORT BSA 72
Eaton/Stephen E. HORT MS 90
Eaton/Stephen E. HORT BSA 80
Eaton/Will James AGBS BSA 95
Eberle/Teresa Marie CLTX BSHE 80
Echels/Dayna A. CLTX BSHE 78
Eck/Clarence S. PHUS MS 48
Eck/Clarence S. GNAG BSA 47
Eden/Darlene Kimbrough BSHE 52
Edgar/Mary Bocquin BSHE 25
Edgmon/Michelle Dawn IFDS BSHES 97
Edmisten/Billy C. BSA 50
Edmisten/Raina IDHS BSHE 89
Edmondson/Andrea CLTX BSHE 87
Edmondson/Jim BSA 33
Edmondson/Justus H. AGRI BSA 47
Edmondson/L. F. BSA 42
Edmondson/R. G. BSA 56
Edmondson/Vance W. AGEC PHD 56
Edmondson/Vance W. AGEC MS 50
Edmondson/Vance W. AGEC BSA 48
Edrington/Thomas Sydney ANSC MS 89
Edwards/Brad AGBS BSA 83
Edwards/Caramie J. AECT 99
Edwards/Emmett BSA 36
Edwards/G. Dean BSA 55
Edwards/Glenda G. CDFR BSHE 76
Edwards/James T. BSA 38
Edwards/Jefferson D. AGEC MS 77
Edwards/Jimmie Lee CDFR MS 73
Edwards/John C. ANSC 99
Edwards/Leonard V. BSA 51
Edwards/Mary Sims BSHE 43
Edwards/Perry C. ENTO MS 61
Edwards/Perry C. ENTO BSA 57
Edwards/Ruth McCord BSHE 38
Edwards/Wilma Charese CLTX BSHE 93
Efird/Paul Kevan ENTO MS 91
Eidson/Gus A. BSA 36
Eifling/Melanie POSC BSA 86
Eisele/Debra L. CTM BSHE 75
Eisenhauer/Carol L. DANB BSA 77
El-Forgany/Mabrouk M. PTSC PHD 83
El-Saffawi/Abdulkadir A. BSA 56
Elam/Carleton Britt GNAG MS 84
Elam/Carleton Britt DANB BSA 81
Elbersen/W. AGRN PHD 94
Elcan/Christine McKissack BSHE 37
Eldridge/Elizabeth N. HEED BSHE 69
Eldridge/Ira Luther CRSC BSA 69
Elenbarger/Brian P. HOUS BSHES 97
Elevazo/Pio Agbanlog AGEC MS 74
Elgin/Sheri CLTX BSHE 85
Elizandro/Marcia HEED BSHE 71
Elkassabany/Nader ENTO PHD 94
Elkins/Cynthia Hamm HOEC MS 93
Elkins/Cynthia Hamm CLTX BSHE 89
Elkins/Cynthia Mackin HEED BSHE 75
Elkins/Joyce Wells BSHE 54
Elkins/Randy ANSC BSA 75
Elkins/Robert L. BSA 54
Ellen/Jodie BSA 50
Ellen/Margaret S. AGBS BSA 80
Ellibee/Margaret A. GNAG MS 88
Ellington/Lewis Stephen PLPA BSA 75
Elliot/D'Ann Downey BSHE 72
Elliott/Blanche Hanks BSHE 24
Elliott/Burton T. FDSC BSA 81
Elliott/Craig C. BSA 39
Elliott/Dianne M. DANB BSA 82
Elliott/Lloyd Carlton BSA 26
Elliott/Margaret A. HOEC MS 92
Elliott/Martha Cox BSHE 43
Elliott/Michael William POSC BSA 90
Elliott/Wayne M. BSA 38
Elliott/Weldon H. BSA 39
Ellis/Frances M. Smith BSHE 51
Ellis/Frank GNAG BSA 50
Ellis/Julia Lemley BSHE 39
Ellis/Julia Rebecca CLTX BSHE 81
Ellis/Mary F. Nettleship BSHE 28
Ellison/Ginger Strong AGBS BSA 85
Ellison/Henry F. BSA 21
Ellison/Lindell B. POSC BSA 82
Elmi/Abdulkadir A. AGRN PHD 92
Elmore/Carolyn J. Cole BSHE 54
Elmore/F. R. BSA 54
Elms/Lawrence H. BSA 38
Elphingstone/Edwin S. BSA 50
Elsner/Gary H. AGEC BSA 62
Elsner/Laura Ann HDFS BSHE 93
Ely/Carol Beth AGRN MS 88
Ely/Patricia Kidd DIET BSHE 54
Emamian/Dorothea Jeske FDNU BSHE 79
Emerson/Barbara J. GHEC BSHE 79
Emerson/Peggy J. HOEC MS 80
Emert/Cathy D. Campbell HEED BSHE 82
Emmerling/Melinda CLTX BSHE 86
Enck/DeNise M. FHNH 98
Endres/Craig Steven AGRN MS 86
Endres/Roland R. AGRI BSA 54
England/Penny S. CLTX BSHE 87
England/Travis Kevin ANSC BSA 89
Engler/Ruby Sharron POSC BSA 88
English/Elbert Hartwell BSA 14
English/Henry W. AGRN PHD 83
English/Henry W. AGRN MS 78
English/Lane L. AGBS BSA 93
English/Merrill H. BSA 61
English/Michael Andrew AGRN BSA 89
Enkvetchakul/Boonprom POSC PHD 94
Ennis/Alicia Renae IDHS BSHE 94
Enoch/Aubrey B. HORT BSA 41
Eoff/Dee R. BSA 31
Eoff/Herbert J. POSC BSA 57
Eoff/Howard Lafayette BSA 34
Epperson/James C. AGRN BSA 57
Epperson/Lee Morris BSA 51
Epperson/Thomas A. BSA 56
Erb/William Alan HORT MS 84
Erickson/Albert ANSC BSA 79
Erlinger/Kimberley Ann CLTX BSHE 94
Ernst/Marianne Zimmerman HOEC MS 67
Erstine/James M. CRSC BSA 71
Erstine/Jon Mark AGEC MS 81
Erstine/Jon Mark AGEC BSA 79
Erstine/Philip D. AGRN MS 79
Ervin/Glenna Skelton HEED BSHE 73
Ervolina/Christine Marie IDHS BSHE 94
Erwin/Patti S. LDUH BSA 82
Espinoza/Hernan Roberto ENTO PHD 94
Essig/Alice Hudson BSHE 53
Essig/H. Werner BSA 53
Estelle/Carolyn Hilliard BSHE 59
Estelle/George H. BSA 51
Estes/Brandy Mechelle IFDS BSHES 98
Estes/Nancy E. HOEC MS 62
Estes/Nancy E. BSHE 47
Estes/Sharon Campbell BSHE 65
Estes/Walter C. BSA 54
Etemad-Kasaian/A. A. FDSC PHD 85
Etheridge/Hampton A. BSA 41
Etheridge/Jill Estes FDNU BSHE 93
Etheridge/Jimmy R. ENTO MS 72
Eubank/Donna Muncy BSHE 65
Eubanks/Clyde HORT BSA 72

Eubanks/Harvey A. AGBS BSA 74
Evans/Betty AGEC BSA 84
Evans/Brad K. AGME BSA 81
Evans/David G. ANSC MS 75
Evans/David G. POSC BSA 73
Evans/Gary Fletcher ANSC BSA 67
Evans/Heidi L. IDHS BSHE 88
Evans/James David AGBS BSA 96
Evans/Jane Matthews FND BSHE 71
Evans/Leo E. BSA 56
Evans/Leon E. BSA 41
Evans/Louise Parkinson BSHE 33
Evans/Mary Henderson BSHE 52
Evans/Michael Benjamin DANB BSA 81
Evans/Raymond Edward BSA 53
Evans/Rocky E. AGME BSA 83
Evans/Rusty N. POSC BSA 82
Evatt/Karen J. ESWS 99
Everett/Ronald E. ENTO MS 69
Everett/Ronald E. BSA 65
Ewan/Jim AGBS BSA 87
Ewan/Patrick Kirby AGBS BSA 91
Ewell/Frank David AGEC PHD 86
Ewing/Helen Walker BSHE 33
Ewing/Lonnie VOAG BSA 81
Ezell/Betty J. HEED BSHE 64
Fader/Thalia HORT MS 79
Fagala/Bill L. CRSC MS 72
Fagala/Bill L. CRSC BSA 70
Failla/John P. BSA 52
Fain/Monte James EXED BSA 94
Fain/Monte James AGED BSA 97
Fain/Monte James AGED MAT 98
Faires/Douglas Grady AGED BSA 85
Fairman/Patricia Kathryn IDHS BSHE 91
Faldon/Teresa A. HEED BSHE 80
Falivene/Moghaddam Hamid ANSC MS 84
Falkner/Deanna L. CLTX BSHE 77
Fallen/Alyssa Jeanine CLTX BSHE 91
Fambrough/William Judon AGEC BSA 68
Famotemi/David Ayodele AGRN MS 79
Fancher/Bryan I. ANSC MS 84
Fancher/Bryan I. POSC BSA 82
Fancher/Burr E. AGEC BSA 53
Fancher/Burr E. ANSC MS 58
Fancher/James Shannon POSC BSA 94
Fancher/Melissa Ruth POSC BSA 97
Fancher/Thomas William BSA 51
Fankhauser/Delmar W. HORT MS 78
Fankhauser/Henry I. HORT BSA 78
Fannah/Sylvanus Joe ENTO MS 84
Farish/Alice Diane FDNU BSHE 68
Farkvam/David Oscar POSC MS 87
Farley/Emily HEED BSHE 87
Farmer/Barney Eugene BSA 51
Farmer/Frank L. RSOC MS 84
Farr/Chuck AGRN BSA 87
Farrar/Clara Ruth Grimes BSHE 47
Farriester/Brian Wayne POSC BSA 94
Farris/Bill AGRN BSA 79
Farris/Camille Decker BSHE 33
Farris/Donald E. GNAG BSA 50
Farris/Donald E. AGEC PHD 58
Farris/Donald E. HORT MS 51
Farris/Sue HOEC MS 66
Farris/Sue HOEC BSHE 54
Farris/William S. BSA 39
Farris/William S. AGEC MS 46
Farzan/Raof POSC BSA 82
Fason/Virginia J. ANSC BSA 88
Fate/Mark A. ANSC MS 88
Fatima/Nuzhat AGEC MS 92
Fatony/Achmad RSOC MS 91
Faubus/Sherry Ann AGBS BSA 93
Faught/William Avenere BSA 40
Faulk/Leah Beth CATX BSHES 98
Faulk/W. Kendall BSA 81
Faulkner/Joyce Plumlee BSHE 63
Faulkner/Martha Virginia BSHE 58
Faupel/Michel A. FDSC MS 92
Faupel/Michel A. PPPM BSA 85
Faur/Saleh M. FDSC BSA 88
Faur/Saleh M. HORT BSA 83
Fausett/John B. ESWS 98
Faust/Edward C. FDSC BSA 93
Favorite/Mindy Dawn IDHS BSHES 98
Fawcett/Mark S. AGBS BSA 92
Fawcett/Richard B. ANSC BSA 79
Fear/Carlos D. HORT MS 78
Feaster/Michael Anthony ENTO MS 96
Featherston/Edward S. ANSC BSA 72
Featherston/Russell M. BSA 54
Featherston/Sam R. ANSC BSA 74
Featherston/William Roy BSA 57
Featherstone/Homer Eugene ANSC BSA 77
Feeney/Glenyce G. FDNU BSHE 69
Feghali/Khalil Elias BSA 54
Fegtly/Debra Sterling Pucci HDID BSHE 80
Feild/Irene Edwards BSHE 41
Feilke/David L. AGBS BSA 76
Feland/John Morgan AGEC BSA 70
Feland/Sarah E. FDNU BSHE 68
Feland/William Price AGBS BSA 74
Fell/Ralph V. POSC BSA 53
Fell/Roger Vincent AGEC MS 80
Feltner/John A. POSC BSA 70
Felton/Trent BSA 74
Fender/David Duke AGBS BSA 88
Fender/Neva Mock BSHE 51
Fender/Theresa Baltz CLTX BSHE 87
Fendley/Kim RSOC MS 85
Ferguson/Alan C. BSA 65
Ferguson/Constance Dore FDSC BSA 88
Ferguson/Danny Lee AGME BSA 93
Ferguson/Dick V. BSA 54
Ferguson/Gail Hewgley ANSC BSA 85
Ferguson/J. B. AGEC MS 65
Ferguson/J. B. BSA 43
Ferguson/Joel A. POSC BSA 85
Ferguson/Joel N. BSA 57
Ferguson/Joel N. GNAG MS 64
Ferguson/John Vance BSA 38
Ferguson/Katherine FDNU BSHE 85
Ferguson/Leon Marvin CRSC MS 71
Ferguson/Leon Marvin GNAG BSA 69
Ferguson/Lou L. K. HEED BSHE 64
Ferguson/Louise F. HOEC MS 75
Ferguson/Mary HOEC MS 78
Ferguson/May McKnight BSHE 41
Ferguson/William S. BSA 50
Ferguson/William W. BSA 66
Fernandez/Mario Rafael HORT MS 85
Ferrer/Claude Gerard AGRN BSA 90
Ferro/Kathleen HEED BSHE 76
Fetters/Rebecca Ann FND BSHE 75
Fidelis/Niba Tafor AGEC MS 81
Fidelis/Niba Tafor AGEC BSA 79
Field/Tammy Kae PLPA MS 90
Field/Tammy Kae PTSC PHD 96
Fielder/Karen HOFS BSHE 78
Fielder/Larry R. ANSC BSA 76
Fields/Henry K. AGRI MS 56
Fields/Henry K. AGRI BSA 55
Fields/Johnny C. POSC BSA 62
Fields/Shawn Allen AGBS BSA 94
Fiely/Mark B. HORT MS 91
Fifer/Irma Ewing BSHE 46
Files/Terry A. AGBS BSA 72
Finch/Michael GNAG MS 87
Finch/Michael PPPM BSA 85
Finch/Sarah S. BSHE 35
Fincher/Angelyn Elayne IDHS BSHES 96
Fincher/Mitchel W. ANSC 99
Fincher/Pamela Graney IDHS BSHE 92
Fingerhut/Jack BSA 50
Finley/Lester Christian BSA 37
Finn/Joe D. ANSC BSA 71
Finn/Mary Ralphe BSHE 55
Finn/Nick R. ANSC BSA 74
Finney/C. Bradley BSA 49
Finney/Joe BSA 60
Finton/Joni Lynn GHEC BSHE 87
Firebaugh/Francille M. BSHE 55
Fischer/Tonya Rose IDHS BSHE 92
Fiscus/Jack D. BSA 43
Fiser/James H. BSAGE 47
Fisher/Carla Jean CLTX BSHE 81
Fisher/James William AGEC MS 73
Fisher/James William AGEC BSA 69
Fisher/Kevin Scott AGRN BSA 95
Fisher/Louis C. MS 54
Fisher/Louis C. BSA 53
Fisher/Mary Virginia M. HDID BSHE 82
Fisher/Neva Lou BSHE 27

Fitch-Hilgenberg/Marjorie FDNU MS 76
Fitch-Hilgenberg/Marjorie BSHE 74
Fitch/Earl Y. BSA 53
Fitts/Bobby A. ANSC BSA 85
Fitts/Harold Lea AGEC MS 88
Fitzgerald/Cleveland Hendricks BS 50
Fitzgerald/John B. HORT MS 80
Flake/Joe A. BSA 56
Flake/Raymond BSA 43
Flanagan/Joseph DASC BSA 73
Flanagin/Randy S. ENTO MS 72
Flanders/Donna J. HOEC MS 84
Flanigan/James E. CRSC BSA 74
Flanigan/Phyllis HOEC MS 74
Flanigan/Phyllis HEED BSHE 73
Fleak/Mabel BSHE 26
Fleet/Susan C. HOEC MS 88
Fleming-Hendricks/Stephanie Jean HDFS BSHE 94
Fleming/Harold W. BSA 50
Fletcher/Christine N. BSHE 40
Fletcher/Daniel E. CLTX BSHE 89
Fletcher/James Robson AGRN MS 53
Fletcher/Marion D. BSA 40
Fletcher/Michael Gordon ENTO MS 84
Fletcher/Roxann Bever ANSC BSA 91
Fleurantin/Gardy AGRN MS 88
Flinn/Heber H. BSA 17
Flint/Jean E. FDNU BSHE 76
Flippo/Miriam Robertson HOEC BSHE 85
Flood/Amos BSA 50
Florer/Douglas E. AGEC BSA 77
Floyd/Belinda POSC BSA 98
Flusche/Vincent Martin HORT BSA 79
Fly/Jerry AGBS BSA 73
Flynn/Archie Glynn AGEC MS 93
Flynn/Craig Joseph AGEC MS 90
Flynn/Debra Ann HEED BSHE 73
Flynn/Michael Cody AGBS BSA 96
Flynn/Michael F. ANNU BSA 71
Flynn/Norman Wayne POSC MS 72
Flynn/Norman Wayne ANSC BSA 70
Fogleman/Franklin G. BSA 49
Fogleman/Scott AGEC BSA 77
Fogleman/W. David AGEC BSA 79
Fondren/Frances Lynch BSHE 37
Foote/Margaret Robinson BSHE 40
Foraker/Cathy Parson HEED BSHE 78
Forbes/Bobbye J. HOEC BSHE 61
Forbess/Richard C. FDSC MS 82
Forbess/Richard C. HORT BSA 79
Ford/Bob E. BSA 51
Ford/Christine Sutton FDNU BSHE 49
Ford/Eleanor Sue BSHE 58
Ford/Gordon L. PHUS MS 64
Ford/Gordon L. BSA 55
Ford/Guy R. BSA 31
Ford/James W. BSA 53
Ford/Jerry J. ENTO MS 58
Ford/Jerry J. BSA 58
Ford/Joseph D. BSA 52
Ford/Loyd Wilburn BSA 53
Ford/Margo K. HEED BSHE 78
Ford/Mark David AGME NODEG 97
Ford/Mark David AGED BSA 97
Ford/Mickey Don AGED BSA 94
Ford/Mona Crawley BSHE 54
Ford/Opal Burton BSHE 35
Ford/Rutledge F. AGRN MS 61
Ford/Rutledge F. BSA 53
Ford/Tonya C. CDFR BSHE 81
Foresee/Mack Carl BSA 43
Forester/Stanley CRSC BSA 66
Formica/Ross R. POSC BSA 74
Forney/Mike A. ANSC BSA 73
Forrest/Howard C. BSA 51
Forrest/Janet HEED BSHE 78
Forrest/Mark C. BSA 37
Forrest/Martha W. FDNU MS 74
Forrest/Randall ANSC BSA 70
Forrester/Cleveland Edward AGRN BSA 54
Forrester/Sidney E. BSA 37
Forrester/Steven Duane ANSC MS 80
Forst/Randy Charles LDUH BSA 88
Forster/Lance Allen ANSC PHD 94
Forte/Robert A. ANSC BSA 84
Fortenberry/Rosemary N. BSHE 47
Fortner/James Bradley ANSC BSA 94
Fortner/Travis K. AGBS 98
Fos/Darrell POSC BSA 87
Foster/Beverly Ann HEED BSHE 75
Foster/Bevley Elishia ANSC BSA 72
Foster/Brantley Lee AGBS BSA 96
Foster/Clifton David BSA 40
Foster/Colleen CLTX BSHE 83
Foster/Dean Allan ANSC BSA 86
Foster/Eunice Fanning AGRN PHD 82
Foster/Fred B. BSA 56
Foster/Harold Edward BSA 50
Foster/Harry Vernon AGEC BSA 68
Foster/John G. ANSC BSA 69
Foster/Riley E. BSA 52
Foster/Sarah Rodgers BSHE 51
Foster/Stephen M. AGBS 98
Foster/Thomas P. ANSC BSA 78
Foul/James William BSA 53
Fountain/Juanita HEED BSHE 63
Fountain/Juanita ELED BSHE 63
Fourmy/Holly Tuttle FDNU BSHE 76
Fowler/Charles J. BSA 53
Fowler/Kimberly Renee IDHS BSHE 91
Fowler/Lehman C. BSA 48
Fowler/Rachelle FDNU BSHE 88
Fowler/Reginald David AGBS BSA 84
Fowler/Shirley BSHE 60
Fowler/Sue Carmical BSHE 40
Fowler/Terry R. HEED BSHE 78
Fox/Calvin L. VOAG BSA 64
Fox/Carol Abercrombie CDFR BSHE 79
Fox/Cheryl Ellis HEED BSHE 69
Fox/Susan Elizabeth FND BSHE 75
Fox/Thomas Meredith DANB BSA 90
Foyil/Viki Lynn Moll CLTX BSHE 76
Frach/Bobby W. ANSC BSA 88
Frachiseur/David E. POSC BSA 69
Frachiseur/Elizabeth Bismark HEED BSHE 70
Frampton/Jane Gregory HORT MS 94
France/Logan L. VOAG BSA 63
Franco-Romero/Fabricio AGEC MS 95
Franco/Fabricio R. DIAM BSA 91
Franco/Gene A. ENTO BSA 71
Franklin/Doris Jean FDNU BSHE 90
Franklin/James R. BSA 47
Franklin/Ralph E. BSA 55
Franklin/Rocky L. AGBS 98
Franklin/Samuel T. AGBS 99
Franklin/Teresa BSHE 77
Franklin/Timothy Don AGED BSA 94
Franks/Floyd P. BSA 41
Franks/Larry E. AHUS BSA 61
Franks/Larry E. AHUS MS 63
Franks/Patricia T. DANB BSA 76
Franks/Royal L. BSA 27
Franks/Ruby White BSHE 27
Frasier/Waldo BSA 24
Fratesi/Benjamin J. BSA 61
Frauenthal/Johnnie Stroud BSHE 42
Free/Bill AGBS BSA 86
Freeland/Thad R. AGRN MS 82
Freeland/Thad R. LDUH BSA 80
Freeman/Carell Thomas ANSC BSA 74
Freeman/Charles Elbert BSA 52
Freeman/Donald W. HOFS MS 76
Freeman/Donald W. HOFS BSA 73
Freeman/Gerald Don VOAG BSA 69
Freeman/John F. AGBS BSA 89
Freeman/Johnny Julius POSC BSA 67
Freeman/Linda Poynter FND BSHE 64
Freeman/Michael Roy AGBS BSA 81
Freeman/Ray L. BSA 53
Freeman/Richard Jay ENTO MS 88
Freeman/Roger Eugene HORT MS 80
Freeze/Dave Marion AGRN MS 87
Freiberger/P. J. AGED BSA 41
Frein/Kathy Williams HOFS BSHE 80
Frein/T. Madge CLTX BSHE 80
French/John A. ENTO MS 74
French/Robert G. BSA 50
French/Tracy AGBS BSA 83
Freyaldenhoven/Donna Burnett ANSC BSA 82
Freyaldenhoven/Mike ANSC BSA 82
Fricke/Ryan Andrew ENSC BSA 98
Friddle/Carolyn Gille AHUS BSA 66
Friddle/Jerry D. BSA 63

Frisby/Jean C. BSHE 41
Frisby/Jim W. AGED BSA 91
Frisby/Monica Morgan AGBS BSA 93
Fritsche/Gary L. AGEC MS 79
Fritts/Christine Angela POUL BSA 98
Fritts/Jerry AGEC BSA 75
Frizzell/Glen W. BSA 84
Frizzell/Johnnie L. BSA 54
Frizzell/Neal AGBS BSA 82
Froman/Nancy Williams BSHE 53
Froud/Darrell Lee POSC BSA 87
Fruechting/Jeffery M. DASC BSA 82
Fry/Larry Lee ANNU BSA 61
Fry/Leslie Randolph ENSC BSA 96
Fryar/Michelle RSOC MS 78
Fudge/Bobbie R. BSA 55
Fudge/Henry O. AGBS BSA 58
Fudge/Kenneth A. BSA 54
Fudge/Thomas G. GNAG MS 70
Fudge/Thomas G. BSA 54
Fugatt/Barry L. HORT MS 70
Fugitt/Terry F. AHUS BSA 64
Fulcher/H. Clay AGRN BSA 77
Fulenwider/Steven A. AGEC BSA 81
Fuller/Billy Wayne POSC BSA 86
Fuller/Charles E. CRSC BSA 72
Fuller/Charles E. AGRN MS 79
Fuller/David L. ANSC BSA 73
Fuller/John Julian ENTO MS 75
Fuller/John Julian ENTO BSA 70
Fuller/John R. PLPA MS 69
Fuller/Juanita HOEC MS 62
Fullerton/John Robert GNAG BSA 68
Fullerton/Terri L. CLTX BSHE 84
Fullerton/Thomas Mankin CRSC PHD 69
Fulmer/Michael Lynn AGED BSA 86
Fulton/Kay FND BSHE 67
Fulton/Neil D. BSA 49
Fulton/Raymond Andy BSA 39
Fultz/David E. DASC BSA 71
Fultz/Latisha Ann AGBS BSA 97
Fultz/Tamara Nelson HDFS BSHE 92
Funderberg/Max BSA 64
Funk/Jerry Richard BSA 58
Funke/Gretchen Elaine CLTX BSHE 88
Fuqua/Ben SOSC PHD 72
Fuqua/Ben SOSC MS 67
Furmanski/Robert Joseph FDNU MS 78
Fuser/James R. ANSC BSA 73
Gabbie/Shelley Wilcoxon HDID BSHE 82
Gabriele/Michele CLTX BSHE 91
Gachot/Laila G. HOEC MS 81
Gachot/Laila G. HOFS BSHE 60
Gadberry/Heather Hackett ANSC MS 96
Gadberry/Michael Shane ANSC BSA 94
Gadbury/Eva Marie FDNU BSHE 88
Gaddis/Joseph Jefferson BSA 49
Gaddy/J. Carroll BSA 25
Gaddy/Jim ANSC BSA 86
Gage/Edward V. ENTO MS 75
Gage/James S. GNAG BSA 79
Gage/Kelly ANSC MS 72
Gage/Kelly ANNU BSA 71
Gage/Richard Kelly AGBS BSA 98
Gairhan/Nathan Earl AGRN BSA 52
Gale/Paula M. AGRN PHD 88
Gale/Paula M. SOCS BSA 83
Gallegly/M. E. GNAG BSA 45
Gallemore/Carrie Lynn BSHE 85
Gallimore/Milton D. CRSC MS 69
Galloway/Carol CLTX BSHE 77
Galloway/Douglas Lee ANSC MS 95
Galloway/Neal AGBS BSA 83
Galvez/Victor H. M. ANSC MS 66
Galvin/Mary Morgan CLTX BSHE 79
Galvin/Terri Bright FDNU BSHE 88
Galyen/Dayne W. ANSC BSA 88
Galyen/Teresa Diane CLTX BSHE 92
Gambacort/Mary Helen AGBS BSA 88
Gamble/Jay Vee VOAG BSA 82
Gamble/Jay Vee GNAG MS 87
Gammill/Cecil A. BSA 49
Gammill/David Robert AGME BSA 85
Gammill/Debbie HEED BSHE 84
Gammill/Michael Alan AGME BSA 90
Gammill/Rick P. AMGT BSA 82
Gandy/Jay ENTO BSA 76
Gann/Larry Ray AGBS BSA 83
Gant/Hilton E. BSA 42
Gant/Marilyn Whitehead BSHE 51
Gant/Phyllis Ann HEED BSHE 84
Garcia-Prendes/Rafael AGME BSA 73
Garcia/Maria Elena HORT MS 82
Garcia/Maria Elena PTSC PHD 97
Garcua-Prendes/Rashed AGME BSA 73
Gardner/Joseph William LDUH BSA 91
Gardner/Mildred Claypool BSHE 27
Garlington/Arthur Roe BSA 23
Garmon/John W. BSA 57
Garneau/Virginia Stapleton BSHE 57
Garner/Billy A. SOSC BSA 65
Garner/Brian Jeffrey AGEC BSA 89
Garner/Carole B. DIET BSHE 67
Garner/Carroll R. AGEC MS 71
Garner/Carroll R. AGED BSA 68
Garner/Charles W. BSA 58
Garner/Darrell E. BSA 58
Garner/Deborah Fiegel CLTX BSHE 83
Garner/Elsie L. HOEC MS 69
Garot/L. J. BSA 33
Garot/Ruth BSHE 34
Garrett/Clara Smith BSHE 72
Garrett/Connie Lynette FDSC MS 87
Garrett/Cynthia Gail HOEC MS 86
Garrett/George H. BSA 50
Garrett/Gibson AGHE BSHE 85
Garrett/Jimmy Lee POSC BSA 87
Garrett/John D. AGRN MS 62
Garrett/John D. BSA 58
Garrett/Justin Lanier BSA 83
Garrett/Nancy Neece HOEC BSHE 58
Garrett/Paul Winston HORT BSA 97
Garrett/Robert Wayne VOAG BSA 71
Garringer/Jason Wade AGBU BSA 98
Garrison/Brenda Greenwood LDUH BSA 95
Garrison/Peggy James BSHE 57
Gartside/Albert L. BSA 40
Gartside/Mary Smith BSHE 43
Gascoigne/Harriett Lyn FDSC MS 83
Gash/Marilyn AGED BSA 95
Gaskill/William F. BSA 47
Gaston/Andrew Thomas AGBS BSA 96
Gaston/Joe L. CRSC BSA 62
Gaston/Margaret L. BSHE 41
Gateley/Audrey HOEC MS 88
Gateley/James T. BSA 53
Gates/Aubrey D. BSA 30
Gates/Mary Musick HDFS BSHE 93
Gates/Patricia Coleman BSHE 57
Gattis/Gay BSHE 28
Gauldin/Deri Lyn HOEC MS 88
Gavin/John FDSC MS 85
Gay/John Lynn HORT BSA 71
Gay/LaLinda CLTX BSHES 96
Gay/Shawn Lynn HORT BSA 96
Gay/Shawn Lynn AGBS NODEG 96
Gaylean/Alice Cordwell BSHE 49
Gazaway/Sharon Horton BSHE 63
Gearhart/Candice HEED BSHE 74
Geary/Charles Watson BSA 27
Geddes/Roderic D. AGRN MS 78
Gee/Garland W. DANB BSA 85
Geels/Linda Foster CLTX BSHE 82
Geels/Steven E. DANB BSA 82
Gehre/Jerry L. AGEC PHD 73
Geller/Jack M. RSOC MS 82
Gentry/Chad K. AGBS 99
Gentry/Jay Harold AGBS BSA 96
Gentry/Leona Estalee BSHE 47
Gentry/Ruff L. BSA 50
Gentry/Terry Joe AGRN BSA 93
George/Barbara A. ANSC 99
George/Cynthia Jean CLTX BSHE 85
George/Jennifer R. POSC 99
George/Joe D. BSA 54
George/John A. VOAG BSA 61
Gerrard/William H. GBUS BSA 75
Gerrard/William Herbert AGBU BSA 97
Geurin/Brenda G. DANB BSA 78
Ghadder/Sami Hassan BSA 80
Ghandorah/Mohammed Omar AGRN PHD 81
Ghorayshi/Majid AGRN MS 79
Gibbs/Clifford AGBS BSA 88
Gibbs/Jerry L. LDUH BSA 82

Gibbs/Laura Woolly IDHS BSHES 97
Gibbs/Sara Jo BSHE 47
Gibbs/William L. VOAG BSA 69
Gibbs/William M. GNAG BSA 49
Gibson/Amy Lynn AGRN BSA 95
Gibson/Buck Canyon DIAM BSA 94
Gibson/Carol Lynn FND BSHE 70
Gibson/Christopher Cline HORT BSA 98
Gibson/Connie W. HEED BSHE 75
Gibson/Crystal Lynne AGBS BSA 82
Gibson/Franklin D. POSC BSA 63
Gibson/Gene Toland BSHE 43
Gibson/Gina Ware CLTX BSHE 90
Gibson/James O. ANSC BSA 54
Giese/James Huling FDSC MS 88
Giese/James Huling FDSC BSA 85
Giese/William G. HORT MS 88
Giese/William G. HORT BSA 82
Giezentanner/Greg LDUH BSA 91
Giffels/Irma F. BSHE 25
Gifford/Carolyn Wiedeman HEED BSHE 73
Gifford/Jana Sue HOEC MS 87
Gift/William L. ANSC BSA 94
Gigoux/Mary J. BSHE 60
Gilbert/Anne Yvonne BSHE 40
Gilbert/Barbara Craig BSHE 69
Gilbert/Evelyn T. BSHE 41
Gilbert/Kenneth L. ENTO BSA 66
Gilbert/Mary Holt BSHE 43
Gilbert/Robin Walthall POSC BSA 88
Gilbey/Philip Douglas AGBS BSA 92
Gilbreath/Dallas E. SOSC BSA 63
Gilbreath/Milton O. BSA 47
Gilby/Elizabeth Stutheit FDNU BSHE 40
Giles/Evelyn Cockerham GNAG BSHE 60
Giles/Laura E. DANB BSA 84
Giles/William L. GNAG BSA 34
Gill/Kathy Williams HEED BSHE 82
Gill/Robert L. BSA 52
Gillean/John A. BSA 39
Gillespie/Lisa J. FDNU BSHE 86
Gillespie/Mary L. BSHE 24
Gilleylen/Thomas Wilson ANSC BSA 71
Gillham/Richard AGEC BSA 77
Gillihan/Carolyn L. Belvins BSHE 59
Gilliland/C. B. GNAG BSA 34
Gilliland/Floyd R. ENTO MS 64
Gilliland/Leno Elwin AGRI BSA 37
Gilliland/Roger Ivan BSA 37
Gillison/William Stacey AGBS BSA 86
Gillum/Johnny W. AGEC BSA 67
Gilmer/Betty R. BSHE 47
Gilmore/George Ivan BSA 30
Gilstrap/Maysel BSHE 42
Gimenez/Tammie J. FDNU BSHE 81
Gingell/Grace Love BSHE 26
Giordano/Frank R. AGRN PHD 75
Girardi/August G. HORT BSA 93
Girma/Bedada AGRN PHD 95
Gish/Glenn Richard HORT BSA 81
Gist/Tom GNAG BSA 58
Givens/Owenita FDNU BSHE 61
Gladden/Andrew C. ANSC BSA 73
Gladden/Baxter Howard AGBS BSA 87
Gladden/Daniel B. ENTO MS 66
Gladden/Daniel B. BSA 65
Gladden/Lorrie Lynne CLTX BSHE 88
Gladfelter/Patricia Watson BSHE 51
Gladney/William J. ENTO MS 64
Glasgow/Irene Gollaher BSHE 27
Glasgow/J. R. BSA 40
Glasgow/Randall T. AGRN BSA 81
Glass/Grace BSHE 27
Glaub/Ted L. AGEC MS 82
Glaub/Ted L. ANSC BSA 71
Glaze/Patti ANSC BSA 84
Glenn/Carl E. AHUS BSA 64
Glenn/Elvira Sellers BSHE 45
Glenn/Kimberly Faye FDSC BSA 95
Glenn/Rupert W. BSA 48
Glenn/Suzanne Marie HORT BSA 98
Gliedt/Douglas W. ANSC PHD 94
Gliedt/Douglas W. ANSC MS 86
Gliedt/Ruby Carol GNAG MS 87
Gliedt/Ruby Carol ANSC BSA 86
Glover/Marina HOFS BSHE 79
Goade/William A. BSA 49
Goatcher/Harold Franklin BSA 49
Goddard/Katrina R. FDNU BSHE 82
Godfrey/Carol BSHE 76
Godfrey/Mary S. Wood HEED BSHE 35
Godkin/Jolynn Bradley GHEC BSHE 81
Godley/John Lee AGRN BSA 80
Godsey/Jason Andrew AGBS BSA 96
Goeppner/Joe BSA 55
Goetsch/Brenda Lee ANSC MS 88
Goetz/Andrew Joseph AGRN PHD 90
Goff/Lloyd N. BSA 51
Goff/Lynn Marie HORT MS 93
Goff/Robert L. AGRN BSA 89
Goforth/Howard Isaac BSA 36
Goforth/Lowell A. BSA 36
Goggans/Greg DIAM BSA 92
Goh/Wee Peng FDSC BSA 90
Goins/Teresa HDFS BSHE 85
Goins/William A. ANSC BSA 82
Golden/Christopher D. ANSC 99
Golden/Donna Marie ANSC BSA 83
Golden/Gary L. ANSC BSA 74
Golden/James F. VOAG BSA 50
Golden/Jamie Lee ANSC BSA 96
Golden/Laron Eldon BSA 49
Golden/Nickie L. HOFS BSHE 80
Goldhammer/David Scott ENTO MS 88
Goldhammer/Frances Marie SOCS BSA 87
Goldmon/Dewayne L. AGRN MS 87
Goldmon/Dewayne L. AGRN BSA 85
Gollon/Edward M. ANSC BSA 69
Golmirzaie/Akbar AGRN PHD 82
Golmirzaie/Ali M. PHD 82
Gomez/Ruben Antonio BSA 59
Gonzalez/Karen Jackson BSHE 87
Good/Robert M. BSA 51
Gooden/Billy F. BSA 54
Gooden/Harold D. BSA 56
Goodgame/Charles Alan AGRN BSA 93
Goodman/Dorothy M. BSHE 47
Goodman/John Grady BSA 65
Goodman/Kathy Lee Early HEED BSHE 73
Goodman/Rayna Babette SOCS BSA 85
Goodman/Ruth Parker HEED BSHE 66
Goodner/Travis O. POSC BSA 69
Goodner/Vick D. ANSC BSA 71
Goodrich/Dalton BSA 49
Goodrich/Sara Sloan HOEC MS 83
Goodrum/Betty Hamilton GHEC BSHE 56
Goodwin/Alice F. BSHE 60
Goodwin/Clariss Obera FDSC BSA 87
Goodwin/Donald R. ANSC BSA 76
Goodwin/Harry Orville BSA 39
Goodwin/John Ashley DIAM BSA 91
Goodwin/Malcolm Cecil BSA 41
Goodwin/Tommy L. POSC MS 60
Goodwin/Tommy L. POSC BSA 58
Gordon/Alfred Y. AGRI BSA 49
Gordon/Carol Robinson CLTX BSHE 81
Gordon/Charlie R. BSA 53
Gordon/Che Michael ENSC BSA 98
Gordon/Dale R. BSA 58
Gordon/Eddie C. AGRN MS 77
Gordon/Eddie C. GNAG BSA 73
Gordon/Gwenith M. AECT 99
Gordon/Lois Davis HEED BSHE 42
Gordon/Ramona HOEC MS 91
Gordon/Raymond E. BSA 54
Gordon/Terrel VOAG BSA 42
Gordon/Willaim R. AECT 99
Gore/Benny G. VOAG BSA 64
Gore/Ulys Roy BSA 25
Gore/Wanda G. FDSC BSA 80
Gore/Wanda J. Whisnant BSHE 76
Gore/Warren A. BSA 52
Gorman/Bob BSA 74
Gosey/Carrie A. CLTX BSHE 88
Gossett/Henry M. GNAG MS 67
Gossett/Henry M. BSA 40
Goto/Kazufumi POSC PHD 81
Gould/Cheryl Ann CLTX BSHE 83
Gower/Scott AGBS BSA 85
Goyne/Danny POSC BSA 77
Grabbe/John A. BSA 60
Grace/Anna Fulton BSHE 41
Grace/Beatrice P. BSHE 41
Grace/Frank BSA 47

Grace/John Allen BSA 52
Grace/Lila B. HOEC MS 81
Grace/Lloyd E. BSA 59
Grace/Walter G. BSA 65
Grace/William G. BSA 53
Graddy/Connie B. BSA 54
Graddy/Jim D. BSA 50
Graddy/John Ed AGBS BSA 87
Graening/Gary Orval BAEG MSBAE 94
Gragg/John R. BSA 48
Graham/Donna L. FND BSHE 70
Graham/Harryette O. BSHE 59
Graham/Heather Leigh HEED BSHES 96
Graham/Henry Gustave BSA 55
Graham/Henry Gustave AGBS BSA 80
Graham/J. Ralph Wayne BSA 41
Graham/M. J. BSA 40
Graham/Rebecca Susan HOEC MS 81
Graham/Rebecca Susan BSHE 80
Graham/Sandye L. FND BSHE 71
Graham/Shirley Glenn BSHE 56
Graham/Stephen J. BSA 57
Graham/Vance Dee GNAG BSA 41
Gramlich/James Vandle AGRN MS 62
Gramlich/James Vandle BSA 61
Gramlich/Paul D. ANSC BSA 74
Grammer/Amy Lynn HDFS BSHES 97
Grammer/Ann Beller HOEC MS 81
Grams/Katherine Elise LDUH BSA 86
Grannemann/Dana Rhodes Antone FDNU BSHE 77
Grassel/Marcia A. IDHS BSHE 89
Graupner/Walter G. ANSC BSA 90
Graves/Angelia CLTX BSHE 87
Graves/D. B. DANB BSA 77
Graves/Dawn Gaye CLTX BSHE 91
Graves/Dawn Marie CLTX BSHE 93
Graves/Frances Marie BSHE 38
Graves/George Martin AGBS BSA 85
Graves/Ginger Seanne HDFS BSHE 86
Graves/Joe Brent AGRN BSA 80
Graves/Joe Brent GNAG MS 82
Graves/Michael J. VOAG BSA 82
Graves/Robert Clinton ANSC BSA 81
Graves/Susan Hamilton HEED BSHE 61
Gravois/Mary Gerke DANB BSA 84
Gray/Ava Anderson BSHE 40
Gray/Bartus M. BSA 40
Gray/Christina M. BSHE 76
Gray/Delmar D. BSA 52
Gray/Gary L. AGED BSA 91
Gray/Jarrell D. BSA 47
Gray/Jonathan Arthur FDST BSA 98
Gray/Karen Sue HORT MS 83
Gray/Patricia A. HOEC BSHE 53
Gray/Priscilla A. ENTO MS 81
Gray/Priscilla A. PLPA BSA 78
Gray/Randall L. BSA 65
Gray/Roma Lisa ANSC BSA 86
Gray/Roy R. AGRN BSA 76
Gray/Steve AGRN BSA 84
Gray/Tami Lynn CLTX BSHE 87
Gray/Thomas H. BSA 50
Green/Bernal L. BSA 58
Green/Bernal L. AGEC MS 61
Green/Dawn Marie HDFS BSHES 96
Green/Garvin BSA 39
Green/Hazel BSHE 57
Green/John BSA 51
Green/Kimberly CLTX BS 86
Green/Tollie Ross AGRN MS 79
Green/Tollie Ross BSA 76
Green/William E. POSC BSA 68
Greene/Billy J. BSA 52
Greene/Daryle E. POSC MS 55
Greene/Daryle E. GNAG BSA 54
Greene/Garland J. BSA 33
Greenfield/William Ed. BSA 57
Greenlee/James Russell POSC BSA 94
Greenwalt/Bert AGEC MS 80
Greenway/Terry ANSC MS 83
Greenwood/James Robert FDNU BSHES 96
Greenwood/Jeanette C. HOME BSHES 97
Greenwood/Karen L. AGBS BSA 85
Greer/Clyde BSA 27
Greer/Thomas B. BSA 27
Greer/Wilma Jane Cearley BSHE 41
Gregg/Margaret BSHE 21
Gregg/Sharon D. Loveland HDID BSHE 82
Gregory/Anson G. BSA 49
Gregory/Elmer R. AGRI BSA 36
Gregory/Larry B. AGEC BSA 72
Gregory/Marguerite BSHE 36
Gregory/Minor BSA 79
Gregory/Nathan AGEC BSA 75
Gregory/Roger DASC BSA 78
Greig/James K. BSA 49
Greig/Star BSA 13
Greig/Sue BSHE 48
Greig/W. Smith BSA 48
Grenwelge/William A. DASC BSA 68
Gresham/Lucille Suggs FDNU BSHE 62
Grice/Virginia M. Ford HEED BSHE 64
Griffin/Charles T. BSA 52
Griffin/Dot E. PTSC PHD 87
Griffin/Dot E. HORT MS 79
Griffin/Howard BSA 54
Griffin/James R. PPPM BSA 84
Griffin/Leah Christine HES BSHES 96
Griffin/Mary Marguerite BSHE 38
Griffin/Terry Wayne AGRN BSA 97
Griffin/Virginia Kay Yager FDSC MS 89
Griffin/William H. BSA 51
Griffith-Tullis/H. June HDID BSHE 79
Griffith/David L. FDSC BSA 76
Griffith/David Leon BSA 52
Griffith/Kenneth M. AGRN MS 82
Griffith/Lance Arden ENSC BSA 98
Griffith/Marcia M. FDSC BSA 75
Griffith/Paul G. GNAG BSA 62
Griffith/Stephen Grant AGED BSA 90
Griffiths/Delta Moore BSHE 40
Grigg/Brandon AGRN MS 91
Grigg/Brandon AGRN BSA 87
Grigg/Dennis Wayne POSC BSA 86
Grigg/Gary M. GNAG BSA 79
Grigg/James E. POSC BSA 71
Grigg/Lyndall L. VOAG BSA 67
Grigg/Willie Dale ANSC BSA 74
Griggs/Barney Ray ENSC BSA 96
Grigsby/John H. BSA 50
Grigsby/Rebecca M. HEED BSHE 72
Grim/David R. BSA 56
Grimes/Dorothy McClellan BSHE 38
Grimes/J. W. BSA 51
Grimes/Mary Moore BSHE 45
Grimmett/J. Byron BSA 53
Grimsley/Clint E. POSC BSA 80
Grimsley/Roy POSC BSA 51
Grinder/Julie R. GHEC BSHE 83
Griscom/Don DASC BSA 63
Grisham/Julie FDNU BSHE 76
Grisham/Laura Larsa IDHS BSHE 92
Grisham/Marlo Elizabeth CLTX BSHE 94
Grisham/Michael P. PLPA MS 73
Grissom/Percy E. BSA 54
Groce/A. Wayne ANSC MS 68
Groce/A. Wayne ANNU BSA 66
Groce/Dennis L. ANNU BSA 68
Groce/Gary D. AGBS BSA 77
Groce/Nolan BSA 48
Groce/William Nolan AGBS BSA 72
Grooms/Bill BSA 65
Grooms/Shirley FND BSHE 64
Gross/Jason Alan FDSC BSA 96
Gross/Merle R. HORT BSA 68
Groves/Russell L. ENTO MS 93
Gruber/Robert L. CDFR BSHE 83
Grummon/Margaret L. HOEC MS 79
Grunwald/Ronald M FDSC MS 82
Guarr/Charles W. BSA 50
Gubler/Walter D. PLPA MS 74
Guerber/Claudia ENTO MS 91
Guerra/Daniel AGRN MS 81
Guerra/Lisa B. CLTX BSHE 84
Guerrero/Tomas HORT BSA 82
Guess/Katherina M. HOEC MS 81
Guest/Charlotte E. CLTX BSHE 89
Guest/Judy BSHE 63
Guffey/Tresa Jane AGBS BSA 88
Guidry/Robert K. HORT MS 77
Guindo/Dore AGRN PHD 93
Guindo/Dore AGRN MS 89
Guinn/Dendin Lilliad ANSC BSA 69

Guinn/Gene AGRN MS 57
Guinn/Gene GNAG BSA 52
Guinn/Shannon Marie DIAM BSA 92
Guise/E. J. BSA 36
Gullett/Lionel Bertrand RSOC MS 81
Gullette/William H. BSA 49
Gulley/Terry L. AGRN MS 86
Gulley/Terry L. AGEC BSA 79
Gump/Dabney IDHS BSHE 89
Gunn/Catsy Collard BSHE 63
Gunsaulis/John ANSC BSA 87
Gunter/Cal ANSC BSA 78
Guo/Changxin AGRN PHD 93
Gurvitz/Tia Akins CLTX BSHE 70
Gustavson/Steven A. ANSC MS 81
Guthrey/Rebecca L. HEED BSHE 71
Guthrie/David Scott AGRN PHD 83
Guthrie/Tammy Lea POSC BSA 95
Guthrie/Thomas Campbell BSA 42
Gutierrez/Mario O. ENTO BSA 69
Guy/Charles Brandon AGRN PHD 86
Guyer/Mary M. FDSC BSA 88
Gwaltney/Mary Gail Anderson BSHE 56
Gwatney/Deborah Kay HEED BSHE 85
Haak/Patricia Horner CLTX BSHE 82
Hacker/Andrea N. FHNH 99
Hackworth/Carolyn Frankenberger FDNU BSHE 71
Hackworth/Max CRSC BSA 71
Haddan/Theresa L. FHNH 99
Haddox/Austin Walsh Webb ENTO MS 74
Hadley/Gary POSC BSA 71
Haegele/Richard BSA 55
Hafer/David P. AGBS 98
Hagan/David Lee ANSC BSA 71
Hagemann/Joel A. ANSC BSA 73
Hagen/Judy B. HLED BSHE 82
Haggard/Brian Edward AGRN MS 97
Haghighat/Nasser G. AGRN PHD 85
Hahn/Edward M. FDSC BSA 71
Hahn/John William POSC MS 97
Haid/John M. BSA 59
Haines/Leslie Dean POSC BSA 89
Haines/Linda Ransom HOEC MS 77
Haines/Linda Ransom FND BSHE 70
Haingaertner/Jennifer Lee IDHS BSHE 91
Hainline/Daniel Milburn POSC BSA 92
Hairr/William T. BSA 63
Haisty/Donna Hunnicutt BSHE 40
Haizlip/Marcia FDSC MS 84
Haizlip/Marcia ENTO MS 79
Hajny/Teresa Taylor FDNU BSHE 74
Hakkak/Mohammad Reza ANSC PHD 86
Halbert/Helen Melissa ANSC BSA 86
Halbert/Steve Manning AGBS BSA 87
Halbrook/Bobbye Jo HOEC MS 63
Halbrook/Waymon A. AGEC MS 64
Halbrook/Waymon A. AGEC BSA 50
Hale/Alfred Clay BSA 22
Hale/Dianne FND BSHE 64
Hale/Eugenia Crawford BSHE 43
Hale/Floyd Everett BSA 57
Hale/Freddy M. AGEC BSA 64
Hale/George CRSC BSA 68
Hale/Irene Johns BSHE 31
Hale/Joe M. ANSC BSA 58
Hale/Kirk K. POSC MS 67
Hale/Kirk K. PHUS BSA 63
Hale/Laurel Ann IDHS BSHE 93
Hale/Linda Lee Goss CLTX BSHE 89
Hale/Lorie FDNU BSHE 86
Hale/Pamala Bumpers CLTX BSHE 87
Hale/Robert W. BSA 51
Hale/Robert W. GNAG MS 66
Haley/Charlotte Jaylon BSHE 70
Haley/Martha Rounsaville CLTX BSHE 82
Haley/Michael S. ANSC BSA 72
Haley/Ronald J. ANSC BSA 72
Haley/Ruth Carson BSHE 41
Halk/Edward L. PLPA BSA 71
Hall/Carmen Edwards CLTX BSHE 88
Hall/Charles V. GNAG BSA 50
Hall/Charles V. HORT MS 53
Hall/Charles Wayne AGBS BSA 89
Hall/Dorothy A. HOEC MS 79
Hall/Duane AGEC BSA 79
Hall/Frances Rose BSHE 41
Hall/George D. BSA 63
Hall/George S. AGEC BSA 82
Hall/Gilbert W. BSA 53
Hall/Glenda D. DPQA BSHE 70
Hall/Jack W. BSA 49
Hall/James Barry HORT BSA 90
Hall/James Robert ANNU BSA 74
Hall/James Robert ANNU MS 76
Hall/Janie Haigh BSHE 29
Hall/Jeffery Thomas ANSC BSA 95
Hall/Jerry Lee DASC BSA 56
Hall/Jesse Doyle BSA 41
Hall/Judy A. HEED BSHE 79
Hall/Julie Diane AGRN MS 83
Hall/Katherine L. ANSC MS 86
Hall/Marie Sullivent BSHE 44
Hall/Mark AGBS BSA 91
Hall/Mary Karen HEED BSHE 91
Hall/Mary White HOEC BSHE 36
Hall/Melvin Richard PTSC PHD 81
Hall/Oather S. BSA 41
Hall/Orville J. BSA 25
Hall/Ralph L. ANSC BSA 85
Hall/Robert S. AGED BSA 93
Hall/Vera Cassat BSHE 35
Hall/Vernon Lloyd BSA 40
Hall/Walter B. BSA 50
Hall/Willis L. BSA 16
Haller/Stevie L. VOAG BSA 79
Hallesy/Barbara FDNU BSHE 78
Halley/John T. ANSC MS 82
Halley/John T. POSC PHD 86
Hallmark/Billy John BSA 57
Hallum/Alan W. ANSC BSA 67
Hallum/Sally Featherston BSHE 59
Halsey/Edward M. BSA 42
Halsted/Mark POSC BSA 86
Halterman/Keith Ray POUL BSA 97
Ham/Cindy Michelle AGED BSA 96
Ham/Everett A. BSA 50
Ham/Joseph Van AGTE BSA 98
Hamann/Rayma W. IDHS BSHE 95
Hamblen/Linda M. GHEC BSHE 78
Hamblen/Mac L. BSA 48
Hamdan/Mohamad S. HORT BSA 82
Hamdan/Mohamad S. HORT MS 85
Hamid/Muhammad Abdul AGRN PHD 79
Hamilton/Cynthia S. FHNH 99
Hamilton/Daisy D. BSHE 36
Hamilton/Emily Mallory BSHE 48
Hamilton/Holli Ann FDNU BSHES 98
Hamilton/James E. BSA 56
Hamilton/John AGBS BSA 97
Hamilton/Michael D. DANB BSA 76
Hamilton/Michael D. GNAG MS 83
Hamilton/Rita Stitt HOEC MS 77
Hamilton/Rita Stitt CDFR BSHE 76
Hamilton/Shane P. AGBS 99
Hamilton/Van E. BSA 42
Hamm/Douglas BSA 57
Hamm/Sarah AGBS BSA 86
Hammans/Charles E. BSA 52
Hammans/Howard S. BSA 56
Hammer/Mark PTSC PHD 92
Hammond/Huie B. BSA 49
Hampton/Richard O. BSA 51
Hampton/Ronald Earl AGRN PHD 90
Han/Tong AGEC MS 96
Hanano/Tanya Louise ENTO MS 96
Hanby/Leigh Ann ANSC BSA 76
Hancock/Cherie A. LDUH BSA 84
Hancock/Floyd Gale AGRN PHD 91
Hancock/Floyd Gale AGRN MS 86
Hancock/June A. AGRN MS 85
Hancock/Marcia Calva AGBS BSA 84
Hancock/Terri Lyn DANB BSA 85
Handayani/Iin Purwati AGRN MS 91
Haney/Robert J. AGRN BSA 51
Haney/Robert J. GNAG MS 63
Hankins/Ada Greer HEED BSHE 35
Hankins/Ann Marie CLTX BSHE 87
Hankins/B. J. BSA 61
Hankins/B. J. CRSC MS 68
Hankins/Cecil BSA 38
Hankins/Clayton ANSC PHD 91
Hankins/Curtis L. BSA 40
Hankins/Herman H. AGEC BSA 34

Hankins/Mark A. AGED BSA 97
Hankins/Meredith Battreal FDNU BSHES 98
Hankins/Mitchell Lynn GNAG MS 92
Hankins/Mitchell Lynn AGED BSA 88
Hankins/Robert Lee BSA 56
Hanks/Randall L. POSC BSA 77
Hanley/Thomas Lovett AGBS BSA 87
Hanna/Larry Don AGME BSA 91
Hanna/Randall W. ENSC BSA 78
Hanna/Robert C. AGEC BSA 81
Hanna/Thad AGBS BSA 81
Hannah/Mitzi Paige CLTX BSHE 89
Hannum/Meredith Coker FND BSHE 74
Hansell/Virginia Palmer HEED BSHE 27
Hansen/Ethan Allen GNAG MS 63
Hansen/H. Clay BSA 37
Hansen/Leah Alice CLTX BSHE 83
Hansen/Louise AGBS BSA 86
Hansen/Teresa HDID BSHE 80
Hansen/Thomas L. AGRN MS 83
Hanson/Beulah Mae BSHE 33
Hanson/Kenneth James POUL BSA 98
Hanson/Margaret BSHE 47
Hanson/Matthew G. AGRN MS 94
Hanson/Peggy Ann BSHE 57
Hanthorn/James Michael AGEC MS 80
Hantz/David K. HORT MS 85
Haralson/Christy Clarene ANSC BSA 94
Haraway/Paula K. ANNU BSA 84
Harbison/Anna Jane BSA 65
Hardage/Gina Linn CLTX BSHE 81
Hardaway/Lola BSHE 46
Hardcastle/Achel E. BSA 49
Harden/Jessica K. FHNH 99
Harder/Sherry Gaye Arnold HEED BSHE 72
Hardin/Bernadine Marie FND BSHE 74
Hardin/Herschel T. BSA 40
Hardin/Jerry Keith AGBS BSA 74
Hardin/Joy AGRN PHD 81
Hardin/Joy AGRN MS 79
Hardison/Patricia L. BSA 82
Hardman/Gordon Brad AGRN BSA 94
Hardman/Louis Christopher AGRN BSA 91
Hardwick/Robert E. BSA 58
Hardy/Brannen Vail HORT BSA 95
Hardy/K. Scott AGBS BSA 82
Hardy/Kathleen Wallis FDSC BSHE 82
Hardy/Kathleen Wallis FDSC MS 86
Hardy/Lorene Sallis BSHE 40
Hardy/Michael G. PLPA BSA 78
Hare/Marvin B. BSA 65
Harger/Thomas R. SOCS BSA 70
Hargis/Betsy BSHE 45
Hargis/Edrie P. Scott HOEC BSHE 32
Hargis/Galyn Frances Wood HEED BSHE 75
Hargraves/Alan D. AGEC BSA 77
Harkin/Jerry Keith AGEC BSAGE 74
Harkreader/Jean C. HEED BSHE 74
Harlan/Donnie Leon VOAG BSA 68
Harmon/Carole L. VOAG BSA 83
Harmon/Dennis D. ANSC BSA 71
Harmon/James C. ANSC BSA 77
Harmon/Sandra Helen HEED BSHE 71
Harms/Edward Henry BSA 51
Harms/R. H. POSC BSA 53
Harp/Arlin D. VOAG BSA 72
Harp/Caren L. AGEC MS 85
Harp/Caren L. AGBS BSA 83
Harp/Christie Vaughan IDHS BSHES 97
Harp/Clinton Alan POSC BSA 92
Harp/Jim D. ANSC MS 78
Harp/Jim D. ANSC BSA 71
Harp/Kendall D. AECT 99
Harp/Larry D. DASC BSA 75
Harp/Robert ANSC BSA 69
Harp/Ronald Lee ANSC BSA 71
Harper/Fred William BSA 38
Harper/Jarvis AGRN PHD 80
Harper/Jarvis AGRN MS 78
Harper/Jayson Kennedy AGEC MS 83
Harper/Kim Moore IDHS BSHE 94
Harper/M. Dean AGRN MS 62
Harper/M. Dean BSA 50
Harral/Christopher Gary POSC BSA 95
Harral/Melissa Kay CLTX BSHE 95
Harrell/Chris A. AGRN BSA 73
Harrell/Debra Jackson HEED BSHE 74
Harrell/Judith Taylor HEED BSHE 64
Harrell/Mary Bunn HOEC MS 64
Harrell/Phyllis DANB BSA 82
Harrell/Reba R. Parker BSHE 40
Harrendorf/Keith ENTO MS 58
Harrendorf/Keith BSA 52
Harriman/Marilyn Jane HEED BSHE 75
Harrington/Alice Ruth BSHE 17
Harrington/Jack T. BSA 56
Harrington/Janet Graf BSHE 58
Harrington/John Adrian BSA 58
Harrington/Lucy Bennett BSHE 22
Harrington/Pat Golden BSHE 56
Harris/Alison R. HEED BSHE 74
Harris/Brenda B. FND BSHE 74
Harris/Carl D. GNAG BSA 48
Harris/Christopher J. POSC 99
Harris/Cynthia D. FDSC BSA 96
Harris/Debra K. AGBS BSA 86
Harris/Donny L. ENTO MS 69
Harris/Gary Curtis ANSC BSA 84
Harris/Jennifer Anne FHNH BSHES 98
Harris/Jerry Joe ANSC BSA 75
Harris/John C. ACCT BSA 63
Harris/John W. ANSC BSA 78
Harris/Joseph E. BSA 50
Harris/Judy BSHE 65
Harris/Louise Woodrum BSHE 48
Harris/Lucille L. IDHS 99
Harris/Marion Lee BSA 75
Harris/Mary G. PLPA MS 81
Harris/O. J. BSA 50
Harris/Phoebe T. BSHE 40
Harris/Ralph L. AGBS BSA 80
Harris/Richard A. BSA 57
Harris/Robert W. POSC BSA 65
Harris/Susan Johnson HEED BSHE 68
Harris/Winfred Don AGRN MS 62
Harris/Winfred Don BSA 60
Harrison/Betty C. HOEC MS 71
Harrison/E. Pearl Strickland BSHE 41
Harrison/Hattie Bell FDNU BSHE 61
Harrison/Henry F. BSA 39
Harrison/Irene Young BSHE 37
Harrison/Kenneth ANSC MS 63
Harrison/Kenneth AHUS BSA 61
Harrison/Peggy Frances Roberts BSHE 46
Harrison/Robert M. BSA 56
Harrison/Stacy Lee POSC BSA 90
Harrison/Steve A. AGRN MS 77
Harrison/Steve A. AGRN BSA 75
Harrod/Fred W. BSA 50
Harrod/William K. BSA 59
Hart-Rommel/Sandi HDFS BSHE 94
Hart/Bradford D. BSA 52
Hart/Christopher L. ANSC BSA 83
Hart/James Wilburn BSA 39
Hart/John Clay ANSC BSA 82
Hart/Karon E. FDNU BSHE 82
Hart/Lelia Finley GHEC BSHE 86
Hart/Thomas Joseph AGBS BSA 92
Hart/Winfred R. BSA 49
Hartford/Edward VOAG BSA 79
Hartin/Joyle Eugene AGBS BSA 92
Hartley/Wesley Murry ANSC BSA 69
Hartman/Donald Ray BSA 56
Hartman/Robert Anthony AGBS BSA 87
Hartnedy/Kathleen Buettner HDFS BSHE 91
Hartrick/Frank D. GNAG BSA 74
Hartsook/Marsha J. HORT BSA 79
Hartwick/Everett E. BSA 49
Hartz/Douglas A. BSA 80
Hartz/Mark E. AGRN BSA 82
Harvey/Bruce Allen AGEC MS 89
Harvey/Matthew Morgan AGBS BSA 92
Harwood/Bettie Sewell BSHE 60
Harwood/David E. POSC MS 87
Hashemi/Masoud ANSC PHD 83
Hashemi/Masoud ANSC MS 81
Hashim/Ridzwan Bin ANSC MS 76
Haskew/Clyde L. BSA 27
Hass/Diana HEED BSHE 73
Hassell/Mary E. Mann BSHE 74
Hatcher/Irma Ellen Vest BSHE 36
Hatcher/Irma Ellen Vest HOEC MS 66
Hatcher/Jon A. POSC BSA 82
Hatcher/Jon A. ANSC MS 83

Hatcher/Kris GHEC BSHE 77
Hatcher/Ora Huff BSHE 36
Hatcher/W. Lester BSA 55
Hatfield/E. E. POSC BSA 42
Hatfield/Ethelyn HEED BSHE 54
Hatfield/Walter B. BSA 27
Hatley/Kathleen Ogden BSHE 68
Hattey/Jeffory A. AGRN PHD 93
Hattey/Jeffory A. AGRN MS 88
Haugen/Dennis A. ENTO MS 82
Hauk/Hunter Sterling AGBU BSA 98
Hausburg/Sally Ann CTM BSHE 75
Haustein/Greta Kristen AGRN BSA 96
Havens/Leslie T. BSA 65
Hawkins/Darren Wayne AGED BSA 89
Hawkins/Gayla D. HEED BSHE 73
Hawkins/Harlon H. BSA 51
Hawkins/James H. BSA 55
Hawkins/Mark GNAG BSA 62
Hawkins/S. D. ANNU BSA 61
Hawkins/William Laughlin AGBS BSA 96
Hawley/Joe BSA 60
Haws/Angela Faulkner FDNU BSHE 89
Hayati/Islah PLPA MS 91
Hayden/Carl E. AGBS BSA 76
Hayden/Joseph S. DASC MS 59
Hayden/Roy T. GNAG BSA 53
Haydon/Cynthia Leigh CLTX BSHE 90
Hayes/Beverly Hicks CLTX BSHE 82
Hayes/George R. ENTO MS 51
Hayes/Grady F. BSA 55
Hayes/Michael S. AGEC MS 82
Hayes/Michael S. BSA 79
Hayes/Pamela Deshay ANSC BSA 97
Hayes/W. E. BSA 38
Haynes/Barbara Buchanan BSHE 57
Haynes/Frank L. AGRN MS 78
Haynes/Marcelle HOEC BSHE 62
Haynes/Michael ANNU BSA 76
Haynes/Paul E. BSA 41
Haynes/Woodrow Wilson BSA 37
Hays/Ann Hutsell BSHE 60
Hays/Annette Hatfield HEED BSHE 73
Hays/Cynthia Lucille FDNU BSHE 88
Hays/David W. GNAG MS 97
Hays/David W. AGED BSA 91
Hays/Deborah Ann ANSC MS 91
Hays/Joan Coleman BSHE 62
Hays/Justin Wayne AGBS BSA 91
Hays/Justin Wayne AGED BSA 90
Hays/Margaret Kelly GHEC BSHE 84
Hays/Margaret Kelly BSHE 80
Haythe/La Shawn CLTX BSHE 90
Hayward/Marie S. BSHE 38
Hazelbaker/John A. BSA 48
Hazelbaker/William Oscar AGEC BSA 39
Hazen/Kenny R. FDNU MS 74
Hazen/Kenny R. POSC BSA 70
Head/Robert Martin AGEC MS 72
Head/Robert Martin CRSC BSA 69
Heagler/Arthur M. AGEC MS 56
Heagler/Arthur M. GNAG BSA 55
Hearnsberger/Angelia Carter CLTX BSHE 87
Hearnsberger/Dilmus H. BSA 50
Heath/Irvin J. AGRI BSA 18
Heck/Patricia L. HOEC MS 67
Heck/Patricia L. BSHE 59
Heckman/Dora Mae BSHE 32
Heckman/Nellie Price BSHE 31
Hedgecock/Carolyn C. HEED BSHE 72
Hedgecock/Ila BSHE 59
Hedges/Stanley Kirk SOSC BSA 87
Hedrick/Betsy Carolyn BSHE 36
Hedrick/John C. ANSC MS 70
Heerwagen/Leo F. BSA 22
Heerwagen/Paul K. AHUS BSA 18
Heflin/William Herschel BSA 48
Hefner/Homer A. BSA 56
Heidloff/Thomas Paul POSC MS 88
Heimsoth/Beverly Anne HDFS BSHE 89
Heins/Marilyn J. HEED BSHE 68
Heird/Charles Edwin ANSC MS 87
Heird/Charles Edwin ANSC BSA 85
Heiss/John Stabe ENTO PHD 82
Heiss/John Stabe ENTO MS 77
Held/Thomas James FDSC BSA 85
Hellman/Edward W. PTSC PHD 82
Hellwig/Dianne Helene ANSC MS 83
Hellwig/Dianne Helene POSC PHD 86
Hellwig/Howard Michael POSC PHD 86
Helm/James Edward AGRN MS 63
Helm/James Edward BSA 58
Helm/Lina Boyce ANNU BSA 64
Helmbeck/Ronald Dale ENSC BSA 96
Helmbeck/Ronald Dale AGRN MS 98
Helmich/Laura E. AGBS BSA 82
Helms/Brian Alan ANSC NODEG 96
Helms/Brian Alan EXED BSA 96
Helms/Ronald Sloan AGRN MS 81
Hembree/H. Lawson ANSC BSA 82
Hemenway/Robert C. ENTO MS 68
Hemphill/John A. BSA 25
Henbest-Pearce/Lucille BSHE 32
Henbest/Kristen Hendrix HDFS BSHE 94
Hendershott/Charles Henry HORT MS 52
Hendershott/Charles Henry GNAG BSA 51
Henderson/Beverly J. HOEC MS 82
Henderson/Brenda Jo Upton HEED BSHE 62
Henderson/Cassandra IDHS BSHE 95
Henderson/Cynthia Spencer HOEC MS 75
Henderson/Cynthia Spencer HEED BSHE 69
Henderson/Darrin G. AECT 99
Henderson/Debra Ann ANSC NODEG 96
Henderson/Debra Ann POSC BSA 96
Henderson/Elizabeth Jennings HEED BSHE 63
Henderson/Gertrude T. HOEC MS 67
Henderson/Jan R. FDSC MS 77
Henderson/Jimmie Jeaneane HOEC MS 74
Henderson/Joe B. BSA 48
Henderson/Karen Denise FDNU BSHE 79
Henderson/Mary Katherine BSHE 57
Henderson/Mildred Alfrey BSHE 43
Henderson/Monty POSC BSA 69
Henderson/Patricia J. Knowles BSHE 59
Henderson/Rebecca A. Thaxton HEED BSHE 73
Henderson/Rick A. POSC BSA 87
Henderson/William D. BSA 67
Hendren/Leslie G. CLTX BSHE 85
Hendren/Mabern F. BSA 43
Hendricks/Jim D. ANSC BSA 77
Hendrickson/Peggy Welch BSHE 44
Hendrickson/Terry M. VOAG BSA 80
Hendrickson/Thayer D. BSA 39
Hendrickson/Walter W. BSA 43
Hendrix-Carper/Jilane L. AGRN BSA 90
Hendrix/Bob POSC BSA 58
Hendrix/John Evan ENSC BSA 96
Hendrix/Lloyd BSA 50
Hendrix/Troy POSC BSA 58
Hendrix/William Hurston ENTO MS 85
Henley/Mark J. DANB BSA 76
Henley/Marlo M. CLTX BSHE 93
Hennard/Sally HEED BSHE 83
Henry/Charles Cornelius BSA 50
Henry/Hubert Douglas BSA 52
Henry/Janice K. FDNU BSHE 76
Henry/Seth E. BSA 50
Henry/William R. BSA 48
Henslee/Lee BSA 51
Hensley/Justin B. AECT 99
Hensley/Rebecca L. ESWS 98
Henson/Blake Douglas POSC BSA 85
Henson/Carmen T. HEED BSHE 72
Henson/Celeste Marie CLFM BSHE 85
Henson/Heather N. HEED BSHES 96
Henson/Lloyd BSA 51
Henson/Michael Chris ANSC PHD 85
Hepler/Alan W. LDUH BSA 81
Hepler/Rachel Reed BSHE 54
Herbaugh/Gerald D. PLPA BSA 68
Herbaugh/Gerald D. PLPA MS 71
Herbaugh/Larry BSA 63
Herlein/Shawna Lee HDFS BSHES 96
Herman/J. Clayton RSOC MS 62
Hermance/Albert Howard BSA 26
Hernandez/Nidia A. ANSC MS 90
Herren/Deborah M. HEED BSHE 71
Herrera-Estrada/Pedra ANSC BSA 78
Herrick/Margaret Mayo BSHE 28
Herrin/Lauri Lynn AGRN MS 84
Herrington/Barbara Stutheit BSHE 65
Herrington/Billy Edmond AGEC MS 79
Hershberger/Lewis E. POSC BSA 50

Herzog/Gary A. ENTO PHD 76
Herzog/Gary A. ENTO MS 72
Herzog/Marianne Jean ANSC MS 85
Herzog/Suzanne HOEC MS 95
Herzog/Suzanne CLTX BSHE 88
Hess/Christine Matelich HOME BSHES 98
Hess/John W. BSA 55
Hess/Thomas William ANSC BSA 86
Hester/A. Ray BSA 49
Hestir/William H. BSA 52
Hethcoat/J. D. BSA 49
Hethcox/Ansil Lance BSA 41
Hetz/Jeanette BSHE 61
Hewitt/Darrin Dewayne AGBS BSA 87
Hickey/Charles R. BSA 52
Hickey/Steven Wayne CLTX BSHE 89
Hickman/Charles T. BSA 52
Hickman/Robert Jason AGME BSA 95
Hicks/Carol Ramona FND BSHE 73
Hicks/Dorlene J. AGRN BSA 87
Hicks/Harold H. BSA 47
Hicks/Harold H. GNAG MSA 64
Hicks/Iva BSHE 47
Hicks/Jay V. LARC BSA 84
Hicks/Jeanne Murphy FDNU BSHE 43
Hicks/Kristi HDFS BSHE 91
Hicks/Olta Jo DeWitt BSHE 60
Hicks/Oscar H. BSA 52
Hicks/Rodney Gene AGBS BSA 89
Hicks/Roy C. BSA 56
Hicks/Wilford BSA 52
Hierholzer/Jennifer Lane ANSC BSA 93
Hierholzer/Robin Edward ANSC MS 92
Higginbotham/G. VOAG BSA 73
Higginbotham/Maurice Platt HDFS BSHE 94
Higginbothom/Mark AGEC BSA 77
Higginbothom/Stephen W. FDNU BSHE 76
Higginbottom/Jewel Ray BSA 52
Higginbottom/L. Frances Hubbard BSHE 48
Higgins/Eddie L. HOEC MS 69
Higgins/J. Dalton ANSC MS 69
Higgins/J. Dalton ANSC BSA 67
Higgins/Mazie S. GHEC BSHE 77
Higgins/Stewart David GNAG MS 95
Higgins/Stewart David PPPM BSA 92
Higgins/Suzy Hollingsworth VTSC MS 83
Higgins/Suzy Hollingsworth ANSC BSA 80
Higgins/Todd R. AGRN PHD 83
High/Jeff BSA 41
Highfill/Bobby E. POSC BSA 69
Highfill/Elizabeth A. HEED BSHE 69
Highfill/J. Virgil BSA 30
Highfill/Leroy BSA 13
Highfill/Nancy Jolene Marshall HEED BSHE 69
Highfill/Ronnie D. VOAG BSA 67
Hight/Cherylina Wages HDFS BSHE 93
Hight/Ferree Brinton BSA 27
Hight/Shana L. CLTX BSHE 79
Hight/William Bryan AGBS BSA 90
Hightower/Bryan HORT BSA 89
Hightower/Kim Sauer ANSC BSA 83
Hightower/Michael B. BSA 59
Hignight/Steven D. ANSC MS 87
Hignight/Steven D. DASC BSA 79
Hignight/Tommy D. AGEC MS 78
Hildebrand/Jeff L. AGBS BSA 85
Hildebrand/Richard Thomas AGBS BSA 86
Hileman/Leslie H. BSA 50
Hilker/Jami Suzanne AGBS BSA 93
Hill/Bradley Alan ANSC BSA 95
Hill/Bryan S. FDSC MS 82
Hill/Catherine A. CDFR BSHE 77
Hill/Charolette Lea Dodson FND BSHE 71
Hill/Dawn Michelle Logue IDHS BSHE 89
Hill/Deborah K. PPPM BSA 83
Hill/Ernest T. BSA 59
Hill/Evelyn Sisson FND BSHE 66
Hill/Gloria A. BSHE 70
Hill/Harold D. BSA 60
Hill/Hugh W. BSA 38
Hill/Jan Luanne CLTX BSHE 93
Hill/Jean Owen FDNU BSHE 61
Hill/Joellen Cunningham BSHE 51
Hill/John Beniamin BSA 41
Hill/Joyce PLPA MS 90
Hill/Katherine Kennedy BSHE 38
Hill/Laura F. HOEC MS 92
Hill/Mary Lee Tapp HOEC MS 01
Hill/Morris T. BSA 48
Hill/Robert AGRN MS 79
Hill/Robert MBIO BSA 72
Hill/Robert A. AGEC BSA 62
Hill/Robert C. POSC PHD 82
Hill/Scott Thomas AGRN BSA 94
Hill/Sharl Delon CLTX BSHE 87
Hill/Terry Randolph HORT MS 80
Hill/Tommy R. POSC BSA 54
Hill/Trudy Ann CLTX BSHE 83
Hill/Vicki Petree HDFS BSHE 86
Hill/Wayland Warren BSA 41
Hill/Wesley B. BSA 28
Hill/Willie E. BSHE 49
Hillard/Courtney M. BSA 68
Hillard/Porter BSA 53
Hillman/Hal AGBS BSA 81
Hillman/Richard Edward AGEC BSA 86
Hillman/Wilma HEED BSHE 61
Himes/Barbara F. HEED BSHE 72
Himstedt/Hickey H. BSA 56
Hinds/Hazel S. BSHE 25
Hinds/Hubert Bynum BSA 20
Hinds/Nanci BSHE 80
Hines/Theresa Lynn AGRN MS 93
Hinkley/R. G. AGED MS 50
Hinkley/R. G. AGRI BSA 31
Hinton/Michael Boyd HDID BSHE 80
Hipp/Gary L. ANSC BSA 76
Hipp/Norma R. HEED BSHE 66
Hitayezu/Felix AGEC BSA 90
Hite/Jill Copeland CLTX BSHE 83
Hitt/Don Melvin BSA 52
Hitt/Stanley G. AGRN MS 84
Hixson/Jess W. BSA 54
Hixson/Sherry K. HEED BSHE 77
Ho/Hin Mee CHEG BSHE 88
Hobbs/Mark S. AGBS BSA 82
Hobbs/Thomas D. POSC BSA 57
Hobson/Judy HEED BSHE 65
Hobson/Wallace B. BSA 51
Hodge/Birchie Clay AGED BSA 86
Hodge/Ruth BSHE 36
Hodges/Arden Kathleen CLTX BSHE 89
Hodges/Carl E. BSA 53
Hodges/F. Elaine CLTX BSHE 88
Hodges/Jane L. CLTX 98
Hodges/Jolyne C. HORT BSA 97
Hodges/Laurie AGRN MS 79
Hodgson/Larry M. AGRN MS 78
Hodgson/Larry M. AGRN PHD 84
Hoelscher/James E. BSA 53
Hoelscher/James E. AGRN MS 83
Hoelscher/Joseph M. HORT BSA 79
Hoelscher/Sheila M. ENTO BSA 79
Hoffman/Margaret J. HOEC MS 78
Hogan/Annette Renee BSHE 86
Hogan/Diane L. CLTX BSHE 81
Hogan/Toby Glen AGEC MS 88
Hoggard/Jasper O. ANNU BSA 61
Hoggard/Robbye N. BSHE 49
Hogins/Jack Newton BSA 49
Hogue/Alesha AGBS BSA 95
Hogue/Beda Sue BSHE 53
Hogue/Charles W. CRSC MS 68
Hogue/Charles W. AGRN BSA 60
Hogue/Emily C. FDNU BSHE 73
Hogue/Gary Alan ANSC BSA 95
Hogue/Jane Evelyn FND BSHE 69
Hogue/Leslie Cook HEED BSHE 70
Hogue/Patrick J. AGRN BSA 80
Hogue/Robert P. ANSC BSA 73
Hogue/William Franklin BSA 52
Hogue/Zoeye L. CATX BSHES 97
Hoke/Diana Lee HEED BSHE 79
Hokett/Laura J. DANB BSA 84
Holcomb/John DIAM BSA 89
Holcombe/Aaron Lee AGBS BSA 94
Holcombe/David Lee VOAG BSA 69
Holcombe/David Sean POSC BSA 90
Holden/Anthony Lee AGED BSA 90
Holder/Linda Strickland BSHE 74
Holder/Shelby H. AGEC MS 59
Holder/Shelby H. BSA 54
Holder/Tammy AGRN MS 92
Holderly/Richard H. BSA 22

Holifield/Edward L. BSA 60
Holifield/Lavern L. AGEC BSA 69
Holiman/James Randy AGBS BSA 93
Holl/Dollie Randleman BSHE 21
Hollabaugh/John BSA 50
Holland/Alsey BSA 40
Holland/Herman Q. DANB BSA 78
Holland/Kenneth L. AGEC BSA 79
Holland/Tom BSA 38
Hollandsworth/Ronnie BSA 65
Hollanger/Ronnie L. AGEC BSA 66
Holleman/Phyllis Judkins BSHE 60
Holley/Andrew Leon BSA 37
Holley/Easter M. FDSC MS 79
Holley/Edmon O. BSA 41
Holliday/David Ray PLPA MS 77
Holliday/David Ray PLPA BSA 74
Holliday/Laura Lindall BSHE 75
Hollingsworth/A. L. AGED BSA 40
Hollis/Charles BSA 60
Hollis/Gilbert R. AHUS BSA 61
Hollis/Katherine Virginia BSHE 40
Holloway/Christopher S. AGBS 98
Holloway/D. L. AGRN BSA 53
Holloway/Keith L. BSA 24
Holloway/Margaret Hawley BSHE 58
Holloway/Rodney L. ENTO MS 69
Holloway/Stacie Rachell AGBS BSA 85
Holloway/Wayland K. VOAG BSA 68
Holloway/Waymon J. BSA 45
Hollowell/Milton W. BSA 52
Holly/Henry Zeylon BSA 41
Holly/Henry Zeylon GNAG MS 63
Holman/Betty Poe BSHE 55
Holman/Dayton BSA 60
Holman/Eugene Merritt AGRN MS 93
Holman/Shannon Michelle RSOC MS 92
Holmes/Karen Suzanne HOEC MS 95
Holmes/Oliver Wendell BSA 41
Holt/Barbara Anne Wells HDFS BSHE 87
Holt/David L. ANSC BSA 83
Holt/Douglas M. ANSC BSA 73
Holt/Katie HEED BSHE 77
Holt/Lloyd C. ANSC BSA 76
Holt/Wayman L. AGRN BSA 83
Holtgrewe/Chris Allen POSC BSA 96
Holthoff/Boren AGRN BSA 80
Holthoff/Susan Blagg HEED BSHE 81
Holthouse/Kim HEED BSHE 75
Holtzclaw/Hanan H. BSA 9
Holtzen/Helen ANSC MS 89
Holtzen/Helen ANSC BSA 82
Holubec/Margaret Morton CLTX BSHE 83
Holzhauer/Fred AGBS BSA 86
Honarmand/Farideh FDSC BSA 91
Honea/Elmer B. BSA 37
Honea/Robert S. BSA 48
Honeycutt/Pat BSA 50
Hongsaprug/Waree ENTO MS 71
Honomichl/Kathi E. CLTX 98
Hooker/Kimberly Wallace IDHS BSHE 93
Hooker/Paul E. BSA 50
Hooks/Charles SOSC MS 66
Hooks/Charles BSA 65
Hooks/Karen Lee ANSC BSA 81
Hooper/Martha Sparkman CLTX MS 68
Hoover/Mary Baker BSHE 51
Hoover/Winford A. BSA 49
Hopfinger/Susan Ann FDNU BSHE 86
Hopkins/Ann Sharum FND BSHE 70
Hopkins/Brett A. ANSC MS 86
Hopkins/Dale Edward AGBS BSA 74
Hopkins/James Alan AGRN MS 86
Hopkins/James Alan AGRN BSA 84
Hopkins/John D. ENTO PHD 83
Hopkins/John William ENTO MS 90
Hopkins/Michael G. ANSC BSA 73
Hopper/Gloria Weatherly HEED BSHE 69
Hopper/John J. BSA 49
Hopson/David E. AHUS BSA 66
Horn/Raymond Curtis ANSC BSA 96
Horn/Roger S. FDSC MS 81
Horn/Roger S. FDSC BSA 79
Horn/Ronnie Thomas AGED BSA 91
Horn/Sherry Lee AGED BSA 89
Horn/Wesley Wright AGBS BSA 96
Horne/Brian Wayne GNAG MS 82
Horne/Gene M. AGME BSA 83
Horne/Keith Miller ENSC BSA 97
Horne/Peggy HEED BSHE 72
Hornor/Edward Tully AGRN BSA 94
Hornsby/Arthur G. SOSC MS 64
Hornsby/Arthur G. SOSC BSA 62
Hornsby/Carolyn Ruth FND BSHE 67
Hornsby/Pete ANSC BSA 77
Hornsby/Quinton ANSC MS 77
Hornsby/Quinton VOAG BSA 72
Horsfall/Annie Laurie Ellis HEED BSHE 37
Horsfall/Frank BSA 24
Horsfall/James G. BSA 25
Horsfall/Sue Overton BSHE 25
Horsfall/W. Robert BSA 28
Horst/Adelyn Deschner HOEC MS 67
Horst/Kathleen Diane ENSC BSA 77
Horton/Billy D. BSA 58
Horton/Carl E. GNAG MS 82
Horton/Dan L. ENTO PHD 82
Horton/Everett S. BSA 43
Horton/J. G. BSA 41
Horton/Joe W. AGBS BSA 74
Horton/Johnnie BSHE 45
Horton/Lucile BSHE 58
Horton/Ricky D. CRSC BSA 71
Horton/Susan Sanders HEED BSHE 68
Horton/Violet Vivian BSHE 40
Hosch/Susan HDFS BSHE 94
Hosford/Melanie FND BSHE 71
Hoskyn/Janita BSHE 58
Houghton/Rebecca Alice HEED BSHE 81
Houldcroft/Donald GNAG MS 82
House/Carol S. BSA 71
House/Frances Marie BSHE 42
House/Jim ANSC BSA 71
House/John L. BSA 48
House/William I. ANSC BSA 77
Houser/William J. BSA 50
Houston/John Samuel BSA 59
Hout/Carroll Gay HDID BSHE 80
Houtchens/Casey Kirk POSC BSA 85
Howard/Amy J. AGRN MS 79
Howard/Amy J. AGBS BSA 77
Howard/Audra Hite HEED BSHE 42
Howard/Charles G. BSA 47
Howard/David A. AGRN BSA 74
Howard/David L. VOAG BSA 70
Howard/Frances C. Mast BSHE 39
Howard/Ginger Lu AGEC BSA 79
Howard/Irene Johnson FND BSHE 64
Howard/Kathryn M. Counts CDFR MS 76
Howard/Kathryn M. Counts HEED BSHE 72
Howard/Kimberly Karol AGBS BSA 80
Howard/Lloyd Crow BSA 50
Howard/Luke R. FDSC PHD 89
Howard/Luke R. FDSC MS 83
Howard/Richard Edwin ENTO MS 75
Howard/Robert G. GNAG BSA 70
Howard/Vicki Diane BSHE 0
Howe/Dawn AGRN MS 82
Howe/Judy Anne BSHE 56
Howe/Otis AGRN MS 85
Howe/Otis AGRN BSA 79
Howell/Charles F. BSA 52
Howell/Dennis B. HORT BSA 95
Howell/Harry N. ENTO BSA 70
Howell/J. Ernest BSA 49
Howell/Jack Olen POSC BSA 54
Howell/James BSA 41
Howell/Suzann GHEC BSHE 86
Howerton/Kathy L. Suratt HDID BSHE 82
Howington/Edward AGBS BSA 73
Howington/Harvey AGRN BSA 80
Howington/Harvey H. AGRI BSA 43
Howle/David S. AGRN PHD 85
Howton/Brent AGRN BSA 77
Howze/Nan CLTX BSHE 88
Hoy/Jeff PLPA MS 78
Hoye/Jo Anne ANSC MS 84
Hoyer/Cynthia A. ANSC BSA 76
Hoyt/Deborah Riggs FDSC MS 98
Hoyt/Deborah Riggs ANSC BSA 87
Hoyt/Tony B. BSA 65
Hsieh/C. F. SOSC MS 71
Huang/Jinping CSEG MSE 98
Huang/Jinping BAEG MSBAE 96

Hubbard/Helen Jones BSHE 43
Hubbard/Janice Curry HOEC MS 91
Hubbard/Janice Curry HDFS BSHE 89
Hubbard/John D. BSA 43
Hubbard/Thurman Doyle BSA 51
Hubbell/Carolyn HEED BSHE 82
Hubbell/Donald S. ANSC BSA 82
Hubbell/Donald S. BSA 58
Hubbs/Craig William AGRN MS 88
Hubbs/Craig William AGRN BSA 85
Hubbs/Randy L. ANSC MS 78
Hubbs/Randy L. ANNU BSA 76
Huck/Alvin L. AGME BSA 83
Huck/Randall William AGME BSA 86
Huck/Richard J. AGBS BSA 74
Huckleberry/Lee Ann FDSC MS 84
Huckleberry/Lee Ann FDSC BSA 83
Huddle/Gina K. CLTX BSHE 79
Huddleston/Byron BSA 49
Huddleston/Harold D. BSA 40
Huddleston/Paul M. ENTO MS 60
Huddleston/Paul M. BSA 59
Hudgens/Robert E. ANSC MS 79
Hudgins/Zem L. GHEC BSHE 83
Hudlow/Jason Earl DIAM BSA 87
Hudlow/Mike ANSC BSA 83
Hudson/Betty R. BSHE 49
Hudson/David N. ANSC BSA 77
Hudson/Edna Sparkman BSHE 56
Hudson/Jennifer Lynne CLTX BSHE 86
Hudson/John M. FDSC PHD 85
Hudson/John M. FDSC MS 81
Hudson/John M. FDSC BSA 78
Hudson/John W. BSA 57
Hudson/Joseph M. ANSC MS 69
Hudson/Joseph M. ANSC BSA 67
Hudson/Judd M. AGEC MS 56
Hudson/Judd M. AGED BSA 39
Hudson/Kenneth Lareece ANSC BSA 88
Hudson/Leola AGBS BSA 86
Hudson/Loyde H. BSA 49
Hudson/Martha F. BSA 33
Hudson/Richard BSA 54
Hudson/Tracy M. HEED BSHE 75
Hudson/Wes ANSC BSA 60
Hudspeth/Bobby CRSC BSA 61
Huenefeld/Donald Flournoy BSA 50
Huenefeld/Erna E. BSHE 22
Huenefeld/Paul R. GNAG MS 64
Huenefeld/Paul R. BSA 54
Huey/Bobby A. SOSC MS 66
Huey/Bobby A. BSA 56
Huey/Eddie Gene POSC BSA 89
Huey/Susan Linebarger BSHE 63
Huff/Charles David ANSC BSA 70
Huff/Debra Sterling HDID BSHE 80
Huff/Wilbur C. BSA 50
Huffaker/Teri A. AGEC MS 86
Huffaker/Teri A. AGBS BSA 82
Huffman/Donna C. HEED BSHE 72
Huffman/Fred R. ENTO PHD 82
Huggins/Geneva Anderson BSHE 26
Hughes/C. Rossell BSA 38
Hughes/Carol R. CLTX BSHE 78
Hughes/Cecilia A. FDNU BSHE 79
Hughes/Charles BSA 52
Hughes/Christal Maylee AGBS BSA 91
Hughes/Cindy FDNU BSHE 82
Hughes/Helen Elvira BSHE 39
Hughes/John Leonard BSA 59
Hughes/Mark F. AGEC BSA 79
Hughes/Mary Jeanette ANSC BSA 86
Hughes/Nancy English HEED BSHE 83
Hughes/Patricia E. CDFR BSHE 74
Hughes/Rosemary HEED BSHE 74
Hughey/Buddy Earl BSA 56
Huie/Gretchen Lynn ENSC BSA 97
Huie/Rita FDNU BSHE 81
Hulen/Elizabeth J. HEED BSHE 79
Hulett/Billy J. BSA 55
Hull/Stephen Jay POSC MS 67
Hull/Stephen Jay BS 66
Hull/Wayne E. ANSC BSA 92
Hulsey/Andrew Howard GNAG BSA 50
Hultz/Jessie Maxine BSHE 53
Humbard/Jacqueline Phillips BSHE 56
Humbard/Jay D. GNAG MS 68
Humbard/Jay D. BSA 56
Hummel/Betty HEED BSHE 61
Hummel/Roy C. POSC MS 66
Hummel/Roy C. GNAG BSA 60
Humphery/David Ford BSA 58
Humphrey/Ellen Henderson BSHE 32
Humphrey/John D. BSA 52
Humphrey/Vernita T. HOEC MS 84
Humphreys/Davidson H. AGRN BSA 72
Humphreys/Ellie Marie CATX BSHES 97
Humphreys/Mary Stevens BSHE 50
Humphreys/Ruth Ellen HEED BSHE 81
Humphries/Acy AGBS BSA 94
Humphries/Mary Lookingbill BSHE 55
Humphries/Milton L. BSA 56
Humphry/Jimmy Byron ENSC BSA 97
Huneycutt/Herbert CRSC MS 69
Huneycutt/Herbert GNAG BSA 48
Huneycutt/Roy G. BSA 51
Hungate/Jayne Scoggins IDHS BSHE 91
Hunnicutt/Deltha BSHE 59
Hunnicutt/Joe H. GNAG BSA 47
Hunnicutt/O. E. BSA 42
Hunt/Anne S. HEED BSHE 71
Hunt/Ben ANSC BSA 71
Hunt/Herman H. BSA 28
Hunt/Lisa Marie GHEC BSHE 86
Hunt/Loye Dawn HEED BSHE 62
Hunter/Alice Faye FND BSHE 71
Hunter/Alice Houston BSHE 45
Hunter/Carl G. BSA 45
Hunter/J. W. BSA 53
Hunter/James S. ENTO MS 78
Hunter/James S. ENTO BSA 77
Hunter/Lou Nicholson HDID BSHE 80
Hunter/Mary Scott BSHE 44
Hunter/Raymond E. BSA 43
Hunter/Richey L. AGRN MS 75
Hunter/Richey L. BSA 65
Hunter/Robert C. BSA 53
Hunter/Robert C. ENTO MS 57
Hunter/Robert W. BSA 53
Hunter/Sammie Rae AGEG BSA 61
Hunter/Sheri Lavelle CLTX BSHE 91
Hunton/Earl J. ANSC BSA 88
Hunton/Jerry AGBS BSA 73
Hurford/Robert Bryon AGED BSA 88
Hurley/Melodie Magnuson IDHS BSHE 94
Hurley/R. Mark HORT MS 83
Hurley/William K. HORT MS 82
Hurst/Ann Milam BSHE 59
Hurst/Harold BSA 57
Hurst/Harold AGRN MS 62
Hurst/Ralph E. POSC BSA 69
Hurst/Scott EXED BSA 97
Hussey/Ross AGME BSA 73
Huston/Mary May BSHE 17
Hutchens/Dilford BSA 42
Hutcheson/Clinton Ross ANSC BSA 96
Hutcheson/Jerri BSHE 87
Hutcheson/Ruth BSHE 37
Hutchinson/Cynthia IDHS BSHE 91
Hutchison/Nancy Elizabeth Boss CTM BSHE 74
Hutchison/Russell D. AGED BSA 92
Hutson/Cecil Dale BSA 47
Hutson/Denver Burke BSA 48
Hutson/Harold Gene AGRN BSA 80
Hutson/Janet Sue BSHE 70
Hutto/Casey Ray AGBS BSA 97
Hyland/Rosalea FDNU BSHE 79
Ibrahim/Babiker Abdalla AGRN PHD 92
Ibrahim/Hamdi Abbas AGEC MS 84
Ikeme/Ikechukwu Alex FDSC MS 80
Im/Byung-Sang AGME BSA 81
Imboden/Richard K. AGME BSA 71
Ingram/Dennis R. POSC BSA 76
Ingrum/Ann Sims BSHE 37
Ingrum/Dan B. BSA 37
Inman/Andrea Ann CLTX BSHE 93
Inman/Fred PLPA BSA 75
Inuwa/Danladi Usman GNAG MS 82
Inuwa/Danladi Usman BSA 81
Inzer/Paul E. BSA 38
Iovino/Anthony Joseph ENTO MS 67
Iqbal/Javed GNAG MS 94
Iraniparast/Hesam POSC BSA 83

Irby/Patricia BSHE 20
Irish/Brian Malcolm HORT BSA 97
Ironside/Joe E. BSA 55
Irwin/Amy Marie HOEC MS 91
Irwin/Bay FDNU BSHE 75
Irwin/Joan Glasgow BSHE 51
Isaacks/John William AGBS BSA 96
Isaacks/Samuel Jackson ANSC BSA 93
Isaacs/John H. ANSC BSA 76
Isgrig/Ben C. BSA 48
Isfalt/Anna S. FHNH 99
Ishmael/Murla Jeanne HOEC MS 71
Islam/S. M. Nazabat CRSC MS 70
Ismail/Mohammad ANSC MS 91
Isom/Barbara Jones BSHE 46
Ison/Mary Miller BSHE 43
Ito/Shoichi AGEC MS 85
Ivester/Carolyn R. HOEC BSHE 63
Ivie/Carol Ann Jones HEED BSHE 75
Ivy/Billy Ray GNAG BSA 75
Ivy/Clifford A. BSA 65
Ivy/E. Victor BSA 43
Ivy/Jacy BSA 60
Ivy/Nancy Wells GHEC BSHE 86
Ivy/Richard POSC MS 69
Ivy/Richard ANSC BSA 67
Izekor/Ehiorobo AGRN MS 89
Jacimore/Laura L. FDNU BSHE 83
Jacks/James F. BSA 51
Jackson/Betsy E. CLTX BSHE 94
Jackson/Beverly Harris Burch BSHE 71
Jackson/Buford O. BSA 45
Jackson/Carolyn HEED BSHE 57
Jackson/Charles E. ENTO MS 61
Jackson/Charles E. VOAG BSA 57
Jackson/Charles E. ANSC MS 77
Jackson/Conley D. BSA 50
Jackson/Conley D. SOSC MS 67
Jackson/Dirk Alan ENSC BSA 95
Jackson/Donna M. HEED BSHE 89
Jackson/Dorothea Rommel BSHE 39
Jackson/Eloise BSHE 46
Jackson/Evelyn Benson BSHE 63
Jackson/Evelyn Benson HOEC MS 66
Jackson/Gova Irene BSHE 65
Jackson/Hilliard AGEC BSA 39
Jackson/Howard Roy AGEC MS 37
Jackson/Janan GHEC BSHE 82
Jackson/Jerry A. BSA 59
Jackson/Kenneth Eldon PLPA MS 72
Jackson/Margaret J. Laprade FND BSHE 71
Jackson/Mark POSC PHD 87
Jackson/Mary Ann BSHE 43
Jackson/Mary Beth Bryson FND BSHE 67
Jackson/Mary Ruth BSHE 33
Jackson/Mildred Elizabeth CLTX BSHE 77
Jackson/Neil E. HUMN BSA 51
Jackson/Paul W. ENTO BSA 65
Jackson/Robert Oliver BSA 53
Jackson/Stephen J. AGRN BSA 76
Jackson/Steven Dean AGBS BSA 80
Jackson/Vicki Michelle FDSC MS 93
Jackson/Wendy Dillahunty ANSC BSA 95
Jackson/William Gregory AGED BSA 86
Jacobs/Bruce AGRN BSA 81
Jacobs/Chad Eric POSC BSA 96
Jacobs/Janice Smith BSHE 58
Jacobs/Jeffrey Neil AGBS BSA 85
Jacobs/Thomas R. AGRN BSA 77
Jacobs/William H. BSA 40
Jacobson/Dan Alan ENTO MS 92
Jacques/Sorel AGRN PHD 91
Jaen/Eudoro BSA 57
Jaggers/Karen Bock FND BSHE 71
Jamell/Lucy M. HOEC BSHE 26
James/Arthur R. CRSC MS 75
James/Arthur R. PLPA BSA 73
James/Jefferson BSA 39
James/Jennifer Hare AGBS BSA 94
James/John H. AGEC BSA 82
James/John H. AGRI BSA 59
James/Margaret Rusher FDNU BSHES 96
James/Raydus BSA 42
James/Tracey IDHS BSHES 98
Jameson/Miles W. BSA 59
Jamison/Jewell Kermit BSA 51
Jamison/Patricia Parette BSHE 60
Jansen/Matthew James DIAM BSA 89
Jantan/Mohd Dan AGEC MS 86
Jarman/Lisa Smith CLTX BSHE 87
Jarrett/Devoe Robert BSA 52
Jarrett/Mary Millar BSHE 72
Jay/Austin Elwood ANSC PHD 70
Jeffery/Will Rogers BSA 60
Jeffrey/Betty Joe Crews BSA 69
Jeffries/Patricia Simmons CLTX BSHE 79
Jeffus/Hazel BSHE 42
Jenkins/Cecily Anne IDHS BSHE 81
Jenkins/Christy Gay FND BSHE 67
Jenkins/David C. POUL BSA 97
Jenkins/David M. ANNU BSA 78
Jenkins/John Powell AGEC BSA 71
Jenkins/Johnie N. AGRN BSA 56
Jenkins/Joyce Marie BSHE 58
Jenkins/Robert Eugene BSA 59
Jenkins/Thomas Gordon ANSC MS 74
Jenkins/Thomas Gordon ANSC BSA 73
Jenkins/Vanessa T. CLTX BSHE 81
Jennings/Cathy Sue HOEC MS 80
Jennings/Cecil R. VOAG BSA 69
Jennings/Cheryl J. HDID BSHE 78
Jennings/Deborah CTM BSHE 74
Jennings/Finis B. BSA 60
Jennings/Greer Marshall HOEC MS 79
Jennings/Jeffery Neil ANSC BSA 95
Jennings/John Alan AGRN MS 85
Jennings/Johnny BSA 53
Jennings/Julie Anna AGBS BSA 88
Jennings/Lonnie D. ANSC BSA 71
Jennings/Margaret B. BSHE 24
Jennings/Martha HORT BSA 79
Jennings/Mary Jo HOEC MS 76
Jennings/Opal S. BSHE 30
Jennings/Ronald Gray DASC BSA 84
Jennings/Samantha M. HDFS 98
Jennings/Troy S. BSA 36
Jennongs/M. Jo Lynn Tate BSHE 71
Jensen/Karen S. ANSC BSA 89
Jernigan/Grover T. BSA 34
Jeske/Trudy Hegwood BSHE 72
Jester/Heidi Hesser IDHS BSHE 93
Jester/Kevin Wayne AGBS BSA 93
Jeter/John W. BSA 41
Jeter/Max Albert BSA 38
Jewell/Irma M. BSHE 43
Jewell/Joe POSC BSA 80
Jewell/Judith Kay CDFR BSHE 69
Jimenez/Antonio A. BSA 56
Jindal/Vineet POSC PHD 94
Johansen/Harold W. BSA 51
Johanson/Debbie HDID BSHE 78
Johnon/Ruben H. GNAG MS 61
Johnon/Ruben H. GNAG BSA 55
Johnson/Alicia A. CLTX BSHE 89
Johnson/Alton Bleh AGRN PHD 93
Johnson/Arthur BSA 33
Johnson/Barbara Marie CLTX BSHE 86
Johnson/Barbara W. BSHE 73
Johnson/Beverly Sanders HEED BSHE 71
Johnson/Boyce ANSC BSA 82
Johnson/Bradley Shane LDUH BSA 94
Johnson/Byron Everette BSA 17
Johnson/Chana W. AGBS BSA 85
Johnson/Cindy M. HEED BSHE 82
Johnson/Clint Elvert AGBU BSA 98
Johnson/David H. AGRN PHD 93
Johnson/David H. AGRN MS 88
Johnson/Deena LaDon CLTX BSHE 88
Johnson/Diana Gail CLTX BSHE 86
Johnson/Diane McKinney FDNU MS 76
Johnson/Diane McKinney FDNU BSHE 69
Johnson/Donald L. BSA 59
Johnson/Donald R. BSA 52
Johnson/Donald R. ENTO MS 71
Johnson/Donald R. ENTO BSA 69
Johnson/Doris Strauss BSHE 52
Johnson/Dwayne R. AGRN BSA 91
Johnson/E. Arnold AHUS BSA 57
Johnson/Elizabeth Jones HOEC MS 81
Johnson/Esther Lane Crutcher BSHE 38
Johnson/Frances Ann CLTX BSHE 76
Johnson/Francille Killion BSHE 44
Johnson/Helen M. GHEC BSHE 79
Johnson/Ida Marie AGBS BSA 96

Johnson/Jason M. ESWS 98
Johnson/Jeannie Ione CLTX BSHE 85
Johnson/Jerry BSA 65
Johnson/Jimmy AGEC MSA 82
Johnson/Jimmy AGEC BSA 80
Johnson/Jimmy R. BSA 63
Johnson/Joan HEED BSHE 91
Johnson/Joan R. MS 55
Johnson/Joe P. BSA 51
Johnson/Joe Paul BSA 51
Johnson/John C. BSA 34
Johnson/John C. BSA 48
Johnson/John Marshall FDNU BSHE 84
Johnson/Johnny Louis BSA 65
Johnson/Joseph Thomas AGRN BSA 94
Johnson/Juanise S. HDID MS 61
Johnson/Judy M. FND BSHE 69
Johnson/Kathy FDNU BSHE 91
Johnson/Kristina Kay HORT BSA 87
Johnson/Larry Dale AGED BSA 88
Johnson/Larsh GNAG BSA 64
Johnson/Lenard R. BSA 54
Johnson/Loran L. BSA 50
Johnson/Madge Bryant BSHE 49
Johnson/Madge E. HOEC BSHE 19
Johnson/Marian Renee CLTX BSHE 91
Johnson/Martha Sue Walker BSHE 53
Johnson/Marvin Dickson BSA 22
Johnson/Mary M. INSC BSHE 29
Johnson/Melinda B. CLTX BSHE 89
Johnson/Melissa Ann IFDS BSHES 97
Johnson/Monte P. ENTO PHD 89
Johnson/Olaf Dean BSA 51
Johnson/Phyllis Whitaker BSHE 43
Johnson/R. Sears BSA 41
Johnson/Rebecca Payne HDID BSHE 81
Johnson/Richard Alan AGEC BSA 81
Johnson/Risa Loreen North IDHS BSHE 90
Johnson/Ruth Craig BSHE 27
Johnson/Samuel P. AGBS BSA 72
Johnson/Sharon E. HOEC MS 92
Johnson/Sharon E. HDFS BSHE 88
Johnson/Steven Lloyd AGED BSA 93
Johnson/Theresa AGBS BSA 83
Johnson/Turner C. BSA 41
Johnson/William Finnis AGRN PHD 95
Johnson/William Gary AGRN PHD 94
Johnson/William Gary AGRN MS 89
Johnson/William Jack ENTO BSA 70
Johnson/William Julian AGRN BSA 81
Johnston/Alice Beasley BSHE 65
Johnston/Anna Jo AGEC BSA 82
Johnston/Carl Raymond BSA 52
Johnston/Carolyn D. CLTX MS 75
Johnston/Charles BSA 60
Johnston/Dick D. BSA 57
Johnston/Frank R. AGED BSA 41
Johnston/Fremont Ellis BSA 51
Johnston/James H. BLL 8
Johnston/James H. BSA 6
Johnston/Kurt D. CLFM BSHE 83
Johnston/Linda D. DIET BSHE 53
Johnston/Maybeth Deese HOEC BSHE 58
Johnston/Mescal Dunn HOEC BSHE 42
Johnston/Mildred G. BSHE 49
Johnston/Morwyn Mae BSHE 59
Johnston/Ray Denison BSA 21
Johnston/Robert Bruce ANSC BSA 72
Johnston/Rogers BSA 57
Johnston/Ruth E. FDNU BSHE 75
Johnston/Stanley Robert POSC BSA 79
Johnston/Tony Vaughn FDSC PHD 95
Johnston/Tony Vaughn FDSC MS 86
Jolliff/D. Jeane BSHE 41
Jones/Alicia Read BSHE 37
Jones/Amy Lynn AGBS BSA 96
Jones/Ann Burcham BSHE 57
Jones/Avis Lane HOFS BSHE 82
Jones/Barbara Ann BSHE 60
Jones/Bernice Heathman BSHE 55
Jones/Bettye Pierce BSHE 42
Jones/Beverly Ann FDNU BSHE 90
Jones/Bill F. ENTO MS 65
Jones/Billy D. ANNU BSA 70
Jones/Bobby R. POSC BSA 56
Jones/Bradley R. AGRN MS 82
Jones/Buffey Michelle IFDS BSHES 98
Jones/Carl B. BSA 57
Jones/Christine Michelle ANSC BSA 96
Jones/Claude R. AGRI BSA 55
Jones/Cora Abington BSHE 59
Jones/Daire O'Neal HEED BSHE 79
Jones/Daniel Webster BSA 17
Jones/Darwin BSA 42
Jones/Deanna Kay FDNU BSHE 83
Jones/Denver B. VOAG BSA 50
Jones/Diane FDNU BSHE 68
Jones/Effie BSHE 42
Jones/Elizabeth GHEC BSHE 80
Jones/Emma Adelaide Stephens BSHE 58
Jones/Flernoy Glenn ENTO MS 61
Jones/Flernoy Glenn BSA 51
Jones/Frances Wall BSHE 44
Jones/Francis GNAG BSA 66
Jones/Gary Martin BSA 59
Jones/Gayle Frances ANSC BSA 88
Jones/Gene Chambers BSHE 64
Jones/Gregory Simpson AGBS BSA 85
Jones/Ira Burton BSA 21
Jones/Iva Lavina BSHE 45
Jones/J. G. AGRN BSA 53
Jones/James M. BSA 57
Jones/James R. AGEC PHD 76
Jones/James W. ENTO BSA 78
Jones/James W. ENTO PHD 92
Jones/James W. ENTO MS 81
Jones/Jane W. FND BSHE 67
Jones/Jeannie L. HEED BSHE 79
Jones/Jeff ANNU BSA 79
Jones/Jerry Lee ANSC BSA 76
Jones/Jerry Marvin AGEC BSA 71
Jones/Jewell O. BSA 56
Jones/Jim ANSC BSA 74
Jones/John Calvin GNAG MS 87
Jones/John Calvin ANSC BSA 84
Jones/John Michael ENTO MS 92
Jones/Juliet Shawn ANSC MS 95
Jones/Keith C. VOAG BSA 79
Jones/Kevin D. ANSC BSA 96
Jones/Kimberly Ann AEED MS 98
Jones/Kimberly Ann AGED BSA 89
Jones/Kimberly Anne DANB BSA 86
Jones/Laura Merck IDHS BSHE 92
Jones/Lewis L. BSA 50
Jones/Lorea Hoback BSHE 33
Jones/Lorin BSA 57
Jones/Louis B. BSA 42
Jones/Lynette Collier GNAG MS 83
Jones/Marilyn Sue HEED BSHE 69
Jones/Marion Baxter BSA 54
Jones/Mark Douglas DANB BSA 77
Jones/Mary E. DANB BSA 82
Jones/Merrie A. Bellmann CTM BSHE 75
Jones/Nancy S. FDNU BSHE 77
Jones/Neal Richard POSC BSA 81
Jones/Penny Hicks BSHE 60
Jones/Rick POSC BSA 88
Jones/Robert Harold ANSC BSA 89
Jones/Russell S. AGRN PHD 85
Jones/Ruth Tannehill FDNU BSHE 76
Jones/Sarah Barham FDNU BSHE 61
Jones/Sharon Kay LDUH BSA 80
Jones/Stephen P. LDUH BSA 79
Jones/Terri Lee CLTX BSHE 81
Jones/Tonja Diane HDFS MS 96
Jones/Tonja Diane HDFS BSHE 94
Jones/Will H. POSC BSA 98
Jordan/Aldy Clarence BSA 53
Jordan/David Lindley AGRN PHD 92
Jordan/Deanna Lynn HDFS BSHE 88
Jordan/Denney BSHE 60
Jordan/Denney HOEC MS 64
Jordan/Hazel Craig AGEC MS 60
Jordan/Hazel Craig BSHE 34
Jordan/Ruby J. FDNU BSHE 75
Joseck/Sydney Hawley BSHE 79
Joslin/Charlie AGJN BSA 82
Jourdan/Scott Wayne AGRN PHD 88
Joyce/Kelley Ann Smith FDSC BSA 87
Joyner/Gaye K. BSHE 69
Joyner/Stanley W. AGRN BSA 73
Juby/Marc Wayne AGED BSA 93
Judd/Jeremiah James POSC BSA 97
Judkins/Billy R. BSA 56

Junek/Jonelle Jeanette FDSC MS 80
Jung/Pil Kynn AGRN MS 80
Jurecka/Marcus AGBS BSA 97
Justus/Jack G. BSA 54
Justus/Joe A. AGME BSA 82
Justus/Norman E. BSA 54
Jyoti/Jawahar Lal ENTO MS 94
Kaboneka/Salvator AGRN MS 93
Kadmin/Bajuri Bin AGRN PHD 83
Kaffka/V. Carol Barnum BSHE 61
Kahsner/Edwin Eugene BSA 42
Kalbach/John Huber HDFS BSHE 86
Kalke/Kenneth LDUH BSA 94
Kalke/Kimberly E. IDHS BSHE 94
Kamputa/Dan David AGRN MS 90
Kanady/Janet Darlene AGED BSA 88
Kanady/Marlon Reese AGBS BSA 88
Kang/Young-Kil AGRN PHD 81
Kanyeka/Zakaria L. AGRN MS 83
Karazawa/Missae AGRN MS 77
Karber/John M. BSA 48
Kardinan/Agus GNAG MS 92
Karlsson/Maria Ulla ANSC BSA 91
Karnes/Rachel N. ANSC BSA 96
Karnes/Stirman W. BSA 54
Karokaro/Setel AGEC MS 89
Kasireddy/Narisimha R. FDSC PHD 83
Kasireddy/Narisimha R. SOSC MS 75
Katkhordeh/Farzaneh FDNU BSHE 82
Katovich/Elizabeth Jean AGRN MS 84
Kattan/Mike FDSC BSA 87
Kattlemann/Kendra Kay HOEC MS 85
Katzen/Sidney Charles BSA 54
Kauble/Reese Everett ANSC BSA 86
Kauffman/John Luther BSA 58
Kaufman/Gretchen Lee CLTX BSHE 81
Kaufman/Judy M. ANSC MS 80
Kaufman/Judy M. DANB BSA 78
Kaufman/Laura Lee FDNU BSHE 77
Kaufman/Ronald E. AGEC BSA 69
Kay/Dorothy Patricia BSHE 55
Kayihura/Francois AGRN BSA 90
Kaylor/Kimberly Ann CLTX BSHE 95
Ke/Zhen-Shou AGRN MS 96
Keacher/Kandi Lynn DANB BSA 77
Keahey/Travis C. AGRN BSA 38
Keahey/Winifred Welborn FDNU BSA 70
Kealy/Richard Dean POSC PHD 67
Kearney Cannon/Lydia CDFR BSHE 75
Kearns/Betty Rogers BSHE 53
Keathley/Gary N. AGEC BSA 67
Keathley/Joe L. GNAG BSA 69
Keathley/Robert C. BSA 40
Keathley/Robert F. AHUS BSA 64
Keaton/Jennifer Rhodes ANSC BSA 96
Keaton/Larry W. ANSC MS 74
Keaton/Larry W. ANSC BSA 70
Keaton/Mark Douglas ANSC MS 86
Keaton/Mark Douglas ANSC BSA 83
Keaton/Vance William AGED BSA 95
Keck/John V. BSA 38
Keck/Lloyd D. POSC BSA 79
Keck/Sheliah Marie AGBS BSA 83
Keegan/Pauline HORT BSA 86
Keel/Peyton AGBS BSA 85
Keeling/Arlen D. BSA 51
Keeling/Howard S. BSA 39
Keeling/Joe E. BSA 51
Keeling/Leonard J. VOAG MS 47
Keeling/Leonard J. AHUS BSA 46
Keeling/Norma Burns BSHE 33
Keeling/Roy Colbert BSA 32
Keen/Gregory W. AGBS BSA 85
Keen/James M. ANSC BSA 88
Keen/John W. ANSC BSA 83
Keene/James H. GNAG BSA 57
Keene/R. Bruce ANNU BSA 61
Keenen/Beverly D. HOEC MS 86
Keenen/Beverly D. HEED BSHE 84
Keener/Elbert L. BSA 42
Keener/J. W. BSA 60
Keeter/Jerry Lee BSA 63
Keeter/Joel B. ANSC BSA 77
Kehn/Russell H. AGRI BSA 51
Keisling/Terry C. SOSC BSA 67
Keith/Allen A. BSA 14
Keith/Ernest A. ANNU MS 75
Keith/Ernest A. ANNU BSA 74
Keith/Herbert Ross SOSC MS 70
Keith/Herbert Ross SOSC BSA 67
Keith/Richard Terrell AGEC MS 79
Keith/Sandra Little FND BSHE 66
Kellam/Annabel L. ANSC PHD 83
Kellam/Catherine Ragsdale BSHE 29
Kellam/Valerie Ann Ford FDNU BSHE 85
Kellar/Johnny D. AGME BSA 75
Keller/Kenneth R. AGEC BSA 72
Keller/Marion G. AGRI BSA 56
Keller/Preston R. ESWS 99
Keller/Sara B. GHEC BSHE 77
Keller/Susan Jackson CLFM BSHE 81
Kelley/Donald E. AGRN MS 77
Kelley/Donald E. AGRN BSA 75
Kelley/Edward C. BSA 53
Kelley/Eric Neal AGRN BSA 96
Kelley/Floyd Littleton BSA 49
Kelley/James W. AGSC MS 77
Kelley/Jana HEED BSHE 61
Kelley/Max P. BSA 59
Kelley/Stephanie Dailey HDID BSHE 79
Kelley/Terry L. AGRN BSA 82
Kelley/Tracy R. AGBS BSA 82
Kelly/Jenelle Derickson CLTX BSHE 84
Kelly/John Thomas AGRN PHD 91
Kelso/Howard C. BSA 50
Kelso/Mildred Spurlock BSHE 51
Keltner/Pat BSHE 59
Kemp/Margaret Helen BSHE 44
Kemp/Robert Victor ANNU BSA 66
Kemp/Susan CLTX MS 75
Kemp/Susan HEED BSHE 73
Kendall/Jessica Ann CLTX BSHE 94
Kendall/Lela Florence Bates BSHE 35
Kendig/John Andrew AGRN PHD 92
Kendig/John Andrew AGRN MS 88
Kendig/Susan Marie PTSC PHD 96
Kendrick/Brenda Ann CLTX BSHE 76
Kendrick/Carol A. FND BSHE 71
Kendrick/Clyde H. BSA 52
Kendrick/Danny POSC BSA 70
Kendrick/Richard POSC BSA 46
Kendrick/Robert L. BSA 57
Kendrick/Robert L. ENTO MS 59
Kendrick/Tracie Nicks BSHE 44
Kennedy/Alvor Jefferson BSA 57
Kennedy/Brandalynn Diane HOEC BS 93
Kennedy/Elizabeth PPPM BSA 89
Kennedy/Florene BSHE 33
Kennedy/Harvey W. BSA 41
Kennedy/Jack M. CRSC PHD 76
Kennedy/Joseph David ANSC BSA 95
Kennedy/Michael R. CRSC MS 75
Kennedy/Robin G. ANSC BSA 79
Kennedy/Rodney Dean AGME BSA 88
Kennedy/Susan Bogle CLTX BSHE 74
Kennedy/Tammy Mizell DANB BSA 85
Kenner/Richard W. VOAG BSA 77
Kennett/Daniel P. BSA 69
Kent/Patsy Sue Nix FDNU BSHE 55
Keogh/John AGRN BSA 76
Kernodle/Jere Glenn AGEC BSA 93
Kerr/John E. BSA 42
Kerr/K. Eugene BSA 53
Kerr/Robert L. POSC BSA 69
Kerr/Sam W. VOAG BSA 67
Kersey/Edwin C. BSA 53
Kersey/Jarrod H. POSC MS 97
Kersey/Jarrod H. POSC BSA 96
Kersh/Garland M. BSA 41
Kersh/Kenneth E. ENSC BSA 81
Kersting/Leaha Roark HDFS BSHE 94
Kester/Cynthia Kay FDSC BSA 94
Kettle-Gay/Rebecca D. HORT BSA 96
Key/Sam BSA 50
Keyser/E. Glenn POSC MS 68
Khan/Mumtz GNAG MS 84
Khilani/Salila FDNU BSHES 96
Khodabandeh/Mohsen AGRN PHD 84
Khodayari/Khosro AGRN PHD 84
Khodayari/Khosro AGRN MS 81
Khusakul/Vichai ENTO MS 70
Kiazolu/James L. AGRN PHD 83
Kice/Ethel Patricia BSHE 43
Kidd/Brenda POSC BSA 98

Kidd/Howard BSA 39
Kidd/Krista Kay GHEC BSHE 93
Kidd/Michael T. POSC BSA 90
Kidd/Michael T. ANSC MS 91
Kidd/Rebecca A. AGBS 98
Kiene/James J. DANB BSA 90
Kiene/Michael J. AGME BSA 82
Kilbourn/Jeffrey Scott ANSC BSA 85
Kilbourn/Keith AGED BSA 89
Kilby/M. Jean FDNU BSHE 77
Kilcrease/John BSA 53
Kilgore-Rorrer/Sylvia FDSC MS 79
Kilgore-Rorrer/Sylvia FDNU BSHE 77
Killian/Dale BSA 48
Killion/Benjamin F. BSA 59
Killough/Kyle HORT BSA 82
Killough/Peggy BSHE 59
Kim/Dong-Geun PTSC PHD 89
Kim/Kyung Soo PLPA PHD 71
Kim/Young Ho PTSC PHD 85
Kimball/Marie B. BSHE 48
Kimbrell/Cynthia HORT BSA 86
Kimbrell/Howdy AGRN BSA 86
Kimbrough/Clinton Eugene POSC BSA 74
Kimbrough/J. J. ENTO MS 70
Kimbrough/J. J. POSC BSA 60
Kimbrough/James R. BSA 55
Kimbrough/Lesley Clinto POSC BSA 85
Kimbrough/Willis Edward BSA 49
Kinard/Shandra Denise IDHS BSHES 97
Kincade/Linda D. CLTX BSHE 77
Kincade/Paula Myrl CLTX BSHE 89
Kincaid/Grover C. AGRN BSA 30
Kincannon/Jill Louise CLTX BSHE 91
Kinder/Lisa POSC MS 98
Kinder/Lisa ANSC BSA 92
Kindrick/Joe C. BSA 60
King/Angela Darby AGBS BSA 93
King/Barbara Bennett HOEC MS 79
King/Barbara Bennett HEED BSHE 66
King/Charles A. AGRN PHD 93
King/Charles A. AGRN MS 88
King/Charles R. BSA 47
King/Christy Lynn IDHS BSHE 91
King/Cory J. AGBS BSA 83
King/Darrell Brooks BSA 54
King/Gerald Ray ANSC BSA 90
King/James R. AGRN BSA 82
King/John Clancy AGBS BSA 83
King/John Henry HORT MS 86
King/Joyce Spinar FDNU BSHE 62
King/Laurie A. AGBS 98
King/Linda L. Moore CDFR MS 76
King/Lita N. HDID BSHE 76
King/Mark A. AGBS BSA 82
King/Pansy Nix BSHE 55
King/Rebecca G. HEED BSHE 78
King/Sam G. BSA 49
King/Stephen Ray AGRN BSA 88
King/Stephen Ray AGRN MS 91
King/Thomas J. BSA 49
King/W. J. BSA 51
King/William Henry ENTO MS 97
Kingrey/Greg Orville VOED BSA 82
Kingsborough/Lydia Lincoln BSHE 60
Kinion/Gail G. FDSC BSA 80
Kinion/Mark A. FDSC BSA 82
Kinkead/Ewing W. BSA 37
Kinnaird/Tommy C. ENSC BSA 81
Kinney/Caroline Roberts BSHE 44
Kinney/Jack Roy AGEC BSA 81
Kinney/Neal Preston BSA 53
Kinser/Barbara Nichols BSHE 51
Kinslow/Eddy AGME BSA 83
Kirby/Elva Hudson BSHE 34
Kirby/Jeffrie Charles AGEC BSA 84
Kirchhoff/Faith HEED BSHE 73
Kirchner/David A. AGEC MS 87
Kirk/Tracy Ann HEED BSHE 86
Kirkland/James D. ANNU BSA 81
Kirkland/Theo L. AHUS BSA 64
Kirkland/Virlean Lofton HOEC MS 78
Kirkland/Virlean Lofton CDFR BSHE 76
Kirkpatrick/Dwight Lance ANSC BSA 97
Kirkpatrick/Gary Dale GNAG MS 92
Kirkpatrick/Grace Durham BSHE 80
Kirkpatrick/Jennifer HORT 98
Kirkpatrick/Matthew Terrence HORT BSA 96
Kirkpatrick/Monroe BSA 42
Kirkpatrick/Patricia K. FND BSHE 72
Kirkpatrick/Terrence L. PLPA MS 78
Kirkpatrick/Terrence L. ANSC BSA 75
Kirksey/Avanelle HOEC BSHE 47
Kirksey/Myron A. FDSC BSA 91
Kirksey/Thomas C. AGRI BSA 43
Kiser/Edsel BSA 51
Kisor/Doris Coger BSHE 64
Kisor/Lucy Beth HOEC BSHE 62
Kitt/Mark John AGRN MS 95
Kittrell/John B. BSA 55
Kizzar/Lee Ann IDHS BSHE 90
Kjeldgaard/John Leonard ANSC BSA 71
Klar/Rebecca Bernard FDSC BSA 89
Klein/Clark Dennis ENTO PHD 93
Klein/Clark Dennis ENTO MS 89
Klein/Kristine ANSC BSA 91
Klerk/Richard A. AGRN MS 83
Kliner/Beverly Ann HEED BSHE 71
Klingaman/Tracy Eugene AGRN PHD 94
Klinger/Christine HOEC MS 78
Klutts/Margaret I. ELED BSHE 38
Knapp/Gretchen Meyer BSHE 45
Knapple/Jack W. BSA 58
Knapple/Linda Lewis BSHE 60
Knapple/Whit SOCS BSA 86
Knesek/Larry R. AGME BSA 72
Knez/Kathy L. HOFS BSHE 82
Knight/Alma Sisk BSHE 35
Knight/Archie C. BSA 48
Knight/Edwin BSA 48
Knight/Evangeline BSHE 38
Knight/Hiron AGRI BSA 54
Knight/Marcus Harold AGBS BSA 96
Knight/Pat DANB BSA 80
Knight/Patsy BSHE 57
Knight/William G. BSA 38
Knighten/Debra Lynne FDNU BSHES 98
Kniseley/James S. HORT BSA 74
Knoepp/Jennifer Donaldson HORT MSA 82
Knoepp/Jennifer Donaldson HORT BSA 80
Knoll/Diane C. HOEC MS 82
Knoll/Janis R. BSHE 49
Knoll/Kathryn R. CTM BSHE 74
Knowles/Harold D. BSA 52
Knowles/Karen Boles HDID BSHE 84
Knowles/Lee Ann CLTX BSHE 87
Knowles/Lou Ella AGEC MS 93
Knowles/Sherman G. BSA 42
Knowles/Vance E. BSA 52
Knowlton/Gail B. CTM BSHE 75
Knox/James A. BSA 49
Knox/John AGEC BSA 75
Knox/John Frank BSA 40
Knox/Lois Talbert BSHE 26
Kobera/John James BSA 82
Koch/Katy Susan HEED BSHE 71
Koehler/Matthew Allen AGRN MS 97
Koehler/Matthew Allen ENSC BSA 96
Koehn/Anita Lou PPPM BSA 85
Koen/Hillman S. POSC BSA 50
Koenig/Clemons A. BSA 52
Koenig/Matthew A. BSA 52
Koerdt/Leslie Marie CLTX BSHES 96
Koger/Roberta W. HEED BSHE 45
Kohler/David V. POSC BSA 84
Kolansinski/William L. BSA 80
Komarmy/Dorothy Mitchell BSHE 52
Konert/Margaret A. FDNU BSHE 78
Koonce/Ira Neely BSA 54
Koone/Carl D. AGED MS 52
Koone/Carl D. AGSC BSA 51
Koone/Harold D. ENTO MS 52
Koone/Harold D. ENTO BSA 51
Koosau/Esteban SOSC MS 71
Kopek/Jo Ann Fulton PLPA MS 79
Kopek/Wakako Yamaguchi ANSC PHD 93
Kopepasah/Donna Summerlin CLTX BSHE 89
Kottkamp/Milton B. ENTO MS 83
Kouakou/Brou ANSC PHD 95
Kousiakis/Dimitrios POSC MS 84
Kowalski/Edward L. ENTO MS 80
Kowalski/Edward L. GNAG BSA 75
Kowk-Fu/Mi-Kwan HOEC MS 80
Kowk-Fu/Mi-Kwan FDNU BSHE 78

Kozup/Joseph FDSC MS 82
Kozup/Rosemary HOFS BSHE 84
Kraft/Virginia Humphreys HEED BSHE 76
Kraim/Amer M. ENTO PHD 83
Kraim/Amer M. ENTO MS 80
Kramer/Mari Lee GHEC BSHE 79
Kreider/David L. ANSC PHD 82
Kreider/David L. ANSC BSA 73
Kreider/Marla Lynn ANSC MS 91
Kreps/Rebecca FDNU BSHE 80
Kroger/Erika Christina FDSC BSA 96
Krueger/Judi L. HDID BSHE 82
Krug/Martha J. HORT BSA 83
Krutz/Larry Jason ENSC BSA 96
Kubistek/Timothy Aaron AGBS BSA 94
Kumpe/Carra Lee BSHE 61
Kumpe/J. O. POSC BSA 33
Kumpe/John O. BS 57
Kunnalaca/Chongkol AGRN MS 77
Kunnalaca/Surajit ENTO MS 78
Kupperman/Kenneth M. HORT MS 93
Kurz/Bernard Ferdinand HORT MS 84
Kurz/Bernard Ferdinand HORT BSA 81
Kurz/Magdalena M. ANSC MS 86
Kurz/Magdalena M. ANNU BSA 85
Kurz/Marilyn Anne HORT MS 86
Kwon/Sam Lyul AGRN PHD 89
Kyle/John R. POSC BSA 75
Kyle/Luther B. BSA 59
Kyler/Alma Alexander BSHE 23
Lachowsky/Brian Albert AGEC BSA 81
Lachowsky/Frank HORT BSA 87
Lachowsky/G. Annette HORT MS 87
Lachowsky/G. Annette HORT BSA 80
Lachowsky/John Edward BSA 50
Lackey/Elizabeth Gail IDHS BSHES 97
Lacour/Carmela Patty ANSC BSA 91
Lacour/John Joseph AGED BSA 90
Lacy/Gay CRSC BSA 70
Lady/Bob J. BSA 51
Lady/Harold L. BSA 50
Lafferty/Daris G. BSA 50
Lafferty/Daris G. AGEC MS 52
Lafferty/Doris Marie FND BSHE 64
Lafferty/Doris Marie FND MS 66
Lafferty/Linda J. FDNU BSHE 69
Lafferty/Linda J. FDNU MS 71
Lagasse/Mike ANSC BSA 89
Lagasse/Stephen ANSC BSA 87
Lairmore/Larry P. BSA 76
Lairmore/Larry Payne GNAG MS 88
Lambert/Deborah Rabin HEED BSHE 78
Lambert/Debra Lee BSHE 85
Lambert/Henry HDFS BSA 46
Lambert/James Lee BSA 50
Lambert/Jordan B. BSA 81
Lambert/Maribelle J. FDNU BSHE 56
Lambert/Martha Kelly BSHE 51
Lambert/Ocrie Melton HOEC MS 66
Lambert/Roy E. BSA 48
Lambert/Sue Spencer HEED BSHE 74
Lambert/William M. AGRN MS 77
Lambert/William M. CRSC BSA 68
Lambeth/Kenneth W. VOAG BSA 77
Lamken/Lisa Anne Causey BSHE 74
Lamkin/Corbet J. AGEC MS 78
Lamkins/Susan McChristian POSC BSA 82
Lammers/Jeffery David AGRN BSA 94
Lancaster/Donn M. HORT BSA 72
Lancaster/Donn M. PTSC PHD 84
Lancaster/Eric B. PPPM BSA 79
Lancaster/J. L. BSA 47
Lancaster/Lynda Ann FND BSHE 66
Lancaster/Virginia P. BSHE 47
Lande/Mansur AGRN PHD 85
Landers/Earl Gillam BSA 33
Landers/Patti S. FDSC PHD 92
Landes/Doelas POSC MS 66
Landes/Doelas BSA 65
Landgraf/Rita Marie HOEC MS 88
Landry/Mona M. LDUH BSA 91
Landsaw/James Robert AGBS BSA 92
Lane/Anne Ledford HEED BSHE 44
Lane/David Earl AGRN BSA 93
Lane/Leota Carole BSHE 63
Lane/Mary Jane BSHE 41
Lane/Tammy HDFS BSHE 86
Lanehart/David A. AGRN BSA 83
Lanehart/Henry C. BSA 53
Laney/Pamela Hollandsworth FDNU BSHE 82
Laney/William H. AGED BSA 85
Lang/Chris CRSC BSA 76
Lang/E. Michele HOEC MS 78
Lang/E. Michele FND BSHE 72
Lange/Carola Angelica CLTX BSHE 85
Langer/Doris C. HOEC MS 79
Langer/Doris C. BSHE 57
Langford/Erma L. BSHE 42
Langlais/Sharon L. Bollinger BSHE 73
Langley/Susan Kay IDHS BSHE 89
Langston/Charles T. BSA 53
Langston/Elijah Ray ACCT BSA 75
Lanier/Fred Booth AGBS BSA 85
Lanier/Michael FDNU MS 79
Laningham/Judith Dawn HDFS BSHES 94
Lankford/Lowell T. POSC MS 52
Lankford/Lowell T. ANSC BSA 49
Lankford/Paul BSA 52
Lanter/Janet M. PLPA MS 81
Lantrip/Dewey BSA 34
Laquey/Janey Carpenter BSHE 59
Larey/Lance L. GNAG MS 63
Largent/Larry Kennon BSA 58
Larimer/Tiffany Nicole IDHS BSHES 97
Larkin/Marcia Puckett BSA 73
Larkin/Mary Earle BSHE 30
Larkin/Steven Dow ANNU BSA 73
Larsen/Richard Craig PLPA PHD 87
Larson/Michelle Marie FDNU BSHE 94
Lasch/Anne M. FDNU BSHE 85
Laser/Ann F. BSHE 65
Lashley/Connie Sue Russow HEED BSHE 71
Lask/Ruth Laverne Aday BSHE 57
Lasley/Agnes J. HEED BSHE 79
Lassiter/Bill E. BSA 53
Lassiter/Billie Flippo BSHE 53
Laster/Charles E. BSA 43
Laster/Janet King CLTX BSHE 86
Laster/Janet King CLST BSHE 86
Laster/Marion ENTO MS 59
Laster/Marion ENTO BSA 58
Lastonica/Ernest A. BSA 73
Latham/Elizabeth Leigh AGRN BSA 90
Latimer/Larry M. AGRN BSA 76
Latorre/Jose R. POSC PHD 86
Latorre/Jose R. POSC MS 83
Latta/Susan M. HEED BSHE 72
Latting/John A. ENTO BSA 66
Latture/James Mitchell AGBS BSA 91
Lauderdale/Hautense Stuckey BSHE 42
Laurent/Amy Diane IDHS BSHES 97
Laurent/Glen SOSC 67
Law/Clymer C. BSA 51
Lawhon/J. N. BSA 43
Lawhon/Onita V. Trawick BSHE 52
Lawhon/Thane Joel AGRC BSA 98
Lawless/J. C. GNAG MS 71
Lawless/J. C. BSA 51
Lawrence/Cecilia W. HDID BSHE 78
Lawrence/Frank M. HEED MD 67
Lawrence/Gary VOAG BSA 79
Lawrence/Norma Harrington BSHE 43
Lawrence/Robert O. ANSC BSA 67
Lawrence/Robert O. ANNU MS 69
Lawrence/Timothy C. AGEC BSA 72
Lawroski/Mary Dewoody BSHE 55
Lawson/D. B. BSA 43
Lawson/Floy B. BSHE 42
Lawson/Fred Avery BSA 43
Lawson/J. Larry BSA 54
Lawson/Kevin Wayne AGRN BSA 95
Lawson/Louanne CLTX BSHE 85
Lawson/Mark ANSC BSA 86
Lax/Alan R. PLPA PHD 83
Laxson/Margaret Terhune HOEC BSHE 52
Lay/Heidi Elene ANSC BSA 94
Layes/Henry J. GNAG MS 83
Layes/Henry J. VOAG BSA 78
Layton/Arthur C. BSA 57
Layton/Marilyn Bonner FND BSHE 70
Lazzaro/Lisa S. DANB BSA 78
LaFerney/Preston E. AGEC MS 60
LaFerney/Preston E. BSA 58
LaPrade/Carole Sue HEED BSHE 74

Le Page/Mary P. CLTX BSHE 82
Leach/Deborah Ann POSC BSA 94
Leach/Robert D. ANSC BSA 79
Leary/William Gay ENTO MS 91
Leavitt/Melinda CLTX BSHE 87
Ledbetter/A. A. BSA 38
Ledbetter/Craig Matthew AGBS BSA 95
Ledbetter/Joy Gregory BSHE 40
Ledbetter/Ricky Darrell AGME BSA 85
Ledford/Denny G. POSC BSA 83
Lee/Arthur Fuller BSA 17
Lee/Ashley Grace CLTX BSHE 94
Lee/Cedric Elton BSA 48
Lee/E. Dave AGEC BSA 73
Lee/Eddie Ray ANSC BSA 72
Lee/Eddy T. FDSC BSA 86
Lee/Edmund R. BSA 49
Lee/Elizabeth Lewis HEED BSHE 68
Lee/Ellen Mae BSHE 50
Lee/Fleet N. PLPA MS 72
Lee/Fleet N. PLPA BSA 69
Lee/Heuisuck POSC MS 86
Lee/James T. ENTO MS 72
Lee/James T. ENTO BSA 70
Lee/Judith FND BSHE 72
Lee/Keith A. LDUH BSA 79
Lee/Mary FDNU BSHE 72
Lee/Meei-Huey FDSC MS 94
Lee/Norman Willis BSA 50
Lee/Paula BSHE 69
Lee/Shelburn BSA 48
Lee/Steve AGRN MS 81
Lee/Steve AGRN BSA 78
Lee/Thomas Collier AGEC BSA 81
Lee/William McGuire BSA 20
Lee/William Roscoe BSA 53
Leeman/Frank W. ANSC BSA 74
Lefors/William McKinley BSA 23
Legg/J. O. AGRN MS 51
Leggett/Joseph E. BSA 60
Leggitt/Rodney H. AGBS 98
Lehigh/Patti Wiese FDSC BSA 88
Lehle/Fredric Roth CRSC MS 75
Lehman/Cheryl L. HOFS BSHE 79
Lehman/J. W. MSA 71
Lehman/J. W. BSA 55
Lehman/Laura S. ANSC BSA 73
Lehnen/Jennifer Ann CLTX BSHE 94
Leising/Tamara A. IDHS BSHE 95
Leister/Stephen L. VOAG BSA 82
Lemons/Doyle Ray ANSC BSA 82
Lemons/E. Wayne ENTO MS 55
Lemons/E. Wayne BSA 52
Lemons/Luther D. BSA 49
Lemser/Lydia Bell BSHE 56
Lemser/Theodore Eugene ANSC BSA 82
Lendermon/William H. BSA 55
Lennon/James H. ANSC BSA 73
Lensing/Cynthia Rushing GHEC BSHE 82
Lenzen/Kathy Beltrani CLTX BSHE 84
Leohner/Rupert R. BSA 54
Leonard/Catherine Louise HOEC MS 84
Leonard/Elston S. VOED BSA 25
Leonard/O. F. BSA 35
Leonard/Tracey Lee ANSC BSA 97
Leonard/Tracey Lee AGBS BSA 97
Leong/Jeffrey Hengtak AGRN MS 86
Lerner/Franklin D. HORT MS 79
Leschen/Richard Alan ENTO MS 88
Leslie/James BSA 50
Leslie/Juanita Hastings HEED BSHE 67
Leslie/Rita CDFR MS 69
Leslie/Thomas E. CRSC MS 70
Leslie/Thomas E. CRSC BSA 67
Leslie/William S. BSA 52
Lessley/Billy V. AGEC MS 60
Lessley/Billy V. AGEC BSA 57
Lester/Connie White ANSC BSA 82
Lester/Melissa Kay GNAG MS 90
Lester/Melissa Kay AGED BSA 88
Lester/Toby Don ANSC MS 93
Lester/Toby Don ANSC BSA 89
Levois/Malcolm N. AGRN BSA 82
Lewallen/Bill Wayne POSC BSA 68
Lewis/Bill F. BSA 49
Lewis/Billy R. BSA 59
Lewis/Eugenia S. BSHE 43
Lewis/Jimmy D. VOAG BSA 70
Lewis/Joseph E. DANB BSA 83
Lewis/Kimberly J. CLTX BSHE 78
Lewis/Larry Allen BSA 63
Lewis/Larvinia D. HOEC MS 81
Lewis/Lawrence A. BSA 37
Lewis/Louise BSHE 38
Lewis/Marian Penrose HEED BSHE 35
Lewis/Mary Harris BSHE 47
Lewis/Mitch BSA 42
Li/Chuan AGEC MS 96
Li/Jian RSOC MS 93
Li/Xin-Yu FDSC MS 93
Licht/Lance L. BSA 80
Lichti/Dennis AGME BSA 87
Lichty/Earnst C. BSA 30
Lichty/Earnst C. AGEC MS 31
Liddell/Steven Brian ENSC BSA 82
Lieblong/Clyde P. VOAG MS 54
Lieblong/Clyde P. VOAG BSA 40
Lieblong/Luther H. BSA 59
Lieven/Martha Kathryn FDNU BSHE 90
Light/Cherry Harkey HEED BSHE 62
Ligon/Chuck AGBS BSA 86
Ligon/Edwin Shaw AGBS BSA 80
Ligon/Fred BSA 50
Ligon/Heather Huffer FDNU BSHE 91
Ligon/Janet Lee CLTX BSHE 80
Ligon/Marilyn E. LDUH BSA 84
Ligon/Micky ANSC BSA 76
Ligon/Sullivan A. BSA 56
Lih/Marita P. ENTO PHD 93
Lih/Marita P. ENTO MS 85
Liles/John Steve ANSC BSA 88
Lilley/Beth Elaine IDHS BSHE 91
Lin/Cheng Yie ANSC PHD 80
Lin/Song-Chang FDSC MS 87
Lin/Yuo Chyung AGEC MS 65
Linam/Carolyn L. BSHE 63
Linane/Marianne GNAG MS 90
Linck/Henry M. BSA 81
Linck/Hollice A. DANB BSA 80
Lincoln/Benjamin Aplin BSA 22
Lincoln/Charles G. BSA 34
Lincoln/Dwight CRSC BSA 70
Lincoln/Jack Morton BSA 35
Lindsey/Charles W. BSA 57
Lindsey/James B. BSA 47
Lindsey/Joseph L. BSA 60
Lindsey/Mary Whiteside BSHE 76
Lineback/Clay Lawrence AGBS BSA 89
Linit/Marc J. ENTO PHD 81
Linker/Susan Elizabeth POSC MS 89
Linker/Susan Elizabeth ANSC MS 87
Linn/Candace A. CLTX BSHE 77
Linn/Conal Odell GNAG MS 64
Linn/Lora Sue IDHS BSHE 94
Linn/T. H. BSA 39
Linton/Rae Danciu HOEC MS 92
Lipsey/Richard Lynn ENTO MS 68
Liscano/Jose F. AGRN BSA 85
Lisko/Frank A. AGME BSA 85
Little/Laura Squyres FDNU BSHE 92
Little/Lucretia Alice BSHE 57
Littlejohn/Donna L. HEED BSHE 60
Littrell/David L. HORT BSA 61
Liu/Randy FDSC BSA 85
Liu/Ting-Feng Michael FDSC MS 80
Liu/Yanrong HOEC MS 95
Livingston/James M. ENTO PHD 78
Livingston/James M. ENTO MS 71
Livingston/John Lee BSA 52
Livingston/Lois Berry BSHE 24
Livingston/Patricia Jean HOEC MS 68
Livingston/Phillip V. POSC BSA 59
Lloyd/Edgar Harold BSA 43
Lloyd/Heywood Bryan BSA 37
Lloyd/James Ernest BSA 32
Lloyd/Lisa CLTX BSHE 83
Lloyd/Regina Renee CLTX BSHE 86
Lloyd/Roy BSA 58
Lobsinger/Christine ANSC BSA 80
Locke/Carol BSHE 59
Locke/Richard D. AGEC BSA 60
Lockhart/Bruce E. ANSC BSA 75
Lockhart/Cyrill D. HEED BSHE 76
Lockhart/Darron Ray AGBS BSA 89

Lockhart/Lucy Shadrach BSHE 43
Lockhart/Randall BSA 77
Lockhart/Ross L. DASC BSA 79
Loe/Carolyn J. HOEC MS 71
Loe/Robert H. AGRN MS 55
Loe/Robert H. GNAG BSA 51
Loe/William C. GNAG BSA 53
Loe/William C. ANNU MS 59
Loewer/Betty S. IDHS BSHE 90
Loewer/Robert C. BSA 57
Loftin/Kelly Mark ENTO MS 87
Loftin/Mary Jane PPPM BSA 87
Loftis/Juanita Cox BSHE 39
Loftis/Katharin Ruth Elmore SOSC BSA 80
Logan/Bess Hodges BSHE 19
Logan/Bonnie Fox HEED BSHE 66
Logan/Kathryn Ann Camp GHEC MS 65
Logan/Michael Dale AGME BSA 87
Loghry/Amy Elizabeth FDNU BSHE 94
Lombardo/Heraclio A. AGRI MS 55
Londagin/George Elvin BSA 55
London/Alvin Clayborne RSOC MS 37
Long/Billy M. BSA 58
Long/Carol Cupples FDNU BSHE 62
Long/Carol D. DANB BSA 82
Long/Courtney Ann FDNU BSHE 88
Long/David Hughes PLPA MS 96
Long/Douglas Jacob AGRN BSA 85
Long/Frank D. ANNU BSA 64
Long/James Henry AGRN PHD 82
Long/James Henry AGRN MS 80
Long/Jane Elmore BSHE 37
Long/Olivia A. DANB BSA 85
Long/Shelly Diane HDFS BSHE 95
Long/Terri A. CLTX BSHE 95
Longgrear/Maurine D. CLTX BSHE 82
Longmate/James A. RSOC BSA 72
Looney-Carle/Christine HDFS BSHE 94
Looney/Cathryn C. CLTX BSHE 78
Looney/Charles R. ANSC MS 80
Looney/Charles R. ANSC BSA 78
Looney/Elma Davis BSHE 33
Looney/Era F. HOEC MS 69
Looney/George W. BSA 40
Looney/J. W. AHUS BSA 66
Looney/Janice Lane CLTX BSHE 90
Looney/Leamon L. GNAG MS 64
Looney/Leamon L. BSA 51
Looney/Rance Franklin EXED BSA 94
Loop/Emma L. Downs BSHE 55
Looper/Charles G. GNAG BSA 58
Looper/Cheryl R. ANSC BSA 91
Looper/Cheryl R. ANSC MS 93
Looper/Mike L. ANSC MS 94
Looper/Mike L. ANSC BSA 92
Loran/Fidelis Joseph ANSC BSA 84
Lord/Layla Miller BSHE 63
Lorenz/Barbara Carnes FDNU BSHE 76
Lorenz/Eric John AGRN MS 83
Lorenz/Gustav Marvel PPPM BSA 82
Lorenz/Gustav Marvel ENTO PHD 95
Lorenz/Gustav Marvel ENTO MS 85
Loss/Elizabeth Angelina BSHE 48
Loucks/George C. ANSC MS 72
Loudermilk/E. W. BSA 38
Loudermilk/James Edwin BSA 58
Love/Albert M. ENTO MS 76
Love/Albert M. ENTO BSA 72
Love/Ann H. FND BSHE 73
Love/Bob E. POSC MS 73
Love/Bob E. ANSC BSA 71
Love/Mosco BSA 53
Lovelady/Patricia BSHE 53
Loveland/Kimberly Lamb BSHE 93
Loveless/Nancy Flusche ENTO MS 83
Lovell/David W. AGBS BSA 80
Lovell/Jessie Louise BSHE 60
Lovell/Jimmy S. ENTO MS 72
Lovell/Jimmy S. ENTO BSA 70
Lovell/Jodina Marie CATX BSHES 98
Lovette/Keitha A. CLTX BSHE 94
Low/Carl C. VOAG BSA 52
Low/Margo BSHE 51
Lowder/Vernon POSC BSA 49
Lowe/Pat McLaughlin CDFR MS 72
Lowe/Pat McLaughlin BSHE 52
Lowe/Vicki HEED BSHE 77
Lowrey/Gene DHUS BSA 62
Lowrey/John BSA 30
Loyd/Janet Lynn Floyd HEED BSHE 87
Loyd/Laura L. Smith HEED BSHE 83
Loyd/Sherry Michelle HDFS BSHE 92
Luallen/Janie BSHE 86
Lubin/Avery AGEC BSA 76
Lucariello/Cynthia M. AGBS BSA 90
Lucariello/Mike POSC BSA 86
Lucariello/Stephanie L. ANSC BSA 88
Lucas/Henry Alexander BSA 19
Lucas/John S. BSA 50
Ludwig/Randy D. AGED BSA 87
Lueken/Patty Watson FDNU BSHE 75
Lueker/Bill ANSC BSA 83
Lueker/Carl ANSC BSA 49
Luers/Daniel J. BSA 82
Lugo/Maria De Lourdes AGRN PHD 93
Lukas/Donna CLTX BSHE 92
Lumpkin/Elizabeth A. HOEC MS 79
Lumsden/Eddie BSA 65
Lund/Carl Frederick BSA 26
Lundergan/Catherine HORT MS 75
Lundhagen/Kathe FDNU BSHE 62
Lusk/Jonathan Scott DANB BSA 77
Lusk/Mary Lou Phillips HDID BSHE 84
Luther/Jacob L. BSA 54
Luther/Lonnie W. POSC BSA 69
Luther/Rod ANSC BSA 71
Luttrell/Randall G. ENTO PHD 82
Luttrell/Randall G. ENTO MS 78
Luttrell/Rick D. FDSC BSA 87
Lyle/Christopher S. ANSC 98
Lyman/Virginia FND BSHE 67
Lynch/Jeffrey Darryl DANB BSA 85
Lynch/Randall BSA 65
Lynch/Ruth BSHE 47
Lynd/J. Quentin BSA 43
Lynn/Ann McHenry BSHE 51
Lynn/Daniel L. ENSC BSA 75
Lynn/Deborah L. HOEC MS 76
Lynn/Deborah L. HEED BSHE 74
Lyon/Amber L. HDFS 98
Lyon/Betty Lou HOEC BSHE 62
Lyon/Geoffrey Scott HORT BSA 83
Lyon/James Maxwell BSA 53
Lyon/Peggy Ross BSHE 58
Mabe/Carol L. FDSC MS 83
Mabray/Craig Jay POSC MS 80
Mabruk/Hassan S. ANSC PHD 81
Mabry/Rhonda F. HEED BSHE 82
Mabry/Wendy Jo HOUS BSHES 98
Macaulay/Laura E. FDSC MS 91
Macedo/Grace Margaret HEED BSHE 95
Machen/Deborah CLTX BSHE 86
Mack/Harriet BSHE 71
Mackey/Suzanne CLTX BSHE 86
Macumber/Lester H. BSA 50
Macy/Larry B. ANSC PHD 90
MacKay/S. Nia ANSC PHD 93
MacKay/S. Nia ANSC MS 90
Madden/Dale C. AGEC MS 70
Maddox/James Gray BSA 27
Maddox/Jefferson Austin GNAG BSA 40
Maddox/Wallace BSA 55
Maddox/Willis G. BSA 50
Madero/Miguel Alberto AGEC BSA 85
Madrigal Rojas/Karla HOEC MS 94
Madrigal Rojas/Sergio A. AGEC MS 95
Madrigal Rojas/Sergio A. POSC MS 94
Maeza/Ali S. POSC PHD 83
Magdefrau/Susan Elizabeth CLTX BSHE 88
Magill/Tamara HDFS BSHE 89
Magness/David A. VOAG BSA 62
Magness/Van Fakes BSHE 88
Maguire/Julia Phillips BSHE 33
Mahan/Cynthia M. HEED BSHE 71
Mahan/Jesse Hoyt ANSC BSA 80
Mahiaini/Julius Maina ANSC BSA 86
Mahmood/Talat PLPA MS 90
Mahmoud/Aslam Mohamed Hosni POSC MS 94
Mahon/Christine Renee HDFS BSHE 86
Maier/Kenneth Ray AGRN BSA 95
Maier/Mark Andrew AGBS BSA 87
Mailloux/Gail M. Brannan ANSC MS 82
Mailloux/Gail M. Brannan ANSC BSA 78
Main/Gary L. FDSC PHD 93

Main/Gary L. FDSC MS 86
Main/Gary L. FDSC BSA 83
Main/Janet M. HOEC MS 74
Main/Janet M. FND BSHE 70
Main/Jeffrey Len AGRN BSA 92
Mainard/Floy C. BSHE 34
Majedi/Mohammad R. AGRN PHD 85
Majedi/Mohammed Reza AGRN PHD 85
Majedi/Mohammed Reza AGRN MS 81
Majors/Rebecca Ann FDNU BSHE 79
Makia/Elner Joyce HOEC MS 88
Mallett/Nathan G. AGEC BSA 51
Mallory/Brenda K. GHEC BSHE 76
Malloy/David BSA 41
Malloy/John R. BSA 60
Malm/David A. GNAG MS 84
Malo/Juan P. AGBS 99
Maloch/D. V. GNAG BSA 37
Maloch/Jim GNAG BSA 61
Malone/Bobby J. GNAG MS 64
Malone/Bobby J. BSA 59
Malone/Darrin SOSC BSA 88
Malone/David H. POSC BSA 68
Malone/Patty GHEC BSHE 77
Malone/Stephen R. AGRN MS 83
Malone/Stephen R. AGRN BSA 82
Manasco/Burl D. BSA 53
Manbeck/Glenda B. BSHE 65
Mancuso/Nora AGRN PHD 90
Mankar-Joshi/Maithili Vijay POSC MS 95
Manning/Marion Heerwagon BSHE 33
Manning/Mark AGRN BSA 80
Manning/Mary ENTO MS 83
Mansell-Kappler/Theresa FDNU BSHE 87
Manson/David H. BSA 54
Mantegani/Gillie Ray BSA 66
Manthani/Krupakar Reddy FDSC MS 85
Mantooth/Calvin T. ANNU BSA 70
Mantooth/Larry W. GBUS BSA 79
Mantooth/Mary Lynn ANSC BSA 87
Manuel/Carrie Beth FDSC BSA 87
Manwan/Ibrahim ENTO MS 64
Maples/Catherine T. HOEC MS 64
Maples/Kenn A. AGEC BSA 70
Maples/Richard L. BSA 50
Marcantel/Carson AGBS BSA 85
Marchand/Kay Elizabeth HEED BSHE 77
Marconi/Timothy E. AGBS BSA 83
Maredia/Karimbhai M. AGRN PHD 85
Marini/Paul J. POSC PHD 72
Marinoni/Paul Albert AGBS BSA 90
Markham/Jerry D. HORT MS 76
Markham/Leslie PLPA MS 79
Marks/Alfred C. FDSC PHD 90
Marks/Betty Fay BSHE 54
Marks/James A. BSA 11
Markum/Dorothy Neil Dean BSHE 56
Marlatt/Melody L. Garland PLPA MS 96
Marler/Jimmie Wayne VOAG BSA 74
Marley/Jeffrey D. ANSC BSA 79
Marple/Linda P. FND BSHE 71
Marquette/Lynn HEED BSHE 74
Marr/Amy L. AGBS BSA 83
Marr/Katherine Heath BSHE 31
Marr/Rita FDNU BSHE 78
Marrs/Dolf C. ANSC BSA 78
Marrs/Dolf C. POSC MS 80
Marrs/James E. AGEC BSA 64
Marrs/Kelly Lynn GHEC BSHE 81
Marsh/Clifford R. BSA 43
Marsh/DeMaris G. BSHE 40
Marsh/Ellie E. RSOC MS 80
Marsh/Riley B. BSA 54
Marsh/Robert W. D. BSA 40
Marsh/William Edward BSA 41
Marshall/Bequita Williams BSHE 44
Marshall/Don A. AGEC MS 47
Marshall/Julia G. GHEC BSHE 77
Marshall/Marion R. CDFR BSHE 74
Marshall/Mary Mange BSHE 51
Marshall/Sue Etta HOEC BSHE 26
Martell/Karen Annette AGRN BSA 88
Martella/Elizabeth Smartt AGEC MS 80
Martin/A. L. BSA 51
Martin/Amos R. AGED BSA 58
Martin/Amy Lynn AGBS BSA 96
Martin/Betty BSHE 68
Martin/Billy E. AGEC MS 61
Martin/Billy E. BSA 59
Martin/Brett Wade POSC BSA 87
Martin/Bruce PLPA MS 79
Martin/Carol Lynn Dillard LDUH BSA 84
Martin/Carole Sue FND BSHE 64
Martin/Celia B. DIET BSHE 68
Martin/Cleo BSA 39
Martin/Derron Lee AGBU BSA 98
Martin/Don W. BSA 60
Martin/Dwayne ANSC BSA 74
Martin/Edwynn VOAG BSA 72
Martin/Elizabeth BSA 86
Martin/Fred Eugene BSA 41
Martin/Harlis Edward BSA 53
Martin/Jackie Jordan ANSC BSA 84
Martin/James W. AHUS BSA 64
Martin/Jennifer Rebecca HDFS BSHES 96
Martin/Jo Anne S. HEED BSHE 68
Martin/Joe B. AGRN BSA 72
Martin/JoAnn Freeman HEED BSHE 60
Martin/Keith Dion POSC BSA 93
Martin/Keith F. PLPA MS 83
Martin/Keith F. PPPM BSA 78
Martin/Leora BSHE 51
Martin/Mark A. AGRN MS 85
Martin/Mark A. AGRN BSA 83
Martin/Mary Jane FDNU BSHE 61
Martin/Merida J. Phillips BSHE 58
Martin/Michael Deion DANB BSA 77
Martin/Michel E. AGBS BSA 84
Martin/Nancy FDNU BSHE 62
Martin/Norma Austin HEED BSHE 73
Martin/R. Gene AGEC MS 69
Martin/R. Gene AGEC BSA 66
Martin/Ray BSA 16
Martin/Rex AGBS BSA 89
Martin/Robert G. PLPA MS 84
Martin/Robert G. PPPM BSA 82
Martin/Todd L. AGBS BSA 86
Martin/Vicki Price HEED BSHE 79
Martin/Victoria HEED BSHE 75
Martin/Walter Neil ANSC BSA 86
Martin/Wendy Lynn CNSL MS 97
Martin/Wendy Lynn HDFS BSHE 95
Martin/William H. BSA 51
Martin/William Russell AGBS BSA 85
Martinez/Ruth E. Trowbridge BSHE 60
Martini/Joseph Albert SOSC MS 58
Martini/Joseph Albert SOSC BSA 57
Marts/Patricia G. HOEC MS 76
Marts/Patricia G. BSHE 54
Maruca/Sheryl Diane ANSC MS 81
Marugg-Wolfe/Marjorie BSHE 57
Marugg/Tami Laris FDNU BSHE 89
Marvel/James N. AGRN BSA 85
Mascaro/Ralph M. POSC BSA 77
Masching/Rebecca Henry FDNU BSHE 77
Mashek/Kathye Caldwell BSHE 65
Mason/Buddy AGEC MS 72
Mason/David Harold PPPM BSA 94
Mason/Donald J. VOAG BSA 80
Mason/Edward P. ENTO MS 69
Mason/Edward P. BSA 63
Mason/James L. BSA 50
Mason/Matthew Paul AGEC MS 93
Mason/Matthew Paul AGBS BSA 91
Mason/Patricia Heck CTM BSHE 73
Mason/Perry Ray AGED BSA 85
Mason/Ralph Odel BSA 50
Mason/Scott Doyle AGEC MS 90
Mason/Scott Doyle AGRN BSA 88
Mason/Suzetta Jane HEED BSHE 69
Massey/Ellen Spears BSHE 40
Massey/Jeffery Wayne AGEC BSA 86
Massey/Joe AGRN MS 90
Massey/Joe AGRN PHD 95
Massey/Walter BSA 41
Massey/Walter ENTO MS 52
Massey/Warren C. ANNU BSA 68
Massie/Kathie Annette BSHE 87
Mast/I. N. BSA 41
Mast/Jawanda HOEC MS 89
Mast/Jawanda GHEC BSHE 84
Masters/Bryce M. BSA 47
Masters/Gene C. POSC BSA 53
Masters/Mildred Bruce BSHE 50

Masterson/Troy James AGBS BSA 86
Mat/Rashid Bin ARCH BSHE 86
Matheny/Sue Ann Plunkett BSHE 60
Mathews/Amy Jeannette HORT BSA 94
Mathias/Bernice J. HEED BSHE 67
Mathias/Harle L. BSA 58
Mathis/Gary AGED BSA 96
Mathis/Kent Ronald AGED BSA 92
Mathis/Lonnie Ray AGBS BSA 94
Mathis/Marion T. BSA 58
Mathis/Treva Martin BSHE 59
Mathis/Vernon W. BSA 38
Mathis/William AGRN PHD 77
Matlock/Bradley Dale AGBS BSA 95
Matlock/Ronald A. GNAG MS 87
Matsler/Phillip Lee POSC MS 96
Matteri/Debra Kaye Gadbury HOEC MS 88
Matteri/Debra Kaye Gadbury HEED BSHE 84
Matteri/Greg James AGED BSA 86
Matteri/Steve D. POSC BSA 90
Matteson/Beverly Hopper BSHE 38
Matteson/George Montgomery AGBS BSA 97
Matteson/Worth L. CRSC BSA 71
Matthews/Boyd E. PPPM BSA 86
Matthews/Edward G. GNAG BSA 56
Matthews/Edwin R. GBUS BSA 77
Matthews/Elizabeth Wagoner CLTX BSHE 94
Matthews/Geraldine Shaw CLTX BSHE 49
Matthews/Jackie L. GNAG BSA 64
Matthews/Joan P. CLTX BSHE 78
Matthews/P. B. AGED MS 50
Matthews/Wallace H. BSA 50
Mattingly/Bob Lynn AGBS BSA 88
Mattingly/Kimberly Lynn CLTX BSHE 86
Mattke/Dana Liane EXED BSA 97
Mauney/James E. HORT BSA 68
Mauney/Ross R. BSA 37
Maxa/Edward L. ANSC PHD 80
Maxwell/Cheryl Nash FND BSHE 67
Maxwell/George R. BSA 40
Maxwell/Howard Edward BSA 28
Maxwell/Joe S. BSA 38
Maxwell/John Woodrow BSA 38
Maxwell/Richard E. AGEC MS 73
Maxwell/Robert E. BSA 40
Maxwell/Ronald S. GNAG BSA 64
May/Camellia Alice HDFS BSHE 88
May/Charles Winston BSA 56
May/Donald C. HORT BSA 87
May/Elaine Arendale BSHE 40
May/James E. BSA 65
May/Kenneth J. CRSC MS 71
May/Kenneth J. AGRN BSA 70
May/Larry Wayne POSC BSA 69
May/Martha HOEC MS 78
May/Mary Anderson ANSC MS 71
May/Mary Anderson ANSC BSA 70
May/Melvin A. POSC BSA 73
May/Michael Lynn AGRN MS 79
May/Michael Lynn ANSC BSA 76
May/Paula Kay Hudgins FDNU BSHE 75
May/Peggy L. HEED BSHE 70
May/Ralph D. AGEC BSA 68
May/Sue Sykes BSHE 57
May/William David BSA 51
Mayes/Julie Berry IDHS BSHE 92
Mayes/Ralph D. ENTO MS 62
Mayes/Ralph D. BSA 55
Mayes/Robert L. ERSC BSA 74
Mayes/Vera Jones BSHE 34
Mayfield/Malcolm Randall AGEC MS 89
Mayfield/Patricia Rotenberry HOEC BSHE 65
Mayhew/Walter Lee AGRN PHD 91
Mayhew/Walter Lee AGRN MS 88
Maynard/Philip Miles ANSC MS 91
Mayner/Sharon A. HOEC MS 70
Mays/Elizabeth Vanhooke CATX BSHES 98
Mays/Jack C. AGBS BSA 91
Mays/Jack C. AGEC MS 94
Mays/Jeffrey Neal FDSC BSA 89
Mays/Jim BSA 50
Mays/Joy A. ESWS 98
Mays/Mary Emily Russell BSHE 24
Mays/Neal AGRN BSA 96
Mazahery/Mehry BSHE 80
Maze/Dennis Lavern POSC BSA 92
Mccourt/Kimberly Ann AGBS BSA 94
Mcdoud/Jeffery Hoyt POSC BSA 82
Mcdoud/Louis E. AGBS BSA 96
McReynolds/Dawn Elizabeth HDFS BSHES 95
McAdams/Johnny GBUS BSA 79
McAdams/Patricia J. FDNU BSHE 61
McAdoo/Anthony AGME BSA 98
McAdoo/Anthony AGED BSA 98
McAdoo/Scott Austin AGEC MS 94
McAdoo/Yamonda Y. LDUH BSA 82
McAlister-Apple/Laurie Marie HOEC MS 91
McAlister-Apple/Laurie Marie CLTX BSHE 87
McAlister/Billy John DASC BSA 68
McAlister/M. K. DANB BSA 81
McAllister/Janet Clare ENTO PHD 93
McAllister/Shirley E. Friend HEED BSHE 62
McBride/Glynn BSA 42
McBride/Joseph Boyd AGEC BSA 70
McBride/Mary Jo HOEC BSHE 45
McBride/Noble J. BS 28
McBride/William M. BSA 59
McCaig/Cari Stephens HDFS BSHE 93
McCain/Glenn AGEC BSA 70
McCain/Gordon William AGED BSA 86
McCain/Isaac Francis BSA 42
McCain/Johnathan L. HORT BSA 79
McCain/Randall A. BSA 49
McCain/Robert H. AGBS BSA 76
McCaleb/Suzanne Dawn IDHS BSHE 92
McCall/Robert David AGRN MS 97
McCall/Robert David AGRN BSA 94
McCape/Beverly Garrett FND BSHE 64
McCarley/Anne HEED BSHE 70
McCarriar/Elane Janssen BSHE 32
McCarroll/Otto Green BSA 26
McCarter/Danny Gene AGEC MS 69
McCarter/Danny M. GNAG MS 85
McCarter/Mike ANSC BSA 79
McCartney/Alton BSA 52
McCartney/Howard GNAG BSA 52
McCartney/Ray GNAG MS 83
McCarty/Anderson Cleve BSA 53
McCarty/Heather Marie IDHS BSHE 94
McCarty/John Timothy AGRN MS 91
McCarty/Mary Hosch HDFS BSHES 95
McCarty/Mary Leslie BSHE 88
McCarty/Michael Paul AGBU BSA 98
McCarver/Clinton H. GNAG BSA 61
McCaskill/Don CRSC BSA 70
McCaul/Louis A. ENTO MS 76
McChristian/Carolyn J. HEED BSHE 64
McChristian/Margery A. DANB BSA 84
McChristian/Rebecca J. ANSC 98
McClain/Amy L. CLTX BSHES 98
McClain/Hugh D. AGBS BSA 82
McClaren/Wilson N. BSA 54
McClellan/David L. AGBS 99
McClelland/Marilyn AGRN MS 77
McClenathan/James R. ANSC BSA 79
McClendon/Hugh Lewis ENSC BSA 96
McClendon/Larry GBUS BSA 75
McClintock/William R. AGRN BSA 41
McClung/William Joseph BSA 35
McClure/Laura BSHE 56
McClure/Nedra Cannon HIED BSHE 81
McClure/Wesley Michael POSC BSA 96
McClure/Wesley Michael AGBS NODEG 96
McClurkin/Herschel H. AGRI BSA 51
McClurkin/Herschel H. AHUS MS 52
McClurkin/John D. BSA 52
McColloch/Carrick L. BSA 24
McCollough/Stacey F. DANB BSA 92
McCollum/Ben D. AGRN BSA 43
McCollum/Elizabeth S. CLTX BSHE 87
McCollum/Joe D. BSA 39
McCollum/Maureen Buckingham CLTX BSHE 82
McCollum/Melle Sewell BSHE 31
McCollum/Patrick Michael ANSC BSA 84
McCollum/Scott Michael AGBS BSA 94
McCone/Shirley Anne FND BSHE 70
McConnell/Edward W. BSA 56
McConnell/John Alison ENTO MS 89
McConnell/Mary Gordon C. FDNU BSHE 82
McConnell/Michael D. POSC 99
McConnell/Reed S. BSA 39

McConoco/Rebecca S. DANB BSA 82
McCool/Deborah Elaine HDFS BSHE 87
McCormick/Daniel R. AGBS BSA 78
McCormick/Herman D. BSA 50
McCormick/Kathryn BSHE 52
McCormick/Myrtle Peek HOEC BSHE 40
McCormick/Nell Staton BSHE 53
McCormick/Ronald L. AGRN BSA 76
McCown/Phyllis Hampel BSHE 54
McCoy/Billy Don AGEC MS 81
McCoy/Helen BSHE 44
McCoy/Kathleen AGED BSA 92
McCoy/Mary Moffitt BSHE 54
McCoy/William M. SOCS BSA 79
McCoy/William O. AGED MS 53
McCoy/William O. AGRI BSA 49
McCracken/Elsie Bernice BSHE 39
McCracken/Lola Rene POSC MS 95
McCrary/Edgar W. BSA 58
McCrary/F. Watkins BSHE 58
McCrary/Moye Lee BSA 30
McCraw/Samuel E. ANSC BSA 83
McCray/Vaughn ANSC BSA 85
McCubbin/William T. BSA 51
McCulloch/Cynthia Ann Gaston HEED BSHE 70
McCullough/Robert Norwood POSC BSA 54
McCurdy/Charles Paul AGRN BSA 94
McCutchen/Billy Eugene BSA 56
McCutchen/Emmett BSA 41
McCutcheon/Elaine Ashley HEED BSHE 77
McCutcheon/Pamela R. HDID BSHE 80
McDaniel/Arthur Devane AGEC BSA 67
McDaniel/Arthur T. HORT MS 63
McDaniel/Arthur T. BSA 48
McDaniel/Jack D. BSA 56
McDaniel/James W. BSA 53
McDaniel/John L. CRSC MS 68
McDaniel/John L. CRSC BSA 66
McDaniel/Linda Ashcraft HEED BSHE 74
McDaniel/M. C. BSA 53
McDaniel/Stanley M. HORT BSA 78
McDaniel/Thomas BSA 37
McDonald/Catherine L. HLED BSHE 81
McDonald/Deborah HEED BSHE 73
McDonald/Ernest D. BSA 49
McDonald/James Earl CRSC BSA 62
McDonald/Jennifer D. AGED BSA 92
McDonald/Jerry Russell ANSC BSA 92
McDonald/Jerry T. VOAG BSA 66
McDonald/Jo McGuire BSHE 57
McDonald/Kathleen Stone BSHE 45
McDonald/Nellie A. BSHE 29
McDonald/R. Henry BSA 37
McDonald/Richard BSA 69
McDonald/Rita Russell HEED BSHE 71
McDonald/Susan Lorraine GNAG MS 88
McDonald/Susan Lorraine ANSC BSA 82
McDonald/William R. AGBS BSA 74
McDougall/Charles W. BSA 57
McDougall/Jo BSHE 57
McDuffie/Jeff D. BSA 31
McElhaney/Curtis W. BSA 59
McElhaney/John C. GNAG MS 85
McElhaney/John C. AGJN BSA 83
McElmurry/Ashley Ann FDNU BSHE 94
McElroy/Ernest M. ANSC BSA 72
McElroy/Michelle Deeann IDHS BSHES 97
McElroy/Mona F. BSHE 48
McElroy/Thomas Clay ANSC BSA 70
McEntire/Amy Janel CLTX BSHE 93
McEntire/James G. VOAG BSA 72
McEntire/Ovaline C. HOEC MS 67
McFall/Kimberly May ANSC BSA 96
McFalls/Ricky Vernon POSC BSA 79
McFalls/Vernon BSA 58
McFarland/C. Edward POSC BSA 94
McFarland/Don D. AGBS BSA 78
McFarlin/J. A. GNAG BSA 76
McFerran/Jack BSA 40
McFerran/Joe BSA 41
McGahee/Marcus Wayne DASC BSA 90
McGahhey/Ervin Daniel BSA 57
McGary/Jerry Linton ANSC BSA 67
McGaughey/John Mark AGRN BSA 76
McGee/Dana Marie ANSC BSA 93
McGee/Leonard W. VOAG BSA 72
McGee/Mary HIED BSHE 83
McGehee/Billy Ray VOAG BSA 64
McGehee/George N. ANSC BSA 84
McGehee/Grace BSHE 52
McGehee/Jesse L. BSA 56
McGhee/Barbara Ann FDNU BSHE 74
McGhee/Gary Thomas GNAG BSA 70
McGhee/James R. BSA 53
McGhee/Ronnie BSA 71
McGhee/Seth Adam AGTE BSA 98
McGibbony/Donna Didier CLTX BSHE 85
McGill/John R. BSA 60
McGill/Josephine T. BSHE 50
McGill/Mary Atkinson BSHE 24
McGill/R. Leighton BSA 28
McGinnis/Alice Crenshaw BSHE 25
McGinnis/Bob BSA 53
McGinnis/Clyde BSA 40
McGinnis/James Paul POSC BSA 86
McGinnis/Martha Trotter BSHE 58
McGinnis/Mary Ann F. BSHE 54
McGinty/Phyllis Kissire HEED BSHE 71
McGlynn/William George FDSC PHD 97
McGlynn/William George FDSC MS 89
McGoodwin/Mary Joyce FDNU MS 63
McGoogan/John Truitt AGRN BSA 88
McGowan/Thomas Watson ANSC BSA 92
McGraw/Fannie R. Harris BSHE 26
McGraw/Ricky AGEC BSA 76
McGraw/Wilmer T. AGRN MS 78
McGregor/Dale BSA 32
McGregor/John Thomas AGRN PHD 86
McGrew/Carl A. BSA 50
McGrew/Patricia E. HEED BSHE 71
McGruder/Diana Bailey CLTX BSHE 81
McGruder/Diana Bailey HOEC MS 84
McGuire/James M. BSA 56
McGuire/Kathleen ANSC BSA 83
McGuire/Lawrence E. DASC BSA 67
McGuire/William S. BSA 47
McIllroy/James Rudolph BSA 50
McIlroy/E. M. BSA 47
McIntosh/Gloria HORT BSA 91
McIntyre/Kenneth E. CRSC BSA 73
McIntyre/Lajuana Beth HEED BSHE 71
McJunkins/James E. GBUS BSA 83
McKay/Cynthia Fanning GHEC BSHE 78
McKee/Charles E. CRSC MS 74
McKee/Charles E. ANSC BSA 69
McKee/Randall L. AGBS 98
McKeever/Jack HORT MS 80
McKelvy/Damon Fletcher AGBS BSA 95
McKelvy/Damon Fletcher AGEC MS 97
McKenzie/Ann Bogle HDID BSHE 82
McKenzie/Bob K. BSA 38
McKenzie/Laura Jeanne FDNU BSHE 72
McKenzie/Laura Jeanne FDNU MS 74
McKie/James McGregor BSA 33
McKim/Joann VOAG BSA 82
McKimmey/Susan L. HDID BSHE 79
McKinney/Nathan Van AGRN BSA 81
McKinney/Nathan Van AGRN MS 83
McKinnon/Ernest Ray BSA 42
McKissack/Terri A. Marshall HDID BSHE 80
McKitrick/Edith Ann HOEC MS 85
McKnight/Christine Louise ANSC BSA 87
McKnight/Howard F. BSA 37
McKnight/John Phillip LDUH BSA 81
McKnight/Marjorie Sue Sims HOEC BSHE 61
McKnight/Mary L. BSHE 47
McKnight/Noel F. BSA 39
McKnight/Robert Atkison BSA 57
McKnight/Susan ENTO MS 73
McKnight/William Bunyan AGRN BSA 87
McKnight/William Bunyan AGRN MS 93
McKown/Guy S. BSA 72
McKown/Stephen J. PLPA MS 68
McKuin/Mary A. FDNU BSHE 62
McLaughlin/Scott Alan ENSC BSA 96
McLean/Marianna D. BSHE 53
McLelland/Robert E. BSA 41
McLemore/Ernestine Faye BSHE 37
McLendon/Frances C. BSHE 40
McLendon/Mack H. BSA 40
McLeod/Ernestine A. BSHE 50
McLeod/Paul J. ENTO PHD 84
McLoud/Douglas J. POSC BSA 83

McLoud/James H. AGRI BSA 60
McLoud/Mary Evelyn HEED BSHE 60
McMahan/Bobby Madison BSA 59
McMahan/Joseph Arch BSA 50
McMahand/Stewart Allen ANSC BSA 84
McMahen/Carmen V. CLTX BSHE 77
McMahen/Elvis N. BSA 38
McMaster/John D. ANSC MS 70
McMaster/John D. POSC BSA 68
McMillan/Alice Wingfield BSHE 63
McMorran/Lesli April Hirby HDFS BSHE 93
McMullen/Richard D. BSA 55
McMurry/Wilda Sharon FND BSHE 66
McNabb/Coy G. BSA 41
McNair/Cindy HDID BSHE 79
McNeal/Charles D. ENTO BSA 68
McNeal/John W. BSA 49
McNeal/Ruth A. Cravens BSHE 48
McNeely/Van Michael AGRN BSA 93
McNeil/Jack AHUS BSA 46
McNeil/Melanie D. AGBS BSA 87
McNeill/Deborah Raye DASC BSA 84
McNew/Beverly M. HOEC MS 75
McNew/Fred BSA 42
McNiece/Cecil L. BSA 41
McNulty/Peter ANSC BSA 88
McPherson/Bill N. POSC BSA 59
McPherson/Ralph R. BSA 14
McQueen/Alan D. AGEC MS 81
McRae/Diana G. HEED BSHE 68
McRae/Phillip E. BSA 29
McSpadden/Christiana HOEC MS 79
McVey/W. A. BSA 44
McWaters/Gene W. BSA 51
McWha/Debra L. Case GHEC BSHE 85
Meacham/Maible S. Simpson BSHE 33
Mead/Paul K. BSA 55
Meadows/Clinton E. BSA 53
Meadows/Howard BSA 49
Meadows/Viola C. BSHE 49
Meahl/Helen E. BSHE 36
Means/Curtis H. AGBS 99
Measel/John W. BSA 35
Measeles/Gilbert M. BSA 31
Meazle/James Leonard VOAG BSA 82
Medal/Julio Cesar ENTO PHD 94
Medley/Orland Dale BSA 58
Medlin/Faye P. BSHE 53
Medlock/Teresa Lynn ANSC BSA 95
Medora/Nilufer P. HOEC MS 79
Meek/Chester Lamar ENTO MS 71
Mehdizadegan/Feridoon ANSC PHD 83
Meier/Deborah J. AGRN PHD 84
Meiner/Bobbie D. Guffey HEED BSHE 61
Melendez/Daniel A. FDSC MS 81
Mellado/Miguel ANSC MS 79
Mellor/Julia I. BSHE 43
Melton/Curtis C. ANSC BSA 60
Melton/Janice Harber HEED BSHE 70
Melton/Mary White BSHE 30
Melton/Nancy Bane FND BSHE 71
Melton/O. J. BSA 33
Melton/William O. BSA 30
Melton/William T. ANSC BSA 84
Mena/JoaQuin AGRN MS 88
Mendenhall/Michael J. AGEC BSA 81
Mendenhall/Myles Elton AGBS BSA 85
Mendenhall/Ruby T. BSHE 17
Mendoza/Joffrey Rodrigo AGEC BSA 95
Mercado/Jose Edgardo BSA 52
Mercer/Robert Keith AGBS BSA 81
Merchant/Charles Ray GBUS BSA 75
Merchant/Deborah L. HEED BSHE 79
Mermoud/Dale Eugene GNAG MS 80
Merrell/Julia Lorraine IDHS BSHE 94
Merrick/Joyce BSHE 58
Merrifield/Mary Nell BSHE 66
Merrill/Bettie Parrott BSHE 59
Merriman/Janet FDNU BSHE 72
Merritt/Dani D. IDHS BSHE 89
Merritt/J. C. Melton BSA 50
Mersiovsky/Edgar AGRN MS 93
Mersky/David AGRN MS 92
Mersky/David AGEC BSA 89
Mertens/William C. LDUH BSA 76
Meshew/Michael Lee POSC BSA 70
Messia/Mboussi a' AGRN MS 96
Metcalf/Brenda Elaine ANSC BSA 84
Metcalf/Janice Ann BSHE 60
Metcalf/Terry ANSC BSA 77
Metcalf/Virgil A. AGEC MS 60
Metcalf/Virgil A. BSA 58
Metheny/Sandra A. AGBS BSA 85
Metheny/Steve AGBS BSA 81
Methvin/Mark POSC BSA 81
Metz/Charles L. ANSC BSA 82
Metz/Danny Ray AGEC BSA 69
Metz/Larry Don ANSC BSA 74
Metz/Mary R. HEED BSHE 77
Metzler/George F. RSOC MS 63
Metzler/George F. HORT BSA 27
Meyer/Alvin Reynold AGBS BSA 96
Meyer/George Burnett BSA 50
Meyer/Jeffery Alan ENTO PHD 81
Meyer/Laura Lee CLTX BSHE 76
Meyer/Nancy Jo BSHE 77
Meyers/Laura HOFS BSA 78
Meyers/O. L. ANSC BSA 83
Mgonja/Andrew Peter PLPA PHD 83
Mgonja/Mary Andrew AGRN MS 83
Miah/Nur M. AGRN PHD 78
Middlekauff/Helen V. BSHE 25
Middleton/Julie Anne AGED BSA 86
Middleton/Kathryn Fitzhugh BSHE 52
Middleton/Robert E. BSA 55
Miesner/Jim CRSC MS 72
Miesner/Jonathan Ray AGRN BSA 96
Mika/Timothy D. AGBS BSA 82
Mikus/Lisa CLTX BSHE 89
Milam/Alfred N. ESWS 99
Miles/Mildred Cash BSHE 46
Miles/Richard D. POSC BSA 71
Miles/Richard J. BSA 49
Miley/Carroll Annette BSHE 65
Miley/Jerry L. POSC BSA 64
Miley/Kenneth A. BSA 58
Milholland/Emma McDonald BSHE 34
Milholland/James G. BSA 35
Milholland/Paul S. BSA 41
Millar/Paul Harwood BSA 17
Millard/Dennis R. ANSC BSA 72
Miller/Alan ANSC MS 82
Miller/Alan POSC BSA 81
Miller/Carl N. AGRI BSA 51
Miller/Carolyn Lou DANB BSA 85
Miller/Debra Smith POSC BSA 86
Miller/Donna HEED BSHE 75
Miller/Doug AGBS BSA 80
Miller/Doyle Lee ANSC BSA 85
Miller/Edra Kay AGBS BSA 90
Miller/Erica M. AGRN MS 83
Miller/Forrest B. BSA 38
Miller/Fredric D. ENTO MS 81
Miller/Gaylon E. BSA 59
Miller/George Franklin ANSC MS 91
Miller/George R. BSA 53
Miller/Glenda J. HEED BSHE 77
Miller/Hattie Mae Dotu BSHE 29
Miller/James Wayman LDUH BSA 88
Miller/Janet Lucretia HEED BSHE 63
Miller/Jeffry Shannon ANSC PHD 91
Miller/John H. BSA 29
Miller/John N. AGEC BSA 70
Miller/John Vorkpanah AGRN BSA 91
Miller/Judy Diane HDFS BSHE 93
Miller/Kelly CLTX BSHE 86
Miller/Kirk William AGBS BSA 93
Miller/Lisa Gale DANB BSA 80
Miller/Lyndon Andrew DASC BSA 87
Miller/Patricia Reinhart ENTO MS 79
Miller/Patricia Reinhart BSA 77
Miller/Philip G. VOAG BSA 68
Miller/Sean Anthony AGED BSA 89
Miller/Virginia Blanshard BSHE 24
Miller/William W. AGRI BSA 54
Milligan/Sarah Fletcher CLTX BSHE 82
Milligan/Scott GNAG MS 90
Milliken/Robert Sean ANSC BSA 97
Mills/Danny G. AGEC BSA 71
Mills/James L. BSA 59
Mills/Joe S. GBUS BSA 77
Mills/John McCollum LDUH 79
Mills/Maurice J. BSA 40
Mills/Timothy D. AGBS BSA 76

Milton/Doris Whittington BSHE 29
Milton/Ed BSA 49
Minchew/Milton L. BSA 54
Mineart/Douglas James ENSC BSA 96
Miner/Arthur BSA 58
Miner/Carolyn Beaty FDNU BSHE 68
Miner/James J. POSC MS 54
Miner/James J. AGRI BSA 53
Minkel/Juliette Smith HOEC MS 91
Minnick/Louise Gardner BSHE 37
Minor/Kathryn L. HOFS BSHE 82
Minton/Wyndal I. BSA 51
Mintz/Angel Santos PLPA MS 91
Mir/Mohammed Sadeghi AGRN PHD 84
Mirabello/Robert Allan HORT MS 97
Mirabello/Robert Allan LDUH BSA 94
Mirbagheri/Firooz AGRN MS 79
Mires/Emma C. BSHE 36
Mirlohi/Aghafakhr F. AGRN PHD 91
Mischo/Ora A. CLTX BSHE 94
Misenheimer/Bob W. BSA 76
Misenheimer/Glenn L. AGRI BSA 61
Misenheimer/Michael Wayne AGED BSA 91
Missey/Tim Lee ANSC BSA 97
Mitcham/Sharon A. HEED BSHE 72
Mitchell/Boss C. BSA 38
Mitchell/Carla Noel CLTX BSHE 89
Mitchell/Carlos N. GNAG BSA 64
Mitchell/Denton Lee AGEC MS 89
Mitchell/Denton Lee PPPM BSA 87
Mitchell/Donna K. FND BSHE 69
Mitchell/E. Underwood BSA 50
Mitchell/Eula HEED MS 56
Mitchell/Evelyn M. BSHE 38
Mitchell/Gretchen Scheurich FDNU BSHE 78
Mitchell/Homer Eldon HORT BSA 87
Mitchell/James E. ANSC BSA 74
Mitchell/Joan E. BSHE 68
Mitchell/Karen HEED BSHE 74
Mitchell/Kelly Dawn LDUH BSA 91
Mitchell/Nancy Moore BSHE 58
Mitchell/Ronald E. VOAG BSA 79
Mitchell/Suzanne Sumner BSHE 74
Mitchell/Timothy H. AGRN BSA 78
Mitchell/Walker A. ENSC BSA 97
Mitchell/William BSA 42
Mitchell/William Isaac POSC BSA 92
Mixon/Bobby J. AGEC MS 79
Mize/Elizabeth A. HOEC BS 94
Mobley/Connie Lee BSA 84
Mobley/Luke P. ANSC BSA 98
Mobley/Matthew Clang ANSC BSA 96
Mobley/Peggy HEED BSHE 70
Mock/Mary Blanche BSHE 28
Modaihsh/Abdullah Saad AGRN PHD 86
Moeljopawiro/Sugiono AGRN PHD 86
Moeller/Bill C. ANSC BSA 72
Moeller/W. J. BSA 64
Moeller/W. J. CRSC MS 66
Moellers/Robert Frank POSC BSA 88
Moery/Clarence B. BSA 64
Moery/Kyle AGEC BSA 85
Moery/Nancy GHEC BSHE 84
Moery/Phillip W. AGEC BSA 70
Moffet/James E. BSA 53
Moffitt/Hugh Price BSA 23
Moffitt/Kate E. ANSC 99
Mohamad Sidin/Azizan AGEC MS 94
Mohamed/Abdalla I. ENTO PHD 83
Mohamed/Abdalla I. ENTO MS 79
Mohan/Rao Balmoori FDSC MS 80
Mohdesa/Yazid Bin ENTO MS 80
Mohdhasim/Norhashiah HOFS BSHE 79
Mohdnoor/Ramli Bin AGRN PHD 79
Moi/Timon Isaac HORT MS 96
Moix/Ernest LDUH BSA 78
Mok/Yoke Ngor FDSC PHD 92
Mok/Yoke Ngor FDSC MS 89
Moline/John Anton PPPM BSA 88
Moll/H. David BSA 78
Moll/Marieta CLTX BSHE 76
Monahan/Nancy Howard BSHE 60
Mones/Kathryn E. FHNH 98
Money/Charles Evans POSC BSA 71
Money/Margaret Anne HOEC MS 85
Money/Margaret Anne CDFR BSHE 76
Monis/Judit PLPA MS 85
Monk/Amy Sue HOEC MS 94
Monks/Charles Dale AGRN MS 85
Monks/David Wilson AGRN PHD 85
Monroe/Marian L. HOEC BSHE 55
Montgomery/F. W. BSA 49
Montgomery/Harold BSA 34
Montgomery/Jefferson R. BSA 59
Montgomery/Jeffery W. AGBS BSA 82
Montgomery/Jon Tyler HTSO BSA 98
Montgomery/N. Catherine HOEC MS 93
Montgomery/Shannon Renner IDHS BSHE 95
Moody/Julius A. BSA 49
Moody/Rosemary Diane Privett FND BSHE 63
Moody/Thomas Dale GNAG MS 91
Moody/Thomas Dale SOSC BSA 84
Moon/Mitchell Lyle DASC BSA 86
Mooney/Chad T. AECT 99
Mooney/Gina Hill BSHE 85
Mooney/John S. AGME BSA 77
Moore/Angela Doren CLTX BSHE 90
Moore/Bernard Jerry BSA 52
Moore/Betty Duncan FDNU BSHE 44
Moore/Bill ENTO BSA 66
Moore/Billy J. BSA 68
Moore/Byron E. BSA 43
Moore/Clay R. BSA 41
Moore/Ed Dye GNAG BSA 51
Moore/Eugenia Taylor CLTX BSHE 80
Moore/Gene W. BSA 38
Moore/George A. GNAG BSA 49
Moore/Glen ENTO MS 75
Moore/James Clinton AGBS BSA 86
Moore/James Howard BSA 31
Moore/James N. BSA 56
Moore/Jimmy J. GNAG MS 91
Moore/Joe Claude AGED BSA 93
Moore/John Burton GBUS BSA 72
Moore/Judy A. HEED BSHE 76
Moore/Karen Elizabeth IDHS BSHE 91
Moore/Karen Grayson HIED BSHE 81
Moore/Kelly L. POSC BSA 89
Moore/Kerry Jay FDSC MS 85
Moore/Kimberly Kirby FDNU BSHE 82
Moore/Leon ENTO MS 59
Moore/Leon BSA 57
Moore/Linda HEED BSHE 64
Moore/Lynn ENTO MS 75
Moore/Mary Cooper FND BSHE 74
Moore/Michael L. ANSC BSA 75
Moore/Misty J. HORT 98
Moore/Natasha S. ESWS 98
Moore/Patricia M. POSC 99
Moore/Patricia G. CLTX BSHE 92
Moore/Patricia Lea CLTX BSHE 89
Moore/Philip A. SOSC BSA 0
Moore/Philip A. AGRN MS 81
Moore/Philip A. SOCS BSA 79
Moore/Rachel Danielle AGRN BSA 90
Moore/Reba Posey BSHE 65
Moore/Russell A. ANSC BSA 74
Moore/Russell Wayne AGME BSA 81
Moore/Steven Hal AGRN PHD 87
Moore/Trude Alley FND BSHE 73
Moore/Vicki Cheryl CLTX BSHE 86
Moore/W. Trueman BSA 51
Moore/Wilma Geraldine Caudle DIET BSHE 49
Moorman/Jason Roy AGBS BSA 94
Moose/Martha Jane BSHE 41
Mooty/Dawn CLTX BSHE 82
Moppin/Mickey Glyn ANSC BSA 93
Moran/Rebecca Christine HDFS BSHES 98
Moran/Renae E. HORT MS 91
Moran/Shelly HEED BSHE 67
Morcos/Vicki Morris IDHS BSHE 89
Morden/Kelly Christine POSC BSA 96
Morelock/Teddy E. HORT MS 68
Morelock/Teddy E. HORT BSA 66
Moreno/Ala'n Jesu's Beracierto AGRN BSA 80
Moreno/Luis H. BSA 55
Moreton/Dan J. ANSC BSA 86
Morgan/David Wayne AGME BSA 95
Morgan/Deborah McKinnis ANNU BSA 78
Morgan/Helen BSHE 30
Morgan/Niven Doyle BSA 30
Morgan/Steven K. POSC BSA 77
Morgan/Tracy Stroud POSC BSA 95
Morgan/Vicki L. HDID BSHE 83

Morphew/Mitchell D. AGBS 98
Morris/Barney P. AGBS BSA 82
Morris/Brad Louis AGBS BSA 88
Morris/Donald B. AGSC MS 61
Morris/Donald B. BSA 49
Morris/Dwight V. BSA 70
Morris/Eugene R. BSA 56
Morris/George L. BSA 57
Morris/James T. VOAG BSA 64
Morris/Johnnie L. GNAG MS 82
Morris/Justin R. BSA 58
Morris/Justin R. HORT MS 61
Morris/Margaret Dean AGEC MS 85
Morris/Marietta BSHE 51
Morris/Mark L. AGBS BSA 81
Morris/Misty Ann HDFS BSHE 95
Morris/Neil AGEC MS 83
Morris/Neil AGEC BSA 80
Morris/Phillip AGRN BSA 86
Morris/Teresa Vick HEED BSHE 92
Morris/Terrie L. CLTX BSHE 79
Morrison/April S. ESWS 99
Morrison/Glenn D. BSA 60
Morrison/Glenn H. AGEC MS 55
Morrison/Glenn H. AGRI BSA 54
Morrison/Harrison L. ANNU BSA 66
Morrison/Harrison L. POSC MS 67
Morrison/James E. BSA 48
Morrison/Nona E. BSHE 60
Morrison/Pat S. BSHE 65
Morrison/Patty FDNU BSHE 61
Morrison/Sarah Parish BSHE 57
Morrison/W. R. AGEC MS 50
Morrissey/Travis Jeremy ENSC BSA 95
Morrow/Beulah Michelle CLTX BSHE 85
Morrow/Carroll S. BSA 32
Morrow/Charlene M. ENSC BSA 78
Morrow/H. Doyle POSC MS 76
Morrow/H. Doyle POSC BSA 71
Morrow/Hugh E. BSA 4
Morrow/Ronald E. ANSC BSA 68
Morrow/Ronald E. ANSC MS 72
Morse/Betty Ruth Etheridge BSHE 54
Morse/Patricia A. FND BSHE 69
Morse/Selmer E. BSA 50
Morse/Susan Brewer FDNU BSHE 73
Morton/Michael William ANSC BSA 88
Morton/William M. HORT BSA 61
Morton/Willie H. BSA 54
Moseley/Melissa Dawn IDHS BSHE 95
Moses/William D. AGEC BSA 70
Mosher/Steven Mitchell AGBS BSA 85
Mosier/Dwight G. AGRN PHD 86
Mosier/Myra Mowery Dexter BSHE 49
Mosley/Mary P. BSHE 55
Mosley/Suzanne CLTX BSHE 89
Mosley/Virginia Payne CDFR BSHE 74
Mosley/Wiley W. AGEC MS 57
Mosley/Wiley W. BSA 55
Moss/Columbus W. GNAG MS 89
Moss/Herman T. AGED MS 50
Moss/Joan C. BSHE 52
Moss/Robert H. MS 48
Moten/James W. BSA 43
Motes/Dennis R. HORT MS 78
Motes/Richard R. AGBS BSA 82
Motichek/George R. HORT MS 85
Moubarak/Ali S. ANSC PHD 82
Moudy/Linton C. BSA 54
Moudy/Paul BSA 53
Mountcastle/Walter Eggleston BSA 27
Mowery/Alonzo C. ANSC BSA 40
Moxley/Shari Coleman HEED BSHE 91
Muda/Abdul Rahim Bin ENTO MS 78
Mudford/Curtis L. GNAG MS 82
Mudford/Curtis L. VOAG BSA 78
Mudford/Steven VOAG BSA 81
Mulhollan/Mary Bess BSHE 56
Mulhollen/Becky BSHE 65
Mulkey/William A. BSA 65
Mullen/Jack G. BSA 51
Mullen/Patrick G. AGRN BSA 79
Mullen/Patrick G. AGRN BSA 78
Mullenix/Otis BSA 53
Mullenix/Phyllis N. FND BSHE 69
Muller/E. Thayne BSA 29
Muller/Joe Edwin GNAG MS 87
Mullikin/Leasea Dawn POUL BSA 98
Mullins/James Andrew ANSC BSA 88
Mullins/Karen Ann FDNU BSHE 81
Mullins/Linda Price BSHE 76
Mullins/Troy BSA 34
Mullis/Stephen Marc AGRN BSA 82
Mulson/Troy Dean FDSC BSA 87
Mumford/Joel Mark POSC BSA 81
Muncey/Paula CTM BSHE 75
Muncy/Henry R. BSA 52
Mundy/Gary E. BSA 69
Mundy/Martha E. HDFS 98
Munir/Christie Lea HOEC MS 93
Munn/Cherlyn Hoyle AGBS BSA 90
Munsey/Elizabeth Ann BSHE 55
Munson/Sheryl Whitby FDNU BSHE 91
Murchison/Benjamin F. BSA 53
Murdoch/Betsy SOSC BSA 80
Murdoch/Charles L. MS 60
Murdoch/Charles L. BSA 59
Murdock/Candace E. Williamson HORT MS 76
Murphy/Albert E. BSA 51
Murphy/Bernard D. POSC MS 79
Murphy/Candice Treat CLTX BSHE 81
Murphy/Carl E. BSA 51
Murphy/Charlie LDUH BSA 90
Murphy/Gary E. AGRN BSA 68
Murphy/Gary E. ANSC MS 77
Murphy/Joseph C. HORT BSA 77
Murphy/Maurice J. BSA 50
Murphy/Pat BSHE 86
Murphy/Ron Yang FDSC MS 89
Murphy/Vicki CLTX BSHE 85
Murr/Dale AGBS BSA 76
Murray/Andrea FDSC MS 84
Murray/Donald BSA 54
Murray/Martha W. BSHE 52
Murry/Cynthia B. CTM BSHE 75
Musbah/Muftah A. POSC PHD 80
Muse/Dorothy Beuse BSHE 33
Musgrave/Amos AGEC MS 30
Musgrave/Otis W. AGED BSA 36
Musial/Penelope Reynolds FND BSHE 75
Musick/Joseph A. AGEC MS 71
Musick/Joseph A. AGEC BSA 69
Muthiah/Selvaraju N. AGRN MS 92
Muwalla/Marwan M. ANSC PHD 83
Myers/Barbara Mazzanti BSHE 60
Myers/Cecil E. BSA 32
Myers/Danny L. POSC BSA 70
Myers/Douglas Wade DASC BSA 85
Myers/Joan Edna BSHE 65
Myers/Kenneth L. BSA 79
Myers/Kent GNAG BSA 78
Myers/Kimberly CLTX BSHE 79
Myers/Larry Bruce ANSC BSA 70
Myers/Larry F. ANSC BSA 98
Myers/Robert D. AGME BSA 78
Myles/Lisa Michelle ANSC BSE 97
Naill/John D. AGBS BSA 73
Nailling/Cecilia Ann FDNU BSHE 71
Nall/George A. BSA 50
Nammuang/Prayong SOSC MS 71
Nance/Candace L. CLTX BSHE 84
Nance/Mauree E. BSHE 33
Nance/Raymon Bryan AGEC MS 85
Nance/Renne Don AGEC MS 85
Naraguma/Joseph PTSC PHD 98
Naraguma/Joseph HORT MS 94
Narramore/Virginia G. Mills BSHE 58
Nash/Donald B. POSC BSA 89
Nash/Lelus B. BSA 58
Nash/Lorraine E. GHEC BSHE 87
Nash/Raymond C. BSA 53
Nash/Ronald VOAG BSA 80
Nash/Rose Wardlaw BSHE 60
Nasre/Esfahani-Manucherhr PTSC PHD 81
Nasre/Esfahani-Manucherhr AGRN MS 79
Nasrollahi/Abbas Ali ENTO BSA 83
Nastasl/Paolo AGRN PHD 85
Nasti/Kimberly Beth ANSC PHD 96
Nation/Mary G. HDID BSHE 82
Nations/Eleanor Bell BSHE 37
Nauman/Hayden V. BSA 51
Navas/Diego E. ENTO MS 57
Navas/Diego E. ENTO BSA 56
Nayigihugu/Venerand ANSC PHD 97

Nayigihugu/Venerand ANSC MS 91
Naylor/Sandra Kay BSA 24
Ndife/Louis Izuchukwu POSC PHD 87
Ndife/Matrid King FDSC PHD 87
Neal/Edith CLTX MS 76
Neal/Phillip T. GNAG BSA 80
Neal/Phillip T. GNAG MS 81
Neal/Virginia Holbrook BSHE 32
Neal/William Oliver BSA 55
Nealon/Francis Richard BSA 51
Nease/Rondall L. DHUS BSA 61
Nederpelt/Sherry Thompson CLTX BSHE 91
Neece/Nikki Alane CLTX BSHE 90
Neel/Joseph B. AGRN BSA 78
Neeley/Billie Jennings BSHE 51
Neely/J. Winston BSA 28
Neely/Michael B. FDSC MS 78
Neff/Bonnie Peterson BSHE 45
Negrotto/David Vincent AGRN MS 88
Neighbor/Noel ANSC MS 91
Neighbors/Gary D. ANSC BSA 77
Neighbors/William Everett BSA 50
Neil/Robert J. BSA 53
Neill/Hoyt BSA 48
Neill/Virginia Cochran BSHE 47
Nelke/Julia Marie FDNU BSHE 77
Nelsen/Steven L. DANB BSA 77
Nelson/Elsie BSHE 59
Nelson/Irene Henrietta BSHE 18
Nelson/James Avey BSA 49
Nelson/James Gregory AGBS BSA 83
Nelson/Laverne HOEC MS 85
Nelson/Matthew Stephen FDSC BSA 88
Nelson/Stephen C. AGEC MS 82
Nelson/Talmadge POSC BSA 52
Nelson/William Leon GNAG MS 86
Nelson/William O. BSA 51
Nelson/William Perry AGEC MS 61
Nelson/William Perry AGRI BSA 58
Nelson/Wynona Jean FDNU BSHE 73
Nesbitt/Jimmie Dow POSC MS 76
Neslage/Laurie Dale LDUH BSA 76
Nester/Paul R. AGRN MS 78
Nester/Paul R. PLPA BSA 75
Nester/Ruel P. BSA 53
Netherton/Barbara Elaine HDFS BSHE 89
Nettles/Charles J. BSA 57
Nettles/Denver E. BSA 50
Nettles/Fred BSA 48
Neu/Kathleen Kay HEED BSHE 68
Neukranz/Jennifer L. HORT MS 85
Neukranz/Jennifer L. LDUH BSA 82
Neumann/Lorena P. FDNU BSHE 47
Neumeier/Mary Ann Brabec BSHE 53
Neumeier/Timothy J. AGED BSA 87
Neumeier/Timothy J. AGME BSA 86
Newbern/Ila BSHE 24
Newberry/George T. POSC BSA 72
Newberry/George T. BSA 57
Newberry/Getulio C. BSA 59
Newberry/Judith A POSC MS 82
Newberry/Lisa Ann POSC PHD 96
Newberry/Lisa Ann ANSC MS 89
Newberry/William L. BSA 47
Newby/Thomas Jack ANSC PHD 83
Newlin/Janice FDNU BSHE 72
Newman/Mark S. BSA 74
Newman/Susan Elisabeth BSA 87
Newquist/Cynthia S. CLTX BSHE 76
Newton/Carolyn FND BSHE 75
Newton/Karen E. GHEC BSHE 85
Newton/Lori Maye HOUS BSHES 98
Newton/Peter W. LDUH BSA 77
Newtown/Barbara Jean BSHE 52
Ngoumou/Titus Nga AGRN MS 88
Nguyen/Dinh Qui POSC MS 63
Nguyen/Dinh Qui ANNU BSA 61
Nguyen/Hong Thi-Thuy CLTX BSHE 92
Nguyen/Thien Da CLTX BSHES 96
Niba/Teresa B. HEED BSHE 81
Nicholas/Dayna Leanne ANSC BSA 85
Nichols/Cherry A. HOEC MS 81
Nichols/D. Jeff SOSC BSA 90
Nichols/Jane Scott CTM BSHE 73
Nichols/Robert Allen VOAG BSA 82
Nichols/Rodney G. ANSC BSA 78
Nicholson/Cynthia Gwaltney CLTX BSHE 81
Nicholson/David C. GNAG BSA 68
Nicholson/Juanita BSHE 43
Nicholson/William Randall GNAG MS 91
Nicholson/William Randall VOAG BSA 79
Nickels/C. W. BSA 38
Nickels/Lucy Williams BSHE 39
Nickels/Wallace E. AGRN BSA 46
Nickerson/James Charles Everett ENTO MS 72
Nickerson/James Charles Everett ENTO BSA 69
Nickles/Melanie Deanice DIAM BSA 89
Nida/Debbie L. PPPM BSA 87
Nieh/Cinping FDSC PHD 91
Nieh/Cinping FDSC MS 86
Niemi/Jyrki Sakari AGEC MS 91
Nik/Abdullah Nik Hashim AGEC MS 83
Nimley/Edwin B. RSOC MS 93
Nipps/Jennifer Barnett CLTX BSHES 97
Nipps/Scott AGBU BSA 97
Nisbett/Charles D. AGBS 99
Niven/Charles F. DASC BSA 35
Niven/James A. BSA 32
Niven/Lauren Kennedy CLTX BSHE 80
Niven/William A. BSA 40
Nix/Elizabeth L. CLTX 99
Nkunzimana/Salvator AGEC MS 94
Nnyagu/Chukwunweogo C. POSC MS 88
Noble/T. Scott ANSC BSA 92
Nobles/Henry A. BSA 39
Nobles/James R. BSA 58
Nobles/Lori Ann ENTO MS 95
Noel/Anita BSHE 54
Noggle/J. C. BSA 52
Noisewater/Janie HEED BSHE 71
Nokes/Charles Lynn HORT BSA 69
Nokes/Linda Kaye HOEC MS 71
Nokes/Linda Kaye HEED BSHE 69
Nokes/Martha Reye ENSC BSA 98
Noland/Betty A. Engler HOEC MS 79
Noland/Betty A. Engler FDNU BSHE 77
Noland/Steve ANSC BSA 72
Noland/Thomas L. PTSC PHD 84
Nolen/Hansel Boyce BSA 51
Norcross/Herrick F. AGEC MS 88
Norcross/Herrick F. AGBS BSA 84
Norcross/Jill H. FDSC MS 85
Norcross/Jill H. FDSC BSA 85
Nordstrom/Connie L. HOFS MS 76
Norman/Logan HDFS BSHE 94
Norman/M. David AGEC MS 82
Norris/Gary F. AHUS BSA 64
Norris/Gini HOEC MS 80
Norris/Jackie Lee AGBS BSA 80
Norsworthy/Jason Keith AGRN MS 97
Norsworthy/Thomas T. AGEC BSA 82
Northcutt/Alan Dane AGRN BSA 92
Northcutt/Imogene Coger GHEC BSHE 40
Northup/James F. BSA 63
Norton/Linda Murphy CLTX BSHE 70
Norton/Michael Chad ANSC MS 93
Norvell/Dona J. HORT MS 81
Norvell/Dona J. LDUH BSA 78
Norwood/Ann Dilliard CLTX BSHE 79
Norwood/Miles I. BSA 50
Norwood/Shirley A. BSHE 72
Nostrand/Catherine A. Lazear GHEC BSHE 79
Nowlin/Kelly Garland CLTX BSHE 95
Nowlin/Lucian Weast BSA 51
Noyes/Mary Jean Sparks HEED BSHE 62
Ntamatungiro/Sixte AGRN PHD 96
Ntamatungiro/Sixte AGRN MS 92
Nunn/John AGRN BSA 79
Nunnally/Joy Morrow DIET BSHE 47
Nurdjanah/Siti FDSC MS 90
Nutt/David W. VOAG BSA 79
Nutt/Davie Vincent BSA 78
Nutt/Frances BSHE 50
Nutt/Hugh BSA 50
Nutt/Sharon J. AGEC BSA 82
Nwoha/Ogbonnaya John AGEC MS 93
Nwoha/Ogbonnaya John ANSC BSA 90
Nyaribo-Roberts/Fanny Boyani AGEC MS 82
O'Bar/Billy D. POSC BSA 68
O'Brien/David R. POSC BSA 74
O'Brien/Gayle L. HEED BSHE 77
O'Brien/Linda Kay ANSC MS 81

O'Bryant/Marshall BSA 49
O'Dell/John P. ABBS 99
O'Donnell/Sherri A. Sample CLTX BSHE 79
O'Hara/Lea Bratton HIED BSHE 81
O'Kelley/Roger Keith AGRN PHD 89
O'Kelly/Joseph Fred BSA 20
O'Kelly/Lois Barrett BSHE 20
O'Leary/Daniel J. PLPA MS 81
O'Leary/Daniel J. PLPA BSA 78
O'Mara/Glenanna K. HDID BSHE 79
O'Neal/Edward K. BSA 40
O'Neal/Linda HEED BSHE 74
O'Neal/M. Kathryn Hogue BSHE 40
O'Neal/Melva BSHE 51
O'Neal/William C. BSA 50
O'Neil/Jennifer Elgina CLTX BSHE 95
O'Quin/Karen Ann CLTX BSHE 87
O'Quin/Morris N. AGRN BSA 85
O'Quin/Morris N. GNAG MS 82
O'Shaughnessy/Marcia ANSC BSA 79
Oakes/Gerald AGED BSA 50
Oakes/Gwen M. HEED BSHE 49
Oakes/Harold GNAG MS 68
Oakes/Harold BSA 49
Oakes/John London HOFS MS 80
Oakes/Ray R. BSA 54
Oakley/Margaret J. BSHE 24
Oakley/Ovita BSHE 40
Oakley/Richard Garland BSA 28
Oakley/Stephen Roland AGRN PHD 85
Oates/David B. GNAG BSA 62
Oates/F. B. BSA 17
Oates/Max Bruce BSA 9
Oates/Robert W. BSA 49
Oates/Rozan HEED BSHE 83
Oates/William O. BSA 51
Obert/Donald Edward AGRN MS 92
Obiagwu/Chukwuma Justin AGRN MS 80
Ocker/Janet Sue CLTX BSHE 82
Odell/Jerry Clark BSHE 86
Odglen/Glen E. ENTO MS 61
Odglen/Glen E. BSA 58
Odglen/Randy POSC BSA 85
Odle/Elaine HOFS BSHE 78
Odom/Doyle O. BSA 51
Odom/J. C. AGRN BSA 86
Oetting/Bryan Clark ANSC BSA 86
Oetting/Bryan Clark ANSC MS 88
Ogden/A. K. BSA 41
Oglesby/Betty J. BSHE 46
Oglesby/Donald L. AHUS BSA 64
Oglesby/Julius Doyle BSA 58
Ogun/Modinat A. K. LDUH BSA 91
Ojukwu/Augustine Ora CLTX BSHE 84
Ojukwu/Edwin C. AGEC MS 83
Ojukwu/Edwin C. AGEC BSA 84
Okereke/Maurice I. PLPA MS 78
Okolo/Sidney O. AGME BSA 86
Okuley/Gail FND BSHE 75
Olcott-Reid/Brenda Lesli PTSC PHD 93
Olive/Floyd R. BSA 36
Olive/William R. BSA 47
Oliver/Andrew Cecil AGRN MS 82
Oliver/Betty F. HOEC MS 67
Oliver/Elizabeth Ann Henson HEED BSHE 70
Oliver/Jane Jeffery HOEC MS 92
Oliver/Jimmy D. FDNU BSHE 91
Oliver/Lawrence R. CRSC MS 66
Oliver/Lawrence R. CRSC BSA 64
Oliver/Linda Keene FND BSHE 70
Oliver/Nancy Laree HEED BSHE 63
Oliver/Thomas P. BSA 31
Oliver/William L. BSA 21
Ollis/Fay Bartlett FDNU BSHE 56
Olsen/Lawrence M. DANB BSA 80
Olson/Christopher Allen ANSC BSA 89
Olson/Helen Harper BSHE 49
Olson/Sharon Harvey HEED BSHE 72
Omakupt/Manu SOSC MS 70
Ontiveros/Rinda R. AGRN MS 83
Ooi/Chee Soon AGEC MS 88
Orange/Robert S. AGEC BSA 69
Orcutt/Dan Livingston AGBS BSA 86
Orfanos/Lisa Stecks CLTX BSHE 81
Orr/Houston W. VOAG BSA 76
Orr/Kenneth Wayne LDUH BSA 86
Orrell/Rhonda J. HOEC MS 89
Orrick/Esther Sharon BSHE 51
Ortiz/Carlos E. AGRN PHD 92
Ortiz/Castro Valentin ANSC BSA 82
Ortiz/Castro Valentin GNAG MS 85
Ortiz/Castro Valentin GNAG MS 85
Ortiz/John H. ENTO MS 80
Osborn/Carl T. AGEC MS 81
Osborn/Carl T. ENSC BSA 80
Osborn/Raymond AGEC BSA 81
Osborne/Alan Wayne ANSC BSA 86
Osborne/Hilton F. HORT MS 66
Osborne/Hilton F. HORT BSA 64
Osborne/Merna HORT BSA 91
Osborne/Terry N. POSC BSA 83
Osburn/Royal R. BSA 58
Osgood/Lucy D. BSHE 31
Osgood/Otis T. BSA 31
Osier/Steven Curtis AGRN BSA 96
Osing/Gerald M. DANB BSA 86
Osman/Nixar G. AGEC MS 65
Osteen/Sissy Roydell HOEC MS 82
Osteen/Sissy Roydell HEED BSHE 80
Ostendorf/Joe H. BSA 40
Ostendorf/Zelma Corley BSHE 37
Oswald/Nolan BSA 59
Oswald/Otto L. FDSC MS 85
Oswald/Otto L. FDSC BSA 83
Ott/John Robert BSA 49
Ott/Velma Clark BSHE 40
Otwell/Brenda J. FDNU BSHE 68
Otwell/Howard David AGBS BSA 93
Otwell/Orel G. BSA 41
Otwell/Richard L. ENTO MS 71
Overbey/George Wilson BSA 52
Overstreet/Helen T. BSHE 35
Overton/Cletis O. BSA 49
Overton/Donna L. Thompson FDNU BSHE 76
Overton/Stephanie CLTX BSHE 90
Overturff/Teddy R. VOAG BSA 80
Oviatt/Catheron E. FND BSHE 73
Owen/Sue FDNU BSHE 61
Owen/Thomas Lawrence ANSC BSA 84
Owen/W. B. BSA 40
Owens/Lisa M. HDFS 98
Owens/Phillip Ray AGRN BSA 93
Owens/Phillip Ray AGRN MS 97
Owens/Rebecca Lynn CLTX BSHE 91
Oxford/Brad G. POSC BSA 83
Oxford/Deryle G. POSC BSA 60
Oxford/Eric Len ANSC BSA 96
Oxner/Michael AGBS BSA 90
Ozment/Georgann Fisher FND BSHE 63
Pabor/Kimberly Miller IDHS BSHES 98
Pace/Bobby J. BSA 60
Pace/Carl E. AGRN PHD 76
Pace/Daryl A. AGBS BSA 84
Pace/Edward L. BSA 50
Pace/Harry W. BSA 69
Pace/Homer BSA 49
Pace/Janet M. FND BSHE 71
Pace/Mark E. HORT BSA 80
Pack/Alan Dee ANNU BSA 69
Pack/Timothy M. ENTO BSA 70
Padgett/Cynthia A. BSHE 61
Padgett/Harold Wayne POSC BSA 64
Padgett/Peggy Lee HEED BSHE 73
Paetzold/Ronald F. SOSC PHD 76
Pagan/Betty Jo HOEC MS 64
Pagan/Betty Jo BSHE 48
Pagan/Joe D. ANNU BSA 79
Pagan/John F. BSA 41
Page/Candace E. HEED BSHE 79
Page/Charles Edwin BSA 51
Page/Clyde R. FDNU BSHE 72
Painter/Leslie Laird BSA 80
Painter/Zachry F. AGBS 98
Paladino/Paul BSA 49
Palarino/Mary Y. FDNU BSHE 86
Palazzi/Kathleen Knowles FND BSHE 71
Palenske/Merhl L. BSA 33
Pallais/Noel Ernesto PTSC PHD 84
Palmer/H. C. PLPA BSA 72
Palumbo/Deborah Rose PLPA MS 83
Panagiotakopoulou/Vassili FDSC BSA 87
Panas/Jeffrey Ted FDSC BSA 96
Pantazides/Athanasios John POSC BSA 68
Pape/Deugkyoung Choe FDSC MS 93

Pardue/Michel David AGEC MS 97
Pardue/Michel David AGBS BSA 95
Parette/Billie E. BSA 58
Parham/Enola Alexander BSHE 42
Parham/Robert W. BSA 60
Parish/J. Austin BSA 41
Parish/Jack M. VOAG BSA 70
Parish/L. B. BSA 47
Parish/Thelma L. BSHE 47
Park/George E. BSA 52
Park/Keun-Kyu ANSC PHD 95
Park/Thomas G. BSA 49
Parker-Gibson/Necia Tomboulian PPPM BSA 80
Parker/A. J. BSA 56
Parker/Anita Tackett BSHE 85
Parker/Billy D. BSA 52
Parker/Carl R. BSA 68
Parker/Charles C. S. AGEC BSA 76
Parker/Colleen FDSC BSA 94
Parker/Doris Ann BSHE 49
Parker/E. Berniece Evans HEED BSHE 43
Parker/Floyd Conelius BSA 53
Parker/Jackie FDSC MS 83
Parker/Jacques AGBS BSA 78
Parker/Joe C. BSA 49
Parker/John C POSC MS 82
Parker/Kathy A. FDNU BSA 79
Parker/Kathy Lynn VOAG BSA 83
Parker/Leslie M. BSA 77
Parker/Martha Stark BSHE 26
Parker/Raymond O. BSA 56
Parker/Russ AGRN BSA 80
Parker/Sarah Frances BSHE 25
Parker/Terrisue FDNU BSHE 72
Parker/Terrisue C. BSHE 72
Parks/Sarah M. POSC 99
Parks/Walter W. BSA 47
Parnell/Chester Max AGED BSA 63
Parnell/Michael Dean AGBS BSA 80
Parochka/Marsheila Jane AGBS BSA 86
Parr/Jimmy V. ANNU MS 65
Parr/Jimmy V. AHUS BSA 57
Parr/Kenneth M. BSA 60
Parrish/Buddy Lee ANSC BSA 93
Parrish/Delbert R. POSC BSA 69
Parsa/Haragopal POSC MS 80
Parsa/Haragopal ANSC MS 79
Parsley/James R. BSA 51
Parsons/Charles E. CRSC BSA 71
Parsons/Ludonna Jay FDNU BSHE 94
Parsons/Robert D. POSC BSA 66
Parsons/Winifred Phayer FND BSHE 70
Partain/Robbie BSHE 47
Partlow/Karen Bradford CLTX BSHE 90
Parvin/Gregg L. AGEC MS 77
Pascale/Laurie F. IDHS BSHE 85
Paschal/Lisa K. LDUH BSA 88
Pasco/Ricardo Antonia BSA 57
Pasierb/Marilyn FDNU BSHE 74
Pasley/Mabel Giles BSHE 48
Patanjali/Rao N. FDSC MS 92
Patanjali/Shoba HOEC MS 93
Patchell/Wesley A. BSA 40
Pate/Adeline BSHE 23
Pate/Johnie Joe BSA 50
Pate/Mark Lee HORT BSA 81
Pate/Mike ENSC BSA 78
Paterson/Robert Gordon HORT MS 87
Patil/Avinash Ramdas ANSC PHD 95
Patrick/Terry CLTX BSHE 86
Patterson-Hardin/Dana Page CLTX BSHE 93
Patterson/Beverly C. CLTX BSHE 85
Patterson/Carl Arthur PLPA MS 89
Patterson/David M. BSA 76
Patterson/Felix E. BSA 61
Patterson/Glenn R. CRSC BSA 73
Patterson/Jennifer Leigh FDNU BSHE 95
Patterson/Maudalene HEED MS 53
Patterson/Paul L. AGME BSA 74
Patterson/Paul Ray ANSC BSA 84
Patterson/W. L. BSA 52
Patterson/William K. HORT MS 79
Patton/Charles C. BSA 51
Patton/Glenn BSA 50
Patton/Hubert B. BSA 47
Patton/Jill Marie AGBS BSA 92
Patton/John Edward BSA 65
Patton/Linda Brown BSHE 65
Patton/Lisa Dane RSOC MS 80
Patton/Rodney Allen AGED BSA 90
Patton/Steven W. VOAG BSA 82
Patton/Vernon S. AGRN BSA 78
Paul/Anita Lucille BSHE 33
Paul/Jack D. BSA 34
Paul/Neva Hill BSHE 37
Paul/Patricia O'Neal FDNU BSHE 76
Pavlis/Gary Charles HORT MS 81
Pawling/Marilyn Kae Hathaway BSHE 60
Pay/David W. BSA 51
Pay/Joe W. BSA 57
Pay/Johnnie BSA 49
Pay/Ralph Thomas BSA 55
Payne/Charles R. ANSC BSA 52
Payne/Darlene Loresa CLTX BSHES 96
Payne/Everett S. ANSC BSA 85
Payne/Jeanette ANSC BSA 87
Payne/John R. POSC BSA 72
Payne/Richard E. POSC BSA 77
Paysinger/Doraine GNAG MS 63
Paysinger/Doraine BSA 49
Pazdera/Billie Lou Endres BSHE 53
Pazdera/F. J. BSA 53
Pazderka/Ehrline Rowden BSHE 33
Peachey/James R. DASC MS 67
Peacock/Dena Denise Kennedy ANNU BSA 84
Peacock/Kenneth Ray AGBS BSA 85
Peacock/Robert H. BSA 56
Pearce/Alan AGBS BSA 96
Pearce/Alan AGEC MS 98
Pearce/Jacky P. BSA 61
Pearce/Lisa Dawn IDHS BSHE 89
Pearce/Richard Edwin AGBS BSA 87
Pearce/Ruth BSHE 29
Pearcy/William O. BSA 65
Pearson/Judy FND BSHE 69
Pearson/Philip AGBS BSA 81
Pearson/Steven Dennis ANSC BSA 85
Pearson/Waymon CRSC BSA 69
Peck/Helen Oakley BSHE 27
Peck/William B. ENTO PHD 68
Pedersen/Paula Renae CLTX BSHE 92
Peebles/Thomas K. GBUS BSA 77
Peek/James M. BSA 39
Peek/Noah S. ANSC BSA 82
Peek/Noah S. AGRN BSA 50
Peel/James S. BSA 51
Peel/Vernon BSA 52
Peeler/Michael Leon AGBS BSA 92
Peer/Alan Carlos SOSC MS 77
Peer/Carolyn Harvel BSHE 42
Peevy/Garry Edward ANSC BSA 71
Pella/Margaret E. Bullard FDNU BSHE 80
Pellecer/Juan Jose ANSC BSA 93
Pelt/Jimmy T. ENTO BSA 67
Pemberton/Cheryl Lynn HLED BSHE 81
Pencoe/Nancy L. ENTO PHD 87
Pendergraft/Lyndall Kyle CNED BSA 89
Pendergraft/Ralph Kendall AGBS BSA 93
Pendergraft/Rusty ANSC BSA 78
Pendergrass/John Paul GBUS BSA 77
Pendleton/Marcy J. CTM BSHE 75
Penix/James Edwin BSA 54
Penix/Susan Ellen PTSC PHD 92
Penn/Allisen GHEC BSHE 87
Penn/Joe B. CRSC BSA 68
Penn/Patricia Jane FND BSHE 69
Penn/Richard L. AGBS BSA 72
Penn/Winona Coleman BSHE 51
Pennell/Sarah Smith HOEC BSHE 62
Pennington/Joe S. AGRN MS 69
Pennington/Joe S. AGRN BSA 65
Pennington/Van BSA 55
Penquite/Don S. AGEC BSA 71
Peoples/Jimmy R. ANSC BSA 76
Pepiot/Sheila FDSC MS 87
Pepper/Betsy Baker CDFR BSHE 76
Pepperkorn/Amy Georgenia BSHE 37
Peratt/Carl F. LDUH BSA 83
Perdue/Mary E. BSHE 45
Perez/Carlos Eduardo GNAG MS 89
Perez/Salvador Gonzalez PTSC PHD 85
Perkins/Edith Langford BSHE 43
Perkins/Glen BSA 50

Perna/Dominique CLTX BSHE 92
Perona/Michael AGEC BSA 91
Perry/Donald Craig POSC BSA 91
Perry/James R. VOAG BSA 67
Perry/Jerry Duane POSC BSA 74
Perry/Joseph R. BSA 39
Perry/Julia Wood HORT MS 81
Perry/Ricky Wayne DANB BSA 80
Perry/Roger M. AGBS BSA 73
Perry/Sallie HEED MS 73
Perry/Sallie BSHE 66
Perry/Shannon IDHS BSHE 95
Perryman/Mike AGRN BSA 85
Persing/Diana Gaye Small BSHE 81
Pertuis/Angela D. FHNH 98
Peters/Dixie Ann IDHS BSHE 92
Peters/Wanda Morgan HEED BSHE 74
Peters/Weldon VOAG BSA 61
Petersen/Jean Perry HOEC MS 79
Petersen/Randy L. ANSC MS 79
Peterson/Betty L. BSHE 59
Peterson/Shawn T. POSC BSA 96
Peterson/Susan I. HOEC MS 78
Peterson/Susan I. CDFR BSHE 77
Petillo/Thomas E. BSA 51
Pettey/Jason J. BSA 58
Pettiet/Joe V. BSA 57
Pettigrew/Donald D. BSA 54
Pettit/Frances Misenhimer BSHE 42
Pettit/James L. AHUS BSA 62
Pettit/Margaret Nell FND BSHE 64
Pettus/Lester G. AGEC BSA 70
Petty/Douglas B. GNAG MS 85
Petty/Douglas B. VOAG BSA 79
Petty/Kelly Joe ANSC BSA 86
Pettyjohn/Earl F. AGEC MS 63
Pettyjohn/Earl F. BSA 39
Peulausk/Thomas C. AGRN BSA 78
Pew/James F. BSA 58
Pfeifer/Ernest E. BSA 50
Pfeifer/Frank BSA 29
Pfeiffer/Beverly J. Long HEED BSHE 70
Pfrimmer/T. R. AHUS BSA 48
Pharr/Cynthia Collins POSC MS 79
Pharr/Cynthia Collins ANSC BSA 76
Pharr/David Mason HORT MS 67
Pharr/David Mason HORT BSA 64
Pharr/Harold Gene POSC BSA 76
Pharr/Joan Dickerman FND BSHE 64
Pharr/L. D. ANNU MS 66
Pharr/L. D. ANNU BSA 64
Pharr/Michael Dale ANSC BSA 91
Pharr/Michael W. BSA 76
Phelps/Carmen E. Jacks HEED BSHE 75
Phelps/Elisabeth Jean HORT MS 86
Phelps/Frank J. ANSC BSA 66
Phelps/James B. BSA 39
Phelps/Louise Johnson BSHE 41
Phelps/Walter B. ANSC BSA 76
Phifer/Iris J. HOEC MS 82
Phillips-Liles/Melinda AGBS BSA 88
Phillips/Ben H. BSA 49
Phillips/Bernice BSHE 26
Phillips/Bettye Martin BSHE 47
Phillips/Blake N. AGEC MS 81
Phillips/Blake N. ANSC BSA 79
Phillips/C. Joann HOEC MS 67
Phillips/C. Joann HOEC BSHE 51
Phillips/Carmen Joann BSHE 51
Phillips/Charles W. GNAG MS 62
Phillips/Charles W. BSA 58
Phillips/Cindy Dabbs GNAG MS 93
Phillips/Cindy Dabbs AGED BSA 87
Phillips/Cynthia Ann IDHS BSHE 91
Phillips/Dale R. HORT BSA 80
Phillips/Debra Faddis POSC BSA 81
Phillips/Freida Katherine HEED BSHE 84
Phillips/Harold F. GNAG MS 90
Phillips/J. D. BSA 61
Phillips/Jack Alden BSA 65
Phillips/Jacob R. ENTO MS 61
Phillips/Jacob R. AGSC BSA 52
Phillips/James Steven AGRN BSA 91
Phillips/Jerry M. AGRN PHD 85
Phillips/Jerry M. AGRN MS 82
Phillips/Juanna Jackson BSHE 54
Phillips/Lance FDSC BSA 85
Phillips/Larry G. BSA 65
Phillips/Pamela G. HORT MS 82
Phillips/Pamela G. LDUH BSA 80
Phillips/Paul J. BSA 50
Phillips/Ralph D. BSA 60
Phillips/Reece W. BSA 47
Phillips/Richie Scott POSC BSA 94
Phillips/Ronald W. ANSC BSA 78
Phillips/Stewart M. ANSC MS 77
Phillips/Stewart M. ANSC BSA 76
Phillips/Thomas E. BSA 57
Phillips/William B. BSA 53
Phillips/William S. BSA 50
Phillips/William W. SOSC MS 67
Phillips/William W. BSA 59
Philpot/John L. BSA 57
Pianalto/Donna Simco CLTX BSHE 84
Pianalto/Margie Yeager AGBS BSA 86
Pianalto/Robert E. BSA 63
Pickard/Claudine T. BSHE 40
Pickens/Ron Anthony AGBS BSA 85
Pickett/Gloria Vasser HOEC MS 81
Pickett/Patrick Ercil ANSC BSA 86
Pickle/Charles E. BSHE 78
Pickle/Shirley Canerday HEED BSHE 64
Pickle/Susan J. GHEC BSHE 78
Pickren/Jim J. BSA 31
Pico-Acosta/Gloria C. GNAG MS 88
Pidcock/Suzanne HEED BSHE 75
Pierce/Joseph W. BSA 60
Pierce/Marissa Ann CLTX BSHE 80
Pierce/Phillip E. ENTO MS 72
Pierce/Shana Rhea IDHS BSHES 97
Pierce/Shirley Sue BSHE 59
Pierce/Vicki Joan Schroeder HORT MS 81
Pierson/Joe Neil DANB BSA 91
Pierson/Sheila Webb AGEC MS 94
Pierson/Sheila Webb AGBS BSA 92
Pike/Leonard M. MS 64
Pike/Leonard M. HORT BSA 62
Pike/Roxy A. HEED BSHE 62
Pile/Camille L. HEED BSHE 66
Pilgrim/Lynda K. HDFS BSHE 82
Pinkerton/Dora Belle BSHE 34
Pinkerton/Jody Gene POSC BSA 96
Pinkerton/Jody Gene AGBS NODEG 96
Pinkston/Greg GNAG BSA 58
Pinson/Glendola BSHE 49
Piper/John B. BSA 41
Pipkin/Nita Vines CDFR BSHE 77
Pippin/A. K. AGEC MS 61
Pippin/A. K. AGEC BSA 60
Pippin/Danny Ray AGEC MS 88
Pishny/Janice Radke HORT BSA 79
Pitchford/Leo A. BSA 50
Pitchford/William H. ANNU BSA 76
Pitman, Justin J. EXED 98
Pitman/Larry H. BSA 58
Pittard/Wanda Puryear BSHE 55
Pittaway/Kenneth Ray AGRN PHD 76
Pittillo/William Chad AGBS BSA 97
Pittman/Christy HOEC MS 87
Pittman/Gail Gateley BSHE 59
Pittman/Sidney L. ANSC BSA 71
Pitts/Paul D. BSA 52
Pitts/Roger Morrow VOAG BSA 61
Pitts/Russell AGBS BSA 88
Pitts/Vikki Whitlock FND BSHE 73
Plafcan/Claire N. IDHS BSHES 97
Plafcan/Frank T. AGEC MSA 75
Plafcan/Frank T. AGBS BSA 70
Plake/Toni Houston HIED BSHE 81
Platt/Roberta Eidson BSHE 59
Playford/Bette Mae HEED BSHE 70
Plugge/Charles Edward AGRN BSA 80
Plugge/Cynthia Marie BSHE 81
Plugge/Dana S. Dickerson CLTX BSHE 82
Plugge/Lori L. Robinson CLTX BSHE 78
Plumlee/Bryan Lawrence POSC MS 97
Plumlee/Hugh W. GNAG MS 66
Plumlee/Hugh W. ANSC BSA 57
Plumlee/Karen BSHE 65
Plummer/Diane Dameron HOEC BSHE 56
Plunkett/Don GNAG MS 86
Plunkett/Edward Ray VOAG BSA 69
Poe/Bryan Wayne LDUH BSA 96
Poe/Buford BSA 35

Poe/Willie E. BSA 26
Polanco-Zea/Luz A. AGBS 99
Polk/Bernard GNAG BSA 31
Polk/Frank D. ANSC BSA 76
Polk/Holland P. Greene HOEC BSHE 32
Pollard/A. E. BSA 50
Polley/Leonard C. POSC 98
Pomerene/Anne H. BSHE 42
Ponder/Jacqueline Young FND BSHE 71
Ponder/Truley S. VOAG BSA 66
Pone/Dieudonne Kamdem ANSC PHD 93
Pone/Dieudonne Kamdem POSC MS 84
Pongsakul/Pichit AGRN MS 77
Pool/Jodye Marie POSC BSA 96
Pool/Juanita Hampton ESEC BSHE 41
Pool/Lera H. BSHE 42
Pool/Lisa Lynn IDHS BSHE 88
Poot/Fausto PPPM BSA 80
Pope/David J. AGBS BSA 91
Pope/Elizabeth Hunt LDUH BSA 81
Pope/R. Todd LDUH BSA 86
Pope/Randall E. AGEC MS 78
Pope/Shannon Lynn AGBS BSA 95
Poplin/Anne K. DIAM BSA 91
Porta/Carmen Manning HEED BSHE 75
Porta/John M. PLPA MS 80
Portch/Stanley S. SOSC PHD 71
Porter/Betty Douglas HEED BSHE 50
Porter/Bob E. ANSC BSA 61
Porter/David B. POSC BSA 82
Porter/Donna E. HORT BSA 86
Porter/James O. BSA 55
Porter/Jim AGRN BSA 75
Porter/Kelli Anne CLTX BSHE 86
Porter/Robert W. BSA 42
Porter/Stewart BSA 52
Porter/Tom B. VOAG BSA 49
Porter/Wilson E. GNAG MS 68
Porter/Wilson E. BSA 39
Porterfield/Lewis T. BSA 59
Posey/Alan F. ENTO MS 69
Posey/John Bradley AGBS BSA 87
Posey/Kelly Ann AGED BSA 90
Posey/Ken AGBS BSA 85
Posey/Rodney Thomas AHUS BSA 61
Post/Mary Gatlin HEED BSHE 70
Post/Samuel C. BSA 58
Pote/Daniel H. AGRN PHD 97
Pote/Daniel H. AGRN MS 93
Pote/Linda Marie ANSC PHD 84
Potter/Donald K. POSC MS 63
Potter/Donald K. ANNU BSA 62
Potter/James D. AGED BSA 83
Potter/Lawrence A. AGEC BSA 66
Potter/Mary C. HEED BSHE 83
Potter/Rissie L. BSA 14
Potts/Doyne F. BSA 55
Potts/Edward S. HORT BSA 92
Potts/Mary Kelley Philbeck BSHE 30
Poulsen-Chao/Normalee FDNU BSHE 76
Poulsen/Gregory C. DANB BSA 76
Pound/Daisy C. BSHE 40
Powell/Curtis Ray BSA 50
Powell/D. RoAnn HDID BSHE 78
Powell/Donnie M. BSA 56
Powell/Donnie M. ENTO MS 58
Powell/Drexel D. AGRI BSA 46
Powell/Gary R. AHUS BSA 66
Powell/Gladys Tallent BSHE 49
Powell/Greg BSA 79
Powell/Horace A. BSA 57
Powell/Jeremy G. TRNS BSBA 98
Powell/Pauline BSHE 39
Powell/Ruth BSHE 29
Powell/Willa Dene BSHE 59
Powell/William HORT BSA 23
Power/Koni Jo Boone BSA 70
Power/Michael J. AGEC BSA 81
Powers/Bruce W LDUH BSA 82
Powers/Jimmy Edd POSC BSA 60
Prager/Joseph S. BSA 55
Pranger/Joe PTSC BSA 77
Prasittikhet/Jirapong AGRN MS 77
Prater/Billy Dean VOAG BSA 79
Prather/Victoria L. CLTX BSHE 87
Pratt/Helen P. FDSC MS 85
Pratt/Helen P. FND BSHE 66
Pratt/Melanie Blankenship DANB BSA 80
Prawito/Priyono AGRN MS 91
Presley/Adrien HORT BSA 79
Presley/Adrien HORT MS 82
Presley/Billy Dwain SOSC MS 68
Pressley/Mary Abbey CTM BSHE 75
Presson/Brenda Lee ANSC MS 84
Presson/Brenda Lee ANSC PHD 86
Prewitt/Wayne R. AGEC MS 85
Price/Alice Bowman BSHE 32
Price/Becky J. FDNU BSHE 82
Price/Claude L. BSA 38
Price/Corinne Beasley BSHE 38
Price/Dale VOAG BSA 61
Price/David Lindsey LDUH BSA 88
Price/Delton E. BSA 50
Price/Frances Morris FDNU BSHE 72
Price/Gregory David ENSC BSA 96
Price/Jacquelyn FDSC BSA 92
Price/Lance Lee POUL BSA 97
Price/Lisa C. FDSC PHD 92
Price/M. Gelene Anderson HEED BSHE 42
Price/Mary E. BSHE 46
Price/Mazo AGRN PHD 77
Price/Sue FDNU BSHE 72
Price/Virginia HOEC BSHE 60
Price/Virginia L. BSHE 60
Prichard/Howard VOAG MS 54
Prichard/Howard AGRI BSA 47
Prickett/Kimberly Allyn BSHE 84
Priebe/Jeffrey R. AGBS BSA 98
Prier/Sarah L. IDHS BSHE 92
Priest/Mark Edward AGBS BSA 86
Priest/Patti Shirleen AGED BSA 95
Prince/Anna HEED BSHE 97
Prisock/Susan A. HEED BSHE 79
Pritchard/Helen Huffman BSHE 61
Pritchett/Davis W. ENTO PHD 81
Pritchett/Davis W. ENTO MS 77
Pritchett/E. H. BSA 39
Pritchett/William L. BSA 41
Privett/Lisa D. HOEC MS 83
Proctor/Andrew FDSC PHD 86
Proctor/Andrew FDSC MS 83
Proctor/Gwendolyn Fudge HEED BSHE 72
Prokakis/George FDSC BSA 83
Prongsirivathana/Chamras CRSC MS 64
Prophet/Kathy L. Cox HOEC MS 81
Prothero/Sara Jane BSHE 57
Pruitt/Bryan AGED BSA 86
Pruitt/Jean Hampton BSHE 57
Pryor/Harold C. BSA 55
Pryor/Robyn Osgatharp HLED BSHE 81
Puckett/Boyd J. BSA 51
Puckett/Daniel Wayne BSA 46
Puckett/Irene BSHE 42
Pugh/Bill T. CRSC BSA 69
Pugh/Kenneth H. AGRN MS 78
Pugh/Robert D. BSA 53
Pullen/Charles P. BSA 41
Pultz/Tony ANSC BSA 68
Purba/Victor AGEC MS 92
Purdy/John William ANSC MS 90
Purnell/Carl D. ANSC BSA 70
Purnell/Jeff S. AGBS BSA 80
Purtle/Carolyn B. BSHE 59
Purvis/Vicki Lee BSHE 80
Putri/Atien Priyanti Sudaryo AGEC MS 93
Pyatt/Jeffrey Alfred AGME BSA 86
Pye/Glen AGRI MS 64
Pye/Glen AGRI BSA 45
Pye/Ruth BSHE 44
Pyeatt/Peter G. DASC BSA 82
Pyland/Steven Earl HORT BSA 81
Pylate/Frank D. BSA 60
Pyle/Frankie R. BSA 65
Pyle/Harold Lee AGBS BSA 93
Pyle/Norvel V. BSA 39
Pyle/Patricia May BSHE 65
Pyles/Wayne E. DHUS BSA 61
Qi/Min FDSC MS 96
Quattlebaum/Shelly D. AGBS BSA 85
Quinton/Roger G. AGEC BSA 75
Raben/Dennis Paul ANSC MS 93
Rabeneck/James N. CRSC BSA 77
Rachal/Bryan Edward FDNU BSHE 83
Radice/Anthony D. PTSC PHD 87

Raffety/Julie L. ANSC BSA 94
Raffety/William Michael FDNU BSHES 96
Ragan/Kay BSHE 55
Ragland/Carl E. ENTO MS 71
Ragland/Carl E. ENTO BSA 68
Ragland/James Bryan BSA 58
Ragland/William W. BSA 65
Raguseo/Reba Cox ANSC BSA 92
Rahal/Mehemmed S. ANSC PHD 81
Raible/Robert AGRN MS 62
Railsback/T. C. AGEC BSA 82
Raines/Paul L. BSA 51
Raines/Sharon W. AGRN MS 77
Rainey/Alfred Seymour AGBS BSA 88
Rainey/Daniel Vincent AGBS BSA 90
Rainey/Leo BSA 54
Rainey/Leo ENTO MS 64
Rainey/Robert P. AGBS BSA 61
Rainey/Ronald Lee AGEC MS 93
Rainey/Ronald Lee AGBS BSA 91
Rainey/Sandra S. FDNU BSHE 79
Rains/James F. BSA 29
Rakes/J. M. ANSC BSA 55
Rakes/J. M. ANSC MS 56
Rakes/Larry L. AGRN BSA 71
Rakes/Terry J. HEED BSHE 72
Raley/Binom J. BSA 43
Raley/Embry VOAG BSA 61
Raley/George H. BSA 65
Ramage/Kelli Diane FDNU BSHE 92
Ramage/Marthell Scoggin BSHE 40
Rambo/Harold A. SOSC BSA 71
Rambo/Joe B. ANSC BSA 74
Ramer/Cordy A. BSA 53
Ramer/Scott D. POSC 98
Ramey/Harmon H. BSA 51
Ramey/Mary Alice ANSC BSA 97
Ramey/Tamara Lynn CLTX BSHE 85
Ramey/Willie Margaret BSHE 41
Ramirez-Oliveras/Gumersindo AGRN PHD 85
Ramirez/Melody Ann HDFS BSHE 94
Ramm/David Paul LDUH BSA 86
Ramsey/Alda J. BSHE 58
Ramsey/Bruce E. POSC BSA 79
Ramsey/E. Dale BSA 60
Ramsey/Lester Charles BSA 53
Ramsey/M. Sharee HDFS BSHES 95
Ramsey/Ronald E. ANSC BSA 68
Ramsfield/Daniel Joseph FDSC BSA 95
Ramthun/Larry E. AGRN MS 77
Ramthun/Larry E. SOCI BSA 74
Ramthun/Sherry HEED BSHE 76
Ranalli/Norbert AGBS BSA 74
Rance/Glenda Fay HOEC MS 81
Randall/Glen Orvice BSA 22
Randolph/Robert Allen PLPA BSA 82
Raney-Harper/Lisa Ann CDFR BSHE 80
Ranger/Donna Jordan FND BSHE 70
Ranglack/Georganna ANSC BSA 70
Rankin/Hubert Wayne BSA 50
Rankin/Jim AGBS BSA 90
Rankin/John J. BSA 50
Rankin/Richard Cameron BSA 22
Rankin/William C. AGEC BSA 76
Ransom/Michael H. POSC BSA 76
Ransom/Michel D. SOSC MS 76
Ransom/Michel D. SOSC BSA 74
Rapley/Leslie K. Tatman CLTX BSHE 78
Rasamivelona/Andre AGRN MS 94
Rasamivelona/Andre' AGRN MS 94
Raseira/Maria Do Bassols PTSC PHD 86
Rashid/Tahir ENTO MS 94
Rask/Cheryl Vassaur IDHS BSHE 91
Rasnic/Marisa J. CLTX BSHE 82
Ratchford/Niagle D. ANSC BSA 78
Ratchford/Whitney Amber ANSC BSA 96
Rathgeber/Bruce Mitchell POSC MS 94
Ratley/Selma Harkey BSHE 41
Raub/Kris AGBS BSA 88
Raub/Kris AGEC MS 90
Raub/Laura Marie GHEC BSHE 85
Raup/Thomas James ANSC MS 89
Rausch/Bennie B. BSA 51
Rausch/Earl L. AGEG BSA 76
Ravi/Jayasree ENTO MS 96
Ray/Brett J. LDUH BSA 90
Ray/James B. BSA 40
Ray/James M. BSA 55
Ray/Larry HOFS MS 74
Ray/Larry HOFS BSA 72
Ray/Maurice L. BSA 42
Ray/Nanette Carolyn AGBS BSA 95
Ray/Phyl A. CDFR BSHE 73
Ray/Sharon Louise HOEC MS 84
Ray/Sharon Louise HLED BSHE 81
Ray/Thurman L. BSA 50
Ray/Tracy VOAG BSA 82
Raybon/Michael Phares LDUH BSA 92
Raycher/Karen Denise HOUS BSHES 97
Rayder/Jan W. BSA 57
Reading/Garlan D. BSA 55
Reading/Virgil L. BSA 50
Reagan/Alice Nagel BSHE 69
Reagan/Rebecca Mitchell HEED BSHE 69
Reames/Carlton E. POSC BSA 57
Reams/Bob F. ANSC BSA 73
Reaper/Jack D. III AGBS 98
Reaves/D. Wade DASC BSA 83
Reavis/Margaret Anne BSHE 37
Reber/Kara L. BSHE 85
Rebsamen/Paul Scott AGBS BSA 80
Reddick/William T. ANSC BSA 76
Reddin/James D. HORT BSA 70
Reddin/James D. HORT MS 77
Reddish/John Mark POSC MS 96
Reddish/John Mark POSC BSA 93
Reddy/Kotha Basi FDSC MS 81
Redfearn/James Norman AGED BSA 91
Redfern/R. E. ENTO MS 61
Redfern/R. E. AGRI BSA 59
Redman/Kobi Jo HDFS BSHES 95
Redpath/Loretha Faye HEED BSHE 85
Redwine/Max R. BSA 52
Reece/Debra Jean CLTX BSHE 89
Reed/A. L. BSA 36
Reed/Benjamin Adam AGRN BSA 93
Reed/Betty Thompson BSHE 43
Reed/Charlene J. FND BSHE 74
Reed/Charlene J. HOEC MS 77
Reed/Dale R. AGME BSA 75
Reed/David B. AGBS BSA 78
Reed/Fred E. BSA 55
Reed/Hazell PTSC PHD 84
Reed/Hollis T. ANNU MS 70
Reed/Hollis T. ANNU BSA 68
Reed/Jack T. ENTO PHD 83
Reed/James David GNAG BSA 70
Reed/Jarvis Joe AGEC MS 32
Reed/Jim F. BSA 50
Reed/Johnnie M. HOEC MS 68
Reed/Josephine Lewis BSHE 36
Reed/Judy Owens HDFS BSHES 96
Reed/Leanne BSHE 58
Reed/Matthew Clinton ANSC BSA 80
Reed/Priscilla Jeffers HEED BSHE 78
Reed/Ronald Lee IDHS BSHES 96
Reed/Sandra K. HOEC BSHE 72
Reed/Stanley E. AGEC BSAGE 73
Reed/T. E. AGED BSA 42
Reed/T. Jack BSA 39
Reed/Theresa A. CLTX BSHE 79
Reed/Tommie Oberley FDNU BSHE 68
Reese/Mary Elizabeth FDNU BSHE 87
Reese/Melissa Gay ANSC BSA 80
Reese/Robert P. ANSC BSA 76
Reeves/Addie Barlow BSHE 43
Reeves/Dot BSHE 50
Reeves/Edward Walker AGEC MS 37
Reeves/Pamela J. HDID BSHE 84
Reeves/Walter Arthur HORT BSA 81
Regan/Carole AGRN MS 79
Reginelli/Teresa G. GHEC BSHE 79
Reiber/Molly Ann POSC BSA 88
Reid/Beverly A. AGJN BSA 82
Reid/Nicole Brandt BAEG MSBAE 96
Reid/Virginia CLTX BSHE 82
Reinhardt/Bida Ellen BSHE 31
Reinhardt/William M. BSA 39
Reis/Lloyd Monroe BSHE 55
Reiter/Jess J. BSA 54
Reitmeir/Cheryll A. HOFS MS 75
Relyea/John F. AGRN PHD 78
Relyea/Walter R. BSA 65
Renfro/Randy AGEC MS 82

Renfrow/Nolen E. BSA 41
Rentfro/Jeanne HEED BSHE 64
Revelle/Daryl K. LDUH BSA 78
Revis/Rosalie BSHE 60
Rexrode/Ronny Dean AGBS BSA 90
Reynolds/Bertahl Dean HORT BSA 71
Reynolds/Betty Pearl Toll BSHE 60
Reynolds/Betty White BSHE 71
Reynolds/Charles M. AGRN PHD 86
Reynolds/Daniel B. AGRN MS 84
Reynolds/Franchon Decker BSHE 35
Reynolds/Kenneth J. AGBS BSA 74
Reynolds/Rebecca HOEC MS 84
Reynolds/Robert L. BSA 60
Reynolds/Tara Lynn CLTX BSHE 93
Reynolds/Thomas Gilbert BSA 48
Reynolds/Toney G. BSA 54
Reynolds/Vickie Diann HEED BSHE 85
Reynolds/Will S. HORT BSA 51
Rhein/Earl A. GNAG BSA 40
Rhoades/Shaun Ray AGEC MS 97
Rhoades/Shaun Ray AGBS BSA 95
Rhodes/Dianne E. HEED BSHE 74
Rhodes/Hugh BSA 49
Rhodes/J. Dwight PLPA MS 85
Rhodes/J. Dwight PLPA BSA 77
Rhodes/Janet Nickerson BSHE 68
Rhodes/Karen Lee HOEC MS 79
Rhodes/Karen Lee FDNU BSHE 76
Rhodes/Mary Kennett Manis HOEC MS 66
Rhodes/Mary Kennett Manis BSHE 47
Rhodes/Ray AGBS BSA 72
Rhodes/Robert H. ANNU MS 68
Rhodes/Robert H. ANSC BSA 66
Rhodes/Sandra Elizabeth FDNU BSHE 73
Rhodes/Stanley E. ANSC BSA 72
Rhudy/Frankie Darlene ANSC BSA 92
Ribera/Luis A. AGBS 98
Rice/Bennie Eugene BSA 31
Rice/Bertha E. BSHE 45
Rice/Clarence BSA 50
Rice/H. B. BSA 40
Rice/Jay Timmons POSC BSA 85
Rice/Kevin Scott GNAG MS 84
Rice/Kevin Scott VOAG BSA 81
Rice/Lanny GNAG MS 83
Rice/Lanny VOAG BSA 75
Rice/Phillip L. ANSC BSA 72
Rice/Robert Eugene BSA 48
Rice/Virginia Anne BSHE 42
Rich/C. Allen BSA 51
Richards/Mark A. ENTO BSA 83
Richardson/Ann N. ANSC BSA 89
Richardson/Ernest L. BSA 39
Richardson/James W. VOAG BSA 72
Richardson/Jimmy K. POSC BSA 70
Richardson/John Thomas AGRN PHD 77
Richardson/John Thomas PLPA BSA 68
Richardson/Judy FND BSHE 64
Richardson/Justin BSA 32
Richardson/Kenneth Kile BSA 55
Richardson/Noveta Pharris BSHE 56
Richardson/Robert F. BSA 52
Richardson/W. Dale BSA 53
Richert/Leslie Dianne FDSC MS 91
Richert/Leslie Dianne FDSC BSA 90
Richey/Mary Margaret Wilson BSHE 36
Richmond/Carol Morgan BSHE 50
Richmond/Carolynn FDNU MS 77
Richmond/Carolynn HOEC BSHE 72
Richmond/Darren Ray ANSC BSA 85
Richmond/Jill R. ANSC BSA 85
Richmond/Todd Allen ANSC BS 93
Richter/Darrall A. BSA 59
Rickard/David A. PLPA MS 70
Rickett/Cyril E. BSA 38
Rickett/Rubelle Roark BSHE 38
Ricketts/Burton Lee AGBS BSA 96
Riddle/Cheryl Welch FDNU BSHE 84
Rider/Archie Allen BSA 54
Rider/Bruce BSA 51
Rider/Mary Ivester BSHE 54
Rider/Mary Lasiter BSHE 47
Ridley/Deniece PLSC BSA 82
Ridling/Little BSA 16
Riebe/Kristina Joy ANSC BSA 91
Rieck/Mary L. HEED BSHE 72
Rieff/Kevin ANSC BSA 85
Rieff/Michelle Ann HEED BSHE 85
Rietman/David F. AGBS BSA 82
Rife/JoAnne Sears BSHE 56
Riggan/Travis H. BSA 57
Riggan/Vencil Lee BSA 60
Riggin/Ethel L. Willcox BSHE 51
Riggs/Clydus Luther BSA 50
Riggs/Robert D. PLPA MS 56
Riggs/Robert D. GNAG BSA 54
Rigsby/John Allen BSA 38
Riley/Lisa Hopkins ENTO MS 83
Riley/Lisa Hopkins PPPM BSA 79
Riley/Mary Kathleen HEED BSHE 76
Riley/Richard S. HORT BSA 77
Riley/Tommy L. SOCI BSA 75
Rimmer/W. J. GNAG MS 67
Rimmer/W. J. BSA 49
Rinehart-Jones/G. Lynn POSC MS 82
Rinehart-Jones/G. Lynn HORT BSA 81
Rinehart/Keith E. ANNU BSA 61
Rinehart/Keith E. POSC MS 64
Rinnert/Gay B. GHEC BSHE 78
Riopel/Richard J. AGBS BSA 82
Ripley/Claudette Bethel FDNU BSHE 82
Risley/Gary Norman ANSC BSA 70
Risley/Mark Alan AGRN PHD 86
Risley/Mark Alan AGRN MS 83
Ritchie/Curt Craig ANSC BSA 92
Ritchie/Ginger CDFR MS 72
Ritchie/Ginger BSHE 68
Ritchie/W. David AGRN BSA 73
Ritter/Emily Ellen Roberts AGEC MS 81
Ritter/Emily Ellen Roberts AGEC BSA 79
Ritter/Martha BSHE 58
Rittman/Deana Kay CLTX BSHE 88
Rizley/Neva F. FDSC MS 79
Roach/Martha E. Combs HOEC MS 67
Roach/Martha E. Combs HEED BSHE 56
Roam/Virgil Douglas CLTX BSHE 80
Roan/Madeline Adele BSHE 41
Roark/C. Hamilton BSA 39
Roark/Harry F. POSC BSA 83
Roark/John H. PLPA MS 79
Robb/Maurice E. BSA 42
Robberson/Delano G. VOAG MS 64
Robberson/Delano G. BSA 58
Robberson/Franklin E. BSA 58
Robberson/Roosevelt B. DHUS BSA 58
Robbins/Frances Hughes BSHE 66
Robbins/Iva Clemmer BSHE 28
Robbins/Jennette D. HEED BSHE 72
Robbins/Kevin Blake AGBS BSA 95
Robbins/Larry S. AGEC MS 77
Robbins/Samuel W. AGEG BSA 63
Roberds/Dorothy McBroom BSHE 30
Roberson/Lisa Dawn FDNU BSHE 84
Roberson/Sara Sue HEED BSHE 68
Roberson/William D. AGRN MS 80
Roberts/Alvin N. POSC BSA 69
Roberts/Bob Franklin POSC BSA 80
Roberts/Caron Smets ENTO BSA 79
Roberts/Craig Arthur AGRN PHD 85
Roberts/Craig Arthur AGRN MS 82
Roberts/D. Trent AGEC MS 82
Roberts/D. Trent AGEC BSA 80
Roberts/Haskell Hugh GNAG BSA 61
Roberts/Haskell Hugh ENTO MS 63
Roberts/James Benson BSA 53
Roberts/James E. ENTO MS 55
Roberts/Jeanette HEED BSHE 67
Roberts/Judith Smith FND BSHE 64
Roberts/Kimberley Dawn CLTX BSHES 96
Roberts/Larry D. ANSC MS 85
Roberts/Larry D. POSC BSA 83
Roberts/Laurie Jo HOEC MS 83
Roberts/Lisa Y. CLTX BSHE 87
Roberts/Lowell David BSA 57
Roberts/Luther H. ENTO MS 69
Roberts/Luther H. AHUS BSA 64
Roberts/Luther Herbert BSA 33
Roberts/Robin D. AGBS BSA 89
Roberts/Roy W. VOAG BSA 21
Roberts/Shirley BSHE 57
Roberts/Timothy Allen AGEC MS 97
Roberts/Timothy Allen AGEC BSA 94
Roberts/Willis L. BSA 48

Robertson/Connie S. BSHE 69
Robertson/Don Dewitt AGBS BSA 90
Robertson/Frances Duggans BSHE 27
Robertson/George Arthur BSA 39
Robertson/Hilliard W. BSA 35
Robertson/Janice Ederington BSHE 71
Robertson/Jenny Claire FDNU BSHES 96
Robertson/Lewis Edwin BSA 39
Robertson/Mark A. HORT MS 91
Robins/Carl M. BSA 50
Robins/Mearl Clayton BSA 50
Robinson/Brenda Sue ANSC BSA 96
Robinson/Charles U. BSA 25
Robinson/Chloera BSHE 25
Robinson/Cynthia A. FND BSHE 72
Robinson/Deborah Stanley CDFR BSHE 73
Robinson/Edwin Freeman BSA 36
Robinson/Harmon N. BSA 41
Robinson/Mary Champion BSHE 26
Robinson/Mary K. S. FND BSHE 69
Robinson/Richard R. AGBS BSA 81
Robinson/Ronald B. VOAG BSA 82
Robinson/Terri Stapleton LDUH BSA 79
Robinson/Truman AGED BSA 36
Robinson/Vicky BSHE 82
Robinson/Vonna Faye Mills BSHE 46
Robinson/Whitnie HDFS BSHE 94
Robison/Carla Ruth ANSC BSA 83
Robison/Marilyn D. PLPA MS 77
Roca/Paul AGME BSA 78
Roca/Paul OMGT BSA 79
Rochelle/Marion J. BSHE 52
Rockenhaus/A. J. BSA 51
Rockwell/Mark David LDUH BSA 92
Roddy/Joe T. BSA 51
Rodery/Steve Kent GNAG MS 87
Rodezno/Hernan Roberto Espinoza ENTO MS 80
Rodezno/Hernan Roberto Espinoza BSA 74
Rodgers/Carlin Lanier BSA 25
Rodgers/Erma HOEC MS 54
Rodgers/John Henry BSA 23
Rodgers/Lynn L. ANSC BSA 77
Rodgers/Sara Cowan BSHE 70
Rodman/J. Denton BSA 48
Rodman/Joe B. BSA 56
Rodriguez/Cynthia Jane PLPA MS 83
Rody/Nancy Crutchfield BSHE 62
Roe/Elwin S. ANSC BSA 86
Roe/Helen BSHE 66
Roe/Roy A. ENTO MS 74
Roe/Roy A. MS 51
Roe/Sherry D. FHNH 99
Rogers/Barry Neal AGME BSA 87
Rogers/Charles B. AGRN PHD 83
Rogers/Dedra Dawn POSC BSA 88
Rogers/Euna Yarbrough BSHE 38
Rogers/Everett ANNU BSA 71
Rogers/Frank Anderson BSA 40
Rogers/Franklin C. BSA 53
Rogers/Harold W. BSA 65
Rogers/Hazel D. HEED BSHE 73
Rogers/Jerold L. CRSC MS 69
Rogers/Jerri Ann FDNU BSHE 93
Rogers/Jerry Lynn ANSC BSA 68
Rogers/Joffre H. BSA 41
Rogers/John N. AGRN BSA 82
Rogers/Juanita Bowling PLPR MS 52
Rogers/L. Nyoka HEED BSHE 70
Rogers/Lee AGBS BSA 88
Rogers/Leslie J. BSA 36
Rogers/Lyle BSA 39
Rogers/Michael Howard AGED BSA 96
Rogers/Noell Kent AGRN PHD 80
Rogers/Ralph Dale BSA 59
Rogers/Royce A. AHUS BSA 64
Rogers/Sammy BSHE 58
Rogers/Travis Wesley MSAGE 96
Rogers/Travis Wesley AGBS BSA 942
Rogers/William M. BSA 42
Rogers/William M. BSA 39
Rolfe/Khristine Turner HOEC MS 94
Rolfe/Khristine Turner CLTX BSHE 90
Roll/Stephen L. POSC BSA 74
Rolland/Cecil A. BSA 53
Rollans/Florence Pitts BSHE 36
Rollans/Wendell BSA 41
Roller/Richard E. AGED BSA 85
Rollins/Paul C. BSA 51
Rollins/Tammie Adams CLTX BSHE 92
Roltsch/William James ENTO MS 83
Rom/Curt HORT BSA 80
Rone/Monica Lee DASC BSA 86
Rone/Ronald W. DASC BSA 76
Rongey/Charles B. BSA 47
Rooks/Susan D. HORT MS 87
Root/Duke M. BSA 24
Root/Peggy C. HOEC MS 67
Root/Robert S. AGEG BSA 71
Rorex/Terry VOAG BSA 81
Rorie/A. J. ANSC MS 85
Rorie/A. J. ANSC BSA 76
Rorie/Deborah Prince FND BSHE 74
Rorie/F. Gay BSA 56
Rorie/Lee Kent ANSC MS 79
Rorie/Lee Kent ANSC BSA 74
Rorie/Rick ANSC MS 79
Rorie/Rick ANSC BSA 77
Rosales/Armando A. BSA 58
Rosborough/Mae Blakely BSHE 22
Rose/Carl Elmer BSA 40
Rose/Charlotte M. ANSC 98
Rose/Denise Lucy FDNU BSHE 86
Rose/Elizabeth C. HEED BSHE 67
Rose/Joy McKinney BSHE 55
Rose/Joyce H. FDSC BSA 92
Rose/Kristi Bradford AGBS BSA 80
Rose/Lois McMillen BSHE 33
Rose/Verlis AGEC MS 56
Rose/Verlis BSA 41
Rose/Veronica Sparkman BSHE 57
Roselius/Charles G. BSA 52
Rosenbaum/Lorita GHEC BSHE 40
Rosencrantz/Franklin C. BSA 16
Ross/Billy C. ANSC BSA 72
Ross/Fern BSHE 34
Ross/Janis Yvonne FND BSHE 73
Ross/Janis Yvonne HOEC MS 78
Ross/Kermit C. BSA 38
Ross/Micah C. AGBS BSA 91
Ross/Sharon L. HEED BSHE 75
Ross/Timothy T. ANSC BSA 73
Ross/Timothy T. ANSC MS 78
Ross/Wilma McKelvey BSHE 37
Rossi/Johnny Edward ANSC BSA 94
Rosson/Kimberly Brooks HIED BSHE 83
Rosson/Roxanne HOEC MS 85
Rosson/Roxanne GHEC BSHE 78
Roten/Paul C. ANSC BSA 72
Roten/Polly Jane CTM BSHE 75
Rouly/Gloria Bednar CLTX BSHE 86
Rountree/Thomas C. DANB BSA 93
Rouse/E. Phil ENTO MS 58
Rouse/E. Phil BSA 48
Routon/John Graydon AGBS BSA 87
Roux/Betty E. HOEC MS 82
Rowbotham/Cara Dawn ENTO MS 88
Rowden/Beth Whelan BSHE 38
Rowe/Lu BSHE 38
Rowe/Phillip HORT MS 63
Rowe/Phillip GNAG BSA 62
Rowe/Stewart F. GNAG BSA 40
Rowe/Van Carroll BSA 41
Rowe/William E. BSA 50
Rowland/Jonathan Saul AGED BSA 93
Rowland/Julie D. HDFS 98
Rowland/Max Howard VOAG BSA 62
Rowland/Virginia T. MS 65
Rowland/Virginia T. BSHE 57
Rowland/William ANSC BSA 81
Rowlett/Robbie Ramey BSHE 41
Rowton/Keith James ANSC BSA 90
Roy/D. C. BSA 38
Roye/Leslie Morelock ANSC MS 93
Roye/Leslie Morelock ANSC BSA 91
Roye/Nick ANSC BSA 92
Rozypal/Tricia Leigh ANSC BSA 92
Rucker/Janis Lynn CLTX BSHE 84
Rucker/William R. AGRI BSA 54
Rudasill/Lisa Spohn AGME BSA 83
Rudd/Jackie Craig AGRN MS 81
Ruddell/Thelma Shannon BSHE 47
Ruden/Patricia HEED BSHE 59
Rudledge/Jim BSA 48

Rudolph/Bradley Kale AGBS BSA 86
Rudolph/James Warren BSA 56
Rudolph/Richard D. CRSC PHD 75
Rudolph/Richard D. CRSC MS 72
Ruestow/Sarah IDHS BSHES 97
Ruff/David Franklin AGRN PHD 93
Ruff/Deborah L. Satterfield HEED BSHE 77
Ruff/Donald G. AGEC BSA 77
Rugerinyange/Jean AGEC MS 90
Rugger/Rosemary A. ANSC BSA 85
Rummel/George A. BSA 43
Rundle/Doug AGEC MS 83
Runsick/Jack BSA 68
Runsick/Jerry Smith SOSC MS 67
Runyan/Louise BSHE 66
Runyan/Roy CRSC MS 68
Runyan/Roy CRSC BSA 66
Ruoff/Marguerite Coleman BSHE 22
Rupe/Wayne R. BSA 52
Rupe/Wayne R. GNAG MS 64
Rusch/Corrie Arkins HDFS BSHES 96
Rush/Sarah L. HDFS BSHE 92
Rushing/Glenda McLaughlin HEED BSHE 62
Rushing/Gregory Scott AGRN MS 95
Rushing/Gregory Scott PPPM BSA 93
Russell/Alicia FND BSHE 67
Russell/Britton Sorelle CLTX BSHES 97
Russell/Carol L. HOEC MS 81
Russell/Chang-Hee Hong IDHS BSHE 89
Russell/Harmon BSA 48
Russell/Herbert Barton GNAG MS 63
Russell/Herbert Barton BSA 37
Russell/Jennifer A. BSHE 65
Russell/Jerles Eldon AGEG BSA 54
Russell/Jimmy G. AGEC BSA 77
Russell/John A. BSA 36
Russell/Julia Beth HDFS BSHE 93
Russell/Lloyd G. BSA 49
Russell/Marion W. ANSC BSA 68
Russell/Michael A. AGEC BSA 82
Russell/Reagan Paige HDFS BSHE 92
Russell/Robin R. ANSC BSA 82
Russell/Steven Glynn HORT BSA 85
Russell/William R. PPPM BSA 89
Russell/William R. GNAG MS 91
Rutherford/Amy Lynn AGEC MS 93
Rutherford/Ben D. AGED BSA 95
Rutherford/Claud F. POSC BSA 68
Rutherford/Darla Faye AGBS BSA 96
Rutherford/Henry O. POSC MS 70
Rutherford/Henry O. BSA 65
Rutherford/William Scott ANSC BSA 90
Rutledge/David Richard BSA 49
Rutledge/E. Moye AGRN MS 63
Rutledge/Floyd L. BSA 51
Rutledge/Lloyd L. AGEC BSA 40
Rutledge-Ryals/Lori A. HDFS 98
Ryan/Aisling Catherine FDNU BSHE 92
Ryan/Archie B. DANB BSA 83
Ryan/Dana HDID BSHE 83
Ryan/Roxann HOEC MS 82
Ryan/Shelly Irene ANSC BSA 96
Ryburn/Renee HOEC MS 79
Rydell/Brian Andrew AGME BSA 89
Rydell/James Alfred ANSC MS 87
Rydell/James Alfred POSC BSA 86
Ryder/John Phillip PPPM BSA 79
Rye/Imogene BSHE 38
Ryles/David H. BSA 60
Sabbatini/Ronald Joseph AGRN BSA 93
Sabbe/Jonell Renae FDSC BSA 90
Sadek/Lutfi S. ENTO MS 82
Saenz/Melody IDHS BSHES 98
Safarpour/Max ANSC PHD 87
Sager/Gary AGBS BSA 85
Sager/Herbert Kenneth BSA 29
Sager/Louise Suzanne Chambers CLTX BSHE 80
Sager/Vera W. BSHE 29
Salako/Exekiel Adebayo PLPA PHD 79
Salazar/Rafael Eduardo AGME BSA 93
Saldivar/Mary Nokes HEED BSHE 74
Sale/Helen G. BSHE 46
Saleh/Ali I. ANSC MS 81
Saleh/Ayman Abd Eldady ANSC PHD 85
Salim/Hussein ANSC PHD 86
Salinas/Luis Miguel AGRN BSA 81
Salom/Scott M. ENTO MS 85
Salomon/Rene J. DIAM BSA 90
Salsbury/Randy Joe POSC BSA 85
Salverson/Charles A. BSA 47
Salzman/Frederick P. AGRN MS 87
Sample/Daniel E. LDUH BSA 83
Sams/Casey Ann FDNU BSHE 91
Sams/Morris Tilman BSA 50
Samudio/Gil M. ANSC BSA 67
Samudio/Vivian Wood HEED BSHE 67
Samuels/Audra Leigh FDNU BSHE 92
Samuels/Glenda Faye IDHS BSHE 93
Samuels/Kristy Morris ANSC BSA 94
Sanchez/Hector Randal PLPA BSA 80
Sandage/Larry Joe CRSC MS 74
Sandage/Larry Joe SOSC BSA 71
Sanderlin/Jeanette C. HEED BSHE 69
Sanderlin/Randy S. PLPA MS 72
Sanders/Beatrice Lane BSHE 59
Sanders/Cary Dale AGEC MS 81
Sanders/Celia Mires BSHE 38
Sanders/Cynthia Karen ANSC BSA 77
Sanders/Emmett E. BSA 60
Sanders/Everette N. AGEC MS 63
Sanders/Everette N. BSA 38
Sanders/Gary G. ANSC MS 75
Sanders/Gregory Wilson AGBS BSA 88
Sanders/James Dale AGME BSA 81
Sanders/James Niles POSC MS 66
Sanders/James Niles BSA 65
Sanders/Joe I. VOAG BSA 62
Sanders/John L. AGRN PHD 78
Sanders/John L. SOSC MS 75
Sanders/Larry Wayne ANSC BSA 71
Sanders/Mark Edward GNAG MS 89
Sanders/Mark Edward VOAG BSA 81
Sanders/Maudine F. BSHE 46
Sanders/O. Blaine ANSC BSA 74
Sanders/O. Blaine VOAG BSA 77
Sanders/Rebecca G. FDNU BSHE 75
Sanders/Ronald Leverette DANB BSA 81
Sanders/Sammy Gill AGME BSA 74
Sanders/Stephanie L. CLTX 98
Sanders/Violet Mullins BSHE 47
Sanderson/Lynn A. GBUS BSA 77
Sandoski/Craig Alan ENTO PHD 87
Sandoski/Craig Alan ENTO MS 83
Sandrin/Todd Ryan PLPA MS 97
Sandvig/Helen King BSHE 45
Sang/Ho Quang AGEC BSA 64
Sanneh/Mohammadu A. AGRN BSA 92
Sanson/David W. ANSC MS 80
Sanson/David W. ANSC BSA 78
Santifer/Charles S. GNAG MS 68
Santifer/Charles S. BSA 59
Sapp/Virgil N. AGED BSA 33
Sargent/Anne Marie HEED BSHE 85
Sarkarung/Surapong AGRN PHD 78
Satterfield/Cheryl L. LDUH BSA 78
Satterfield/Cleda Oldham BSHE 40
Satterfield/Greg AGEC BSA 78
Satterfield/Lloyd O. BSA 58
Saulsberry/Daryll B. AGRN MS 80
Saulsberry/Daryll B. SOCI BSA 77
Saunders/Patricia Ann AGRN MS 80
Savage/Jim POSC BSA 43
Savage/Michael Glen POSC BSA 913
Savage/Richard Francis AGBS BSA 81
Sawatski/Treva E. HEED BSHE 73
Sawyer/Jordan BSA 40
Sawyer/Mary Etheridge BSHE 43
Sawyer/Thomas G. AGEC MS 76
Sax/Charles Edward GNAG BSA 62
Sayles/Tamara L. HDFS BSHE 94
Sbanotto/Harry J. POSC BSA 77
Sbanotto/Pete ANSC BSA 71
Scanlon/James Clinton POUL BSA 97
Scanlon/Paulena P. HEED BSHE 67
Scaramuzza/Carole HEED BSHE 61
Scaramuzza/Frank POSC MS 68
Scaramuzza/Frank AHUS BSA 63
Scarborough/William F. BSA 22
Scearce/Janna Luebkemann AGEC MS 86
Scearce/Janna Luebkemann AGBS BSA 83
Schaal/Gerald A. POSC BSA 70
Schaal/Henry D. POSC BSA 72
Schaefer/Clayton F. AGRC BSA 97

Schaefers/Leonard GNAG MS 68
Schaefers/Leonard AHUS BSA 62
Schaefers/Patricia BSHE 63
Schaeffer/Anne Phillips HDFS BSHE 87
Scharf/Debra CLTX BSHE 85
Scharhag/Scott Alexander POSC BSA 95
Scharlau/Karen FDNU BS 84
Schermerhorn/James A. AGRN BSA 92
Scherrey/Clara Spencer BSHE 23
Schiefer/Bernard A. ENTO MS 70
Schiefer/Bernard A. AGRN MS 61
Schiefer/Bernard A. BSA 60
Schieffler/Amanda CLTX BSHE 86
Schilling/George Silas BSA 24
Schilling/Kurt PTSC PHD 84
Schilling/Kurt PLPA MS 81
Schlachman/Ruby Lipe BSHE 36
Schleifer/Samuel B. BSA 39
Schmelzer/Donna Driver BSHE 41
Schmidt/Kayla Middleton IDHS BSHE 93
Schmidt/Lance Albert AGRN BSA 96
Schmidt/Lance Albert AGBS NODEG 96
Schmitt/Carolyn L. CDFR MS 72
Schmitt/Carolyn L. HEED BSHE 64
Schmitz/Harold Herman FDSC BSA 87
Schnakenberg/Charles Timothy AGRN MS 90
Schnebelen/Stephanie Lyn IDHS BSHE 94
Schneider/Duane Palmer ANNU MS 67
Schneider/Jessie Hill BSHE 41
Schneider/Judy BSHE 85
Schnelle/Rebecca J. GHEC BSHE 78
Schoeffler/Sara J. FND BSHE 73
Schoenenberger/Marsha G. CLTX BSHE 75
Scholze/Herman E. E. CRSC BSA 64
Schreiner/Sue G. BSHE 49
Schroeder/R. William BSA 37
Schroeder/Steven H. HORT MS 80
Schuck/Shona Laree FDNU BSHES 98
Schudy/Clinton Earl CLTX BSHE 89
Schuler/Steve F. AGRN PHD 92
Schulteis/David T. SOSC PHD 72
Schumacher/Thomas J. HOEC MS 91
Schuman/Donald Bruce AGRN MS 80
Schuman/Jeffery Lynn RSOC MS 84
Schuncke/M. Candace LDUH BSA 78
Scifres/Coy E. BSA 37
Scifres/Cullen D. AGRI BSA 38
Scifres/Odell BSA 37
Scisson/James P. BSA 48
Sckerl/Max M. CRSC PHD 68
Scobee/Mary Jean BSHE 44
Scogin/Vernon L. POSC MS 81
Scogin/Vernon L. ANSC BSA 79
Scott/Beulah Ann FDNU BSHES 96
Scott/Brad BSA 27
Scott/Darren D. FDSC BSA 91
Scott/David D. SOSC MS 73
Scott/David D. SOSC BSA 71
Scott/Heide Anna Garlinger HEED BSHE 86
Scott/Herman A. BSA 65
Scott/Imogene M. HOEC MS 67
Scott/James Thomas POSC BSA 80
Scott/Jayne Ellyn GHEC BSHE 81
Scott/Jennifer M. LDUH 98
Scott/Joe Benton HORT MS 83
Scott/Karla CLTX BSHE 81
Scott/Kenneth R. BSA 59
Scott/Loretta Moon BSHE 55
Scott/Maya Millicent ANSC BSA 96
Scott/Michael Leo ANSC MS 90
Scott/Milton B. GNAG BSA 51
Scott/Reece BSA 50
Scott/Robert Douglas AGEC MS 83
Scott/Robin M. HDFS 98
Scott/Robin McVey GHEC BSHE 79
Scott/Susan Rogers HOEC MS 81
Scott/Susan Rogers HEED BSHE 79
Scott/W. Miller PPPM BSA 80
Scott/Waymette Gambill BSHE 74
Scribner/Joseph E. DANB BSA 79
Scroggin/Michael P. BSA 47
Seaman/Dina Lynn DANB BSA 91
Searcy/Dorothy Sandford VOED BSHE 26
Searcy/Mickey A. CLTX BSHE 78
Sears/Michael L. POSC BSA 88
Sears/Michele L. LDUH BSA 87
Seay/Robert POSC MS 74
Seay/Robert VOAG BSA 72
Sebushahu/Tharcisse ANSC MS 91
See/Elizabeth Bright BSHE 61
Sego/Janet S. Woodruff BSHE 64
Segraves/Danny CRSC BSA 71
Segraves/Don S. BSA 51
Seidensticker/Jeffrey Robert HORT BSA 80
Seitz/Kathy Reynolds FDNU BSHE 82
Selby/Robert D. ENTO BSA 64
Self/Clara Mae Hunton HEED BSHE 75
Self/Robert Dewey BSA 49
Sellers/Donita HDFS BSHE 96
Sellers/Edward D. CRSC BSA 69
Sellers/Richard POSC MS 79
Selman/Betty A. Guion Hull BSHE 48
Selph/Roger A. ANSC MS 68
Selph/Roger A. AHUS BSA 66
Selvey/Dallas Baker HDFS BSHE 92
Semidey-Laracuente/Nelson AGRN PHD 92
Senkevech/Kevin Adam AGBS BSA 90
Senkevech/Tara Marlo FDNU BSHE 88
Senko/Jo Ann FND BSHE 71
Senko/Michael J. AGEC BSA 75
Senseman/Scott Allen AGRN PHD 94
Senseman/Scott Allen AGRN MS 90
Sergeant/Debra Bowser HOEC MS 85
Serra/Alexander Biasbas ANSC MS 83
Sese/Abras I. SOSC MS 76
Sessions/Betty GNAG MS 85
Sethpakdee/Ravie PTSC PHD 87
Settlage/Dan AGBS BSA 97
Setzler/Rebecca HEED BSHE 73
Setzler/Rudolph D. BSA 32
Sewell/Lesley LDUH BSA 81
Sexton/Elizabeth Dorsey BSHE 72
Sexton/Jeffery Gill POUL BSA 98
Sexton/Kelly J. POSC MS 84
Sexton/Michele Diane HDID BSHE 83
Sexton/Winifred Champion BSHE 31
Seymore/Dennis N. ANSC BSA 73
Seymour/Carter P. BSA 49
Shaddox/William R. BSA 56
Shadrick/Jennifer Lynn AGBS BSA 96
Shafer/Richard Everett RSOC MS 81
Shaffer/Dana Brown ANSC BSA 87
Shaffer/Velma BSHE 28
Shafizadeh/Dariush AGME BSA 79
Shaikh Abdullah/Shaikh Awadz Bin AGRN MS 79
Shakespear/Carl E. PLPA MS 75
Shamoun/Shamoun F. PLPA PHD 90
Shaner/J. F. BSA 65
Shank/Karen Easson HOEC MS 85
Shank/Karen Easson HEED BSHE 74
Shanks/Carolyn Wray BSHE 57
Shannon/Eardie Roy BSA 33
Shannon/Elizabeth AGEC BSA 86
Shannon/Gary Gene ANSC BSA 69
Shannon/Jerry W. ENTO MS 76
Shannon/Jerry W. ANSC BSA 73
Shannon/Judy Burton FND BSHE 71
Shannon/Polly Sweetser HOEC MS 67
Shannon/Polly Sweetser HEED BSHE 62
Sharby/Thomas F. POSC PHD 73
Shariffa/Azimunisa Hamid LDUH BSA 86
Sharma/Jyotsna HORT BSA 95
Sharp/Brenda Joyce FDNU BSHE 78
Sharp/Carolyn Q. HEED BSHE 64
Sharp/Maurce C. MS 54
Sharp/Roy Neil ANNU BSA 68
Sharp/Roy Neil FDSC PHD 81
Sharp/Serenity Kaye IDHS BSHES 97
Sharp/Thomas W. AGEC BSA 79
Sharp/Timothy C. AGRN MS 80
Shassere/Mary A. DANB BSA 76
Shateri/Ahmad Mirabadi AGSC BSA 84
Shaver/Allen A. BSA 53
Shaver/Marcus Carl ENSC BSA 96
Shaver/Paul Hayden FDSC BSA 72
Shaver/Robert V. BSA 65
Shaw/Deanna L. HEED BSHE 69
Shaw/Elisabeth HOEC MS 77
Shaw/Glenn Walter HORT MS 67
Shaw/James Harvey BSA 35
Shaw/Luther O. BSA 52
Shaw/Martha Phillips BSHE 50
Shaw/Melissa B. GHEC BSHE 82

Shaw/Sidney Stanford BSA 61
Shaw/William E. GNAG BSA 36
Shay/Junior Ralph BSA 39
Shearer/Frances Lawrence HOEC BSHE 66
Shearer/Max A. BSA 59
Sheets/Ted AGBS BSA 91
Shelby/Carole Adams FDNU BSHE 62
Shelby/Tina HORT BSA 80
Shell/Jerry BSA 79
Shell/Jerry GNAG MS 85
Shellhorn/Jennifer Lynne IDHS BSHE 92
Shelton/Dean Dorwin BSA 50
Shelton/Donald Joseph AGME BSA 85
Shelton/J. B. BSA 55
Shelton/Loretta Lynn HORT MS 80
Shelton/William D. BSA 53
Shen/Patrick Fu-Li ANSC PHD 82
Shepard/James H. BSA 50
Shepard/Nathaniel Leonard BSHE 22
Shepard/Thad Edward ENSC BSA 96
Shepherd/John H. HORT BSA 75
Shepherd/Keith Allan AGBS BSA 87
Shepherd/Leslie Eugene POSC BSA 94
Shepherd/Robert R. ANSC BSA 82
Shepherd/Sharon Kay Simmons HEED BSHE 74
Sheppard/Angela C. POSC 99
Sherchand/Laxman GNAG MS 90
Sheridan/Richard M. BSA 44
Sheriff/Lorna Kathleen HLED BSHE 81
Sherwin/Ronald D. AGME BSA 85
Sheu/Gueyhwa FDSC MS 86
Shideler/Eva A. HOEC BSHE 45
Shields/Lyle G. GNAG MS 78
Shields/Lyle G. ANNU BSA 72
Shinn/Aileen McCoy BSHE 18
Shinn/Anne Davis BSHE 43
Shipley/Geraldine BSHE 56
Shipley/Holly P. HDID BSHE 82
Shipley/Janis Hawkins HOEC MS 81
Shipley/Janis Hawkins BSHE 52
Shipley/Kerry L. HDID BSHE 82
Shipp/Mimi FDNU BSHE 76
Shirazi/Abbas M. LDUH BSA 80
Shirley/Vincent Raif ANSC BSA 94
Shivazad/Mahmood POSC PHD 84
Shiver/Linda Stone FND BSHE 66
Shiver/Mary Buechley BSHE 25
Shobowale/Adewale Shakiur AGEC MS 81
Shock/Anthony A. BSA 50
Shock/Melissa M. HEED BSHE 70
Shockley/Powell A. GNAG BSA 56
Shoemake/Lyman CRSC BSA 61
Shoemaker/Carolyn D. HORT BSA 81
Shoemaker/Elizabeth Stone BSHE 86
Shoemaker/Sara Chester BSHE 39
Shofner/Alice BSHE 51
Shofner/Bob ANSC BSA 82
Shofner/Janice HEED BSHE 82
Shofner/William K. ANSC BSA 76
Shofner/William K. BSA 51
Shook/Helen Lafern BSHE 51
Shook/James G. PHUS BSA 63
Shopfner/Bruce Allen POSC BSA 89
Shores/Jana M. HEED BSHE 72
Short/J. Rick BSA 79
Short/M. Portia HOEC MS 78
Short/M. Portia BSHE 65
Shower/David Lee ANSC BSA 75
Shrable/Eric Jonathan ANSC MS 84
Shrable/Mary Margaret GHEC BSHE 84
Shreve/Henry Benson POSC BSA 94
Shriver/Leslie Ann ANSC MS 83
Shriver/Leslie Ann DASC BSA 81
Shrout/Kaye P. HOEC MS 86
Shrum/Shannon L. CLTX BSHE 90
Shrum/Shawn Paul ANSC BSA 95
Shuck/Tami AGBS BSA 85
Shuffield/Glen D. ANSC BSA 76
Shull/Callie Ann King FDNU MS 67
Shumaker/Racheal Dunn BSHE 34
Shumate/Alton E. ENTO MS 65
Shumate/Alton E. BSA 52
Shumate/Richard A. POSC BSA 73
Shumate/William Melton BSA 43
Shurgar/Thomas Edward AGBS BSA 97
Sibert/Victor Jay BSA 41
Siddique/Mda SOSC MS 70
Sides/Donald R. HORT BSA 78
Sides/James D. MGMT BSA 72
Sides/Robert E. HORT BSA 82
Siebenmorgen/David Gerard POSC BSA 89
Siebenmorgen/Dennis C. AGED BSA 89
Siebert/Jane Leeann HOEC MS 88
Siegler/Elise Dietz CLTX BSHE 82
Siems/Christa Lane IDHS BSHE 92
Siever/Dick BSA 77
Sigman/Byron Keith AGBS BSA 90
Sigua/Gilbert Castaneda AGRN MS 83
Sikes/Rusty Nathan POSC BSA 91
Silber/Wilma Scott BSHE 31
Silberstein/Steven Philip HORT MS 82
Siler/Alva Ray AGRN MS 62
Siler/Alva Ray GNAG BSA 60
Siler/Cinda Elizabeth FDSC MS 96
Siler/Cinda Elizabeth FDSC BSA 95
Silva/Massie B. GNAG MS 72
Silvey/Thomas J. BSA 37
Silzell/Sally Ann EXED BSA 96
Simco/Debbie Lynn CLTX BSHE 86
Simco/Joseph S. BSA 53
Simco/Joseph S. ENTO MS 62
Simco/Robert F. AGBS BSA 90
Simco/Thomas N. AGEC BSA 82
Simelane/David Okhi ENTO MS 96
Simkins/Hiram D. AGEC BSA 78
Simkins/Janice BSHE 55
Simmons/Brandy Lynn CLTX BSHE 94
Simmons/Edith Gosnell BSHE 40
Simmons/H. H. BSA 49
Simmons/James J. BSA 49
Simmons/Jody L. AGBS 99
Simmons/Rita Susan BSHE 49
Simon/Casey Andrew AGBS BSA 96
Simon/Daniel Philip GNAG MS 93
Simon/Daniel Philip AGED BSA 87
Simon/Jason William AGRC BSA 98
Simon/Kendall HDFS BSHES 98
Simon/Louis G. FDSC MS 82
Simon/Louis G. AGRN BSA 79
Simon/Matthew Joseph ANSC BSA 96
Simon/Michael BSA 65
Simon/Timothy ANSC BSA 94
Simpson/Arthur Marion CRSC MS 75
Simpson/Arthur Marion CRSC BSA 74
Simpson/Elaine HEED BSHE 61
Simpson/Gregory David AGRN BSA 86
Simpson/Gregory K. AGBS BSA 82
Simpson/James D. ANSC MS 85
Simpson/Jeff IDHS BSHE 92
Simpson/Miles D. POSC MS 74
Simpson/Miles D. POSC BSA 69
Simpson/Pamela S. CLTX BSHE 77
Simpson/Steven W. POSC MS 76
Simpson/Steven W. POSC BSA 75
Simpson/Virginia Edwards HOEC BSHE 44
Simpson/Wilburn A. AGBS BSA 76
Simpson/William B. BSA 51
Sims/Arling L. BSA 52
Sims/Barry Dale AGRN PHD 86
Sims/Charles A. FDSC PHD 86
Sims/Charles A. FDSC MS 82
Sims/Charles A. FDSC BSA 80
Sims/James R. BSA 60
Sims/Karen Sue Webster PLPA BSA 79
Sims/Katrina L. FDSC 99
Sims/Natalie Rae HDID BSHE 82
Singleton/James Allen AGED BSA 85
Singleton/Jimmy F. POSC BSA 73
Singleton/Mark E. AGRN BSA 91
Sink/Richelle LDUH BSA 97
Sinor/Carla Jane ANSC BSA 85
Sipes/Joseph Paul BSA 50
Siragusa/Gregory R. FDSC PHD 89
Sisk/Clyde Aaron BSA 31
Sisson/Walter W. BSA 43
Sistrunk/Callen Patrick FDSC BSA 76
Sistrunk/Laurence AGEG BSA 69
Sites/Clyde BSA 57
Sites/Daniel E. POSC BSA 80
Sites/Jerry William AGBS BSA 87
Sitlington/Billy AGBS BSA 83
Sitton/Dorothy Price HOEC BSHE 46
Sitton/Lloyd I. BSA 59

Six/Charlotte KAy HEED BSHE 70
Skaggs/Karen K. HOEC MS 77
Skaggs/Pamela Olienyk FDSC MS 94
Skaggs/Pamela Olienyk AGBS BSA 91
Skaggs/William W. DIAM BSA 92
Skarda/Pat AGRN BSA 78
Skeets/Karen L. Light ANSC BSA 93
Skeets/Mac Hewett AGEC BSA 93
Skeith/Shirleen HORT BSA 81
Skelton/Louis Carl AGBS BSA 95
Skelton/Mark Allen AGBS BSA 85
Skelton/Ronald Quentin ANSC BSA 83
Skies/James B. BSA 58
Skinner/Andrew Ray HTSO BSA 98
Skinner/James B. AHUS MS 53
Skinner/James B. AGRI BSA 52
Skinner/James Thomas ANSC PHD 92
Skinner/Sidney Lynn AGRN BSA 85
Skinner/Vaughn AGRN BSA 82
Skordal/Katherine A. FDNU BSHE 78
Skulman/Briggs FDSC PHD 86
Slack/Steven A. PLPA MS 71
Slack/Steven A. PLPA BSA 69
Slade/Ilma Johnson BSHE 32
Slade/Kivi Kivia BSA 17
Slagter/Peter J. G. POSC MS 80
Slaton/Nathan Alan AGRN PHD 98
Slaton/Nathan Alan AGRN MS 89
Slaton/Patricia Clement HEED BSHE 69
Slaughter/Linda Kaye CTM BSHE 75
Slaughter/Roy BSA 80
Slaughter/Susan HEED BSHE 77
Slaughter/Vera BSHE 23
Slaven/Joe J. BSA 41
Slavens/Frank D. BSA 50
Slay/Debra Raley HOEC MS 78
Slay/Debra Raley BSHE 76
Slayman/Kathi R. CLTX BSHE 76
Sledge/Delia Y. HEED BSHE 78
Sloan/Anne HEED BSHE 64
Sloan/B. Pauline Faucett BSHE 51
Sloan/Betty T. BSHE 45
Sloan/Charles T. BSA 51
Sloan/Danny G. AGEG BSA 81
Sloan/Don R. VOAG MS 70
Sloan/Don R. VOAG 3SA 67
Sloan/Frank F. BSA 54
Sloan/J. Hope BSHE 41
Sloan/Jack M. GNAG BSA 51
Sloan/Linda B. FDNU BSHE 78
Sloan/Margaret Adele BSHE 43
Sloan/Martha Ann Huey HDID BSHE 79
Sloan/Robert H. BSA 49
Sloan/Winston I. BSA 58
Sloate/Jennifer Suzette HORT BSA 87
Small/Jeffrey L. ANSC BSA 84
Smalley/Michelle AGBS BSA 96
Smartt/Alfred B. BSA 48
Smay/Stephen E. FDSC BSA 70
Smedley/Robert P. AGED BSA 92
Smiley/Samuel C. ESWS 98
Smith/Almer C. BSA 39
Smith/Audrey L. FND BSHE 70
Smith/Betty Faye BSHE 51
Smith/Billy G. BSA 52
Smith/Boyd R. LDUH BSA 82
Smith/Bruce Laron BSA 39
Smith/Byron T. BSA 21
Smith/Carl A. BSA 25
Smith/Carolyn Marie HDFS BSHE 92
Smith/Carr BSA 17
Smith/Charlotte A. GHEC BSHE 89
Smith/Clarence Turner BSA 23
Smith/Claudia Ann HOEC MS 95
Smith/Clifford L. BSA 36
Smith/Clovis P. BSHE 65
Smith/Daniel N. GNAG BSA 50
Smith/David B. AGBS BSA 82
Smith/Dewitt McKinley BSA 23
Smith/Don M. AGBS BSA 80
Smith/Donnetta Gooch CLTX BSHE 85
Smith/Donnie Alvie AGEC MS 76
Smith/Donnie Ray CRSC MS 68
Smith/Donnie Ray BSA 65
Smith/Doyne ENTO BSA 64
Smith/Duffie L. BSA 50
Smith/Dwayne Austin GNAG BSA 62
Smith/Earl T. BSA 40
Smith/Earl W. BSA 40
Smith/Earl Webster BSA 16
Smith/Eddie B. AGEC MS 60
Smith/Eddie B. BSA 57
Smith/Edman A. BSA 52
Smith/Elmer Dee AGRN BSA 89
Smith/Eugene H. BSA 38
Smith/Evelyn Holt BSHE 38
Smith/Evelyn Shawn HEED BSHE 85
Smith/Floyd J. BSA 50
Smith/Frances McElroy BSHE 36
Smith/Fred Alfred BSA 25
Smith/Fred H. ANSC BSA 73
Smith/Freddy Allan ANSC MS 69
Smith/Garner H. BSA 35
Smith/Glenn W. AGRI BSA 78
Smith/Harlan E. BSA 49
Smith/Harold E. BSA 58
Smith/Harold R. CRSC MS 69
Smith/Harold R. BSA 65
Smith/Hollie Alane HIST BA 84
Smith/Holly P. IDHS BSHE 92
Smith/Homer L. BSA 59
Smith/J. Ritchie BSA 41
Smith/James Alan DASC BSA 86
Smith/James C. ANSC BSA 83
Smith/James F. ANSC MS 88
Smith/James Floyd ANSC BSA 81
Smith/James K. AGBS BSA 74
Smith/James L. GNAG BSA 74
Smith/James M. ANNU BSA 67
Smith/James Victor AGEC BSA 81
Smith/Janis Lynn HOEC BSHE 87
Smith/Jeffery J. ANSC BSA 82
Smith/Jennifer M. PLPA BSA 89
Smith/Jerry McKinley GNAG MS 67
Smith/Jerry McKinley BSA 57
Smith/Joanna W. CLTX BSHE 84
Smith/John Bee AGED BSA 93
Smith/John Boyd ANSC BSA 94
Smith/John C. GNAG BSA 67
Smith/John P. ENTO MS 79
Smith/John P. ENTO BSA 77
Smith/John R. BSA 49
Smith/Julene Sells BSHE 79
Smith/Karen Lynn CDFR 82
Smith/Karen Lynn CDFR BSHE 80
Smith/Karen Lynn Jarvis HOEC BSHE 82
Smith/Karon Bennett BSHE 79
Smith/Kathleen Rose HOEC MS 93
Smith/Kelly Don AGBS BSA 85
Smith/Kenneth A. BSA 56
Smith/Kenneth B. BSA 53
Smith/Kenneth Leon BSA 42
Smith/Kimberly Jean HDID BSHE 80
Smith/Kirk Dixon PPPM BSA 87
Smith/Kit BSA 56
Smith/Lance E. AGBS BSA 94
Smith/Larry D. PLPA MS 75
Smith/Larry J. GNAG BSA 72
Smith/Larry W. GNAG MS 85
Smith/Lealdon BSA 36
Smith/Lela Maude BSHE 25
Smith/Lewis Dawson GNAG MS 84
Smith/Linda E. FND BSHE 69
Smith/Lucy Stephenson CLTX BSHE 87
Smith/Lydle Paris BSA 20
Smith/Lynda Lee HEED BSHE 75
Smith/Lynn Luman BSA 26
Smith/Mabel HOEC BSHE 29
Smith/Malory Leanne IDHS BSHE 93
Smith/Marjorie E. BSHE 44
Smith/Marshal Cade AGRN MS 96
Smith/Martin Cleoh BSA 47
Smith/Mary A. Person PLPA BSHE 69
Smith/Mary Lois Foster BSHE 33
Smith/Maxine GHEC MS 60
Smith/Maxine GHEC BSHE 46
Smith/Melba McKenzie BSHE 46
Smith/Michael J. ANNU BSA 76
Smith/Michelle Ann HEED BSHE 8C
Smith/Norman K. POSC PHD 87
Smith/Ocie Lee AGEC MS 59
Smith/Odessa Pearce BSHE 23
Smith/Oren R. POSC BSA 49
Smith/Oscar D. BSA 16

Smith/Owen E. AGEC MS 74
Smith/Pearle A. BSA 25
Smith/Phil Stacy ANSC BSA 86
Smith/Phillip POSC PHD 95
Smith/Phillip POSC MS 81
Smith/Phillip ANSC BSA 79
Smith/R. Douglas AGBS BSA 88
Smith/Ralph J. BSA 40
Smith/Randall W. GNAG MS 88
Smith/Randall W. AGRN BSA 83
Smith/Randy K. AGEC MS 89
Smith/Randy K. ANSC BSA 82
Smith/Richard Grady ANSC BSA 88
Smith/Robert G. BSA 50
Smith/Robert G. GNAG MS 68
Smith/Robert M. ANSC BSA 58
Smith/Rocquelle Deneise CLTX BSHES 96
Smith/Roland Edward ANSC BSA 75
Smith/Ronald Leslie AGRN BSA 80
Smith/Ronald S. LDUH BSA 77
Smith/Russell McKinney AGRN BSA 86
Smith/Russell Wayne DANB BSA 80
Smith/Sandra H. HOEC 78
Smith/Sandra H. FND BSHE 70
Smith/Scott Allen ANSC BSA 90
Smith/Shelby M. BSA 55
Smith/Sondra Hall FDNU BSHE 60
Smith/Tamara G. DANB BSA 89
Smith/Thomas R. DASC MS 76
Smith/Thomas R. DASC BSA 71
Smith/Timothy Shelby AGEC MS 87
Smith/Timothy Shelby AGBS BSA 86
Smith/Todd AGRN BSA 85
Smith/Velma BSHE 19
Smith/W. Maurice BSA 42
Smith/Watson I. BSA 47
Smith/William Charles PPPM BSA 81
Smith/William G. BSA 59
Smith/William Grady ANSC MS 60
Smith/William Grady ANSC BSA 59
Smith/William W. ANSC MS 77
Smith/Winifred Rudolph BSHE 25
Smitherman/William M. BSA 39
Smithey/Vicki A. AGEC MS 86
Smithey/Vicki A. AGBS BSA 83
Smithson/Jill Lyall CLTX BSHE 95
Smithson/Kathlyn Jeffery FND BSHE 63
Smithson/Larry J. AHUS BSA 62
Smittle/Burrell J. ENTO MS 56
Smittle/Burrell J. BSA 55
Smittle/Doyle Allen HORT BSA 61
Smittle/Robert BSA 59
Smykla/Misti Acuff AGBS BSA 95
Sneed/Fred D. HORT MS 69
Sneed/Fred D. INEG BSA 68
Snider/Marge GHEC BSHE 81
Snow/Caleb L. AGBS BSA 83
Snow/Gertrude BSHE 39
Snow/Paula Fay GNAG MS 85
Snow/Paula Fay HORT BSA 80
Snow/Renita Paige FDNU BSHE 82
Snowden/Carol Erich HOEC MS 74
Snowden/Susanne FDNU BSHE 76
Snyder/Clifford S. AGRN MS 80
Snyder/Clifford S. SOCS BSA 78
Snyder/Frank Owen BSA 50
Snyder/Mary Ann Stone BSHE 50
Snyder/Ralph Harold BSA 38
Snyder/William W. BSA 51
Soast/Frances Harper BSHE 34
Sogut/Bunyamin ANSC PHD 96
Soller/Ruth S. CLFM BSHE 78
Solomon/James Doyle ENTO MS 60
Solomon/James Doyle GNAG BSA 56
Somayyajula/Srikanth A. ENEG MSENE 94
Sooksai/Sanay ENTO MS 77
Sorensen/Katherine Rhea CLTX BSHE 95
Sorrells/Richard D. AGEC MS 72
Sorrells/Richard D. AGEC BSA 64
Sorrels/George W. BSA 37
Sotelo/Judy Gail BSHE 69
Soulti/Lydia FDNU BSHE 94
Southerland/Ernest Leroy BSA 52
Southern/Jerald Ray ANSC BSA 92
Southern/Minnie Irene Cole BSHE 31
Souza/Vania Beatriz V. PTSC PHD 89
Spanel/Mitchell J. HORT BSA 79
Spann/Eric Christopher CLTX BSHE 88
Sparrow/Justin Wray PPPM BSA 95
Spaulding/Marion BSA 54
Spayd/Sara Elizabeth FDSC PHD 80
Spayd/Sara Elizabeth FDSC MS 77
Spears/Bob Lee POSC BSA 62
Spears/Karen Sue HDFS BSHE 86
Speck/Thomas H. AGBS BSA 81
Speer/Dixie W. HOEC BSHE 41
Speight/Julie D. AGED 98
Speight/Michael L. AGED 98
Spells/Lori Leigh CLTX BSHE 84
Spence/Clara S. HEED MS 52
Spencer/Chris A. AGEC BSA 82
Spencer/Edith E. Speir BSHE 33
Spencer/Edward D. BSA 50
Spencer/G. B. BSA 49
Spencer/Gregory Kenneth POSC MS 84
Spencer/James A. PLPA MS 62
Spencer/James A. BSA 54
Spencer/Joe W. BSA 51
Spencer/Lois M. BSHE 54
Spencer/Matthew Lowe AGEC MS 85
Spencer/Stanley BSA 42
Sperduto/Edith Ray BSHE 46
Spicer/Barbara L. HEED BSHE 73
Spicer/Mary Laura BSHE 49
Spigner/Rhonda Carol AGEC BSA 88
Spinks/Garland Keith ANSC BSA 91
Spinks/Michael Douglas ANSC BSA 90
Spitze/Hazel Taylor HEED BSHE 43
Spitze/Robert G. F. AGRI BSA 47
Spivey/John W. BSA 49
Spivey/Kim Regina HDFS BSHE 85
Spivey/Wanda R. BSHE 41
Spivey/William F. AGED BSA 35
Spotts/James W. SOSC MS 70
Spradley/Plez PLPA MS 82
Spradlin/Paul D. BSA 50
Spradling/Anna Lea HDFS BSHES 97
Spray/Beverly J. ANSC PHD 89
Spreen/Scott POSC MS 86
Spreen/Scott DANB BSA 85
Springer/Jennifer Darlene HORT BSA 95
Springer/Mary Loutitia HDID BSHE 84
Springfiled/Amada G. CLTX 99
Sprott/Joseph M. AGEC MS 65
Sprott/Joseph M. BSA 63
Spurgeon/Dale W. ENTO PHD 92
Spurgeon/Winifred Ishmeal BSA 56
Spurlin/Jim W. AGME BSA 82
Spurlin/Susan McKinney CLTX BSHE 82
Spurlock/Brian D. AGEC MS 74
Spurlock/Charlotte Ann FND BSHE 70
Spurlock/Darby L. POSC 99
Spurlock/Dennis C. POSC BSA 75
Spurlock/John O. BSA 49
Spurlock/Leo Ernest BSA 42
Spurlock/William Greg AGBS BSA 85
Spyres/Sean Mitchell BSA 94
Squier/May Washington BSHE 44
Squire/Deri Lyn HDFS BSHE 86
Squires/Keith Allen POSC BSA 86
Srisomboon/Sumon SOSC MS 70
Srivastava/Puneet BAEG MSBAE 95
Sriyotai/Kasem AGEC MS 76
St. John/C. W. BSA 73
St. John/Robert Eugene VOAG BSA 82
St. Onge/Mark Roger DANB BSA 90
Stacey/Alvie Leon ENTO PHD 78
Stacey/Alvie Leon AGEC MS 74
Stacey/Joseph Scott AGBU BSA 97
Stacpoole/Lee Ann Starr GHEC BSHE 77
Stadthagen-Vogl/Roberto David BSA 65
Stafford/Dixie E. BSHE 37
Stafford/Phillip AGRN MS 79
Stafford/Phillip SOSC BSA 75
Staggs/Adriane F. IDHS 99
Stalker/Scott M. AGBS BSA 88
Stallcup/Anne FDNU BSHE 82
Stallcup/Odie T. BSA 43
Stallcup/Susan E. FDNU BSHE 82
Stallings/Alan E. BSA 41
Stallings/Brenda D. HOEC MS 78
Stallings/Brenda D. FND BSHE 75
Stamper/Marjorie King BSHE 38
Stamps/Ashley Charlene IDHS BSHE 91

Stanberry/Thomas W. POSC BSA 71
Standefer/Charles BSA 53
Standefer/William Haskell BSA 50
Standerfer/Mavis Hale MS 53
Standfill/Thomas Jeff BSA 54
Standridge/Philip Harold ANSC BSA 85
Standridge/Rodney Wayne POSC BSA 87
Standridge/William E. BSA 42
Stanfield/Pamela C. HORT BSA 83
Stanford/Eloise Martha BSHE 29
Stanford/Malcolm F. BSA 26
Stanford/Timothy Leroy ANSC BSA 73
Stanford/William L. BSA 48
Stanley/A. D. POSC BSA 52
Stanley/Lynn AGRN MS 83
Stanley/Lynn AGRN BSA 81
Stanley/Ronald Eugene AGEC MS 83
Stanley/Ronald Eugene AGBS BSA 80
Stanley/William L. BSA 56
Stanton/Derik Wicks ANSC BSA 88
Stanton/Harold Stanley AGEG MS 53
Stapleton/W. A. GNAG BSA 53
Stark/Bill Frederick BSA 50
Stark/Pamela Maureen ENTO MS 83
Stark/Susan Carol POSC BSA 86
Starnes/Debra POSC BSA 81
Starnes/George Richardson BSA 54
Starr/Betty R. BSHE 62
Starr/Scott HORT BSA 97
Starwalt/Barbara A. HEED BSHE 72
Stasiak/Matt J. HORT MS 90
Staton/Billy Carlton BAEG MSBAE 95
Staton/Billy Carlton AGRN BSA 91
Staton/Henry Carlos CRSC MS 74
Stauber/Larry Gene AGRN BSA 87
Stauber/Martin Stevens BSA 58
Stauber/Robert S. AGBS BSA 91
Staudinger/Felicia Denise ANSC BSA 91
Staudinger/Felicia Denise ANSC MS 94
Stavely/Edwin L. ANSC BSA 74
Stavely/Herman Edwin BSA 50
Stearns/Bryan BSA 19
Stebbins/Albert H. BSA 60
Stecher/David A. LDUH BSA 80
Stecks/Gregory Edward AGBS BSA 92
Steed/Charles Ray BSA 39
Steel/Vestal Montgomery BSHE 39
Steele/Dorothy Guilliam HEED BSHE 37
Steele/Erma B. BSHE 38
Steele/Janie Haley FND BSHE 67
Steele/Michelle Ann HDFS BSHE 95
Steely/Barbara HEED BSHE 61
Steen/Katherine Stewart HOEC MS 73
Steffens/Elizabeth Pipek FDNU BSHE 76
Steiger/Fritz Speier AGEC BSA 81
Steiner/Robert AGME BSA 87
Steiner/Robert BSA 58
Steinhart/Valorie Rogers CLTX BSHE 77
Steinsiek/J. Bill BSA 82
Steinsiek/J. Bill AGRN MS 82
Steinsiek/Scott Stuart POSC BSA 86
Stemple/Dale AGEC BSA 79
Stendel/Joan Ellen GHEC BSHE 91
Stendel/Marvin B. ANSC BSA 52
Stephens/Cindy Shaw CLTX BSHE 80
Stephens/David H. AGBS 98
Stephens/Fraser AGRN BSA 40
Stephens/Gary Don CRSC BSA 74
Stephens/Harold Hall BSA 39
Stephens/Jerry D. AGED BSA 92
Stephens/Jerry L. GNAG BSA 53
Stephens/John H. AGEC BSA 31
Stephens/John H. AGEC BSA 64
Stephens/Madge BSHE 41
Stephens/Madge HOEC MS 66
Stephens/Mary Lemley Gardner BSHE 65
Stephens/Maurice L. BSA 40
Stephens/Mollie Wilkerson BSHE 41
Stephens/Phoebe L. HORT BSA 85
Stephens/Ruth Rudolph BSHE 29
Stephens/Steven Dale FDSC BSA 92
Stephens/Tracy Lee AGBS BSA 97
Stephens/Wanda B. BSHE 54
Stephens/William Asbury BSA 55
Stephenson/Mary E. FND BSHE 72
Stephenson/Susan E. FDSC BSA 80
Stephenson/Susan E. ELED BSA 80
Stephenson/Thaddus J. BSA 52
Sterne/Rickie E. PLPA MS 74
Steuber/Dennis R. BSA 76
Stevens/Barbara Cotton BSHE 57
Stevens/Betty Files BSHE 61
Stevens/Curtis B. BSA 37
Stevens/George Robert ANSC BSA 80
Stevens/Jane H. FDNU BSHE 60
Stevens/John Madison BSA 40
Stevens/Kimberly K. CLTX BSHE 87
Stevens/Mae Schirmer BSHE 41
Stevens/Melvin Henryd BSA 48
Stevenson/Dale M. AGBS BSA 83
Stevenson/James Edwin BSA 17
Stevenson/Regina K. CLTX BSHE 83
Stevenson/Weldon AGED BSA 51
Steward/Billy E. BSA 50
Steward/Victor Bruce ENTO MS 86
Stewart/Anita L. GHEC BSHE 82
Stewart/Billy R. POSC MS 66
Stewart/Billy R. BSA 65
Stewart/Florence T. BSHE 56
Stewart/Jerry W. ENTO MS 69
Stewart/Judith Chivers BSHE 61
Stewart/Lenard U. BSA 50
Stewart/Mary Jo HOEC MS 67
Stewart/Randall H. ANSC BSA 79
Stewart/Rebecca Lynn CLTX BSHE 93
Stewart/Roberta Switzer BSHE 63
Stewart/Sanlen Samuel BSA 43
Stewart/Tracy Portis Thompson CLTX BSHE 81
Stewart/Virginia L. CLTX BSHE 76
Stidham/William Douglas ANSC PHD 83
Stilborn/Heather Lynn POSC MS 87
Stiles/Philip G. BSA 53
Stiles/Samuel P. BSA 53
Still/Jack GNAG MS 86
Still/Jack VOAG BSA 79
Still/Jennifer Renee IDHS BSHE 91
Still/Kaye Lynne HEED BSHE 85
Stillwell/Donna Ruth GHEC BSHE 93
Stine/Kimberly A. ANSC 98
Stinnett/James R. AGEC BSA 71
Stipe/John R. BSA 53
Stipe/Martin E. BSA 57
Stipe/Mary Cato BSHE 55
Stitz/Leah Davis FDSC BSA 85
Stivers/Odell Nolan BSA 40
Stockalper/Ralph Bennet AGED BSA 86
Stogsdill/Christopher C. ANSC BSA 74
Stogsdill/Deborah CDFR MS 75
Stogsdill/Deborah CDFR BSHE 73
Stokenbury/William Dolph BSA 52
Stokes/Mamie Olive BSHE 36
Stolz/Wesley VOAG BSA 64
Stone/Barbara BSHE 60
Stone/Boyce L. GNAG BSA 52
Stone/Catherine Kay CLTX BSHE 85
Stone/Fonda Hale HDFS BSHE 86
Stone/Helen K. FDNU BSHE 74
Stone/Jeffrey Dean AGEC BSA 85
Stone/Jerry R. CRSC BSA 61
Stone/Jesse R. BSA 41
Stone/Russell Blanton BSA 48
Storlie/Ellen M. HOEC MS 94
Storlie/Ellen Margaret HDFS BSHE 88
Storm/Kenny W. POSC BSA 96
Storms/Penney Michaelis FND BSHE 75
Story/John D. POSC PHD 79
Story/Samuel T. AGBS BSA 79
Stouffer/Caran Meadors DANB BSA 85
Stout/Diana R. CLTX BSHE 78
Stout/Samuel Robman BSA 15
Stover/Betty Jean Files BSHE 61
Stover/Bryan J. HORT BSA 95
Stover/Jane Longino BSHE 51
Strabala/Richard Shane ENSC BSA 97
Strahl/Meredith A. HOFS BSHE 78
Strain/Ricky David AGEC BSA 81
Strain/Susan DANB BSA 80
Strang/James Adelbert BSA 48
Stratemann/Betty Ann FDNU BSHE 68
Stratton/Katharine J. IDHS BSHE 90
Stratton/Royce L. AGEC MS 75
Strebe/Timothy Allen AGRN MS 95
Streett/Sue Heidgen FDNU BSHE 62

Strickland/George G. BSA 13
Strickland/Harold Wayne ANSC BSA 70
Strickland/Slade LDUH BSA 81
Strickland/Susan CLTX BSHE 81
Striegler/Carol Throop FDSC MS 81
Striegler/Carol Throop FDSC BSA 80
Striegler/Curtis T. POSC MS 66
Striegler/Curtis T. PHUS BSA 62
Striegler/Keith R. FDSC BSA 80
Striegler/Keith R. FDSC MS 81
Stringfellow/C. M. AGME BSA 82
Stringfellow/Evelyn Freeman HEED BSHE 50
Stringfield/Mildred T. BSHE 32
Strombeck/Cynthia Ann POSC BSA 87
Stroub/Joe T. BSA 64
Stroud/Don C. AGBS BSA 82
Stroud/Gloria J. BSHE 53
Strozyk/Traci HEED BSHE 89
Struebing/J. Jan Godfrey BSHE 65
Stuart/Carolyn G. FND BSHE 72
Stuart/Cecil Harlan BSA 57
Stuart/Charles C. BSA 50
Stuart/Clyde AGEC MS 69
Stuart/Joe P. AGED BSA 76
Stubblefield/Frances G. BSHE 33
Stubblefield/Ralph E. HORT BSA 27
Stuckey/Fred Gilmore AGRN BSA 81
Stuckey/Samuel H. CRSC BSA 77
Studebaker/Glenn Earl ENTO MS 89
Stueven/Kimberly Wiseman AGBS BSA 83
Stumbaugh/Catherine HEED BSHE 73
Stumbaugh/Kimberly Kay CLTX BSHE 86
Stutte/Gerald Wade AGEC MS 87
Stuttle/Judy D. IDHS BSHE 91
Styles/James Doyle GNAG BSA 62
Suchsland/Julia L. CLTX BSHE 77
Sudarsono/Hamim ENTO MS 89
Sudarsono/Hamim ENTO PHD 92
Sudmeyer/Danny Fay BSA 71
Sudmeyer/Donny R. BSA 71
Suffridge/Lana Snawder CLTX BSHE 92
Sugg/Ann Wiggans HEED BSHE 52
Sugg/Jeffrey Leon ANSC MS 91
Sugg/Jeffrey Leon ANSC BSA 89
Sukhontasing/Somchai HORT MS 75
Sulaiman/Ropeah H. AGRN MS 79
Suliman/Adan Sallam AGRN PHD 86
Sullivan/James P. ENTO BSA 61
Sullivan/James Timothy AGRN BSA 86
Sullivan/Kathleen M. BSA 77
Sullivan/Kipp B. BSA 47
Sullivan/Michael P. AGRN BSA 83
Sullivan/Scott Allen AGRN BSA 89
Sullivan/Thomas W. POSC MS 56
Sullivant/A. Raybon BSA 28
Summa/Julie LDUH BSA 80
Summerford/Joe Alan ENTO BSA 80
Summerford/Lisa Bell CLTX BSHE 82
Summers/Janette SOSC BSA 86
Summers/Katherine M. HEED BSHE 75
Summers/Kipplyn Shankle HDFS BSHE 94
Summers/Linda Elizabeth BSHE 85
Sumner/Rosanna HEED BSHE 48
Sun/Weilin ANSC MS 92
Susanto/Dwi AGEC MS 89
Sutherland/James W. VOAG BSA 71
Suthmer/Kenneth D. BSA 51
Sutterfield/Marvin O. POSC BSA 73
Sutton/Elizabeth Aeberli GNAG MSA 92
Sutton/Elizabeth Aeberli LDUH BSA 90
Sutton/Lou Ann HEED BSHE 71
Sutton/Roy E. AGBS BSA 72
Sutton/Russell Lane AGRN MS 87
Sutton/Russell Lane AGRN BSA 83
Swaim/G. B. BSA 50
Swaim/Loyal W. BSA 51
Swain/Christy Sherene HDFS BSHE 94
Swain/Michael J. POSC BSA 78
Swan/Rachell A. CLTX BSHE 80
Swantek/Jason Michael AGRN MS 97
Swaty/Albert M. GNAG MS 62
Swaty/Albert M. BSA 51
Swaty/Mark MS 62
Swaty/Mark BSA 5
Swayne/David E. DANB BSA 80
Swearingen/Becky HOEC MS 81
Swearingen/Becky HDID BSHE 79
Swearingen/Gordon J. BSA 39
Sweeden/Marshall Brian ENTO PHD 96
Sweeden/Marshall Brian ENTO MS 92
Sweeney/Michael J. VOAG BSA 66
Sweeney/Michael J. POSC MS 69
Sweet/Michelle L. ESWS 99
Sweetin/Howard Lee AGRN BSA 85
Sweetin/Randall G. AGBS BSA 83
Sweetser/Linda S. FND BSHE 74
Swiderski/Terry ANSC BSA 90
Swift/Emily Gaughan BSHE 43
Swift/James B. BSA 38
Swift/Warren G. BSA 42
Swilling/Fran L. HEED BSHE 72
Swilling/George P. BSA 49
Swim/Jason Wayne ENSC BSA 96
Swindle/Dale Lloyd PHD 72
Swindle/Dale Lloyd MS 69
Swindle/Mike Paul LDUH BSA 86
Swink/Karen Lynn CLTX BSHE 80
Sye/Melody Luise CLTX BSHE 89
Sykes/Rosetta Johnson BSHE 55
Symes/Stella Mae Fitch BSHE 31
Szenher/Jessica Cowart FDNU BSHE 76
Sziszak/Teresa L. ANSC MS 86
Sziszak/Teresa L. ANSC BSA 86
Tabb/Otis AGBS BSA 88
Tabler/Sam ANSC BSA 80
Tabler/Tommy GNAG MS 93
Tabler/Tommy POSC BSA 79
Tabor/Elizabeth Jane HEED BSHE 71
Tackett/Caye ANSC BSA 96
Tackett/Cecil BSA 51
Tackett/J. C. BSA 52
Taggart-Loss/Tracy Elaine IDHS BSHE 88
Taki/Reiko HOEC MS 85
Talbert/Pauline Bigham BSHE 49
Talbert/Philip Wayne AGME BSA 86
Talbert/Robert O. BSA 50
Talbot/M. P. BSA 50
Talburt/Dale C. BSA 53
Taliaferro/Carl E. BSA 40
Talison/Dina Talebizadeh CLTX BSHE 85
Tallent/Beverly A. HEED BSHE 70
Talley/Beth S. HDFS 98
Talley/William T. HORT MS 78
Tallman/Richard AGBS BSA 72
Tanaka/Ryuta AGEC MS 91
Tang/Tianjia AGRN PHD 92
Tanifum/Ambe A. AGEC MS 83
Tanner-Marr/Debra Lynn CLTX BSHE 89
Tanner/Charles David ENSC BSA 95
Tanner/David Gerald ANSC BSA 70
Tanner/James William AGED BSA 94
Tapp/William N. VOAG BSA 82
Tarn/Chung-Yiao ANSC PHD 96
Tarn/Chung-Yiao ANSC MS 92
Tate/B. Carmen CLTX BSHE 80
Tate/G. Carolyn BSHE 57
Tate/Leigh A. LDUH BSA 85
Tatum/Teresa Sue AGRN BSA 81
Taubman/Arthur BSA 42
Tavasoli/Gity FDSC MS 91
Taylor/Aaron T. BSA 52
Taylor/Amie FDSC BSA 95
Taylor/Anita Shafer BSHE 47
Taylor/Barksdale H. AGRN MS 76
Taylor/Barksdale H. HORT BSA 75
Taylor/Brett Ray POSC BSA 97
Taylor/Carol Randolph HEED BSHE 84
Taylor/Charles W. BSA 63
Taylor/Christopher Bryan ENSC BSA 95
Taylor/Corey A. AECT 99
Taylor/D. Quinn CLTX BSHE 87
Taylor/Earl Lynn AGEC MS 88
Taylor/Emily Lynette Gulley BSHE 64
Taylor/Francis A. BSA 41
Taylor/G. D. BSA 40
Taylor/Gayla Ann Brewer HOEC BS 82
Taylor/Geneva Crouch BSHE 60
Taylor/George E. BSA 32
Taylor/James E. AGBS BSA 91
Taylor/Janice Binam FDSC BSA 92
Taylor/Jerry William ANSC BSA 85
Taylor/Jessie Mitchell BSHE 37
Taylor/John D. BSA 51
Taylor/Joyce Hudgins BSHE 43

Taylor/Keith BSA 53
Taylor/Kelli Renea AGED BSA 94
Taylor/Kimberly A. CDFR BSHE 78
Taylor/Leatha HOEC MS 69
Taylor/Maxsie H. BSA 55
Taylor/Michael B. AGME BSA 77
Taylor/Michael D. AGRN MS 78
Taylor/Michael D. CRSC BSA 75
Taylor/Paul A. BSA 29
Taylor/Paula K. BSHE 58
Taylor/Robert Love BSA 52
Taylor/Shane Edward AGRN BSA 93
Taylor/Virginia Littlefield HOEC MS 62
Taylor/Virginia Littlefield BSHE 56
Taylor/Wendell M. VOAG BSA 64
Taylor/Wilford W. BSA 49
Taylor/William J. PLPA MS 80
Taylor/William J. HORT BSA 74
Teague/Hercle C. BSA 55
Teague/Kenny M. ANSC BSA 69
Teague/Paul W. AGEC MS 81
Teague/Tina Gray ENTO MS 81
Tedder/Johnna Kay CLTX BSHE 95
Tedder/Sandra A. ENTO PHD 91
Teems/Deborah Kay Zohn ANSC BSA 87
Teeter/Bill GNAG BSA 59
Teeter/G. L. BSA 25
Teeter/James R. BSA 51
Teeter/Janice Turpin BSHE 49
Tempelmeyer/Jennifer Kelly HDFS BSHES 97
Templeton/George E. PLPA MS 54
Templeton/George E. AGRI BSA 53
Tenborg/Crystol Campbell BSHE 36
Tencleve/Bruce Charles AGBS BSA 89
Tennant/Forrest M. BSA 52
Tennant/Fred G. AGEC BSA 82
Tennant/Jeffrey Evan AGEC BSA 79
Tennant/Mike V. AGBS BSA 82
Tennison/Donna Lynn HEED BSHE 74
Tennison/Jimmie E. BSA 55
Tennison/Patsy Ann BSHE 52
Teo/Yan Hock AGRN PHD 94
Teo/Yan Hock AGRN MS 91
Teppen/Brian John AGRN PHD 94
Teppen/Brian John AGRN BSA 90
Teresa/Nelia Luers BSHE 54
Terhune/Eugene AGRN MSA 84
Terhune/Eugene AGRN BSA 79
Terminella/Karen Kimrey HORT BSA 89
Terminella/Monica Cox CLTX BSHE 91
Terrell/Angela Marnette CLTX BSHE 91
Terry/Andrew FDSC MS 81
Terry/Gwendolyn Spivey CLTX BSHE 82
Test/Peggy A. ANSC MS 75
Test/Peggy A. ANSC BSA 72
Teter/Bobby Gene BSA 56
Tevebaugh/Dwayne POSC BSA 77
Thacker/James Hosea BSA 48
Thamhain/Petra HOEC MS 86
Tharel/Janice Lynn FND BSHE 69
Tharel/Lance M. AGRN PHD 81
Tharp/Gary Alan ANSC BSA 89
Tharp/Jimmy D. ANSC BSA 77
Tharp/Melanie H. ANSC BSA 79
Thatcher/Jean D. BSA 48
Theodore/Melissa CLTX BSHE 84
Theodore/Melissa HOEC MS 86
Thetford/Alphonso BSA 9
Theus/Sue Ann ANSC BSA 85
Thibault/David BSA 40
Thieme/Helen Zumwalt BSHE 41
Thieme/Romona Lea Duncan CLTX BSHE 80
Thiesse/Branon R. AGRN MS 84
Thiesse/Branon R. SOSC BSA 80
Thigpen/Benny L. AGBS BSA 72
Thoma/David Patrick AGRN BSA 91
Thomas-Quarcoo/Olumbunmi GNAG MS 81
Thomas/Carol Lynn BSHE 59
Thomas/Caroline F. CDFR BSHE 77
Thomas/Cecil Amo BSA 35
Thomas/Christine J. HOFS BSHE 82
Thomas/Gloria J. FDNU BSHE 86
Thomas/James A. ENSC BSA 74
Thomas/James D. CRSC MS 74
Thomas/James D. AGRN BSA 72
Thomas/James J. BSA 47
Thomas/Janie Bartlett DANB BSA 78
Thomas/Jerry AGBS BSA 80
Thomas/Jimmy Edward POSC BSA 57
Thomas/Jimmy Edward PHUS MS 64
Thomas/John P. BSA 54
Thomas/Laura B. CLTX BSHE 83
Thomas/Lena M. HEED BSHE 70
Thomas/Lynn AGBS BSA 88
Thomas/Mara Douglas FDSC BSA 91
Thomas/Olen BSA 56
Thomas/Phillip Derell AGED BSA 92
Thomas/Rebecca Bevill HOEC MS 78
Thomas/Rebecca Bevill BSHE 77
Thomas/Russell Alan POSC MS 88
Thomas/Russell Alan ANSC BSA 87
Thomas/Shauna C. FHNH 99
Thomas/Susan Annette Young CLTX BSHE 84
Thomas/Travis R. BSA 25
Thomas/Willard M. BSA 50
Thomason/Dayton AGEC MS 64
Thomason/Dayton BSA 52
Thomason/Dewey S. BSA 22
Thomason/Joshua Carlas AGBS BSA 95
Thomasson/Marilyn S. HEED BSHE 70
Thompson/A. B. AGEC BSA 55
Thompson/Alan K. ANSC MS 83
Thompson/Alan K. ANSC BSA 79
Thompson/Amy M. HORT 98
Thompson/Annette CATX BSHES 98
Thompson/Brian T. LDUH BSA 83
Thompson/Burl BSA 29
Thompson/Cecelia HEED BSHE 71
Thompson/Cinde Bauer HORT MS 77
Thompson/Dale E. VOAG BSA 70
Thompson/David A. AGBS BSA 82
Thompson/Diane Grizzell HEED BSHE 77
Thompson/Elizabeth F. DIET BSHE 23
Thompson/Freddie W. BSHE 34
Thompson/Gary D. ENTO MS 77
Thompson/Glenn A. BSA 69
Thompson/Graham W. AGRN BSA 94
Thompson/Horace Eugene BSA 27
Thompson/Howell Tommy AGME BSA 71
Thompson/J. W. BSA 50
Thompson/Jarry L. BSA 76
Thompson/Jerry Jefferson PHYS MS 51
Thompson/John C. AGBS BSA 73
Thompson/Johnie Dee BSA 56
Thompson/Linda Richter HEED BSHE 67
Thompson/Lois BSA 52
Thompson/Marcella RSOC MS 66
Thompson/Marsha Shorts FND BSHE 71
Thompson/Marshall R. BSA 52
Thompson/Mary Munsch GHEC BSHE 86
Thompson/Melanie Anne CLTX BSHE 87
Thompson/Melanie Suzanne FND BSHE 73
Thompson/Mildred Weir BSHE 37
Thompson/Milton Lee ANSC BSA 83
Thompson/Oma P. BSHE 32
Thompson/Paul G. HORT MS 75
Thompson/Paul G. BSA 61
Thompson/Paul James POSC BSA 55
Thompson/Philip A. HORT MS 86
Thompson/Randall Layne AGBS BSA 92
Thompson/Randy Earl POSC BSA 92
Thompson/Raymond G. ENTO PHD 79
Thompson/Reba F. FDNU BSHE 94
Thompson/Ricky L. GNAG MS 86
Thompson/Robert Grant HORT BSA 86
Thompson/Robert P. LDUH BSA 82
Thompson/Samuel Edward AGED BSA 88
Thompson/Shirley Ann Brock BSHE 58
Thompson/Steve Lynn AGEC MS 87
Thompson/Steve Lynn AGBS BSA 85
Thompson/Tina Kathleen CLTX BSHE 93
Thompson/Van VOAG BSA 54
Thomson/Jared Scott POUL BSA 97
Thornley/Charles W. AGRI BSA 49
Thornton/Alyson Brown PLPA MS 94
Thornton/Alyson Brown LDUH BSA 91
Thornton/Celia Camille CLTX BSHE 92
Thornton/Donna CLTX BSHE 83
Thrash/Tonya HDFS BSHE 94
Threet/James D. BSA 56
Threlkeld/Douglas ANSC BSA 75
Threlkeld/Janna West ANSC BSA 86
Threlkeld/Kevin C. AGRN MS 86
Threlkeld/Kevin C. ANSC BSA 82

Throgmorton/Glenburn L. BSA 53
Thurber/Lisa Dawn AGED BSA 93
Thurlkill/Roy C. BSA 52
Thurman/Carl G. BSA 40
Thurman/Elva L. BSA 39
Thurman/Judy Hogue HEED BSHE 67
Thurman/Lori Hutton ANSC BSA 97
Thurman/Virginia Jo Hudlow BSHE 53
Thurston/Ronald J. ANSC MS 69
Tidd/Bettye Edwards BSHE 59
Tidwell/Neils March POSC MS 88
Tidwell/Tony FDSC BSA 94
Tierney/Michael John AGRN MS 96
Tilley/Donna Jennings HOEC MS 78
Tilley/Donna Jennings FND BSHE 70
Tilley/William C. ANSC MS 82
Tilley/William C. ANSC BSA 80
Tillman/Ann Davis BSHE 45
Tillman/Joan Noller CDFR BSHE 71
Tillman/Mary Penix BSHE 40
Tillman/Paul Benjamin POSC PHD 87
Tilmon/Charles E. BSA 29
Tilmon/Wayne BSA 36
Timberlake/David W. BSA 54
Timby/Gabrielle G. ESWS 99
Timme/Linda FDSC MS 84
Tinker/Allyson Andrews FDNU BSHE 83
Tisdale/Andrea D. HDFS 99
Titus/Joan Beasley BSHE 59
Todd/Anita Sloan BSHE 58
Todd/George R. DASC BSA 67
Todd/Marion Wade AGED BSA 88
Todd/Mary Beth HEED BSHE 66
Todd/Nell McKay BSHE 49
Toler/Juanita E. BSHE 43
Toler/Robert W. BSA 50
Tollett/David Q. BSA 63
Tollett/Eugene Elmore BSA 60
Tollett/F. Joann BSHE 58
Tollett/James T. MSA 58
Tollett/James T. BSA 57
Tollett/Leland E. POSC MS 60
Tollett/Leland E. POSC BSA 58
Tollett/Vergil E. POSC MS 71
Tollett/Vergil E. POSC BSA 67
Tomlinson/Hortense Theresa BSHE 28
Tomlinson/Mary O'Brien HOEC MS 66
Tomlinson/Mike A. FDSC BSA 86
Tompaw/Paul Gustavus AGRI BSA 61
Tompkins/James F. BSA 34
Toney/Robert L. BSA 56
Tontyaporn/Sanchai PLPA MS 70
Toon/Donald L. GNAG BSA 61
Torberson/Krista Maren FHNH BSHES 98
Torbett/Chester C. BSA 38
Torian/Emmet P. BSA 65
Tornek/Allen Vernon GNAG BSA 40
Tougaw/Thomas William BSA 54
Townsend/Allen Lee POSC BSA 89
Townsend/Bill BSA 65
Townsend/Jane Ivester BSHE 58
Townsend/Judy Brinkley HEED BSHE 80
Townsend/Monica Lynn ENTO MS 92
Townsend/Sharon Dianne HOEC MS 91
Trace/Timothy J. ANSC BSA 73
Tracer/Glenda G. CLTX BSHE 76
Trammell/Henry D. BSA 56
Trammell/Vicki HEED BSHE 73
Traylor/Kimberly AGBS BSA 98
Treadwell/Melissa Kaye IDHS BSHE 92
Treat/Barbara Lawrence BSHE 58
Treat/Bob POSC MS 70
Treat/Bob ANSC BSA 69
Treat/Clifford M. POSC BSA 55
Treat/Doyle W. FDSC MS 73
Treat/Doyle W. ANNU BSA 71
Treat/Jay L. AGRI BSA 49
Treat/John A. POSC BSA 50
Treat/Ralph A. GNAG BSA 52
Treat/Ralph A. POSC MS 75
Treat/Randy R. AGED BSA 92
Treece/Eugene B. BSA 32
Treece/Jeweldean BSHE 65
Treece/Neill Hamilton AGRN MS 62
Treece/Neill Hamilton BSA 60
Trenary/Lea Ann HOEC MS 95
Tribble/Brian E. AGTE BSA 97
Tribble/George S. BSA 40
Trice/Arlus W. VOED BSA 70
Trice/Kalven Lanell AGEC MS 86
Trice/Richard L. BSA 57
Trimble/Julie A. Griffin CDFR BSHE 82
Trimble/Sativa A. HDFS 98
Triplett/Cara L. CLTX BSHE 79
Trotter/Claudia Denise BSA 80
Trotter/Danny R. AGEC MS 83
Trotter/Danny R. AGBS BSA 80
Trout/Carol Lee PTSC PHD 95
Trout/Carol Lee PLPA MS 91
Troutman/Barry C. AGRN PHD 78
Trudeau/Ruth I. HOEC MS 68
Trudeau/Ruth I. BSHE 65
Truelove/Edward ANSC BSA 77
Truitt/Glen A. POSC BSA 67
Truitt/Joel E. POSC BSA 64
Truitt/Scott Alan AGBS BSA 94
Truitt/Tracy Aron AGBS BSA 95
Trujillo/Eduardo E. BSA 56
Trull/Ann P. FDNU BSHE 68
Truong/Anh HDFS BSHES 98
Tsai/Hsiu-Chuan PTSC PHD 0
Tschirhart/Janet E. CLTX BSHE 87
Tucker/Bobby J. BSA 58
Tucker/Christopher A. ENTO PHD 92
Tucker/Christopher A. ENTO MS 89
Tucker/Curtis R. AGBS BSA 92
Tucker/Dennis L. BSA 50
Tucker/Edwin Sallu AGRN MS 81
Tucker/Justin R. AGEC MS 35
Tucker/Justin R. BSA 9
Tucker/Kermit N. BSA 39
Tucker/Lillian BSHE 42
Tucker/Lou Ann BSHE 55
Tucker/Melvin C. GNAG BSA 46
Tucker/Milton R. SOSC MS 69
Tucker/Milton R. SOSC PHD 74
Tucker/Moseley C. BSA 14
Tucker/Roberta Cummings BSHE 40
Tucker/Willie HEED BSHE 72
Tuggle/Jon A. AGBS 98
Tugwell/Donnie R. ANSC MS 74
Tugwell/Donnie R. ANSC BSA 70
Tugwell/Philip BSA 58
Tullos/Melvin L. BSA 63
Tullos/Walter L. BSA 57
Turbeville/Teresa Dettenheim BSA 82
Turchi/Jim ANSC BSA 58
Turentine/Tommy POSC BSA 86
Turley/Toni Usher HDID BSHE 80
Turner-Finn/Tesa FDNU BSHE 84
Turner/Carolyn B. BSHE 65
Turner/Ewanta Taylor HOEC MS 85
Turner/George E. AGRN MS 82
Turner/Hazel Swindler HOEC BSHE 36
Turner/James B. ANSC 98
Turner/Jennifer AGED BSA 96
Turner/Jim GNAG BSA 54
Turner/Lance Regan AGBS BSA 85
Turner/Lehman R. VOAG BSA 64
Turner/Nancy Rebecca HOEC MS 67
Turner/Nancy Rebecca HEED BSHE 66
Turner/Patricia A. HORT BSA 79
Turner/Patricia S. FDNU MS 72
Turner/Patricia S. BSHE 70
Turner/Robert Wallace GNAG MS 86
Turner/Ruth C. BSHE 59
Turner/Ruth Silvey BSHE 42
Turner/Terry AGBS BSA 72
Turner/Tommy Lowell AGRN BSA 83
Turney/H. A. ENTO MS 56
Turney/H. A. AGRI BSA 52
Turney/Mary Elizabeth CLTX BSHE 83
Turney/Reedy Olen BSA 41
Turpin/Kellie McCaughey FDNU BSHE 86
Tyer/Jason Duane POSC BSA 95
Tyler/Joseph R. BSA 54
Tyler/Virginia Corley BSHE 40
Tyree/Orvis G. BSA 44
Tyree/Roger E. POSC BSA 73
Tyree/Windell Lee BSA 55
Tyrrell/Jennifer Lynn CLTX BSHE 92
Tyson/Harvey J. BSA 14
Tyson/Rose Maloch HEED BSHE 64
Tzul/Feliz POSC BSA 91

Udey/Edwin C. BSA 36
Uhl/Edith BSHE 24
Ulmer/George E. AGEG BSA 69
Umberson/Donald C. POSC BSA 66
Underwood/Allen K. ENTO MS 77
Underwood/Allen K. ENTO BSA 70
Underwood/Amos H. BSA 42
Underwood/Howard J. BSA 55
Underwood/M. Don AGBS BSA 82
Unger/Charles M. BSA 50
Updegraff/Erica Simmons CLTX BSHE 95
Urcuyo/Vicky FND BSHE 70
Urrey/Garland Eugene BSA 49
Urwiler/Michael J. AGRN PHD 85
Urwiler/Michael J. AGRN MS 82
Usrey/Byron Dell ANSC BSA 97
Usrey/Clara E. BSHE 41
Vafakos/William N. BSA 58
Vail/Gordon D. AGRN BSA 87
Vail/Harlan D. AGBS BSA 79
Valdez/J. Vincente SOSC MS 71
Valencia/Manuel Emilio ANSC MS 92
Valentine/Cindy BSHE 82
Valentine/John E. BSA 60
Valentine/John E. POSC MS 62
Valentine/Ruth HEED BSHE 61
Valverde/Rodrigo A. PTSC PHD 83
Van Asche/Martha FDNU BSHE 71
Van Brunt/Omer Gerald ANSC MS 74
Van Brunt/Omer Gerald ANSC BSA 71
Van Der Horn/Teresa R. FDNU BSHES 97
Van Epps/Mamie Hope BSHE 37
Van Es/Kelley Yvonne FDNU BSHES 97
Van Iersel/Marc Wilhelmus AGRN PHD 94
Van Marion/Catherine Thompson BSHE 44
Vanderboom/Joanie BSHE 70
Vanderford/Stephenie Sorrells IDHS BSHE 92
Vandergriff/Anne BSHE 43
Vandergrift/Libby Johnson HOEC MS 75
Vandervort/Kenneth D. AGRI BSA 54
Vandervort/Terry Joe AHUS BSA 62
Vandivere/William R. AGRN MS 69
Vaness/Kelli Ann CLTX BSHE 90
Vanmiddlesworth/James John ANSC PHD 80
Vanmiddlesworth/Sheryl Ann ANSC BSA 82
Vann/William Wade AGED BSA 91
Vanwinkle/Brandie Paige ENSC BSA 96
VanDalsem/William Paul BSA 36
VanHook/Wayne AGBS BSA 72
VanPatter/Anne L. ENSC BSA 97
VanVlack/Melva B. BSHE 32
VanVlack/Melva B. HOEC MS 66
Vargas/Carlos HORT BSA 87
Vargas/Lisa HOEC BSHE 85
Varner/Charleen L. M. GHEC MS 58
Varner/Joe C. BSA 58
Varner/Linda BSHE 59
Varner/W. L. ANSC BSA 56
Varvil/Stuart Wayne AGME BSA 74
Vathakos/Michael G. PLPA MS 78
Vaughan/A. Mark AGRN BSA 87
Vaughan/Jimmy Dean ANSC BSA 87
Vaughan/P. Audwin AGBS BSA 86
Vaughan/Peggy Lee HEED BSHE 71
Vaughan/Ronald Frank AGME BSA 80
Vaughan/Stephanie Wells CLTX BSHE 88
Vaughn/Martha Romayne Tate BSHE 40
Vaughn/Susan J. FDNU BSHE 79
Vaught/Carla Jean ANSC BSA 85
Vaught/Carla Jean GNAG MS 91
Vaught/Daniel A. AGEC MS 86
Vaught/Daniel A. AGEC BSA 85
Vaught/Don AGRN BSA 61
Vaught/Eric Keith ANSC MS 88
Vaught/Eric Keith DANB BSA 85
Vaught/William G. ANSC BSA 72
Veazey/Darrell POSC BSA 86
Vecchiarelli/Melissa R. CTM BSHE 85
Vega/Juan Carlos AGBS BSA 97
Velez/Jorge L. Rodriguez AGRN PHD 82
Venable/John L. BSA 55
Vending/Ann M. BSHE 65
Vendrell/Paul F. AGRN PHD 88
Vera-Gonzalez/Doralisa ANSC MS 83
Verkamp/John Paul AGBS BSA 88
Verkamp/Kevin M. AGBS BSA 83
Verkamp/Mark Joseph AGME BSA 85
Verkamp/Timothy W. ANSC BSA 81
Verrall/Thelma McCatherine BSHE 25
Verser/Michael W. AGRN BSA 75
Verser/Mollie Trimble BSHE 47
Vest/Alvin F. BSA 49
Vest/Terry S. AGBS BSA 72
Vestal/Cheryl Jeanne LDUH BSA 82
Vestal/Joe David ANSC MS 77
Vestal/Joe David ANSC BSA 75
Vester/Charles AGRN BSA 67
Via/Linda Irene DANB BSA 76
Vick/Christopher C. SOSC BSA 66
Vick/Christopher C. SOSC MS 71
Vick/Dorris C. HOEC MS 62
Vickers/Loy Dwain AGME BSA 90
Vieira/Jose Maria AGRN PHD 81
Viles/Molly Ann HOEC MS 75
Viles/Molly Ann HEED BSHE 69
Villines-Carr/Mary Ann AGRN BSA 81
Villines/Greg J. POSC BSA 82
Villines/Jacquelyn Dawn POSC BSA 93
Villines/Lewis Britt AGED BSA 93
Villines/Marion Beacher POSC BSA 82
Villines/Steven T. ENTO BSA 79
Vincent/Charles BSA 59
Vincent/James Welch ANSC BSA 81
Vincent/Thomas W. BSA 55
Viner/Lila BSHE 54
Vines/Austin BSA 33
Vines/Dwight Theodore ANSC MS 72
Vines/Dwight Theodore ANSC BSA 68
Vines/Iva Harness Margaret BSHE 36
Vines/Linda R. HORT BSA 85
Vining/Thelma Rogers BSHE 37
Vinson/Brad Allen AGBS BSA 95
Vinson/Donna Priebe HEED BSHE 66
Vinson/Sheryl CLTX BSHE 86
Virden/Mary E. HOFS BSHE 79
Vitelli/Veronica AGEC MS 68
Vitoonvitialak/Kitti AGRN PHD 86
Vodrazka/Adolph BSA 51
Vodrazka/Frank BSA 54
Voeller/Jack E. AGRN MS 77
Vogel/Jerome Henry BSA 58
Vogele/Terry HOFS BSHE 84
Vogelsang/Peggy Free BSHE 45
Voldeng/Katherine Tobin BSHE 85
Volentine/John C. BSA 38
Von Der Luft/Sue Wooddy BSHE 17
Vongsawatdi/Chamniern ENTO MS 71
Vorsanger/Daphne Papizan HOFS BSHE 78
Vose/Henry Hinkle BSA 49
Votaw/Mildred C. BSHE 27
Vowell/Amy Elizabeth LDUH BSA 96
Vrionaguena/Catherine Artz ANSC MS 81
Waddell/Douglas M. BSA 51
Waddle/Wanda L. FDNU BSHE 68
Wade/Keith M. ANSC BSA 74
Wade/Vera Drake BSHE 30
Wadkins/Jesse AGBS BSA 83
Wages/Patricia Lea HOEC MS 76
Wages/Patricia Lea BSHE 61
Waggoner/Jane Gattis HOEC BSHE 65
Waggoner/Jane Gattis HOEC MS 78
Waggoner/Robert E. BSA 58
Wagner/Catherine D. IDHS 99
Wagner/Grant W. PLPA MS 66
Wagoner/Chris Randall ANSC BSA 93
Wagoner/Eddie Joe ANSC MS 93
Wagoner/Eddie Joe ANSC BSA 91
Wagoner/Jennifer Ellen IDHS BSHE 90
Wagoner/Stacy Klock ANSC BSA 93
Wait/Heather Walker IDHS BSHES 95
Waits/Madden Fain LDUH BSA 95
Waits/Wallace F. BSA 35
Walden/Betty G. MS 71
Walden/Michael J. FDSC 99
Waldrip/Angela Eason CLTX BSHE 79
Waldrip/Byron T. AGRN BSA 39
Waldrip/Joseph H. ANSC BSA 72
Waldrip/Marcus M. AGBS BA 77
Waldron/Jeanette Marie HOEC MS 83
Walker/Brian DIAM BSA 89
Walker/Carolyn Files FDNU BSHE 59
Walker/Danny O. AGRN BSA 78
Walker/Danny O. AGRN MS 80
Walker/Darrell W. POSC BSA 66

Walker/David R. AGEC BSA 71
Walker/Dennis Robin AGME BSA 84
Walker/Dennis W. AGRN MS 87
Walker/Donald POSC BSA 74
Walker/Doug DIAM BSA 89
Walker/Elizabeth Veola BSHE 40
Walker/Ella Mae HOEC MS 69
Walker/Frank L. AGRN BSA 79
Walker/Gary D. AGEC MS 75
Walker/Gary D. AGEC MS 75
Walker/George Rea AGRN BSA 65
Walker/Jo Ann BSHE 65
Walker/Joe E. SOSC MS 33
Walker/Joe E. AGRN BSA 31
Walker/Kevin K. ANSC BSA 74
Walker/Louise S. HOEC MS 84
Walker/Marc FDNU BSHES 96
Walker/Mary G. HEED BSHE 72
Walker/Maxi Ann CLTX BSHE 83
Walker/Robert Dean HLSC MS 98
Walker/Robert Dean ANSC BSA 95
Walker/Robyn E. HDFS 98
Walker/Russell E. ANSC BSA 74
Walker/Susan Denise HOEC MS 93
Walker/Teresa Lynn FDSC BSA 94
Walker/Terry K. CRSC MS 71
Walker/Terry K. CRSC BSA 70
Walker/Todd W. ENTO MS 80
Walker/Todd W. ENTO BSA 77
Walker/William C. AGBS 99
Walkingstick/Tamera Lynne ENSC BSA 84
Wall/Carey Wagoner POSC MS 97
Wall/Carey Wagoner ANSC BSA 89
Wall/David M. ANSC BSA 76
Wall/Gregory John POSC BSA 89
Wall/Howard C. ENTO MS 61
Wall/Howard C. BSA 56
Wall/Marvin L. ENTO MS 64
Wall/Marvin L. ENTO BSA 62
Wallace/Amanda Lee DANB BSA 94
Wallace/Darrell Leon DHUS BSA 62
Wallace/Floyd W. VOAG MS 49
Wallace/Floyd W. GNAG BSA 46
Wallace/Jack D. GNAG MS 83
Wallace/James W. AGEC BSA 68
Wallace/Jean E. HORT BSA 79
Wallace/Jim Lee BSA 63
Wallace/Joe Terrell AGRN BSA 80
Wallace/John A. BSA 51
Wallace/John Rieves AGBS BSA 91
Wallace/Margaret Lorena BSHE 41
Wallace/Mark Edward POSC BSA 93
Wallace/Mason D. PPPM BSA 93
Wallace/Patricia M. BSHE 61
Wallace/Ronald C. AGRN BSA 76
Wallace/William T. AGRN MS 61
Wallace/William T. GNAG BSA 58
Waller/Cheryl Ann HEED BSHE 71
Waller/Tom ANNU BSA 81
Waller/Trudy Ozment CLTX BSHE 89
Wallgren/Gregory Paul ANSC BSA 84
Wallin/Shanda C. AGRN BSA 79
Wallinder/Cheryl J. AGRN MS 84
Wallis/Betty FDNU BSHE 66
Wallis/Elizabeth HEED BSHE 76
Wallis/Gerald W. ENTO MS 64
Wallis/Gerald W. ENTO BSA 62
Wallis/Jimmy D. BSA 70
Wallis/Rod AGBS BSA 91
Walls/Carroll E. SOSC MS 56
Walls/Carroll E. AGRN BSA 55
Walrod/Albert L. ENTO MS 70
Walsh/Bill POSC MS 63
Walsh/Bill ANNU BSA 61
Walsh/David Anthony POSC BSA 95
Walsh/Kenneth Douglas ENSC BSA 97
Walsh/Marvin BSA 49
Walsh/Pamela Bequette CLTX BSHE 88
Walt/Lee AGRN BSA 88
Walt/Martin L. BSA 60
Walter/Doris M. BSHE 34
Walter/Mary E. FDNU BSHE 44
Walters/Hazel Craig HOEC MS 63
Walters/Robert Eugene BSA 52
Walters/Thurmond L. ECON BSA 37
Walters/Trudy FND BSHE 71
Walters/Warren Earl BSA 40
Walton/Cleo A. BSA 32
Walton/Earnestine B. HOEC MS 70
Walton/Kimberly Robin Morton IDHS BSHE 87
Walton/Ola Mae BSHE 31
Walts/Irene Mae BSHE 31
Wang/Bing ANSC MS 95
Wang/Hong FDSC MS 93
Wang/Hwa-Ying Diana AGEC MS 90
Wang/Rueen-Chiou FDNU MS 83
Wang/Shiying POSC MS 88
Wang/Sui-Ying POSC PHD 97
Wang/Sui-Ying ANSC MS 93
Wang/Tong FDSC MS 92
Wang/Yang BAEG PHD 95
Wann/Carlton Victor GNAG BSA 62
Wann/Elbert V. BSA 59
Ward/Bruce F. POSC BSA 71
Ward/Carolyn B. BSHE 58
Ward/David A. AGEC MS 91
Ward/Donald Arthur DHUS BSA 62
Ward/Georgette L. LDUH BSA 91
Ward/Grover BSA 50
Ward/James Everett BSA 80
Ward/Jamie Linn AGED BSA 96
Ward/Jim ENSC BSA 82
Ward/Jimmy POSC BSA 75
Ward/John BSA 24
Ward/Larry B. BSA 68
Ward/Larry E. AGEC MS 73
Ward/Mary Evelyn Cate BSHE 38
Ward/Regina Newman BSHE 85
Ward/Ronald Fred BSA 59
Ward/Sherry FDNU BSHE 72
Ward/Shirl W. BSA 53
Ward/Sue Carol HEED BSHE 73
Wardlaw/Sallye Lowrey DIET BSHE 65
Wardlow/Renette Elizabeth HOEC MS 86
Wardlow/Renette Elizabeth HDFS BSHE 85
Ware/George W. BSA 24
Ware/George W. ENTO MS 52
Ware/George W. BSA 51
Ware/Mason T. BSA 50
Warford/Ruth Ellen Guynn FND BSHE 70
Warmack/Lori CLTX BSHE 95
Warner/Sidney Catherine FDNU BSHE 77
Warnock/Cecil L. BSA 48
Warnock/Ward L. BSA 49
Warren/Ann G. HEED BSHE 76
Warren/Catherine G. Dodd FND BSHE 68
Warren/Donald E. AGRN MS 53
Warren/Donald E. AGRI BSA 51
Warren/Eugene L. BSA 50
Warren/Herbert Thomas AGED BSA 95
Warren/Leon B. AGRI MS 63
Warren/Lisa Engelmann BSHE 85
Warren/Melissa Rose ANSC BSA 88
Warren/Molly Williams FDSC MS 87
Warren/Sharon Ashley HEED BSHE 78
Warren/Stacy L. CLTX 98
Warriner/Richard W. III AGBS 99
Washburn/Dana BSHE 59
Washburn/Jack B. BSA 56
Washington/Doris Eskridge AGRN BSA 89
Wasilwa/Lusike A. HORT MS 92
Wasson/Carolyn FDSC MS 80
Wasson/Carolyn FDNU BSA 77
Waters/Gaylon BSA 55
Waters/Karen Young CLTX BSHE 94
Waters/Lillar Mae BSHE 39
Waters/Loyd E. BSA 34
Waters/Mary Iman AGRN BSA 82
Waters/Thomas L. POSC BSA 82
Waters/Wilbur L. BSA 56
Watkins/Anna L. HOEC MS 79
Watkins/Gregory B. AGRN BSA 88
Watkins/James H. AGRN BSA 85
Watkins/John H. BSA 57
Watkins/John W. BSA 56
Watkins/Julian BSA 56
Watkins/Kenton Bradley AGEC MS 90
Watkins/Kenton Bradley AGBS BSA 88
Watkins/Kitty Wills BSHE 49
Watkins/Rebecca Dawn ENSC BSA 97
Watkins/Sarah Barrett HOEC MS 60
Watkins/Sarah Barrett HEED BSHE 43
Watkins/Susan E. POSC PHD 97
Watkins/Susan E. ANNU BSA 84

Watkins/Susan E. POSC MS 83
Watkins/William L. AGJN BSA 82
Watson/Belinda Yen ANSC PHD 71
Watson/Bobby D. BSA 61
Watson/C. Carolyn McCullough GHEC BSHE 40
Watson/Harold Berry POSC BSA 97
Watson/Jeffrey E. AGBS BSA 81
Watson/Jennifer Dawn AGBS BSA 97
Watson/Jimmy E. CRSC BSA 72
Watson/L. Z. BSA 49
Watson/Laura L. CLTX BSHE 91
Watson/Lavon V. BSA 43
Watson/Linda Carol HEED BSHE 69
Watson/Mary Bandeen HEED BSHE 73
Watson/Melissa Sink FDNU BSHE 76
Watson/Melvern BSA 47
Watson/Rebecca Bock GNAG MS 92
Watson/Robert H. BSA 58
Watson/Robert Lee ENTO MS 63
Watson/Robert Lee BSA 56
Watson/Terri Renee ANSC BSA 82
Watson/Theo F. ENTO BSA 53
Watson/Theo F. ENTO MS 58
Watson/Thomas Eugene AGBS BSA 85
Watson/Trudy R. HEED BSHE 72
Watson/W. S. SOSC MS 64
Watson/W. S. BSA 49
Watts/Gwendolyn Kay AGRN BSA 81
Watts/Harley A. BSA 51
Watts/John R. VOAG BSA 79
Watts/Leslie Alice CLTX BSHE 84
Watts/Tilford M. BSA 70
Watts/William Michael AGEC MS 88
Watts/William Michael AGEC BSA 85
Waymack/Lester B. BSA 60
Wayman/Agnes Anne HEED BSHE 68
Weare/Robert H. BSA 51
Weatherford/Paul Douglas AGBS BSA 82
Weatherford/Terry K. AGBS BSA 80
Weathers/Barbara J. HOEC MS 68
Weathers/Fred A. BSA 57
Weathers/James T. BSA 53
Weathers/Jerry William POSC BSA 86
Weathers/Linda C. HEED BSHE 72
Weathersbee/Albert A. ENTO PHD 91
Weaver/Payton POSC BSA 96
Weaver/Rhona Joy GNAG MS 84
Weaver/Rochelle E. IFDS BSHES 98
Weaver/Ronnie Dale ANSC BSA 90
Weaver/Steven Ray AGED BSA 86
Webb/Billy Bob ANSC BSA 93
Webb/Billy R. BSA 57
Webb/Billy R. AGEC MS 60
Webb/Christopher Paul DIAM BSA 90
Webb/Clarence W. BSA 50
Webb/Harley D. BSA 65
Webb/Harley D. DASC MS 67
Webb/Jenifer Dawn AGBS BSA 94
Webb/Jerry E. POSC BSA 67
Webb/Jerry E. POSC MS 69
Webb/John D. AGME BSA 79
Webb/Leslie Lee AGBS BSA 89
Webb/Linda Rozan BSHE 61
Webb/Pamela Kaye AGBS BSA 88
Webb/Phillip Ross ANSC MS 92
Webb/Ralph BSA 21
Webb/Randall W. FND BSA 70
Webb/Richard Lee HLSC BHE 97
Webb/Stanley R. ENTO MS 71
Webb/Tarvin F. BSA 49
Webb/William Stanley ANSC BSA 85
Webster/Billie Janice BSHE 51
Webster/Lucy C. BSHE 34
Webster/Susan Tiemann FDNU BSHE 77
Weeks/Ronald AGED BSA 86
Weems/Charles L. POSC BSA 51
Weems/Chris HOEC MS 62
Wegert/Sidney J. BSA 57
Wehunt/E. J. BSA 54
Weiker/Steven C. ANSC MS 82
Weiker/Viola Pflantz ANSC PHD 82
Weiland/Robert Timothy AGRN PHD 80
Weir/Coranne BSHE 59
Weir/Leslie B. ANSC BSA 52
Weir/Marjorie L. HEED BSHE 37
Weir/Max G. BSA 36
Weis/Cynthia Lee DANB BSA 89
Weise/Oswald BSA 52
Weiss/Roberta Lynn HOEC MS 93
Welch/Eva Morton BSHE 41
Welch/Ewell F. PLPA BSA 47
Welch/Ewell Ray BSA 76
Welch/Ewell Ray AGEC MS 77
Welch/Frances Campbell BSHE 68
Welch/Jeff B. AGEC MS 77
Welch/Jeff B. ENTO MS 77
Welch/Jeff D. BSA 40
Welch/Mae Butler BSHE 41
Welch/Quintin B. ANNU BSA 57
Welch/Robert Dale DANB BSA 82
Welch/William Paul AGEC MS 84
Welch/William Paul GBUS BSA 77
Weldon/Matthew Wayne HORT BSA 98
Wells/Earl Wayne FDSC MS 85
Wells/Gary Lee ANSC BSA 71
Wells/Helen L. BSHE 43
Wells/Joe P. BSA 53
Wells/Robert Gene AGRN BSA 80
Wenzelburger/Mark Edward AGRN MS 83
Were/Lilian Mirembe FDSC MS 98
Were/Lilian Mirembe FDSC BSA 95
Wesson/Jennifer Courtney CLTX BSHE 91
Wesson/Mary Lewis HOEC BSHE 54
West/Barth Allen PPPM BSA 93
West/Dean V. BSA 52
West/Donald M. ESWS 99
West/Hazel K. FND BSHE 63
West/Jerry Wesley POSC BSA 94
West/Larry Thomas AGRN MS 78
West/Larry Thomas AGRN BSA 73
West/Mary L. FDNU BSHE 76
West/Matthew Clay ENSC BSA 98
West/Melvis C. ANSC BSA 70
West/Wiley S. SOSC MS 76
Westberg/Dan Eric AGRN MS 87
Westberg/Dan Eric AGRN BSA 86
Westbrook/Carol Wallace CDFR BSHE 80
Westbrook/Cole J. BSA 37
Westbrook/Jim DANB BSA 73
Westbrook/Lloyd T. GNAG MS 64
Westbrook/Lloyd T. GNAG BSA 55
Westbrook/Stephen Lynn AGRN BSA 80
Wester/Frances H. BSHE 45
Westerfield/Mary Cotton BSHE 58
Westfall/Eric Anthony AGBS BSA 95
Westfall/Ethan A. ANSC BSA 77
Westfall/Fred ANSC MS 74
Westfall/Fred POSC BSA 72
Westfall/Paul W. ANSC MS 84
Westfall/Paul W. ANSC BSA 78
Westfall/Sharon Flynn HOFS BSHE 84
Westlake/David L. AGBS BSA 76
Weston/Andrew M. ANSC MS 80
Westridge/Wanda Smith DIET BSHE 43
Weyl/Aaron Scott ANSC BSA 95
Wheeler/Carol Beth CLTX BSHE 90
Wheeler/Charles E. GNAG MS 65
Wheeler/Karen Lee FDNU BSHES 98
Wheeler/Patricia Ann HDFS BSHES 96
Wheeler/Pauline Friddle BSHE 36
Wheetley/Johnny Wayne AGRN BSA 83
Whelan/Vera BSHE 31
Whelchel/David Lee BSA 74
Whiley/Jimmy Charles AGEC BSA 71
Whillock/Anita BSHE 43
Whisker/Sue Ann ANSC BSA 94
Whistle/Fred S. AGEG BSA 81
Whitaker/C. Layton BSA 47
Whitaker/Mary Lou DIET BSHE 61
Whitaker/Patricia Parish BSHE 55
Whitaker/Sam B. AGRN BSA 87
Whitbeck/Bonnie Sue HEED BSHE 73
White/Beulah Rogers BSHE 32
White/Billie Ollar BSHE 55
White/Billy R. BSA 54
White/Bradley W. ANSC BSA 85
White/Calvin Eddy ANSC PHD 79
White/Charles L. ANNU MS 66
White/Charles L. POSC BSA 65
White/Corbit ANSC MS 53
White/Corbit ANSC BSA 48
White/Cyndi A. ANSC BSA 83
White/David L. DASC BSA 81
White/David W. BSA 66

White/Dean BSA 55
White/Elmer B. BSA 26
White/Gladys Tullis BSHE 30
White/Helen BSHE 36
White/Henry L. GNAG BSA 67
White/Jacob Nelson AGBU BSA 98
White/James Harold BSA 52
White/James Harold AGEC MS 53
White/Jamie L. CLTX BSHE 82
White/Jerry W. GNAG BSA 71
White/John William BSA 35
White/Julie Ann GHEC BSHE 83
White/Lloyd J. BSA 32
White/Loretta McClennahan BSHE 53
White/Lyle V. BSA 54
White/Margaret Baxter BSHE 27
White/Mary Jane Blythe BSHE 76
White/Michelle Renee ANSC BSA 96
White/Ollie Eugenia RSOC MS 84
White/Paula E. ESWS 99
White/Peggy Leal BSHE 48
White/Rebecca L. FDNU BSHE 79
White/Robert L. BSA 76
White/Robert M. BSA 50
White/Ruth Nadine FDNU BSHE 74
White/Shari M. CLTX BSHE 82
White/Thomas F. AGEC BSA 84
White/Thomas F. AGEC MS 86
White/Thomas H. BSA 49
White/Thomas Newton LDUH BSA 94
White/Tuell A. BSA 26
White/Victor Lance POSC BSA 85
White/William A. AGBS BSA 82
Whited/Danny L. GBUS BSA 77
Whited/Debbie HEED BSHE 75
Whitehead/Dale L. Huskins CDFR BSHE 76
Whitehead/James David ENTO MS 63
Whitehead/James David DHUS BSA 61
Whitehurst/Harry B. AGRN MS 82
Whitelaw/Alma Lois AGBS BSA 89
Whiteley/Teresa L. HEED BSHE 71
Whiteley/Terry A. BSHE 65
Whitfield/W. H. Bill AGEC MS 79
Whiting/Earl P. BSA 30
Whiting/Thomas Scott POSC PHD 88
Whitley/Betty L. BSHE 71
Whitlow/James M. POSC BSA 77
Whitlow/John R. POSC BSA 66
Whitmore/James H. POSC BSA 56
Whitmore/Ralph G. BSA 36
Whitmore/Ray B. BSA 34
Whitmore/Sherry Ann CLTX BSHE 88
Whittenton/Eloise Irving BSHE 40
Whitting/Cheryl HLED BSHE 82
Whittington/Paul F. BSA 56
Whittington/Sandra E. HEED BSHE 73
Whittle/Shirley J. BSHE 81
Whitwam/Helen Kleine BSHE 43
Whitworth/Robin Harvey BSHE 39
Wichert/Rex Alan AGRN PHD 93
Wichert/Rex Alan AGRN MS 88
Wickizer/James R. AGME BSA 79
Wickizer/Sandra L. HORT MS 78
Wicklund/F. E. BSA 50
Wicklund/Mary Hudson BSHE 48
Widick/Joe Darell CRSC MS 69
Widner/James C. ANSC BSA 74
Wiebe/Bonnie Dew HOEC MS 69
Wiederkehr/Brenda L. LDUH BSA 78
Wiederkehr/Gail M. CDFR BSHE 74
Wiederkehr/Garrett H. HORT BSA 72
Wiese/Kurt L. FDSC PHD 86
Wiese/Kurt L. FDSC BSA 80
Wiggins/Patricia Ann HEED BSHE 62
Wikman/Michele Paulin HDFS BSHES 95
Wilber/Mickey B. AGEC BSA 67
Wilbers/Russell A. ENTO BSA 90
Wilcox/Stephanie CLTX BSHE 89
Wildy/David ANSC BSA 75
Wildy/Earl BSA 38
Wiley/Don N. GNAG MS 88
Wilkerson/Don C. HORT MS 79
Wilkerson/James E. AGBS BSA 96
Wilkerson/Sunny L. AGRN BSA 97
Wilkinson/Alise BSHE 52
Wilkinson/Cordia FDNU BSHE 76
Wilkinson/Liani Susan FDSC BSA 96
Wilkinson/Thomas M. AGRI BSA 52
Wilkison/John Kevin DIAM BSA 94
Wilks/Cyndy HOFS MS 72
Wilks/Cyndy FDNU BSA 70
Willborg/Marcus O. ANSC BSA 72
Willcox/Jesse J. GNAG MS 59
Willcox/Jesse J. GNAG BSA 57
Willems/Harry Justin AGBS BSA 85
Willett-Rice/Guyla S. BSHE 75
Willett/Edward E. BSA 49
Willey/Mary Ann Pich BSHE 54
Williams-Graham/Dolores BSHE 48
Williams/A. J. ENTO MS 51
Williams/A. J. BSA 49
Williams/Anne Wallace BSHE 26
Williams/Becky Lynn POSC BSA 85
Williams/Bess G. BSHE 35
Williams/Betty Jean Hill HOEC MS 69
Williams/Bobby O. AGEC MS 80
Williams/Bobby O. AGEC BSA 78
Williams/Brad AGED BSA 93
Williams/Catherine Irene ANSC MS 93
Williams/Catherine Irene ANSC BSA 91
Williams/Chester F. BSA 37
Williams/Christine HEED BSHE 71
Williams/Clyde Kendrick BSHE 31
Williams/Curtis CRSC PHD 70
Williams/Donald Leon VOAG BSA 62
Williams/Douglas Leon AGBS BSA 86
Williams/Dwight ANSC BSA 77
Williams/Dwight ENTO MS 80
Williams/E. Louise BSHE 51
Williams/Erma Jean HOEC MS 74
Williams/Erma Jean HEED BSHE 73
Williams/Francis J. BSA 49
Williams/Francis J. ENTO MS 51
Williams/Gerald W. AGBS BSA 79
Williams/Harold B. AGEC MS 77
Williams/Harrod J. AHUS BSA 62
Williams/Janis HOEC MS 82
Williams/Janis CTM BSHE 80
Williams/Jeff Ray ANSC BSA 85
Williams/Jerry W. ENTO MS 69
Williams/Jerry W. GNAG BSA 67
Williams/Jesse E. BSA 42
Williams/Jimmy L. DASC BSA 66
Williams/Jo Ann Price BSHE 53
Williams/Joe M. BSA 56
Williams/John Henry BSA 34
Williams/John Kenneth BSA 76
Williams/John W. FDSC MS 78
Williams/Kate Hardister BSHE 32
Williams/Kay CLTX BSHE 83
Williams/Kenneth R. GNAG BSA 71
Williams/Kenneth R. ENTO MS 73
Williams/Kim Anderson BSA 81
Williams/L. D. BSA 58
Williams/Leland H. BSA 65
Williams/Livy Holman ENTO MS 85
Williams/Lola BSHE 27
Williams/Loraine A. BSHE 31
Williams/Marcus BSA 39
Williams/Margo Jane FDNV BSHE 62
Williams/Marietta Stanford BSHE 33
Williams/Martha Patton BSHE 39
Williams/Neoma L. BSHE 46
Williams/Paul ANSC BSA 79
Williams/Paul Waner BSA 48
Williams/R. W. VOAG BSA 42
Williams/Ray AGEC BSA 51
Williams/Robert Charles ANSC MS 81
Williams/Roger Keith DIAM BSA 90
Williams/S. Janice Whiley HEED BSHE 72
Williams/Shundra Latricia AGEC MS 93
Williams/Terry Lee VOAG BSA 71
Williams/Thelma Jones BSHE 48
Williams/Theresa K. FDNU BSHE 95
Williams/Thomas L. ANSC BSA 72
Williams/Timothy J. IDHS BSHE 91
Williams/Wade Allan AGBS BSA 87
Williams/William H. BSA 56
Williams/Willie Byers BSHE 41
Williams/Winifred S. MSHE 56
Williams/Wyona S. BSHE 56
Williamson/Gary L. ANSC BSA 68
Williamson/Holly HORT BSA 82
Williamson/Jennifer Kay FNBK BSHE 89

Williamson/Mary D. HOFS BSHE 79
Willis/Bradley Harold DANB BSA 87
Willis/Calvin ANSC MS 78
Willis/Calvin ANSC BSA 76
Willis/Dick E. ANSC BSA 69
Willis/Garlen T. BSA 58
Willis/Linda E. BSHE 69
Willis/Robert B. BSA 17
Willis/Robert Hamilton BSA 42
Willis/Rose Marie HOEC BSHE 62
Willis/Shelia G. Rhodes HOEC MS 78
Willis/Shelia G. Rhodes GHEC BSHE 77
Willman/Lloyd C. GNAG BSA 41
Wills/Vernon C. BSA 40
Willshire/Vera Fay Morrow BSHE 35
Wilmoth/Helen B. HOEC MS 66
Wilmoth/Jody L. HDID BSHE 80
Wilmoth/Mack R. BSA 58
Wilmoth/Shirley Ward BSHE 58
Wilmuth/Virginia BSHE 40
Wilson/Billy Lynn AHUS BSA 62
Wilson/Candes A. HEED BSHE 75
Wilson/Charles E. AGRN PHD 92
Wilson/Charles E. AGRN MS 88
Wilson/Charles Kent LDUH BSA 79
Wilson/Constant Perkins BSA 6
Wilson/Cynthia Paige CLTX BSHE 92
Wilson/Daniel Earl ANSC BSA 97
Wilson/Deborah Dianne BSHE 72
Wilson/Deborah Dianne HOEC MS 85
Wilson/Denise Deaton CLTX BSHE 92
Wilson/Dick BSA 59
Wilson/Doris Ann HOEC MS 68
Wilson/Doris J. FDNU BSHE 55
Wilson/Earl W. ANSC MS 76
Wilson/Earl W. DHUS BSA 64
Wilson/Eric Gregory POSC BSA 94
Wilson/Erma Lucille BSHE 45
Wilson/Flossie E. BSHE 31
Wilson/Floy BSHE 29
Wilson/Gerald R. ENTO MS 68
Wilson/Gerald R. AGRN BSA 65
Wilson/Glen D. BSA 65
Wilson/Glenn Venson AGRN PHD 86
Wilson/Gretchen R. HDFS BSHE 89
Wilson/Herbert R. BSA 52
Wilson/Herman H. BSA 49
Wilson/James R. BSA 55
Wilson/James Royce ANSC BSA 71
Wilson/Janifer Denet BSHE 88
Wilson/Jesse P. BSA 47
Wilson/John W. LDUH BSA 81
Wilson/Julie Ann HEED BSHE 74
Wilson/Julie Masterson HOEC MS 80
Wilson/Jwan R. HEED BSHE 75
Wilson/Kevin H. AGBS BSA 89
Wilson/Kristi L. AGBS 98
Wilson/Letha Bridges HOEC MS 68
Wilson/Lynn FDNU BSHE 76
Wilson/Manie A. Palmer BSHE 27
Wilson/Mary Jane HOEC MS 66
Wilson/Mildred Lucille BSHE 27
Wilson/Nina BSHE 43
Wilson/Olen Clarence BSA 51
Wilson/Patsy Poteete BSHE 58
Wilson/Ronnie F. ENTO MS 72
Wilson/Ruth A. Dodson FDNU BSHE 68
Wilson/Ruth Manon BSHE 86
Wilson/Sally BSHE 60
Wilson/Stephen A. CRSC BSA 70
Wilson/Steven Curtis HORT BSA 82
Wilson/Thomas H. ENTO MS 68
Wilson/Thomas W. AGBS BSA 72
Wilson/Valerie CLTX BSHE 84
Wilson/Vicki Anne CLTX BSHE 81
Wilson/Virginia Lee HDID BSHE 85
Wilson/William BSA 18
Wilson/William T. AGEC MS 31
Wilson/William T. BSA 25
Wilson/William W. BSA 42
Wilson/Woodrow M. BSA 42
Wiltse/Kathy Micken GHEC BSHE 81
Wimpy/Tommy S. CRSC MS 68
Wimpy/Tommy S. CRSC BSA 66
Winchester/Roberta E. BSHE 31
Windborne/Janice Kelly LDUH BSA 79
Windham/Lesley Elise IDHS BSHE 91
Windham/William James BSA 50
Windland/Robert E. BSA 57
Wineland/Don A. AGEC BSA 68
Winemiller/Donna Lederman BSHE 83
Winfrey/George W. BSA 18
Winfrey/Hugh BSA 41
Winfrey/Ralph E. BSA 30
Wingert/Stanley J. ANSC MS 70
Wingert/Stanley J. ANSC BSA 69
Wingfield/Melvin T. ENTO MS 68
Wingfield/Melvin T. GNAG BSA 64
Wingfield/Patty FDNU BSHE 77
Winn/Kevin Scott AGBS BSA 89
Winn/William Clayton AGRN BSA 91
Winningham/Kendra Dawn ANSC BSA 95
Winningham/Ray M. AGRN MS 76
Winningham/Ray M. DASC BSA 71
Winstead/Christina S. POSC MS 83
Winstead/Christina S. DANB BSA 82
Winters/Myles M. AGEC BSA 69
Winters/Scott A. HORT MS 87
Winton/Kimberly Teresa AGRN PHD 86
Wisdom/Forrest B. GNAG BSA 49
Wisdom/Milinda Mae GHEC BSHE 91
Wise/David H. VOAG BSA 81
Wise/James S. BSA 10
Wise/Max R. BSA 61
Wise/Willa McCuistion BSHE 62
Wiseman/Donna Jeanneien ANSC BSA 86
Wiser/Franklin D. BSA 60
Wiser/William J. BSA 41
Wishon/Carie L. ANSC 99
Wisley/Judith Ann HEED BSHE 62
Wisley/Paul Gene ANNU BSA 62
Withers/Beatrice Davis BSHE 39
Witherspoon/Michael Shawn ANSC BSA 88
Withrow/Michael Wayne AGRN BSA 94
Witt/David Allen AGBS BSA 94
Witt/Elizabeth Ann CLTX BSHE 87
Witt/Elizabeth Greenway BSHE 87
Witt/Jacqueline Rosewell HOEC MS 62
Witt/Jacqueline Rosewell BSHE 57
Witt/Jan E. CLTX BSHE 80
Witt/Vanessa Nadine LDUH BSA 96
Wittorff/Edward Karl DASC BSA 85
Wittorff/Edward Karl ANSC MS 87
Wittorff/Walter Edwin DASC BSA 81
Woelk/Megan Louise HORT BSA 95
Wofford/Boyce HORT BSA 85
Wofford/Daniel James BSA 42
Wofford/Kelly Horton ANSC BSA 85
Wohlford/Victor BSA 31
Wojciechowski/Teresa Marie DANB BSA 91
Woldert/Jerra Odum CLTX BSHE 81
Wolf/Daniel B. ENSC BSA 81
Wolf/Dianne Carolyn AGRN MS 90
Wolf/Donna Lynne FND BSHE 70
Wolf/John Clarence AGBU BSA 97
Wolf/Nancy A. AGRN MS 85
Wolf/Nancy A. SOSC BSA 82
Wolf/Shirley Jo LDUH BSA 81
Wolfe/Barbara L. FND BSHE 72
Wolfe/Jim P. ANSC BSA 71
Wolfe/Joseph L. ANSC BSA 89
Womack/David E. AGRN MS 62
Womack/David E. BSA 59
Womack/Nelda Taylor BSHE 47
Womack/Wanda L. BSHE 56
Wong/May Ling Kelland FDNU BSHE 87
Wong/Sybil BSA 54
Wood Blake/Sue BSHE 59
Wood-Huels/Nanci ANSC MS 88
Wood-Huels/Nanci DANB BSA 83
Wood/Bobby D. AGRI BSA 64
Wood/Charles E. BSA 52
Wood/David K. DASC BSA 85
Wood/Dennis Lynn AGME BSA 85
Wood/Edgar C. BSA 40
Wood/Forrest M. AGME BSA 72
Wood/Heather Hartrick BSA 70
Wood/James A. SOSC BSA 74
Wood/James R. BSA 49
Wood/Lisa S. AGRN MS 88
Wood/Lisa S. AGRN BSA 84
Wood/Martha Lou BSHE 45
Wood/Monica Maria HEED BSHE 84
Wood/Nancy Edmondson HOEC BSHE 66

Wood/Roy G. BSA 15
Wood/Scott M. ANSC BSA 75
Wood/Thurman BSA 42
Wood/Timothy John AGBS BSA 86
Wood/Todd AGBS BSA 88
Woodall/William E. CRSC MS 66
Woodard/Larry J. BSA 59
Woodard/Patricia Ann FDNU BSHE 76
Woodbridge/Jay E. AHUS BSA 53
Woodbridge/Jay E. POSC MS 54
Woodbridge/Susan Marie DASC BSA 85
Woodell/Lloyd Joseph BSA 39
Woodhead/Suzan Hayes PLPA MS 73
Woodley/Marion Allen POSC BSA 97
Woodruff/Arlon L. BSA 69
Woodruff/Harry BSA 29
Woodruff/Jasper Wilson BSA 40
Woods/Alisha Antonette HDFS BSHES 96
Woods/Cynthia Faye ANSC MS 95
Woods/Gene Richey DANB BSA 78
Woods/Joe Bryce BSA 41
Woods/Michael D. AGEC MS 78
Woodson/Randy HORT BSA 79
Woodward/Polly Rouse BSHE 36
Woody/Elmer Randall AGEC MS 79
Woody/Stacy D. HDFS BSHE 94
Woody/Timothy Joel POSC MS 68
Woody/Timothy Joel ANNU BSA 67
Wooldridge/Louise McGaugh BSHE 26
Wooley/Jerry L. GNAG MS 86
Wooley/Jerry L. ANSC BSA 75
Wooley/Johnny Roy AGRN BSA 80
Wooley/Laura BSHE 42
Wooley/Rita HDID BSHE 80
Woolsey/Bonnie J. ANSC BSA 97
Woolsey/John Paul BSA 51
Woolsey/Madge Bowlin BSHE 42
Woolsey/Mark Allen ANSC BSA 82
Woosley/Patsy R. HOEC MS 76
Wooten/Carol Sue ANSC MS 71
Wooten/Thomas Lloyd DANB BSA 85
Worawisitthumrong/Adul PLPA MS 63
Word/Harry D. BSA 59
Word/Samuel Wesley POSC MS 74
Word/Ulyss G. BSA 50
Workman/Drew Stacy FDSC MS 88
Workman/Drew Stacy FDSC BSA 86
Workman/Herman Edmond BSA 53
Workman/Richie Neil AGRC BSA 97
Worley/Glenn B. ENTO MS 66
Worthington/Euna H. BSHE 34
Worthington/John C. ANSC MS 75
Wray/Donald E. POSC BSA 59
Wray/Gladys A. BSHE 47
Wren/Carolyn S. CLTX BSHE 82
Wren/Henry Hudson BSA 29
Wren/Mary Jim BSHE 29
Wren/Tere ANSC BSA 76
Wright/Amy Lynn CDIS BSE 97
Wright/Audrey Louise FDNU BSHE 51
Wright/Barbara Louise AGJN BSA 74
Wright/Betty Lou BSHE 58
Wright/Brett Alan POSC BSA 97
Wright/Carol Carter BSHE 42
Wright/Chris S. ANSC BSA 96
Wright/Debbie CDFR BSHE 76
Wright/Douglas W. ANSC PHD 90
Wright/E. Bradford BSA 53
Wright/Floyd E. PLPA MS 71
Wright/Jesse W. BSA 47
Wright/Karen Newman HEED BSHE 81
Wright/Leslie Erin AGED BSA 96
Wright/Linda K. FND BSHE 67
Wright/M. Deborah CTM BSHE 74
Wright/Margaret L. FDSC BSA 90
Wright/Marigene Howell HOEC BSHE 41
Wright/Mark D. DANB BSA 73
Wright/Mitchel AGED BSA 85
Wright/Orville E. ANSC BSA 65
Wright/Peggy Brink HEED BSHE 64
Wright/Peggy Franks BSHE 60
Wright/Randy AGED BSA 83
Wright/Robert Jenkins AGRN MS 93
Wright/Ronnie AGEC MS 90
Wright/Ronnie ANSC BSA 85
Wright/Shelley Ann ANSC BSA 96
Wright/Walter A. BSA 51
Wright/William F. AGRI BSA 40
Wrona/Jennifer A. ANSC 98
Wullschleger/Stan AGRN PHD 90
Wyatt/Andy Jack POSC BSA 49
Wyatt/Jimmy E. HORT MS 67
Wyche/Robert C. POSC MS 71
Wylie-Rosett/Judith A. DIET BSHE 66
Wylie/Leo D. BSA 31
Wylie/Lorraine Burns HOEC MS 62
Wylie/Lorraine Burns BSHE 41
Wylie/Marie Horne BSHE 43
Wylie/W. D. AGRI BSA 37
Wyman/David Jay POSC BSA 89
Wynn/Billy R. AGEC MS 76
Wynn/Dallas Ashley POSC BSA 88
Wynn/Freddie C. ANSC BSA 79
Wynn/Holland Dwight AGED BSA 91
Wynne/Wendell B. AGBS 98
Wynne/Willie P. HEED BSHE 43
Wyrick/Hal Atherton BSA 51
Xia/Jun Qiang PTSC PHD 91
Xie/Bensong AGRN PHD 93
Xie/Wuzi AGRN MS 89
Xiong/Hua BAEG MSBAE 96
Yan/Wengui AGRN PHD 92
Yancey/Carolyn BSHE 34
Yancey/Edward Jacob EXED BSA 97
Yancey/Rachel Tschabold BSHE 42
Yandell/Charles AGBS BSA 64
Yandell/Fern HEED BSHE 64
Yandell/James Owen POSC BSA 63
Yanez/Lourdes FDSC MS 95
Yang/Luoqing FDSC PHD 90
Yang/Rosa Hsin FDST BSA 98
Yang/Yufen AGEC MS 92
Yarborough/Natalie Renee CLTX BSHE 94
Yarbrough/Brian Todd ANSC BSA 94
Yarbrough/Glenda Sue POSC BSA 93
Yarbrough/Kimberly Bowen CLTX BSHE 95
Yarbrough/William C. AGEC BSA 65
Yarbrough/William D. BSA 59
Yates/Elizabeth H. FDNU BSHE 74
Yates/Jerome D. POSC MS 59
Yates/Jerome D. ANNU BSA 58
Yates/Jerry W. AGED BSA 86
Yates/Mary Miller BSHE 30
Yates/Randy D. POSC 99
Yates/Warren H. BSA 52
Yazdansepas/Amir GNAG MS 83
Yearian/William C. ENTO MS 61
Yearian/William C. BSA 60
Yeatman/Marjorie Ann Thompson BSHE 57
Yeo/Lay Keow HDFS BSHE 89
Yerger/Sandra S. Pate BSHE 63
Yielding/Charles E. ANSC BSA 69
Yingst/Lennie D. POSC MS 70
Yingst/Lennie D. BSA 68
Yocum/David CRSC BSA 74
Yocum/Odell BSA 48
Yoes/Alicia Oldham BSHE 39
Yoho/Douglas Eric ANSC BSA 94
Yonan/Joanne Sue Waldorph AGEC MS 87
York/Dan ANSC BSA 75
York/Harver A. BSA 19
York/Hollie Richelle IDHS BSHES 97
York/Sarah R. BSHE 50
Youmans/Cletus AGRN PHD 85
Young/Barbara Kay CLTX BSHE 83
Young/Debera GNAG MS 90
Young/Don AGBS BSA 83
Young/Douglas Charles AGBS BSA 87
Young/Douglas S. GNAG MS 90
Young/Helen HEED BSHE 70
Young/Helen Appleby BSHE 35
Young/Howard S. BSA 54
Young/Jack F. HORT MS 80
Young/James B. BSA 58
Young/Jim POSC MS 66
Young/Jim BSA 65
Young/Kenneth BSA 52
Young/Kenneth Dale GBUS BSA 77
Young/Lawrence Nall GNAG BSA 62
Young/Lonnie Eugene AGBS BSA 88
Young/Lonnie Eugene AGEC MS 90
Young/Margaret S. HDID BSHE 82
Young/Mary Katherine FDSC BSA 85
Young/Mary Lou BSHE 46

Young/Patricia A. CLTX BSHE 82
Young/Patrisha M. CLTX BSHE 90
Young/Ronald Allen AGEC MS 88
Young/Santford C. ENTO MS 73
Young/Susan C. HORT BSA 82
Young/Wilford Clarence AEED MS 65
Young/Wilford Clarence AGRI BSA 49
Youngblood/Elton Ray POSC BSA 81
Younger/Steve AGBS BSA 88
Younkin/Ada A. HEED BSHE 84
Yowell/Ronnie L. ANSC BSA 71
Yowell/Sara HOEC BSHE 62
Yu/Su-May PTSC PHD 84
Yunos/Ishak Bin AGEC MS 96
Yusof/Mohammed Zain AGBS BSA 85
Yust/Karen A. ENTO MS 81
Zabinski/Mark Steven HORT BSA 85
Zaccanti/Alfred E. POSC BSA 75
Zachariah/Dorothy BSHE 36
Zachry/Nell BSHE 23
Zarlengo/Paul Joseph AGRN MS 93
Zawadowski/Peter S. HORT MS 85
Zayas/Roberto A. AGEC MS 60
Zeglin/Suzanne BSHE 51
Zeglin/Suzanne HOEC MS 70
Zeigler/Helen Marie HOEC BSHE 53
Zeigler/Rick Fred AGED BSA 95
Zelenka/Sara J. ANSC BSA 83
Zepeda/F. Cesar A. HORT BSA 75
Zerman/Cortney A. CLTX BSHE 84
Zhang/Bin AGEC MS 90
Zhang/Jun Ping ENTO PHD 94
Zhao/Duli AGRN PHD 97
Zhu/Baolong ANSC MS 89
Zierak/Mary Ann BSHE 59
Zimmerman/George E. BSA 50
Zimmerman/Janet Beck FDSC MS 82
Zini/Janette Denise ANSC MS 85
Zini/Janette Denise BSA 82
Zini/Kelly Warren POSC BSA 85
Zodrow/Steven F. ANSC BSA 77
Zoecklein/Laurie Joan ANSC MS 81
Zollinger/Claudia AGBS BSA 86
Zollner/Sarah Jane AGBS BSA 95
Zook/Catherine FDNU BSHE 94
Zornes/Albert Scott ANSC BSA 86
Zou/Yiqiao ENTO MS 94
Zumfelde/Mary Ellen Q. McManus BSHE 57

INDEX

About the Author

Gary Zellar

Gary Zellar is on the faculty of the Department of History at Sam Houston State University in Huntsville, Texas. Born in Denver, Colorado, and raised in Indiana, he received his B.A. and M.A. in history from Sam Houston State and is completing his Ph.D. from the University of Arkansas, Fayetteville. His dissertation, "'If I A'int One, There A'int One in This Territory:' Race, Citizenship, Identity and Land, the Creek Freedmen Experience in the Indian Territory and Oklahoma, 1830-1910," is under the direction of Dr. Elliott West. Besides Agricultural history, Mr. Zellar's research areas include African American and Native American history, the Civil War and Reconstruction eras, history of the American West and Southern history. In addition to articles published in *Agricultural History* and the *Chronicles of Oklahoma*, Mr. Zellar is co-author, with Luti Salisbury, of *Agriculture and Rural Life in Arkansas, 1820-1945: A Bibliography*. He lives in Huntsville, Texas, with his wife, daughter and son.